T0267342

Graph Theory and Its Applications, Third Edition

Textbooks in Mathematics

Series editors:
Al Boggess and Ken Rosen

A CONCRETE INTRODUCTION TO REAL ANALYSIS, SECOND EDITION

Robert Carlson

MATHEMATICAL MODELING FOR BUSINESS ANALYTICS

William P. Fox

ELEMENTARY LINEAR ALGEBRA

James R. Kirkwood and Bessie H. Kirkwood

APPLIED FUNCTIONAL ANALYSIS, THIRD EDITION

J. Tinsley Oden and Leszek Demkowicz

AN INTRODUCTION TO NUMBER THEORY WITH CRYPTOGRAPHY, SECOND EDITION

James R. Kraft and Lawrence Washington

MATHEMATICAL MODELING: BRANCHING BEYOND CALCULUS

Crista Arangala, Nicolas S. Luke and Karen A. Yokley

ELEMENTARY DIFFERENTIAL EQUATIONS, SECOND EDITION

Charles Roberts

ELEMENTARY INTRODUCTION TO THE LEBESGUE INTEGRAL

Steven G. Krantz

LINEAR METHODS FOR THE LIBERAL ARTS

David Hecker and Stephen Andrilli

CRYPTOGRAPHY: THEORY AND PRACTICE, FOURTH EDITION

Douglas R. Stinson and Maura B. Paterson

DISCRETE MATHEMATICS WITH DUCKS, SECOND EDITION

Sarah-Marie Belcastro

BUSINESS PROCESS MODELING, SIMULATION AND DESIGN, THIRD EDITION

Manual Laguna and Johan Marklund

GRAPH THEORY AND ITS APPLICATIONS, THIRD EDITION

Jonathan L. Gross, Jay Yellen and Mark Anderson

Graph Theory and Its Applications, Third Edition

Jonathan L. Gross
Jay Yellen
Mark Anderson

CRC Press
Taylor & Francis Group
Boca Raton London New York

CRC Press is an imprint of the
Taylor & Francis Group, an **informa** business

CRC Press
Taylor & Francis Group
6000 Broken Sound Parkway NW, Suite 300
Boca Raton, FL 33487-2742

© 2019 by Taylor & Francis Group, LLC
CRC Press is an imprint of Taylor & Francis Group, an Informa business

No claim to original U.S. Government works

International Standard Book Number-13: 978-1-4822-4948-4 (Hardback)

Visit the Taylor & Francis Web site at
http://www.taylorandfrancis.com

and the CRC Press Web site at
http://www.crcpress.com

Printed and bound in Great Britain by
TJ International Ltd, Padstow, Cornwall

Jonathan dedicates this book to Alisa.

Jay dedicates this book to the memory of his brother Marty.

Mark dedicates this book to Libertad.

CONTENTS

PREFACE

Graphs have proven to be useful models for a wide variety of problems, arising in computer science, operations research, and in the natural and social sciences. This text targets the need for a comprehensive approach to the theory, integrating a careful exposition of classical developments with emerging methods, models, and practical needs. It is suitable for classroom presentation at the introductory graduate or advanced undergraduate level, or for self-study and reference by working professionals.

Graph theory has evolved as a collection of seemingly disparate topics. The intent of the authors is to present this material in a more cohesive framework. In the process, important techniques and analytic tools are transformed into a unified mathematical methodology.

Emphasis throughout is conceptual, with more than 600 graph drawings included to strengthen intuition and more than 1500 exercises ranging from routine drill to more challenging problem solving. Applications and concrete examples are designed to stimulate interest and to demonstrate the relevance of new concepts.

Algorithms are presented in a concise format, and computer science students will find numerous projects inviting them to convert algorithms to computer programs. Software design issues are addressed throughout the book in the form of computational notes, which can be skipped by those not interested in implementation. These design issues include the efficient use of computational resources, software reusability, and the user interface.

This book is a collection of topics drawn from the second edition of *Graph Theory and Its Applications*, written by the first two authors of this book. The topics chosen provide a strong foundation in graph theory.

Summary of Contents

Chapters 1 through 6 concentrate on graph representation, basic properties, modeling, and applications. Graphical constructions here are concrete and primarily spatial. When necessary, we introduce abstractions in a supportive role.

Chapter 7, which is devoted to planarity and Kuratowski's theorem, presents some of the underpinnings of topological graph theory. Chapter 8 is about graph colorings, including vertex- and edge-colorings, map-colorings, and the related topics of cliques, independence numbers, and graph factorization.

The material in Chapters 9 and 10, special digraph models and network flows, overlaps with various areas in operations research and computer science.

Chapter 11 explores the connections between a graph's automorphisms (its symmetries) and its colorings.

The appendix provides a review of a variety of topics which a student would typically encounter in earlier math classes.

To the Instructor

The book is well-suited for a variety of one-semester courses. A *general course* in graph theory might include many of the topics covered in the first eight chapters. A course oriented toward *operations research/optimization* should include most or all of the material in Chapters 4 (spanning trees), 5 (connectivity), 6 (traversability), 8 (colorings), 9 (digraph models), and 10 (flows). A course emphasizing the role of *data structures and algorithms* might include more material from Chapter 3 (trees). A course designed to have strong ties to *geometry, linear algebra, and group theory*, might include parts of Chapters 4 (graphs and vector spaces), 7 (planarity), 8 (graph colorings), and 11 (graph symmetry).

The definitions and results from all of Chapter 1, most of Chapter 2, and from §3.1 are used throughout the text, and we recommend that they be covered in any course that uses the book. The remaining chapters are largely independent of each other, with the following exceptions:

- §3.2 is used in §4.1.

- §4.1 and §4.2 are used in §5.4 and §9.5.

- §4.5 is referred to in §5.3 and §6.1.

- Parts of Chapter 5 are used in §10.3.

Features of This Book

- *Applications.* The Index of Applications lists 70 applications, distributed throughout the text.

- *Algorithms.* The Index of Algorithms points to 51 algorithms, outlined in the text with enough detail to make software implementation feasible.

- *Foreshadowing.* The first three chapters now preview a number of concepts, mostly via the exercises, that are more fully developed in later chapters. This makes it easier to encourage students to take earlier excursions in research areas that may be of particular interest to the instructor.

- *Solutions* and *Hints.* Each exercise marked with a superscript^{S} has a solution or hint appearing in the back of the book.

- *Supplementary Exercises.* In addition to the section exercises, each chapter concludes with a section of supplementary exercises, which are intended to develop the problem-solving skills of students and test whether they can go beyond what has been explicitly taught. Most of these exercises were designed as examination questions for students at Columbia University.

Websites

Suggestions and comments from readers are welcomed and may be sent to the authors' website at *www.graphtheory.com*. Thanks to our founding webmaster Aaron Gross for his years of maintaining this website. The general website for CRC Press is *www.crcpress.com*. As with our previous graph theory books, we will post the corrections to all known errors on our website.

Jonathan L. Gross, Jay Yellen, and Mark Anderson

ABOUT THE AUTHORS

Jonathan L. Gross is Professor of Computer Science at Columbia University. His research in topology, graph theory, and cultural sociometry has earned him an Alfred P. Sloan Fellowship, an IBM Postdoctoral Fellowship, and various research grants from the Office of Naval Research, the National Science Foundation, and the Russell Sage Foundation.

Professor Gross has created and delivered numerous software development short courses for Bell Laboratories and for IBM. These include mathematical methods for performance evaluation at the advanced level and for developing reusable software at a basic level. He has received several awards for outstanding teaching at Columbia University, including the career Great Teacher Award from the Society of Columbia Graduates. His peak semester enrollment in his graph theory course at Columbia was 101 students.

His previous books include *Topological Graph Theory*, coauthored with Thomas W. Tucker; *Measuring Culture*, coauthored with Steve Rayner, constructs network-theoretic tools for measuring sociological phenomena; and *Combinatorial Methods with Computer Applications*.

Prior to Columbia University, Professor Gross was in the Mathematics Department at Princeton University. His undergraduate work was at M.I.T., and he wrote his Ph.D. thesis on 3-dimensional topology at Dartmouth College.

Jay Yellen is Professor of Mathematics at Rollins College. He received his B.S. and M.S. in Mathematics at Polytechnic University of New York and did his doctoral work in finite group theory at Colorado State University. Dr. Yellen has had regular faculty appointments at Allegheny College, State University of New York at Fredonia, and Florida Institute of Technology, where he was Chair of Operations Research from 1995 to 1999. He has had visiting appointments at Emory University, Georgia Institute of Technology, and Columbia University, University of Nottingham, UK, and KU Leuven, Belgium.

In addition to the *Handbook* of *Graph Theory*, which he co-edited with Professor Gross, Professor Yellen has written manuscripts used at IBM for two courses in discrete mathematics within the Principles of Computer Science Series and has contributed two sections to the *Handbook* of *Discrete and Combinatorial Mathematics*. He also has designed and conducted several summer workshops on creative problem solving for secondary-school mathematics teachers, which were funded by the National Science Foundation and New York State. He has been a recipient of a Student's Choice Professor Award at Rollins College.

Dr. Yellen has published research articles in character theory of finite groups, graph theory, power-system scheduling, and timetabling. His current research interests include graph theory, discrete optimization, and the development of course-scheduling software based on graph coloring.

Mark Anderson is Professor of Mathematics and Computer Science at Rollins College. He received his doctorate from the University of Virginia for work in algebraic and differential topology. After arriving at Rollins College, he switched his emphasis to graph theory and also earned a masters degree in Computer Science from the University of Central Florida. He has publications in the areas of algebraic topology (cobordism), topological graph theory (crossing number and Cayley maps), graph coloring (using voltage graphs), as well as graph security (eternal domination). He is currently working on problems in robust graph coloring, which combines ideas from graph coloring with  those of graph security. Undergraduate students have done collaborative research with Dr. Anderson in topological graph theory, graph coloring, and software testing.

Methods included in this book appear in a wide range of courses that Professor Anderson has designed. In his Artificial Intelligence course, students explore graph search algorithms; in Data Compression, they discover Huffman codes; in Coding Theory, they work with the geometry of the n-dimensional hypercube. Courses that use algebra to answer geometric questions include Algebraic Topology and Transformational Geometry. The Geometry of Islamic Patterns involves imbeddings of graphs on surfaces as a way of characterizing planar tilings. Courses he has developed for students with majors outside math and computer science include The Mathematics of Games and Gaming and Zero, A Revolutionary Idea.

Chapter 1

INTRODUCTION TO GRAPH MODELS

INTRODUCTION

Configurations of nodes and connections occur in a great diversity of applications. They may represent physical networks, such as electrical circuits, roadways, or organic molecules. And they are also used in representing less tangible interactions as might occur in ecosystems, sociological relationships, databases, or the flow of control in a computer program.

Formally, such configurations are modeled by combinatorial structures called *graphs*, consisting of two sets called *vertices* and *edges* and an incidence relation between them. The vertices and edges may have additional attributes, such as color or weight, or anything else useful to a particular model. Graph models tend to fall into a handful of categories. For instance, the network of one-way streets of a city requires a model in which each edge is assigned a direction; two atoms in an organic molecule may have more than one bond between them; and a computer program is likely to have loop structures. These examples require graphs with directions on their edges, with multiple connections between vertices, or with connections from a vertex to itself.

In the past these different types of graphs were often regarded as separate entities, each with its own set of definitions and properties. We have adopted a unified approach by introducing all of these various graphs at once. This allows us to establish properties that are shared by several classes of graphs without having to repeat arguments. Moreover, this broader perspective has an added bonus: it inspires computer representations that lead to the design of fully reusable software for applications.

We begin by introducing the basic terminology needed to view graphs both as configurations in space and as combinatorial objects that lend themselves to computer representation. Pure and applied examples illustrate how different kinds of graphs arise as models.

1.1   GRAPHS AND DIGRAPHS

We think of a *graph* as a set of points in a plane or in 3-space and a set of line segments (possibly curved), each of which either joins two points or joins a point to itself.

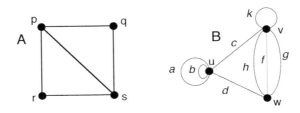

Figure 1.1.1   Line drawings of a graph A and a graph B.

Graphs are highly versatile models for analyzing a wide range of practical problems in which points and connections between them have some physical or conceptual interpretation. Placing such analysis on solid footing requires precise definitions, terminology, and notation.

DEFINITION: A **graph** $G = (V, E)$ is a mathematical structure consisting of two finite sets V and E. The elements of V are called **vertices** (or **nodes**), and the elements of E are called **edges**. Each edge has a set of one or two vertices associated to it, which are called its **endpoints**.

TERMINOLOGY: An edge is said to **join** its endpoints. A vertex joined by an edge to a vertex v is said to be a **neighbor** of v.

DEFINITION: The (**open**) **neighborhood** of a vertex v in a graph G, denoted $N(v)$, is the set of all the neighbors of v. The **closed neighborhood** of v is given by $N[v] = N(v) \cup \{v\}$.

NOTATION: When G is not the only graph under consideration, the notations V_G and E_G (or $V(G)$ and $E(G)$) are used for the vertex- and edge-sets of G, and the notations $N_G(v)$ and $N_G[v]$ are used for the neighborhoods of v.

Example 1.1.1: The vertex- and edge-sets of graph A in Figure 1.1.1 are given by

$$V_A = \{p, q, r, s\} \quad \text{and} \quad E_A = \{pq, pr, ps, rs, qs\}$$

and the vertex- and edge-sets of graph B are given by

$$V_B = \{u, v, w\} \quad \text{and} \quad E_B = \{a, b, c, d, f, g, h, k\}$$

Notice that in graph A, we are able to denote each edge simply by juxtaposing its endpoints, because those endpoints are unique to that edge. On the other hand, in graph B, where some edges have the same set of endpoints, we use explicit names to denote the edges.

Simple Graphs and General Graphs

In certain applications of graph theory and in some theoretical contexts, there are frequent instances in which an edge joins a vertex to itself or multiple edges have the same set of endpoints. In other applications or theoretical contexts, such instances are absent.

DEFINITION: A **proper edge** is an edge that joins two distinct vertices.

DEFINITION: A **self-loop** is an edge that joins a single endpoint to itself.[†]

DEFINITION: A **multi-edge** is a collection of two or more edges having identical endpoints. The **edge multiplicity** is the number of edges within the multi-edge.

DEFINITION: A **simple graph** has neither self-loops nor multi-edges.

DEFINITION: A **loopless graph** (or **multi-graph**) may have multi-edges but no self-loops.

DEFINITION: A (**general**) **graph** may have self-loops and/or multi-edges.

Example 1.1.1 continued: Graph A in Figure 1.1.1 is simple. Graph B is not simple; the edges a, b, and k are self-loops, and the edge-sets $\{f, g, h\}$ and $\{a, b\}$ are multi-edges.

TERMINOLOGY: When we use the term *graph* without a modifier, we mean a *general graph*. An exception to this convention occurs when an entire section concerns simple graphs only, in which case, we make an explicit declaration at the beginning of that section.

TERMINOLOGY NOTE: Some authors use the term *graph* without a modifier to mean simple graph, and they use *pseudograph* to mean general graph.

Null and Trivial Graphs

DEFINITION: A **null graph** is a graph whose vertex- and edge-sets are empty.

DEFINITION: A **trivial graph** is a graph consisting of one vertex and no edges.

Edge Directions

An edge between two vertices creates a connection in two opposite senses at once. Assigning a direction makes one of these senses *forward* and the other *backward*. In a line drawing, the choice of forward direction is indicated by placing an arrow on an edge.

DEFINITION: A **directed edge** (or **arc**) is an edge, one of whose endpoints is designated as the **tail**, and whose other endpoint is designated as the **head**.

TERMINOLOGY: An arc is said to be **directed from** its tail to its head.

NOTATION: In a general digraph, the head and tail of an arc e may be denoted $head(e)$ and $tail(e)$, respectively.

DEFINITION: Two arcs between a pair of vertices are said to be **oppositely directed** if they do not have the same head and tail.

DEFINITION: A **multi-arc** is a set of two or more arcs having the same tail and same head. The **arc multiplicity** is the number of arcs within the multi-arc.

DEFINITION: A **directed graph** (or **digraph**) is a graph each of whose edges is directed.

[†]We use the term "self-loop" instead of the more commonly used term "loop" because loop means something else in many applications.

DEFINITION: A digraph is **simple** if it has neither self-loops nor multi-arcs.

NOTATION: In a simple digraph, an arc from vertex u to vertex v may be denoted by uv or by the ordered pair $[u, v]$.

Example 1.1.2: The digraph in Figure 1.1.2 is simple. Its arcs are uv, vu, and vw.

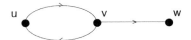

Figure 1.1.2   A simple digraph with a pair of oppositely directed arcs.

DEFINITION: A **mixed graph** (or **partially directed graph**) is a graph that has both undirected and directed edges.

DEFINITION: The **underlying graph** of a directed or mixed graph G is the graph that results from removing all the designations of *head* and *tail* from the directed edges of G (i.e., deleting all the edge-directions).

Example 1.1.3: The digraph D in Figure 1.1.3 has the graph G as its underlying graph.

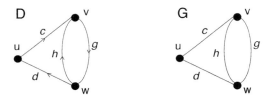

Figure 1.1.3   A digraph and its underlying graph.

Simple and non-simple graphs and digraphs all commonly arise as models; the numerous and varied examples in §1.3 illustrate the robustness of our comprehensive graph model.

Formal Specification of Graphs and Digraphs

Except for the smallest graphs, line drawings are inadequate for describing a graph; imagine trying to draw a graph to represent a telephone network for a small city. Since many applications involve computations on graphs having hundreds, or even thousands, of vertices, another, more formal kind of specification of a graph is often needed.

The specification must include (implicitly or explicitly) a function *endpts* that specifies, for each edge, the subset of vertices on which that edge is incident (i.e., its endpoint set). In a simple graph, the juxtaposition notation for each edge implicitly specifies its endpoints, making formal specification simpler for simple graphs than for general graphs.

DEFINITION: A **formal specification of a simple graph** is given by an **adjacency table** with a row for each vertex, containing the list of neighbors of that vertex.

Example 1.1.4: Figure 1.1.4 shows a line drawing and a formal specification for a simple graph.

$$
\begin{aligned}
p &: \quad q \quad r \quad s \\
q &: \quad p \quad s \\
r &: \quad p \quad s \\
s &: \quad p \quad q \quad r
\end{aligned}
$$

Figure 1.1.4   A simple graph and its formal specification.

DEFINITION: A **formal specification of a general graph** $G = (V, E, endpts)$ consists of a list of its vertices, a list of its edges, and a two-row **incidence table** (specifying the *endpts* function) whose columns are indexed by the edges. The entries in the column corresponding to edge e are the endpoints of e. The same endpoint appears twice if e is a self-loop. (An isolated vertex will appear only in the vertex list.)

Example 1.1.5: Figure 1.1.5 shows a line drawing and a formal specification for a general graph.

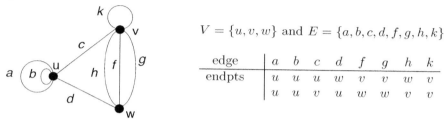

$V = \{u, v, w\}$ and $E = \{a, b, c, d, f, g, h, k\}$

edge	a	b	c	d	f	g	h	k
endpts	u	u	u	w	v	v	w	v
	u	u	v	u	w	w	v	v

Figure 1.1.5   A general graph and its formal specification.

DEFINITION: A **formal specification of a general digraph** or **a mixed graph** $D = (V, E, endpts, head, tail)$ is obtained from the formal specification of the underlying graph by adding the functions $head : E_G \to V_G$ and $tail : E_G \to V_G$, which designate the *head* vertex and *tail* vertex of each arc.

One way to specify these designations in the incidence table is to mark in each column the endpoint that is the head of the corresponding arc.

Remark: A mixed graph is specified by simply restricting the functions *head* and *tail* to a proper subset of E_G. In this case, a column of the incidence table that has no mark means that the corresponding edge is undirected.

Example 1.1.6: Figure 1.1.6 gives the formal specification for the digraph shown, including the corresponding values of the functions *head* and *tail*. A superscript "h" is used to indicate the *head* vertex.

Remark: Our approach treats a digraph as an *augmented* type of graph, where each edge e of a digraph is still associated with a subset $endpts(e)$, but which now also includes a mark on one of the endpoints to specify the head of the directed edge.

This viewpoint is partly motivated by its impact on computer implementations of graph algorithms (see the computational notes), but it has some advantages from a mathematical perspective as well. Regarding digraphs as augmented graphs makes it easier to view certain results that tend to be established separately for graphs and for digraphs as a single result that applies to both.

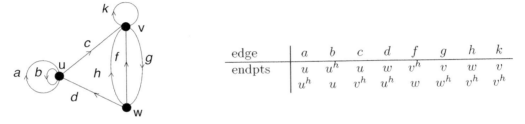

edge	a	b	c	d	f	g	h	k
endpts	u	u^h	u	w	v^h	v	w	v
	u^h	u	v^h	u^h	w	w^h	v^h	v^h

$$head(a) = tail(a) = head(b) = tail(b) = head(d) = tail(c) = u;$$
$$head(c) = head(h) = head(f) = tail(g) = head(k) = tail(k) = v;$$
$$head(g) = tail(d) = tail(h) = tail(f) = w.$$

Figure 1.1.6   A general digraph and its formal specification.

Also, our formal incidence specification permits us to reverse the direction of an edge e at any time, just by reversing the values of $head(e)$ and $tail(e)$. This amounts to switching the h mark in the relevant column of the incidence table or reversing the arrowhead in the digraph drawing.

COMPUTATIONAL NOTE 1: These formal specifications for a graph and a digraph can easily be implemented with a variety of programmer-defined data structures, whatever is most appropriate to the application. A discussion of the comparative advantages and disadvantages of a few of the most common computer information structures for graphs and digraphs appears at the end of Chapter 2.

COMPUTATIONAL NOTE 2: (*A caution to software designers*) From the perspective of object-oriented software design, the ordered-pair representation of arcs in a digraph treats digraphs as a different class of objects from graphs. This could seriously undermine *software reuse*. Large portions of computer code might have to be rewritten in order to adapt an algorithm that was originally designed for a digraph to work on an undirected graph.

The ordered-pair representation could also prove awkward in implementing algorithms for which the graphs or digraphs are *dynamic* structures (i.e., they change during the algorithm). Whenever the direction on a particular edge must be reversed, the associated ordered pair has to be deleted and replaced by its reverse. Even worse, if a directed edge is to become undirected, then an ordered pair must be replaced with an unordered pair. Similarly, the undirected and directed edges of a mixed graph would require two different types of objects.

COMPUTATIONAL NOTE 3: For some applications (network layouts on a surface, for instance), the direction of flow around a self-loop has practical importance, and distinguishing between the ends of a self-loop becomes necessary. This distinction is made in Chapters 8 and 16 but not elsewhere.

Mathematical Modeling with Graphs

To bring the power of mathematics to bear on real-world problems, one must first *model* the problem mathematically. Graphs are remarkably versatile tools for modeling, and their wide-ranging versatility is a central focus throughout the text.

Example 1.1.7: The mixed graph in Figure 1.1.7 is a model for a roadmap. The vertices represent landmarks, and the directed and undirected edges represent the one-way and two-way streets, respectively.

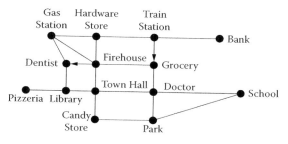

Figure 1.1.7   Road-map of landmarks in a small town.

Example 1.1.8: The digraph in Figure 1.1.8 represents the hierarchy within a company. This illustrates how, beyond physical networks, graphs and digraphs are used to model social relationships.

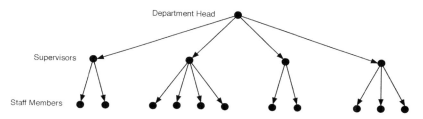

Figure 1.1.8   A corporate hierarchy.

Degree of a Vertex

DEFINITION: **Adjacent vertices** are two vertices that are joined by an edge.

DEFINITION: **Adjacent edges** are two distinct edges that have an endpoint in common.

DEFINITION: If vertex v is an endpoint of edge e, then v is said to be **incident** on e, and e is incident on v.

DEFINITION: The **degree** (or **valence**) of a vertex v in a graph G, denoted $deg(v)$, is the number of proper edges incident on v plus twice the number of self-loops.[†]

TERMINOLOGY: A vertex of degree d is also called a d-**valent vertex**.

NOTATION: The smallest and largest degrees in a graph G are denoted $\delta_{\min}$ and $\delta_{\max}$ (or $\delta_{\min}(G)$ and $\delta_{\max}(G)$ when there is more than one graph under discussion). Some authors use δ instead of $\delta_{\min}$ and Δ instead of $\delta_{\max}$.

DEFINITION: The **degree sequence** of a graph is the sequence formed by arranging the vertex degrees in non-increasing order.

[†]Applications of graph theory to physical chemistry motivate the use of the term *valence*.

Example 1.1.9: Figure 1.1.9 shows a graph and its degree sequence.

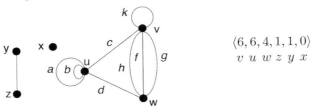

$$\langle 6, 6, 4, 1, 1, 0 \rangle$$
$$v \ u \ w \ z \ y \ x$$

Figure 1.1.9   A graph and its degree sequence.

Although each graph has a unique degree sequence, two structurally different graphs can have identical degree sequences.

Example 1.1.10: Figure 1.1.10 shows two different graphs, G and H, with the same degree sequence.

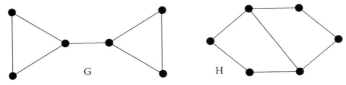

Figure 1.1.10   Two graphs whose degree sequences are both $\langle 3, 3, 2, 2, 2, 2 \rangle$.

The following result shows that the degree sequence of a simple graph must have at least two equal terms. This has an interesting interpretation in a sociological model that appears in Section 1.3 (see Application 1.3.2).

Proposition 1.1.1: *A non-trivial simple graph G must have at least one pair of vertices whose degrees are equal.*

Proof: Suppose that the graph G has n vertices. Then there appear to be n possible degree values, namely $0, \ldots, n - 1$. However, there cannot be both a vertex of degree 0 and a vertex of degree $n - 1$, since the presence of a vertex of degree 0 implies that each of the remaining $n - 1$ vertices is adjacent to at most $n - 2$ other vertices. Hence, the n vertices of G can realize at most $n - 1$ possible values for their degrees. Thus, the *pigeonhole principle* implies that at least two of the n vertices have equal degree. $\diamond$

The work of Leonhard Euler (1707–1783) is regarded as the beginning of graph theory as a mathematical discipline. The following theorem of Euler establishes a fundamental relationship between the vertices and edges of a graph.

Theorem 1.1.2: [***Euler's Degree-Sum Theorem***] *The sum of the degrees of the vertices of a graph is twice the number of edges.*

Proof: Each edge contributes two to the degree sum. $\diamond$

Corollary 1.1.3: *In a graph, there is an even number of vertices having odd degree.*

Proof: Consider separately, the sum of the degrees that are odd and the sum of those that are even. The combined sum is even by Theorem 1.1.2, and since the sum of the even degrees is even, the sum of the odd degrees must also be even. Hence, there must be an even number of vertices of odd degree. $\diamond$

Remark: By Theorem 1.1.2, the sum of the terms of a degree sequence of a graph is even. Theorem 1.1.4 establishes the following converse: any non-increasing, nonnegative sequence of integers whose sum is even is the degree sequence of some graph. Example 1.1.11 illustrates the construction used in its proof.

Example 1.1.11: To construct a graph whose degree sequence is $\langle 5, 4, 3, 3, 2, 1, 0 \rangle$, start with seven isolated vertices $v_1, v_2, \ldots, v_7$. For the even-valued terms of the sequence, draw the appropriate number of self-loops on the corresponding vertices. Thus, v_2 gets two self-loops, v_5 gets one self-loop, and v_7 remains isolated. For the four remaining odd-valued terms, group the corresponding vertices into any two pairs, for instance, v_1, v_3 and v_4, v_6. Then join each pair by a single edge and add to each vertex the appropriate number of self-loops. The resulting graph is shown in Figure 1.1.11.

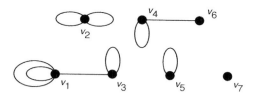

Figure 1.1.11   Constructing a graph with degree sequence $\langle 5, 4, 3, 3, 2, 1, 0 \rangle$.

Theorem 1.1.4: *Suppose that $\langle d_1, d_2, \ldots, d_n \rangle$ is a sequence of nonnegative integers whose sum is even. Then there exists a graph with vertices $v_1, v_2, \ldots, v_n$ such that $deg(v_i) = d_i$, for $i = 1, \ldots, n$.*

Proof: Start with n isolated vertices $v_1, v_2, \ldots, v_n$. For each i, if d_i is even, draw $\frac{d_1}{2}$ self-loops on vertex v_i, and if d_i is odd, draw $\frac{d_1-1}{2}$ self-loops. By Corollary 1.1.3, there is an even number of odd d_i's. Thus, the construction can be completed by grouping the vertices associated with the odd terms into pairs and then joining each pair by a single edge.   ◇

Graphic Sequences

The construction in Theorem 1.1.4 is straightforward but hinges on allowing the graph to be non-simple. A more interesting problem is determining when a sequence is the degree sequence of a *simple* graph.

DEFINITION: A non-increasing sequence $\langle d_1, d_2, \ldots, d_n \rangle$ is said to be **graphic** if it is the degree sequence of some simple graph. That simple graph is said to **realize** the sequence.

Remark: If $\langle d_1, d_2, \ldots, d_n \rangle$ is the degree sequence of a simple graph, then, clearly, $d_1 \leq n - 1$.

Theorem 1.1.5: *Let $\langle d_1, d_2, \ldots, d_n \rangle$ be a graphic sequence, with $d_1 \geq d_2 \geq \ldots \geq d_n$. Then there is a simple graph with vertex-set $\{v_1, \ldots, v_n\}$ satisfying $deg\,(v_i) = d_i$ for $i = 1, 2, \ldots, n$, such that v_1 is adjacent to vertices $v_2, \ldots, v_{d_1+1}$.*

Proof: Among all simple graphs with vertex-set $\{v_1, v_2, \ldots, v_n\}$ and $deg(v_i) = d_i$, $i = 1, 2, \ldots, n$, let G be one for which $r = |N_G(v_1) \cap \{v_2, \ldots, v_{d_1+1}\}|$ is maximum. If $r = d_1$, then the conclusion follows. If $r < d_1$, then there exists a vertex $v_s, 2 \leq s \leq d_1 + 1$, such that v_1 is not adjacent to v_s, and there exists a vertex $v_t, t > d_1 + 1$ such that v_1 is adjacent to v_t (since $deg(v_1) = d_1$). Moreover, since $deg(v_s) \geq deg(v_t)$, there exists a vertex v_k such that v_k is adjacent to v_s but not to v_t. Let $\tilde{G}$ be the graph obtained from G by replacing the edges $v_1 v_t$ and $v_s v_k$ with the edges $v_1 v_s$ and $v_t v_k$ (as shown in Figure 1.1.12). Then the degrees are all preserved and $v_s \in N_{\tilde{G}}(v_1) \cap \{v_3, \dots, v_{d_1+1}\}$. Thus, $|N_{\tilde{G}}(v_1) \cap \{v_2, \dots, v_{d_1+1}\}| = r + 1$, which contradicts the choice of G and completes the proof. $\diamond$

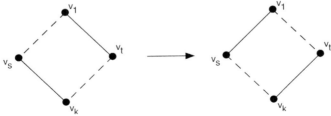

Figure 1.1.12   Switching adjacencies while preserving all degrees.

Corollary 1.1.6: [*Havel (1955) and Hakimi (1961)*] *A sequence* $\langle d_1, d_2, \dots, d_n \rangle$ *of nonnegative integers such that* $d_1 \leq n-1$ *and* $d_1 \geq d_2 \geq \dots \geq d_n$ *is graphic if and only if the sequence* $\langle d_2 - 1, \dots, d_{d_1+1} - 1, d_{d_1+2}, \dots, d_n \rangle$ *is graphic.*

$\diamond$ *(Exercises)*

Remark: Corollary 1.1.6 yields a recursive algorithm that decides whether a non-increasing sequence is graphic.

Algorithm 1.1.1:   **GraphicSequence** $(\langle d_1,\ d_2, \dots, d_n \rangle)$

Input: a non-increasing sequence $\langle d_1,\ d_2, \dots, d_n \rangle$.
Output: TRUE if the sequence is graphic, FALSE if it is not.

If $d_1 \geq n$ or $d_n < 0$
    Return FALSE
Else
    If $d_1 = 0$
        Return TRUE
    Else
        Let $\langle a_1,\ a_2, \dots, a_{n-1} \rangle$ be a non-increasing permutation
            of $\langle d_2 - 1, \dots, d_{d_1+1} - 1, d_{d_1+2}, \dots, d_n \rangle$.
        Return GraphicSequence($\langle a_1,\ a_2, \dots, a_{n-1} \rangle$)

Remark: An iterative version of the algorithm GraphicSequence based on repeated application of Corollary 1.1.6 can also be written and is left as an exercise (see Exercises). Also, given a graphic sequence, the steps of the iterative version can be reversed to construct a graph realizing the sequence. However many zeroes you get at the end of the forward pass, start with that many isolated vertices. Then backtrack the algorithm, adding a vertex each time. The following example illustrates these ideas.

Example 1.1.12: We start with the sequence $\langle 3, 3, 2, 2, 1, 1 \rangle$. Figure 1.1.13 illustrates an iterative version of the algorithm GraphicSequence and then illustrates the backtracking steps leading to a graph that realizes the original sequence. The hollow vertex shown in each backtracking step is the new vertex added at that step.

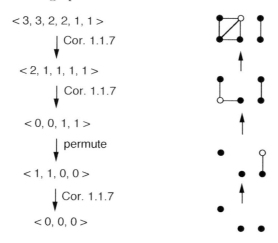

$$< 3, 3, 2, 2, 1, 1 >$$
$\downarrow$ Cor. 1.1.7
$$< 2, 1, 1, 1, 1 >$$
$\downarrow$ Cor. 1.1.7
$$< 0, 0, 1, 1 >$$
$\downarrow$ permute
$$< 1, 1, 0, 0 >$$
$\downarrow$ Cor. 1.1.7
$$< 0, 0, 0 >$$

Figure 1.1.13   Testing and realizing the sequence $\langle 3, 3, 2, 2, 1, 1 \rangle$.

Indegree and Outdegree in a Digraph

The definition of vertex degree is slightly refined for digraphs.

DEFINITION: The **indegree** of a vertex v in a digraph is the number of arcs directed to v; the **outdegree** of vertex v is the number of arcs directed from v. Each self-loop at v counts one toward the indegree of v and one toward the outdegree.

vertex	u	v	w
indegree	3	4	1
outdegree	3	2	3

Figure 1.1.14   The indegrees and outdegrees of the vertices of a digraph.

The next theorem is the digraph version of Euler's Degree-Sum Theorem 1.1.2.

Theorem 1.1.7: *In a digraph, the sum of the indegrees and the sum the outdegrees both equal the number of edges.*

Proof: Each directed edge e contributes one to the indegree at $head(e)$ and one to the outdegree at $tail(e)$. ◇

EXERCISES for Section 1.1

In Exercises 1.1.1 through 1.1.3, construct a line drawing, an incidence table, and the degree sequence of the graph with the given formal specification.

1.1.1$^\text{S}$
$$V = \{u, w, x, z\}\,; \quad E = \{e, f, g\}$$
$$endpts(e) = \{w\}; \; endpts(f) = \{x, w\}; \;\; endpts(g) = \{x, z\}$$

1.1.2
$$V = \{u, v, x, y, z\}\,; \quad E = \{a, b, c, d\}$$
$$endpts(e) = \{u, v\}; \; endpts(b) = \{x, v\}; \; endpts(c) = \{u, v\}; \; endpts(d) = \{x\}$$

1.1.3
$$V = \{u, v, x, y, z\}\,; \quad E = \{e, f, g, h, k\}$$
$$endpts(e) = endpts(f) = \{u, v\}; endpts(g) = \{x, z\}; endpts(h) = endpts(k) = \{y\}$$

In Exercises 1.1.4 through 1.1.6, construct a line drawing for the digraph or mixed graph with vertex-set $V = \{u, v, w, x, y\}$, edge-set $E = \{e, f, g, h\}$, and the given incidence table.

1.1.4$^\text{S}$

edges	e	f	g	h
endpts	y	x^h	v^h	v^h
	w^h	u	x	u

1.1.5

edges	e	f	g	h
endpts	x	v^h	v	v
	w^h	u	u^h	u^h

1.1.6

edges	e	f	g	h
endpts	u	x^h	v	v
	w^h	u	y	u

In Exercises 1.1.7 through 1.1.9, give a formal specification for the given digraph.

1.1.7$^\text{S}$

1.1.8

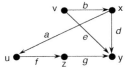

1.1.9

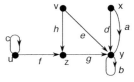

In Exercises 1.1.10 through 1.1.12, give a formal specification for the underlying graph of the digraph indicated.

1.1.10$^\text{S}$ The digraph of Exercise 1.1.7.

1.1.11 The digraph of Exercise 1.1.8.

1.1.12 The digraph of Exercise 1.1.9.

1.1.13 Draw a graph with the given degree sequence.

   a. $\langle 8, 7, 3 \rangle$ b. $\langle 9, 8, 8, 6, 5, 3, 1 \rangle$

1.1.14 Draw a simple graph with the given degree sequence.

   a. $\langle 6, 4, 4, 3, 3, 2, 1, 1 \rangle$ b. $\langle 5, 5, 5, 3, 3, 3, 3, 3 \rangle$

For each of the number sequences in Exercises 1.1.15 through 1.1.18, either draw a simple graph that realizes it, or explain, without resorting to Corollary 1.1.6 or Algorithm 1.1.1, why no such graph can exist.

1.1.15^S a. $\langle 2, 2, 1, 0, 0 \rangle$ b. $\langle 4, 3, 2, 1, 0 \rangle$

1.1.16   a. $\langle 4, 2, 2, 1, 1 \rangle$ b. $\langle 2, 2, 2, 2 \rangle$

1.1.17   a. $\langle 4, 3, 2, 2, 1 \rangle$ b. $\langle 4, 3, 3, 3, 1 \rangle$

1.1.18   a. $\langle 4, 4, 4, 4, 3, 3, 3, 3 \rangle$ b. $\langle 3, 2, 2, 1, 0 \rangle$

1.1.19 Apply Algorithm 1.1.1 to each of the following sequences to determine whether it is graphic. If the sequence is graphic, then draw a simple graph that realizes it.

   a. $\langle 7, 6, 6, 5, 4, 3, 2, 1 \rangle$ b. $\langle 5, 5, 5, 4, 2, 1, 1, 1 \rangle$
   c. $\langle 7, 7, 6, 5, 4, 4, 3, 2 \rangle$ d. $\langle 5, 5, 4, 4, 2, 2, 1, 1 \rangle$

1.1.20 Use Theorem 1.1.5 to prove Corollary 1.1.6.

1.1.21 Write an iterative version of Algorithm 1.1.1 that applies Corollary 1.1.6 repeatedly until a sequence of all zeros or a sequence with a negative term results.

1.1.22^S Given a group of nine people, is it possible for each person to shake hands with exactly three other people?

1.1.23 Draw a graph whose degree sequence has no duplicate terms.

1.1.24^S What special property of a function must the *endpts* function have for a graph to have no multi-edges?

1.1.25 Draw a digraph for each of the following indegree and outdegree sequences, such that the indegree and outdegree of each vertex occupy the same position in both sequences.

   a. in: $\langle 1, 1, 1 \rangle$ out: $\langle 1, 1, 1 \rangle$
   b. in: $\langle 2, 1 \rangle$ out: $\langle 3, 0 \rangle$

DEFINITION: A pair of sequences $\langle a_1, a_2, \ldots, a_n \rangle$ and $\langle b_1, b_2, \ldots, b_n \rangle$ is called **digraphic** if there exists a simple digraph with vertex-set $\{v_1, v_2, \ldots, v_n\}$ such that $outdegree(v_i) = a_i$ and $indegree(v_i) = b_i$ for $i = 1, 2, \ldots, n$.

1.1.26 Determine whether the pair of sequences $\langle 3, 1, 1, 0 \rangle$ and $\langle 1, 1, 1, 2 \rangle$ is digraphic.

1.1.27 Establish a result like Corollary 1.1.6 for a pair sequences to be digraphic.

1.1.28 How many different degree sequences can be realized for a graph having three vertices and three edges?

1.1.29 Given a list of three vertices and a list of seven edges, show that 3^7 different formal specifications for simple graphs are possible.

1.1.30 Given a list of four vertices and a list of seven edges, show that $\binom{7}{2}5^6 2^{10}$ different formal specifications are possible if there are exactly two self-loops.

1.1.31 Given a list of three vertices and a list of seven edges, how many different formal specifications are possible if exactly three of the edges are directed?

1.1.32[S] Does there exist a simple graph with five vertices, such that every vertex is incident with at least one edge, but no two edges are adjacent?

1.1.33 Prove or disprove: There exists a simple graph with 13 vertices, 31 edges, three 1-valent vertices, and seven 4-valent vertices.

DEFINITION: Let $G = (V, E)$ be a graph and let $W \subseteq V$. Then W is a **vertex cover** of G if every edge is incident on at least one vertex in W. (See §10.4.)

1.1.34 Find upper and lower bounds for the size of a minimum (smallest) vertex cover of an n-vertex connected simple graph G. Then draw three 8-vertex graphs, one that achieves the lower bound, one that achieves the upper bound, and one that achieves neither.

1.1.35 Find a minimum vertex cover for the underlying graph of the mixed graph shown in Figure 1.1.7.

1.1.36[S] The graph shown below represents a network of tunnels, where the edges are sections of tunnel and the vertices are junction points. Suppose that a guard placed at a junction can monitor every section of tunnel leaving it. Determine the minimum number of guards and a placement for them so that every section of tunnel is monitored.

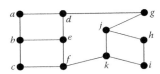

DEFINITION: An **independent set of vertices** in a graph G is a set of mutually non-adjacent vertices. (See §2.3.)

1.1.37 Find upper and lower bounds for the size of a maximum (largest) independent set of vertices in an n-vertex connected graph. Then draw three 8-vertex graphs, one that achieves the lower bound, one that achieves the upper bound, and one that achieves neither.

1.1.38 Find a maximum independent set of vertices in the underlying graph of the mixed graph shown in Figure 1.1.7.

1.1.39[S] Find a maximum independent set of vertices in the graph of Exercise 1.1.36.

DEFINITION: A **matching** in a graph G is a set of mutually non-adjacent edges in G. (See §8.3 and §10.4.)

1.1.40 Find upper and lower bounds for the size of a maximum (largest) matching in an n-vertex connected graph. Then draw three 8-vertex graphs, one that achieves the lower bound, one that achieves the upper bound, and one that achieves neither.

1.1.41 Find a maximum matching in the underlying graph of the mixed graph shown in Figure 1.1.7.

1.1.42^S Find a maximum matching in the graph of Exercise 1.1.36.

DEFINITION: Let $G = (V, E)$ be a graph and let $W \subseteq V$. Then W **dominates** G (or is a **dominating set** of G) if every vertex in V is in W or is adjacent to at least one vertex in W. That is, $\forall v \in V, \exists w \in W, v \in N[w]$.

1.1.43 Find upper and lower bounds for the size of a minimum (smallest) dominating set of an n-vertex graph. Then draw three 8-vertex graphs, one that achieves the lower bound, one that achieves the upper bound, and one that achieves neither.

1.1.44 Find a minimum dominating set for the underlying graph of the mixed graph shown in Figure 1.1.7.

1.1.45^S Find a minimum dominating set for the graph of Exercise 1.1.36.

1.2   COMMON FAMILIES OF GRAPHS

There is a multitude of standard examples that recur throughout graph theory.

Complete Graphs

DEFINITION: A **complete graph** is a simple graph such that every pair of vertices is joined by an edge. Any complete graph on n vertices is denoted K_n.

Example 1.2.1: Complete graphs on one, two, three, four, and five vertices are shown in Figure 1.2.1.

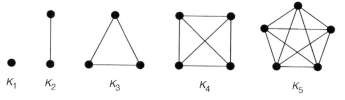

K_1 K_2 K_3 K_4 K_5

Figure 1.2.1   The first five complete graphs.

Bipartite Graphs

DEFINITION: A **bipartite** graph G is a graph whose vertex-set V can be partitioned into two subsets U and W, such that each edge of G has one endpoint in U and one endpoint in W. The pair U, W is called a **(vertex) bipartition** of G, and U and W are called the **bipartition subsets**.

Example 1.2.2: Two bipartite graphs are shown in Figure 1.2.2. The bipartition subsets are indicated by the solid and hollow vertices.

Figure 1.2.2   Two bipartite graphs.

Proposition 1.2.1: *A bipartite graph cannot have any self-loops.*

Proof: This is an immediate consequence of the definition.   ◇

Example 1.2.3: The smallest possible simple graph that is not bipartite is the complete graph K_3, shown in Figure 1.2.3.

Figure 1.2.3   The smallest non-bipartite simple graph.

DEFINITION: A **complete bipartite graph** is a simple bipartite graph such that every vertex in one of the bipartition subsets is joined to every vertex in the other bipartition subset. Any complete bipartite graph that has m vertices in one of its bipartition subsets and n vertices in the other is denoted $K_{m,n}$.[†]

Example 1.2.4: The complete bipartite graph $K_{3,4}$ is shown in Figure 1.2.4.

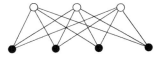

Figure 1.2.4   The complete bipartite graph $K_{3,4}$.

Regular Graphs

DEFINITION: A **regular** graph is a graph whose vertices all have equal degree. A k-**regular** graph is a regular graph whose common degree is k.

DEFINITION: The five regular polyhedra illustrated in Figure 1.2.5 are known as the **platonic solids**. Their vertex and edge configurations form regular graphs called the **platonic graphs**.

DEFINITION: The **Petersen graph** is the 3-regular graph represented by the line drawing in Figure 1.2.6. Because it possesses a number of interesting graph-theoretic properties, the Petersen graph is frequently used both to illustrate established theorems and to test conjectures.

[†]The sense in which $K_{m,n}$ is a unique object is described in §2.1.

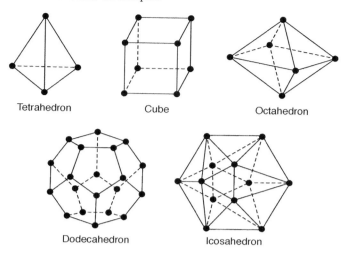

Tetrahedron   Cube   Octahedron

Dodecahedron   Icosahedron

Figure 1.2.5   The five platonic graphs.

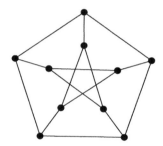

Figure 1.2.6   The Petersen graph.

Example 1.2.5: The oxygen molecule O_2, made up of two oxygen atoms linked by a double bond, can be represented by the 2-regular graph shown in Figure 1.2.7.

Figure 1.2.7   A 2-regular graph representing the oxygen molecule O_2.

Bouquets and Dipoles

One-vertex and two-vertex (non-simple) graphs often serve as building blocks for various interconnection networks, including certain parallel architectures.

DEFINITION: A graph consisting of a single vertex with n self-loops is called a **bouquet** and is denoted B_n.

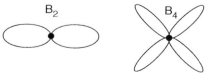

Figure 1.2.8   Bouquets B_2 and B_4.

DEFINITION: A graph consisting of two vertices and n edges joining them is called a **dipole** and is denoted D_n.

Example 1.2.6: The graph representation of the oxygen molecule in Figure 1.2.7 is an instance of the dipole D_2. Figure 1.2.9 shows the dipoles D_3 and D_4.

Figure 1.2.9   Dipoles D_3 and D_4.

Path Graphs and Cycle Graphs

DEFINITION: A **path graph** P is a simple graph with $|V_P| = |E_P| + 1$ that can be drawn so that all of its vertices and edges lie on a single straight line. A path graph with n vertices and $n - 1$ edges is denoted P_n.

Example 1.2.7: Path graphs P_2 and P_4 are shown in Figure 1.2.10.

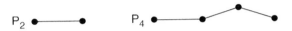

Figure 1.2.10   Path graphs P_2 and P_4.

DEFINITION: A **cycle graph** is a single vertex with a self-loop or a simple graph C with $|V_C| = |E_C|$ that can be drawn so that all of its vertices and edges lie on a single circle. An n-vertex cycle graph is denoted C_n.

Example 1.2.8: The cycle graphs C_1, C_2, and C_4 are shown in Figure 1.2.11.

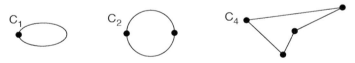

Figure 1.2.11   Cycle graphs C_1, C_2, and C_4.

Remark: The terms *path* and *cycle* also refer to special sequences of vertices and edges within a graph and are defined in §1.4 and §1.5.

Hypercubes and Circular Ladders

DEFINITION: The **hypercube graph** Q_n is the n-regular graph whose vertex-set is the set of bitstrings of length n, such that there is an edge between two vertices if and only if they differ in exactly one bit.

Example 1.2.9: The 8-vertex cube graph that appeared in Figure 1.2.5 is a hypercube graph Q_3 (see Exercises).

DEFINITION: The **circular ladder graph** CLn is visualized as two concentric n-cycles in which each of the n pairs of corresponding vertices is joined by an edge.

Example 1.2.10: The circular ladder graph CL_4 is shown in Figure 1.2.12.

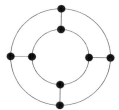

Figure 1.2.12   Circular ladder graph CL_4.

Circulant Graphs

DEFINITION: To the group of integers $Z_n = \{0, 1, \ldots, n-1\}$ under addition modulo n and a set $S \subseteq \{1, \ldots, n-1\}$, we associate the **circulant graph** $circ(n : S)$ whose vertex set is Z_n, such that two vertices i and j are adjacent if and only if there is a number $s \in S$ such that $i + s = j \mod n$ or $j + s = i \mod n$. In this regard, the elements of the set S are called **connections**.

NOTATION: It is often convenient to specify the connection set $S = \{s_1, \ldots, s_r\}$ without the braces, and to write $circ(n : s_1, \ldots, s_r)$.

Example 1.2.11: Figure 1.2.13 shows three circulant graphs.

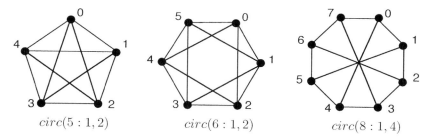

$circ(5 : 1, 2)$ $circ(6 : 1, 2)$ $circ(8 : 1, 4)$

Figure 1.2.13   Three circulant graphs.

Remark: Notice that circulant graphs are simple graphs. Circulant graphs are a special case of **Cayley graphs**, which are themselves derived from a special case of *voltage graphs*.

Intersection and Interval Graphs

DEFINITION: A simple graph G with vertex-set $V_G = \{v_1, v_2 \ldots, v_n\}$ is an **intersection graph** if there exists a family of sets $F = \{S_1, S_2, \ldots, S_n\}$ such that vertex v_i is adjacent to v_j if and only if $i \neq j$ and $S_i \cap S_j \neq \emptyset$.

DEFINITION: A simple graph is an **interval graph** if it is an intersection graph corresponding to a family of intervals on the real line.

Example 1.2.12: The graph in Figure 1.2.14 is an interval graph for the following family of intervals:

$$a \leftrightarrow (1,3) \quad b \leftrightarrow (2,6) \quad c \leftrightarrow (5,8) \quad d \leftrightarrow (4,7)$$

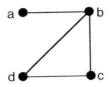

Figure 1.2.14   An interval graph.

Application 1.2.1:   *Archeology* Suppose that a collection of artifacts was found at the site of a town known to have existed from 1000 BC to AD 1000. In the graph shown below, the vertices correspond to artifact types, and two vertices are adjacent if some grave contains artifacts of both types. It is reasonable to assume that artifacts found in the same grave have overlapping time intervals during which they were in use. If the graph is an interval graph, then there is an assignment of subintervals of the interval $(-1000, 1000)$ (by suitable scaling, if necessary) that is consistent with the archeological find.

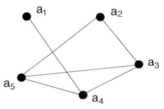

Figure 1.2.15   A graph model of an archeological find.

Line Graphs

Line graphs are a special case of intersection graphs.

DEFINITION: The **line graph** $L(G)$ of a graph G has a vertex for each edge of G, and two vertices in $L(G)$ are adjacent if and only if the corresponding edges in G have a vertex in common.

Thus, the line graph $L(G)$ is the intersection graph corresponding to the endpoint sets of the edges of G.

Example 1.2.13: Figure 1.2.16 shows a graph G and its line graph $L(G)$.

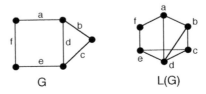

Figure 1.2.16   A graph and its line graph.

EXERCISES for Section 1.2

1.2.1^S Find the number of edges for each of the following graphs.

    a. K_n b. $K_{m,n}$

1.2.2 What is the maximum possible number of edges in a simple bipartite graph on m vertices?

1.2.3 Draw the smallest possible non-bipartite graph.

1.2.4^S Determine the values of n for which the given graph is bipartite.

    a. K_n b. C_n c. P_n

1.2.5 Draw a 3-regular bipartite graph that is not $K_{3,3}$.

In Exercises 1.2.6 to 1.2.8, determine whether the given graph is bipartite. In each case, give a vertex bipartition or explain why the graph is not bipartite.

1.2.6^S   **1.2.7**   **1.2.8**

1.2.9 Label the vertices of the cube graph in Figure 1.2.5 with 3-bit binary strings so that the labels on adjacent vertices differ in exactly one bit.

1.2.10 Prove that the graph Q_n is bipartite.

1.2.11 For each of the platonic graphs, is it possible to trace a *tour* of all the vertices by starting at one vertex, traveling only along edges, never revisiting a vertex, and never lifting the pen off the paper? Is it possible to make the tour return to the starting vertex?

1.2.12^S Prove or disprove: There does not exist a 5-regular graph on 11 vertices.

DEFINITION: A **tournament** is a digraph whose underlying graph is a complete graph.

1.2.13 a. Draw all the 3-vertex tournaments whose vertices are u , v, x.
         b. Determine the number of 4-vertex tournaments whose vertices are u, v, x, y.

1.2.14 Prove that every tournament has at most one vertex of indegree 0 and at most one vertex of outdegree 0.

1.2.15^S Suppose that n vertices $v_1, v_2, \ldots, v_n$ are drawn in the plane. How many different n-vertex tournaments can be drawn on those vertices?

1.2.16 Chartrand and Lesniak [ChLe04] define a pair of sequences of nonnegative integers $\langle a_1, a_2, \ldots a_r \rangle$ and $\langle b_1, b_2, \ldots b_t \rangle$ to be **bigraphical** if there exists a bipartite graph G with bipartition subsets $U = \{u_1, u_2, \ldots u_r\}$ and $W = \{w_1, w_2, \ldots w_t\}$ such that $deg\,(u_i) = a_i, i = 1, 2, \ldots, r$, and $deg(w_i) = b_i, i = 1, 2, \ldots, t$. Prove that a pair of non-increasing sequences of nonnegative integers $\langle a_1, a_2, \ldots, a_r \rangle$ and $\langle b_1, b_2, \ldots, b_t \rangle$ with $r \geq 2$, $0 < a_1 \leq t$, and $0 < a_1 \leq r$ is bigraphical if and only if the pair $\langle a_2, \ldots, a_r \rangle$ and $\langle b_1 - 1, b_2 - 1, \ldots, b_{a_1} - 1, b_{a_1+1}, b_{a_1+2}, \ldots, b_t \rangle$ is bigraphical.

1.2.17 Find all the 4-vertex circulant graphs.

1.2.18 Show that each of the following graphs is a circulant graph.

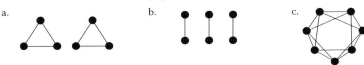

1.2.19 State a necessary and sufficient condition on the positive integers n and k for $circ(\ n : k)$ to be the cycle graph C_n.

1.2.20 Find necessary and sufficient conditions on the positive integers n and k for $circ(n : k)$ to be the graph consisting of $n/2$ mutually non-adjacent edges.

1.2.21 Determine the size of a smallest dominating set (defined in §1.1 exercises) in the graph indicated.

    a. K_n b. $K_{m,n}$ c. C_n d. P_n e. CL_n

1.2.22 Determine the size of a smallest vertex cover (defined in §1.1 exercises) in the graph indicated.

    a. K_n b. $K_{m,n}$ c. C_n d. P_n e. CL_n

1.2.23 Determine the size of a largest independent set of vertices (defined in §1.1 exercises) in the graph indicated.

    a. K_n b. $K_{m,n}$ c. C_n d. P_n e. CL_n

1.2.24 Determine the size of a maximum matching (defined in §1.1 exercises) in the graph indicated.

    a. K_n b. $K_{m,n}$ c. C_n d. P_n e. CL_n

1.2.25[S] Show that the complete graph K_n is an interval graph for all $n \geq 1$.

1.2.26 Draw the interval graph for the intervals $(0,2), (3,8), (1,4), (3,4), (2,5), (7,9)$.

1.2.27[S] Show that the graph modeling the archeological find in Application 1.2.1 is an interval graph by using a family of subintervals of the interval $(-1000, 1000)$.

1.2.28 Prove that the cycle graph C_n is not an interval graph for any $n \geq 4$.

1.2.29 Draw the intersection graph for the family of all subsets of the set $\{1, 2, 3\}$.

1.2.30 Prove that every simple graph is an intersection graph by describing how to construct a suitable family of sets.

1.3   GRAPH MODELING APPLICATIONS

Different kinds of graphs arise in modeling real-world problems. Sometimes, simple graphs are adequate; other times, non-simple graphs are needed. The analysis of some of the applications considered in this section is deferred to later chapters, where the necessary theoretical methods are fully developed.

Models That Use Simple Graphs

Application 1.3.1:   *Personnel-Assignment Problem*   Suppose that a company requires a number of different types of jobs, and suppose each employee is suited for some
of these jobs, but not others. Assuming that each person can perform at most one job
at a time, how should the jobs be assigned so that the maximum number of jobs can
be performed simultaneously? In the bipartite graph of Figure 1.3.1, the hollow vertices
represent the employees, the solid vertices the jobs, and each edge represents a suitable
job assignment. The bold edges represent a largest possible set of suitable assignments.
Chapter 10 provides a fast algorithm to solve large instances of this classical problem in
operations research.

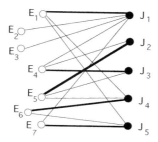

Figure 1.3.1   An optimal assignment of employees to jobs.

Application 1.3.2:   *Sociological-Acquaintance Networks*   In an **acquaintance network**, the vertices represent persons, such as the students in a college class. An edge
joining two vertices indicates that the corresponding pair of persons knew each other
when the course began. The simple graph in Figure 1.3.2 shows a typical acquaintance
network. Including the Socratic concept of *self-knowledge* would require the model to
allow self-loops. For instance, a self-loop drawn at the vertex representing Slim might
mean that she was "in touch" with herself.

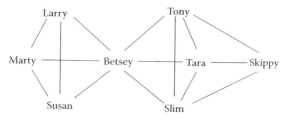

Figure 1.3.2   An acquaintance network.

By Proposition 1.1.1, every group of two or more persons must contain at least two who
know the same number of persons in the group. The acquaintance network of Figure
1.3.2 has degree sequence $\langle 6, 4, 4, 4, 3, 3, 3, 3 \rangle$.

Application 1.3.3:   *Geographic Adjacency*   In the geographical model in Figure
1.3.3, each vertex represents a northeastern state, and each adjacency represents sharing
a border.

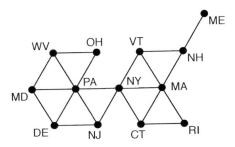

Figure 1.3.3   Geographic adjacency of the northeastern states.

Application 1.3.4:   *Geometric Polyhedra*   The vertex and edge configuration of any polyhedron in 3-space forms a simple graph, which topologists call its **1-skeleton**. The 1-skeletons of the platonic solids, appearing in the previous section, are regular graphs. Figure 1.3.4 shows a polyhedron whose 1-skeleton is not regular.

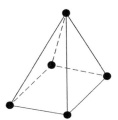

Figure 1.3.4   A non-regular 1-skeleton of a polyhedron.

Application 1.3.5:   *Interconnection Networks for Parallel Architectures*
Numerous processors can be linked together on a single chip for a multi-processor computer that can execute parallel algorithms. In a graph model for such an interconnection network, each vertex represents an individual processor, and each edge represents a direct link between two processors. Figure 1.3.5 illustrates the underlying graph structure of one such interconnection network, called a *wrapped butterfly*.

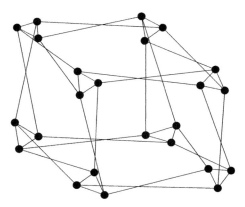

Figure 1.3.5   A wrapped-butterfly-interconnection-network model.

Application 1.3.6: *Assigning Broadcasting Frequencies*   When the radio transmitters in a geographical region are assigned broadcasting frequencies, some pairs of transmitters require different frequencies to avoid interference. A graph model can be used for the problem of minimizing the number of different frequencies assigned.

Suppose that the seven radio transmitters, $A, B, \ldots, G$, must be assigned frequencies. For simplicity, assume that if two transmitters are less than 100 miles apart, they must broadcast at different frequencies. Consider a graph whose vertices represent the transmitters, and whose edges indicate those pairs that are less than 100 miles apart. Figure 1.3.6 shows a table of distances for the seven transmitters and the corresponding graph on seven vertices.

	B	C	D	E	F	G
A	55	110	108	60	150	88
B		87	142	133	98	139
C			77	91	85	93
D				75	114	82
E					107	41
F						123

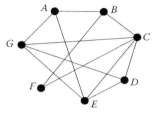

Figure 1.3.6   A simple graph for a radio-frequency-assignment problem.

The problem of assigning radio frequencies to avoid interference is equivalent to the problem of *coloring* the vertices of the graph so that adjacent vertices get different colors. The minimum number of frequencies will equal the minimum number of different colors required for such a coloring. This and several other graph-coloring problems and applications are discussed in Chapter 8.

Models Requiring Non-Simple Graphs

Application 1.3.7: *Roadways between States*   If in the Geographic-Adjacency Application 1.3.3, each edge joining two vertices represented a road that crosses a border between the corresponding two states, then the graph would be non-simple, since pairs of bordering states have more than one road joining them.

Application 1.3.8: *Chemical Molecules*   The benzene molecule shown in Figure 1.3.7 has double bonds for some pairs of its atoms, so it is modeled by a non-simple graph. Since each carbon atom has valence 4, corresponding to four electrons in its outer shell, it is represented by a vertex of degree 4; and since each hydrogen atom has one electron in its only shell, it is represented by a vertex of degree 1.

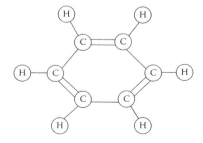

Figure 1.3.7   The benzene molecule.

Models That Use Simple Digraphs

For each of the next series of applications, a link in one direction does not imply a link in the opposite direction.

Application 1.3.9:   *Ecosystems*   The feeding relationships among the plant and animal species of an ecosystem are called a *food web* and may be modeled by a simple digraph. The food web for a Canadian willow forest is illustrated in Figure 1.3.8.[†] Each species in the system is represented by a vertex, and a directed edge from vertex u to vertex v means that the species corresponding to u feeds on the species corresponding to v.

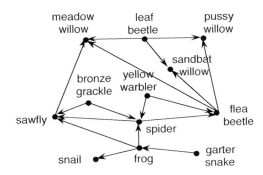

Figure 1.3.8   The food web in a Canadian willow forest.

Application 1.3.10:   *Activity-Scheduling Networks*   In large projects, often there are some tasks that cannot start until certain others are completed. Figure 1.3.9 shows a digraph model of the *precedence relationships* among some tasks for building a house. Vertices correspond to tasks. An arc from vertex u to vertex v means that task v cannot start until task u is completed. To simplify the drawing, arcs that are implied by transitivity are not drawn. This digraph is the *cover diagram* of a partial ordering of the tasks. A different model, in which the tasks are represented by the arcs of a digraph, is studied in Chapter 9.

	Activities
1	Foundation
2	Walls and Ceilings
3	Roof
4	Electrical wiring
5	Windows
6	Siding
7	Paint interior
8	Paint exterior

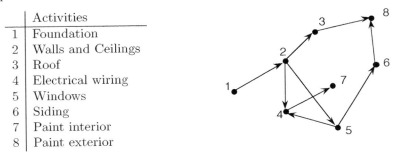

Figure 1.3.9   An activity digraph for building a house.

[†]This illustration was adapted from [WiWa90], p. 69.

Application 1.3.11:   *Flow Diagrams for Computer Programs*   A computer program is often designed as a collection of *program blocks*, with appropriate flow control. A digraph is a natural model for this decomposition. Each program block is associated with a vertex, and if control at the last instruction of block u can transfer to the first instruction of block v, then an arc is drawn from vertex u to vertex v. Computer flow diagrams do not usually have multi-arcs. Unless a single block is permitted both to change values of some variables and to retest those values, a flow diagram with a self-loop would mean that the program has an infinite loop.

Models Requiring Non-Simple Digraphs

Application 1.3.12:   *Markov Diagrams*   Suppose that the inhabitants of some remote area purchase only two brands of breakfast cereal, O's and W's. The consumption patterns of the two brands are encapsulated by the *transition matrix* shown in Figure 1.3.10. For instance, if someone just bought O's, there is a 0.4 chance that the person's next purchase will be W's and a 0.6 chance it will be O's.

In a *Markov process*, the *transition probability* of going from one state to another depends only on the current state. Here, states "O" and "W" correspond to whether the most recent purchase was O's or W's, respectively. The digraph model for this Markov process, called a *Markov diagram*, is shown in Figure 1.3.10. Each arc is labeled with the transition probability of moving from the state at the tail vertex to the state at the head. Thus, the probabilities on the outgoing edges from each vertex must sum to 1. This Markov diagram is an example of a *weighted graph* (see §1.6). Other examples of Markov diagrams appear in Chapter 9.

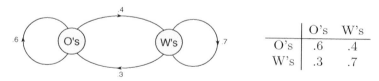

	O's	W's
O's	.6	.4
W's	.3	.7

Figure 1.3.10   A Markov diagram and its transition matrix.

Application 1.3.13:   *Lexical Scanners*   The source code of a computer program may be regarded as a string of symbols. A *lexical scanner* must scan these symbols, one at a time, and recognize which symbols "go together" to form a syntactic *token* or *lexeme*. We now consider a single-purpose scanner whose task is to recognize whether an input string of characters is a valid *identifier* in the C programming language. Such a scanner is a special case of a *finite-state recognizer* and can be modeled by a labeled digraph, as in Figure 1.3.11. One vertex represents the *start* state, in effect before any symbols have been scanned. Another represents the *accept* state, in which the substring of symbols scanned so far forms a valid C identifier. The third vertex is the *reject* state, indicating that the substring has been discarded because it is not a valid C identifier. The arc labels tell what kinds of symbols cause a transition from the tail state to the head state. If the final state after the input string is completely scanned is the accept state, then the string is a valid C identifier.

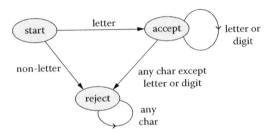

Figure 1.3.11   Finite-state recognizer for identifiers.

EXERCISES for Section 1.3

1.3.1[S] Solve the radio-frequency-assignment problem in Application 1.3.6 by determining the minimum number of colors needed to color the vertices of the associated graph. Argue why the graph cannot be colored with fewer colors.

1.3.2 What is wrong with a computer program having the following abstract flow pattern?

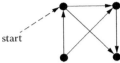

1.3.3[S] Referring to the Markov diagram in Application 1.3.12, suppose that someone just purchased a box of O's. What is the probability that his next three purchases are W's, O's, and then W's?

1.3.4 Modify the finite-state recognizer in Application 1.3.13 so that it accepts only those strings that begin with two letters and whose remaining characters, if any, are digits. (Hint: Consider adding one or more new states.)

1.3.5[S] Which strings are accepted by the following finite-state recognizer?

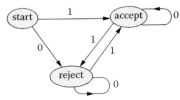

In Exercises 1.3.6 through 1.3.13, design appropriate graph or digraph models and problems for the given situation. That is, specify the vertices and edges and any additional features.

1.3.6[S] Two-person rescue teams are being formed from a pool of n volunteers from several countries. The only requirement is that both members of a team must have a language in common.

1.3.7 Suppose that meetings must be scheduled for several committees. Two committees must be assigned different meeting times if there is at least one person on both committees.

1.3.8 Represent the "relative strength" of a finite collection of logical propositional forms. For example, the proposition $p \wedge q$ is at least as strong as $p \vee q$ since the first implies the second (i.e., $(p \wedge q) \Rightarrow (p \vee q)$ is a *tautology*).

1.3.9 Suppose there are three power generators to be built in three of the seven most populated cities of a certain country. The distances between each pair of cities is given in the table shown. One would like to situate the generators so that each city is within 50 miles of at least one generator.

	B	C	D	E	F	G
A	80	110	15	60	100	80
B		40	45	55	70	90
C			65	20	50	80
D				35	55	40
E					25	60
F						70

1.3.10 Suppose there are k machines and l jobs, and each machine can do only a subset of the jobs. a) Draw a graph to model this situation. b) Express in terms of your graph model, the problem of assigning jobs to machines so that the maximum number of jobs can be performed simultaneously.

1.3.11 A bridge tournament for five teams is to be scheduled so that each team plays two other teams.

1.3.12 Let R be a binary *relation* on a set S. (Relations are discussed in Appendix A.2.)

   a. Describe a digraph model for a binary relation on a finite set.

   b. Draw the digraph for the relation R on the set $S = \{1, 2, 3, 4, 5\}$, given by

$$R = \{(1,2), (2,1), (1,1), (1,5), (4,5), (3,3)\}.$$

1.3.13^S Describe in graph-theory terms, the digraph properties corresponding to each of the following possible properties of binary relations.

   a. reflexive;   b. symmetric;   c. transitive;   d. antisymmetric.

1.4   WALKS AND DISTANCE

Many applications call for graph models that can represent traversal and distance. For instance, the number of node-links traversed by an email message on its route from sender to recipient is a form of distance. Beyond physical distance, another example is that a sequence of tasks in an activity-scheduling network forms a *critical path* if a delay in any one of the tasks would cause a delay in the overall project completion. This section and the following one clarify the notion of walk and related terminology.

Walks and Directed Walks

In proceeding continuously from a starting vertex to a destination vertex of a physical representation of a graph, one would alternately encounter vertices and edges. Accordingly, a *walk* in a graph is modeled by such a sequence.

DEFINITION: In a graph G, a **walk** from vertex v_0 to vertex v_n is an alternating sequence

$$W = \langle v_0, e_1, v_1, e_2, \ldots, v_{n-1}, e_n, v_n \rangle$$

of vertices and edges, such that $endpts(e_i) = \{v_{i-1}, v_i\}$, for $i = 1, \ldots, n$. If G is a digraph (or mixed graph), then W is a **directed walk** if each edge e_i is directed from vertex v_{i-1} to vertex v_i, i.e., $tail(e_i) = v_{i-1}$ and $head(e_i) = v_i$.

In a simple graph, there is only one edge between two consecutive vertices of a walk, so one could abbreviate the representation as a vertex sequence

$$W = \langle v_0, v_1, \ldots, v_n \rangle$$

In a general graph, one might abbreviate the representation as an edge sequence from the starting vertex to the destination vertex

$$W = \langle v_0, e_1, e_2, \ldots, e_n, v_n \rangle$$

TERMINOLOGY: A walk (or directed walk) from a vertex x to a vertex y is also called an x-y **walk** (or x-y **directed walk**).

DEFINITION: The **length** of a walk or directed walk is the number of edge-steps in the walk sequence.

DEFINITION: A walk of length 0, i.e., with one vertex and no edges, is called a **trivial walk**.

DEFINITION: A **closed walk** (or **closed directed walk**) is a walk (or directed walk) that begins and ends at the same vertex. An **open walk** (or **open directed walk**) begins and ends at different vertices.

Example 1.4.1: In Figure 1.4.1, there is an open walk of length 6,

$$< \text{OH, PA, NY, VT, MA, NY, PA} >$$

that starts at Ohio and ends at Pennsylvania. Notice that the given walk is an inefficient route, since it contains two repeated vertices and retraces an edge. §1.5 establishes the terminology necessary for distinguishing between walks that repeat vertices and/or edges and those that do not.

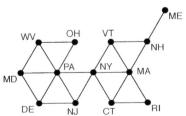

Figure 1.4.1   Geographic adjacency of the northeastern states.

Example 1.4.2: In the Markov diagram from Application 1.3.12, shown in Figure 1.4.2, the choice sequence of a cereal eater who buys O's, switches to W's, sticks with W's for two more boxes, and then switches back to O's is represented by the closed directed walk

$$< \text{O, W, W, W, O} >$$

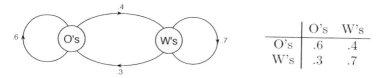

Figure 1.4.2   Markov process from Application 1.3.12.

The product of the transition probabilities along a walk in any Markov diagram equals the probability that the process will follow that walk during an experimental trial. Thus, the probability that this walk occurs, when starting from O's equals $.4 \times .7 \times .7 \times .3 = 0.0588$.

Example 1.4.3: In the lexical scanner of Application 1.3.13, the identifier *counter12* would generate an open directed walk of length 9 from the start vertex to the accept vertex.

DEFINITION: The **concatenation** of two walks $W_1 = \langle v_0, e_1, \ldots, v_{k-1}, e_k, v_k \rangle$ and $W_2 = \langle v_k, e_{k+1}, v_{k+1}, e_{k+2}, \ldots, v_{n-1}, e_n, v_n \rangle$ such that walk W_2 begins where walk W_1 ends, is the walk

$$W_1 \circ W_2 = \langle v_0, e_1, \ldots, v_{k-1}, e_k, v_k, e_{k+1}, \ldots, v_{n-1}, e_n, v_n \rangle$$

Example 1.4.4: Figure 1.4.3 shows the concatenation of a walk of length 2 with a walk of length 3.

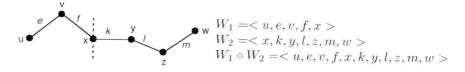

$$W_1 = < u, e, v, f, x >$$
$$W_2 = < x, k, y, l, z, m, w >$$
$$W_1 \circ W_2 = < u, e, v, f, x, k, y, l, z, m, w >$$

Figure 1.4.3   Concatenation of two walks.

DEFINITION: A **subwalk** of a walk $W = \langle v_0, e_1, v_1, e_2, \ldots, v_{n-1}, e_n, v_n \rangle$ is a subsequence of consecutive entries $S = \langle v_j, e_{j+1}, v_{j+1}, \ldots, e_k, v_k \rangle$ such that $0 \le j \le k \le n$, that begins and ends at a vertex. Thus, the subwalk is itself a walk.

Example 1.4.5: In the figure below, the closed directed walk $\langle v, x, y, z, v \rangle$ is a subwalk of the open directed walk $\langle u, v, x, y, z, v, w, t \rangle$.

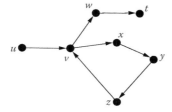

Distance

DEFINITION: The **distance** $d(s, t)$ from a vertex s to a vertex t in a graph G is the length of a shortest $s - t$ walk if one exists; otherwise, $d(s, t) = \infty$. For digraphs, the **directed distance** $\vec{d}(s, t)$ is the length of a shortest directed walk from s to t.

Example 1.4.6: In Figure 1.4.4, the distance from West Virginia to Maine is five. That is, starting in West Virginia, one cannot get to Maine without crossing at least five state borders.

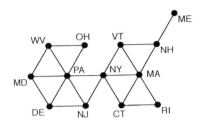

Figure 1.4.4   Geographic adjacency of the northeastern states.

A shortest walk (or directed walk) contains no repeated vertices or edges (see Exercises). It is instructive to think about how one might find a shortest walk. Ad hoc approaches are adequate for small graphs, but a systematic algorithm is essential for larger graphs (see §4.3).

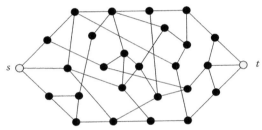

Figure 1.4.5   How might you find a shortest walk from s to t in this graph?

Eccentricity, Diameter, and Radius

DEFINITION: The **eccentricity** of a vertex v in a graph G, denoted $ecc(v)$, is the distance from v to a vertex farthest from v. That is,

$$ecc(v) = \max_{x \in V_G} \{d(v,x)\}$$

DEFINITION: The **diameter** of a graph G, denoted $diam(G)$, is the maximum of the vertex eccentricities in G or, equivalently, the maximum distance between two vertices in G. That is,

$$diam(G) = \max_{x \in V_G} \{ecc(x)\} = \max_{x,y \in V_G} \{d(x,y)\}$$

DEFINITION: The **radius** of a graph G, denoted $rad(G)$, is the minimum of the vertex eccentricities. That is,

$$rad(G) = \min_{x \in V_G} \{ecc(x)\}$$

DEFINITION: A **central vertex** v of a graph G is a vertex with minimum eccentricity. Thus, $ecc(v) = rad(G)$.

Example 1.4.7: The graph of Figure 1.4.6 has diameter 4, achieved by the vertex pairs u, v and u, w. The vertices x and y have eccentricity 2 and all other vertices have greater eccentricity. Thus, the graph has radius 2 and central vertices x and y.

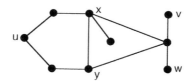

Figure 1.4.6   A graph with diameter 4 and radius 2.

Example 1.4.8: Let G be the graph whose vertex-set is the set of all people on the planet and whose edges correspond to the pairs of people who are acquainted. Then according to the *six degrees of separation* conjecture, the graph G has diameter 6.

Connectedness

DEFINITION: Vertex v is **reachable from** vertex u if there is a walk from u to v.

DEFINITION: A graph is **connected** if for every pair of vertices u and v, there is a walk from u to v.

DEFINITION: A digraph is **connected** if its underlying graph is connected.

TERMINOLOGY NOTE: Some authors use the term *weakly connected* digraph instead of connected digraph.

Example 1.4.9: The non-connected graph in Figure 1.4.7 is made up of connected pieces called *components*, and each component consists of vertices that are all reachable from one another. This concept is defined formally in §2.3.

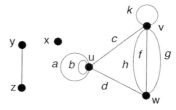

Figure 1.4.7   A non-connected graph with three components.

Strongly Connected Digraphs

DEFINITION: Two vertices u and v in a digraph D are said to be **mutually reachable** if D contains both a directed u-v walk and a directed v-u walk.

DEFINITION: A digraph D is **strongly connected** if every two of its vertices are mutually reachable.

Example 1.4.10: The digraph shown in Figure 1.4.8 is strongly connected.

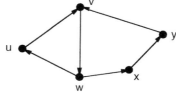

Figure 1.4.8   A strongly connected digraph.

Example 1.4.11: Suppose the graph in Figure 1.4.9 represents the network of roads in a certain national park. The park officials would like to make all of the roads one-way, but only if visitors will still be able to drive from anyone intersection to any other.

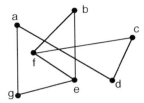

Figure 1.4.9   A national park's system of roads.

This problem can be expressed in graph-theoretic terms with the following definition.

DEFINITION: A graph is said to be ***strongly orientable*** if there is an assignment of directions to its edges so that the resulting digraph is strongly connected. Such an assignment is called a ***strong orientation***.

Chapter 9 includes a characterization of strongly orientable graphs, which leads to the design of a polynomial-time algorithm for determining whether a graph is strongly orientable.

Application of Connectedness to Rigidity

Application 1.4.1:   *Rigidity of Rectangular Frameworks*[†]   Consider a 2-dimensional framework of steel beams connected by joints that allow each beam to swivel in the plane. The framework is said to be ***rigid*** if none of its beams can swivel. The rectangle shown below is not rigid, since it can be deformed into a parallelogram.

Adding one diagonal brace would make the framework rigid, since the brace keeps the horizontal and vertical edges perpendicular to each other.

[†] No subsequent material depends on this application.

A rectangular framework with some diagonal braces might be rigid or non-rigid. For example, framework B in Figure 1.4.10 is rigid, but framework A is not.

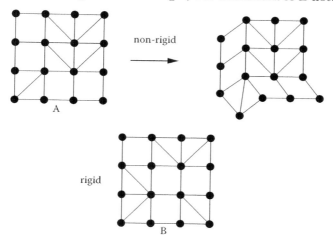

Figure 1.4.10   Two rectangular frameworks.

Surprisingly, the problem of determining whether a given framework is rigid can be reduced to a simple problem concerning connectivity in a bipartite graph.

Transforming the Problem: Suppose that an imaginary line is drawn through the middle of each row and each column of rectangles in the framework, as illustrated in the figure below. The framework is rigid if and only if each row-line r_i stays perpendicular to each column-line c_j, $1, 2 \ldots$.

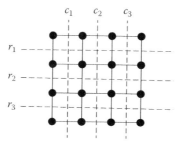

Illustration: Consider the following rigid 2-by-2 framework.

The brace in row 1, column 2 keeps $r_1 \perp c_2$. Similarly, the other two braces keep $c_2 \perp r_2$ and $r_2 \perp c_1$. Thus, each pair of consecutive entries in the sequence r_1, c_2, r_2, c_1 is perpendicular. This implies, by simple plane geometry, that $r_1 \perp c_1$, even though the $(1, 1)$ rectangle is not braced. Therefore, the framework is rigid.

More generally, a framework is rigid if for each i and j there is a perpendicularity sequence that begins with r_i and ends with c_j. This observation leads to a simple method for determining whether a given rectangular framework is rigid.

Construct a bipartite graph by first drawing two sets of vertices, $r_1, r_2, \ldots$ and $c_1, c_2, \ldots$, one set corresponding to the rows of the framework, and the other set corresponding to the columns. Then draw an edge between vertices r_i and c_j if there is a brace in the row i, column j rectangle of the framework.

By the observation above regarding perpendicularity sequences, the framework is rigid if and only if there is a walk in the bipartite graph between every pair of vertices, r_i and c_j. This simple criterion takes the form of the following theorem.

Theorem 1.4.1: *A rectangular framework is rigid if and only if its associated bipartite graph is connected.* ◇

Illustration: Figure 1.4.11 shows the two frameworks A and B from Figure 1.4.11 along with their associated bipartite graphs. Notice that for the non-rigid framework A, each deformable rectangle corresponds to a pair of vertices in the bipartite graph that has no walk between them. On the other hand, for the rigid framework B, the bipartite graph is connected.

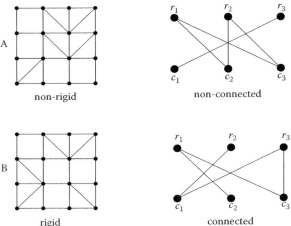

Figure 1.4.11   Sample applications for Theorem 1.4.1 on rigid frameworks.

EXERCISES for Section 1.4

1.4.1^{S} Determine which of the following vertex sequences represent walks in the graph below.

    a. $\langle u, v \rangle$;   b. $\langle v \rangle$;   c. $\langle u, z, v \rangle$;   d. $\langle u, v, w, v \rangle$

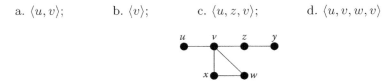

1.4.2 Find all walks of length 4 or 5 from vertex w to vertex r in the following graph.

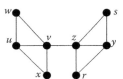

1.4.3 Find all directed walks from x to d in the following digraph.

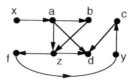

1.4.4^S Which of the following incidence tables represent connected graphs?

edge	a	b	c	d
endpts	u	v	w	x
	u	w	x	v

edge	a	b	c	d
endpts	u	v	v	x
	v	w	x	v

edge	a	b	c	d
endpts	u	v	w	x
	v	v	x	x

1.4.5 Express the walk $\langle u, v, w, x, v, z, y, z, v\rangle$ as the concatenation of a series of walks such that each walk has no repeated vertices.

In Exercises 1.4.6 through 1.4.8, determine whether the given incidence table represents a strongly connected digraph. Assume that there are no isolated vertices.

1.4.6^S

edges	a	b	c	d
endpts	u	v	v	x
	v^h	w^h	x^h	v^h

1.4.7

edges	a	b	c	d
endpts	u	v^h	w	x
	v^h	w	x^h	v^h

1.4.8

edges	a	b	c	d	e
endpts	u	v	w^h	v	w
	v^h	u^h	x	x^h	v^h

1.4.9 Explain why all graphs having at least one edge have walks of arbitrarily large length. Can you make the same statement for digraphs?

1.4.10^S Draw a simple digraph that has directed walks of arbitrarily large length.

1.4.11 For each of the following graphs, either find a strong orientation or argue why no such orientation exists.

1.4.12^S Find the distance between vertices x and y in the following graph.

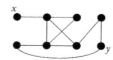

1.4.13 Is there an even-length walk between two antipodal vertices (i.e., endpoints of a long diagonal) of a cube?

1.4.**14** Draw an 8-vertex connected graph with as few edges as possible.

1.4.**15**$^\text{S}$ Draw a 7-vertex connected graph such that the removal of any one edge results in a non-connected graph.

1.4.**16** Draw an 8-vertex connected graph with no closed walks, except those that retrace at least one edge.

1.4.**17** Draw a 5-vertex connected graph that remains connected after the removal of any two of its vertices.

In Exercises 1.4.18 through 1.4.28, determine the diameter, radius, and central vertices of the graph indicated.

1.4.**18**$^\text{S}$ The graph in Exercise 1.4.1.

1.4.**19** The graph in Exercise 1.4.2.

1.4.**20** Path graph $P_n, n \geq 3$ (from §1.2).

1.4.**21** Cycle graph $C_n, n \geq 4$ (from §1.2).

1.4.**22** Complete graph K_n , $n \geq 3$.

1.4.**23** Complete bipartite graph $K_{m,n}$, $m \geq n \geq 3$ (from §1.2).

1.4.**24** The Petersen graph (from §1.2).

1.4.**25** Hypercube graph Q_3; can you generalize to Q_n? (from §1.2).

1.4.**26** Circular ladder graph CL_n , $n \geq 4$ (from §1.2).

1.4.**27** Dodecahedral graph (from §1.2).

1.4.**28** Icosahedral graph (from §1.2).

1.4.**29** Determine the radius and diameter of each of the following circulant graphs (from §1.2).

      a. $circ(5:1,2)$;   b. $circ(6:1,2)$;   c. $circ(8:1,2)$.

1.4.**30** Describe the diameter and radius of the circulant graph $circ(n:m)$ in terms of integer n and connection m.

1.4.**31** Describe the diameter and radius of the circulant graph $circ(n:a,b)$ in terms of integer n and the connections a and b.

1.4.**32**$^\text{S}$ Let G be a connected graph. Prove that

$$rad\,(G) \leq diam\,(G) \leq 2 \cdot rad\,(G)$$

1.4.**33** Prove that a digraph must have a closed directed walk if each of its vertices has nonzero outdegree.

In Exercises 1.4.34 through 1.4.36, use a bipartite graph model to determine whether the given rectangular framework is rigid.

1.4.**34**^{S}

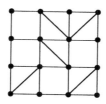

1.4.**35**

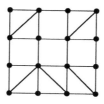

1.4.**36**

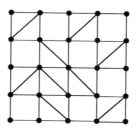

1.4.**37** Show that the framework shown below is not rigid. Can moving one brace make the framework rigid?

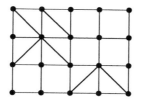

1.4.**38**^{S} Let u and v be any two vertices of a connected graph G. Prove that there exists a $u-v$ walk containing all the vertices of G.

1.4.**39** Prove that a shortest walk between two vertices cannot repeat a vertex or an edge.

1.4.**40** Let D be a digraph. Prove that mutual reachability is an equivalence relation on V_D.

1.4.**41**^{S} Let x and y be two different vertices in the complete graph K_4. Find the number of x-y walks of length 2 and of length 3.

1.4.**42** Let x and y be two different adjacent vertices in the complete bipartite graph $K_{3,3}$. Find the number of x-y walks of length 2 and of length 3.

1.4.43 Let x and y be two different non-adjacent vertices in the complete bipartite graph $K_{3,3}$. Find the number of x-y walks of length 2 and of length 3.

1.4.44 Prove that the distance function d on a graph G satisfies the *triangle inequality*. That is,

$$\text{For all } x, y, z \in V_G, \quad d(x, z) \leq d(x, y) + d(y, z).$$

1.4.45 Let G be a connected graph. Prove that if $d(x, y) \geq 2$, then there exists a vertex w such that $d(x, y) = d(x, w) + d(w, y)$.

1.4.46^S Let G be an n-vertex simple graph such that $\deg(v) \geq \frac{(n-1)}{2}$ for every vertex $v \in V_G$. Prove that graph G is connected.

1.5   PATHS, CYCLES, AND TREES

By stripping away its extraneous *detours*, an x-y walk can eventually be reduced to one that has no repetition of vertices or edges. Working with these shorter x-y walks simplifies discussion in many situations.

Trails and Paths

DEFINITION: A **trail** is a walk with no repeated edges.

DEFINITION: A **path** is a trail with no repeated vertices (except possibly the initial and final vertices).

DEFINITION: A walk, trail, or path is **trivial** if it has only one vertex and no edges.

TERMINOLOGY NOTE: Unfortunately, there is no universally agreed-upon terminology for walks, trails, and paths. For instance, some authors use the term *path* instead of walk and *simple path* instead of path. Others use the term *path* instead of trail. It is best to check the terminology when reading articles about this material.

Example 1.5.1: For the graph shown in Figure 1.5.1, $W = \langle v, a, e, f, a, d, z \rangle$ is the edge sequence of a walk but not a trail, because edge a is repeated, and $T = \langle v, a, b, c, d, e, u \rangle$ is a trail but not a path, because vertex x is repeated.

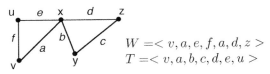

$$W = <v, a, e, f, a, d, z>$$
$$T = <v, a, b, c, d, e, u>$$

Figure 1.5.1   Walk W is not a trail, and trail T is not a path.

Remark: Directed trails and directed paths are directed walks that satisfy conditions analogous to those of their undirected counterparts. For example, a directed trail is a directed walk in which no directed edge appears more than once. When it is clear from the context that the objects under discussion are directed, then the modifier *directed* will be omitted.

Deleting Closed Subwalks from Walks

DEFINITION: Given a walk $W = \langle v_0, e_1, \ldots, v_{n-1}, e_n, v_n \rangle$ that contains a nontrivial, closed subwalk $W' = \langle v_k, e_{k+1}, \ldots, v_{m-1}, e_m, v_k \rangle$, the **reduction of walk W by subwalk W'**, denoted $W - W'$, is the walk

$$W - W' = \langle v_0, e_1, \ldots, v_{k-1}, e_k, v_k, e_{m+1}, v_{m+1}, \ldots, v_{n-1}, e_n \rangle$$

Thus, $W - W'$ is obtained by deleting from W all of the vertices and edges of W' except v_k. Or, less formally, $W - W'$ is obtained by **deleting W' from W**.

Example 1.5.2: Figure 1.5.2 shows a simple graph and the edge sequences of three walks, W, W', and $W - W'$. Observe that $W - W'$ is traversed by starting a traversal of walk W at vertex u and avoiding the detour W' at vertex v by continuing directly to vertex w.

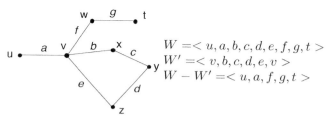

$$W = \langle u, a, b, c, d, e, f, g, t \rangle$$
$$W' = \langle v, b, c, d, e, v \rangle$$
$$W - W' = \langle u, a, f, g, t \rangle$$

Figure 1.5.2 Walk W is not a trail, and trail T is not a path.

Remark: The reduction of a walk by a (nontrivial) closed subwalk yields a walk that is obviously shorter than the original walk.

DEFINITION: Given two walks A and B, the walk B is said to be a **reduced walk** of A if there exists a sequence of walks $A = W_1, W_2, \ldots, W_r = B$ such that for each $i = 1, \ldots, r - 1$, walk W_{i+1} is the reduction of W_i by some closed subwalk of W_i.

   The next three results show that the concepts of distance and reachability in a graph can be defined in terms of paths instead of walks. This often simplifies arguments in which the detours of a walk can be ignored.

Lemma 1.5.1: *Every open x-y walk W is either an x-y path or contains a closed subwalk.*

Proof: If W is not an x-y path, then the subsequence of W between repeated vertices defines a closed subwalk of W.   ◇

Theorem 1.5.2: *Let W be an open x-y walk. Then either W is an x-y path or there is an x-y path that is a reduced walk of W.*

Proof: If W is not an x-y path, then delete a closed subwalk from W to obtain a shorter x-y walk. Repeat the process until the resulting x-y walk contains no closed subwalks and, hence, is an x-y path, by Lemma 1.5.1.   ◇

Corollary 1.5.3: *The distance from a vertex x to a reachable vertex y is always realizable by an x-y path.*

Proof: A shortest x-y walk must be an x-y path, by Theorem 1.5.2.   ◇

Cycles

DEFINITION: A nontrivial closed path is called a **cycle**.

DEFINITION: An **acyclic graph** is a graph that has no cycles.

DEFINITION: A cycle that includes every vertex of a graph is call a **Hamiltonian cycle.**[†]

DEFINITION: A **Hamiltonian graph** is a graph that has a Hamiltonian cycle. (§6.3 elaborates on Hamiltonian graphs.)

Example 1.5.3: The graph in Figure 1.5.3 is Hamiltonian. The edges of the Hamiltonian cycle $\langle u, z, y, x, w, t, v, u \rangle$ are shown in bold.

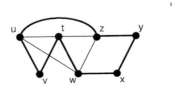

Figure 1.5.3   A Hamiltonian graph.

The following result gives an important characterization of bipartite graphs.

Theorem 1.5.4: *A graph G is bipartite if and only if it has no cycles of odd length.*

Proof: *Necessity* ($\Rightarrow$): Suppose that G is bipartite. Since traversing each edge in a walk switches sides of the bipartition, it requires an even number of steps for a walk to return to the side from which it started. Thus, a cycle must have even length.

Sufficiency ($\Leftarrow$): Let G be a graph with $n \geq 2$ vertices and no cycles of odd length. Without loss of generality, assume that G is connected. Pick any vertex u of G, and define a partition (X, Y) of V as follows:

$$X = \{x \mid d(u, x) \text{ is even}\}$$
$$Y = \{y \mid d(u, y) \text{ is odd}\}$$

If (X, Y) is not a bipartition of G, then there are two vertices in one of the sets, say v and w, that are joined by an edge, say e. Let P_1 be a shortest $u - v$ path, and let P_2 be a shortest $u - w$ path. By definition of the sets X and Y, the lengths of these paths are both even or both odd. Starting from vertex u, let z be the last vertex common to both paths (see Figure 1.5.4).

Since P_1 and P_2 are both shortest paths, their $u \to z$ sections have equal length. Thus, the lengths of the $z \to v$ section of P_1 and the $z \to w$ section of P_2 are either both even or both odd. But then the concatenation of those two sections with edge e forms a cycle of odd length, contradicting the hypothesis. Hence, (X, Y) is a bipartition of G. $\diamond$

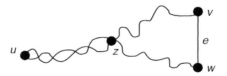

Figure 1.5.4   Figure for sufficiency part of Theorem 1.5.4 proof.

[†]The term "Hamiltonian" is in honor of the Irish mathematician William Rowan Hamilton.

The following proposition is similar to Theorem 1.5.2.

Proposition 1.5.5: *Every nontrivial, closed trail T contains a subwalk that is a cycle.*

Proof: Let T' be a minimum-length, nontrivial, closed subwalk of T. Its minimum length implies that T' has no proper closed subwalks, and, hence, its only repeated vertices are its first and last. Thus, T' is a cycle. $\diamond$

Remark: The assertion of Proposition 1.5.5 is no longer true if T is merely a closed walk (see Exercises).

DEFINITION: A collection of edge-disjoint cycles, $C_1, C_2, \ldots, C_m$, is called a **decomposition** of a closed trail T if each cycle C_i is either a subwalk or a reduced walk of T and the edge-sets of the cycles *partition* the edge-set of trail T.

Theorem 1.5.6: *A closed trail can be decomposed into edge-disjoint cycles.*

Proof: A closed trail having only one edge is itself a cycle. By way of induction, assume that the theorem is true for all closed trails having m or fewer edges. Next, let T be a closed trail with $m+1$ edges. By Proposition 1.5.5, the trail T contains a cycle, say C, and $T - C$ is a closed trail having m or fewer edges. By the induction hypothesis, the trail $T - C$ can be decomposed into edge-disjoint cycles. Therefore, these cycles together with C form an edge-decomposition of trail T. $\diamond$

Example 1.5.4: The three cycles C_1, C_2, and C_3 form a decomposition of the closed trail T in Figure 1.5.5. The edge-based abbreviated representation of walks is used here to emphasize the edge partition of the decomposition.

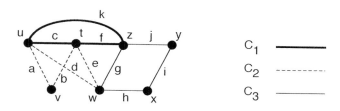

$$T = <u, a, b, c, d, e, f, g, h, i, j, k, u>$$

$$C_1 = <z, k, c, f, z>; \quad C_2 = <u, a, b, e, d, u>; \quad C_3 = <w, h, i, j, g, w>$$

Figure 1.5.5   A closed trail T decomposed into three cycles.

Eulerian Trails

DEFINITION: An **Eulerian trail** in a graph is a trail that contains every edge of that graph.

DEFINITION: An **Eulerian tour** is a closed Eulerian trail.

DEFINITION: An **Eulerian graph** is a connected graph that has an Eulerian tour.[†]

[†]The term "Eulerian" is in honor of the Swiss mathematician Leonhard Euler.

Example 1.5.5: The trail $T = \langle u, a, b, c, d, e, f, g, h, i, j, k, u \rangle$ in Figure 1.5.6 is an Eulerian tour. Its decomposition into edge-disjoint cycles, as in Figure 1.5.5, is a property that holds for all Eulerian tours and is part of a classical characterization of Eulerian graphs given in §4.5.

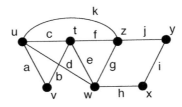

Figure 1.5.6   An Eulerian graph.

Application 1.5.1:   *Traversing the Edges of a Network*   Suppose that the graph shown in Figure 1.5.6 represents a network of railroad tracks. A special cart equipped with a sensing device will traverse every section of the network, checking for imperfections. Can the cart be routed so that it traverses each section of track exactly once and then returns to its starting point?

The problem is equivalent to determining whether the graph is Eulerian, and as seen from Example 1.5.5, trail T does indeed provide the desired routing.

Algorithms to construct Eulerian tours and several other applications involving Eulerian graphs appear in Chapter 6.

Girth

DEFINITION: The **girth** of a graph G with at least one cycle is the length of a shortest cycle in G. The girth of an acyclic graph is undefined.

Example 1.5.6: The girth of the graph in Figure 1.5.7 below is 3 since there is a 3-cycle but no 2-cycle or 1-cycle.

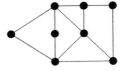

Figure 1.5.7   A graph with girth 3.

Trees

DEFINITION: A **tree** is a connected graph that has no cycles.

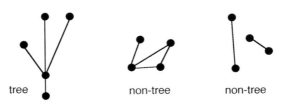

Figure 1.5.8   A tree and two non-trees.

Trees are central to the structural understanding of graphs, and they have a wide range of applications, including information processing. Trees often occur with additional attributes such as roots and vertex orderings. Chapters 3 and 4 are devoted entirely to establishing and applying many of the properties of trees. The following examples provide a preview of some of the material on trees and illustrate how they arise naturally.

Application 1.5.2:   *Physical Chemistry*   A hydrocarbon is a chemical molecule composed of hydrogen and carbon atoms. It is said to be *saturated* if it includes the maximum number of hydrogen atoms for its number of carbon atoms. Saturation occurs when no two carbon atoms have a multiple bond, that is, when the hydrocarbon's structural model is a simple graph. We may recall from elementary chemistry that a hydrogen atom has one electron, and thus it is always 1-valent in a molecule. Also, a carbon atom has four electrons in its outer orbit, making it 4-valent. The saturated hydrocarbons butane and isobutane both have the same chemical formula, i.e., C_4H_{10}, as shown in Figure 1.5.9, so they are said to be *isomers*.

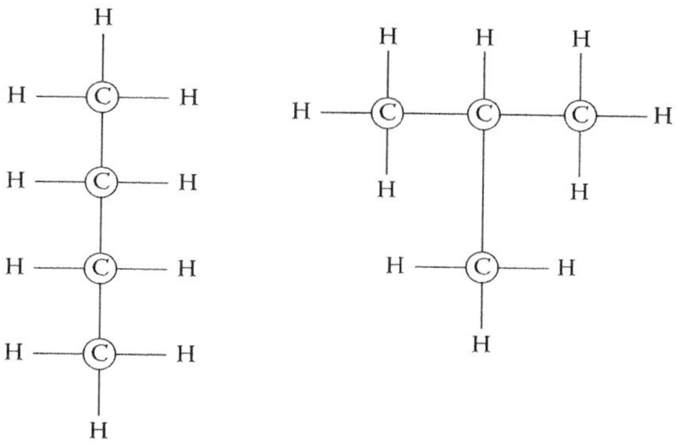

Figure 1.5.9   Two saturated hydrocarbons: butane and isobutane.

Application 1.5.3:   *Operating-System Directories*   Many information structures of computer science are based on trees. The directories and subdirectories (or folders and subfolders) containing a computer user's files are typically represented by the operating system as vertices in a *rooted tree* (see §3.2), as illustrated in Figure 1.5.10.

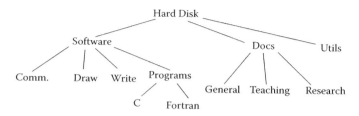

Figure 1.5.10   A directory structure modeled by a rooted tree.

Application 1.5.4:   *Information Retrieval*   *Binary trees* (see §3.2) store ordered data in a manner that permits an efficient search. In particular, if n nodes of a binary tree are used appropriately to store the *keys* of n data elements, then a search can be completed in running time $O(\log n)$. Figure 1.5.11 shows a set of 3-digit keys that are stored in a binary tree. Notice that all the keys in the *left subtree* of each node are smaller than the key at that node, and that all the keys in the *right subtree* are larger. This is called the *binary-search-tree property*.

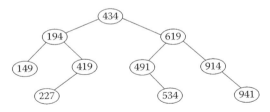

Figure 1.5.11   A binary-search tree storing 3-digit keys.

Application 1.5.5:   *Minimal Connected Networks*   Suppose a network is to be created from n computers. There are $\binom{n}{2} = \frac{n^2-n}{2}$ possible pairs of computers that can be directly joined. If all of these pairs are linked, then the n computers will certainly be connected, but many of these links will be unnecessary. Figure 1.5.12 shows a network connected with a minimum number of edges. Notice that these edges form a tree, and that the number of edges in the tree is one less than the number of vertices.

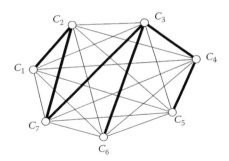

Figure 1.5.12   Connecting seven computers using a minimum number of edges.

EXERCISES for Section 1.5

1.5.1^S Which of the following vertex sequences represent a walk in the graph shown below? Which walks are paths? Which walks are cycles? What are the lengths of those that are walks?

a. $\langle u, y, v, w, v \rangle$
b. $\langle u, y, u, x, v, w, u \rangle$
c. $\langle y, v, u, x, v, y \rangle$
d. $\langle w, v, x, u, y, w \rangle$

1.5.2 Which of the following vertex sequences represent a directed walk in the graph shown below? Which directed walks are directed paths? Which directed walks are directed cycles? What are the lengths of those that are directed walks?

a. $\langle x, v, y, w, v \rangle$
b. $\langle x, u, x, u, x \rangle$
c. $\langle x, u, v, y, x \rangle$
d. $\langle x, v, y, w, v, u, x \rangle$

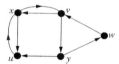

1.5.3^S In the graph shown at right, find

a. a closed walk that is not a trail.
b. a closed trail that is not a cycle.
c. all the cycles of length 5 or less.

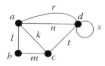

1.5.4 In the graph shown at right, find

a. a walk of length 7 between vertices b and d.
b. a path of maximum length.

1.5.5^S Determine the number of paths from w to r in the following graph.

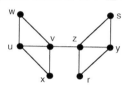

1.5.6 Draw a copy of the Petersen graph (from §1.2) with vertex labels. Then find

a. a trail of length 5.
b. a path length 9.
c. cycles of lengths 5, 6, 8, and 9.

1.5.7 Consider the digraph shown at the right. Find

a. an open directed walk that is not a directed trail.
b. an open directed trail that is not a directed path.
c. a closed directed walk that is not a directed cycle.

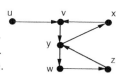

1.5.8^S Give an example of a nontrivial, closed walk that does not contain a cycle.

1.5.9 Let x and y be two different vertices in the complete graph K_5. Find the number of x-y paths of length 2 and of length 3.

1.5.10 Let x and y be two different vertices in the complete graph K_n, $n \geq 5$. Find the number of x-y paths of length 4.

1.5.11^S Let x and y be two adjacent vertices in the complete bipartite graph $K_{3,3}$. Find the number of x-y paths of length 2, or length 3, and of length 4.

1.5.12 Let x and y be two different non-adjacent vertices in the complete bipartite graph $K_{3,3}$. Find the number of x-y paths of length 2, of length 3, and of length 4.

1.5.13 Let x and y be two adjacent vertices in the complete bipartite graph $K_{n,n}$, $n \geq 3$. Find the number of x-y paths of length 2, of length 3, and of length 4.

1.5.14 Let x and y be two different non-adjacent vertices in the complete bipartite graph $K_{n,n}$, $n \geq 3$. Find the number of x-y paths of length 2, of length 3, and of length 4.

In Exercises 1.5.15 through 1.5.22, determine the girth of the graph indicated.

1.5.15^S Complete bipartite graph $K_{3,7}$.

1.5.16 Complete bipartite graph $K_{m,n}$, $m \geq n \geq 3$.

1.5.17 Complete graph K_n.

1.5.18 Hypercube graph Q_5. Can you generalize to Q_n?

1.5.19 Circular ladder graph CL_6. Can you generalize to CL_n?

1.5.20 The Petersen graph.

1.5.21 Dodecahedral graph.

1.5.22 Icosahedral graph.

1.5.23 Determine the girth of each of the following circulant graphs (from §1.2).
    a. $circ(5:1,2)$;   b. $circ(6:1,2)$;   c. $circ(8:1,2)$.

1.5.24 Describe the girth of the circulant graph $circ(n:m)$ in terms of integer n and connection m.

1.5.25 Describe the girth of the circulant graph $circ(n:a,b)$ in terms of integer n and the connections a and b.

1.5.26 Find necessary and sufficient conditions for the circulant graph $circ(n:a,b)$ to be Hamiltonian.

1.5.27 Determine whether the hypercube graph Q_3 is Hamiltonian.

1.5.28 Determine whether the Petersen graph is Hamiltonian.

1.5.29 Determine whether the circular ladder graph CL_n, $n \geq 3$, is Hamiltonian.

1.5.30^S Prove that if v is a vertex on a nontrivial, closed trail, then v lies on a cycle.

1.5.31 Prove or disprove: Every graph that has a closed walk of odd length has a cycle.

1.5.32^S Prove that in a digraph, a shortest directed walk from a vertex x to a vertex y is a directed path from x to y.

1.5.33 Suppose G is a simple graph whose vertices all have degree at least k.
    a. Prove that G contains a path of length k.
    b. Prove that if $k \geq 2$, then G contains a cycle of length at least k.

1.5.34 State and prove the digraph version of Theorem 1.5.2.

1.5.35 State and prove the digraph version of Proposition 1.5.5.

1.5.36 State and prove the digraph version of Theorem 1.5.6.

1.5.37 Find a collection of cycles whose edge-sets partition the edge-set of the closed trail in Example 1.5.4 but that is not a decomposition of the closed trail (i.e., at least one of the cycles in the collection is neither a subwalk nor a reduced walk of the closed trail).

1.5.38 a. In the following binary tree, store the ten 3-digit keys of Figure 1.5.11, in adherence to the binary-search-tree property.

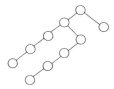

b. Is this tree easier or harder to search than the one in Figure 1.5.11?

1.5.39 Find an Eulerian tour in the following graph.

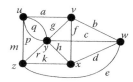

1.5.40[S] Describe how to use a rooted tree to represent all possible moves of a game of tic-tac-toe, where each node in the tree corresponds to a different board configuration.

1.5.41 Which of the platonic graphs are bipartite?

1.5.42 Prove that if G is a digraph such that every vertex has positive indegree, then G contains a directed cycle.

1.5.43[S] Prove that if G is a digraph such that every vertex has positive outdegree, then G contains a directed cycle.

1.6   VERTEX AND EDGE ATTRIBUTES: MORE APPLICATIONS

The all-purpose model of a graph includes attribute lists for edges and vertices. In viewing digraphs as a species of graphs, direction is an edge attribute. The Markov-diagram and lexical-scanner applications, introduced in §1.3, gave a first glimpse of graph models that require edge labels, another kind of edge attribute.

Four Classical Edge-Weight Problems in Combinatorial Optimization

DEFINITION: A **weighted graph** is a graph in which each edge is assigned a number, called its **edge-weight**.

Edge-weights are among the most commonly used attributes of graphs, especially in combinatorial optimization. For instance, the edge-weight might represent transportation cost, travel time, spatial distance, power loss, upper bounds on flow, or inputs and outputs for transitions in a finite state machine.

The definitions of length and distance given in Section §1.4 extend naturally to weighted graphs.

DEFINITION: The **length** of a path in a weighted graph is the sum of the edge-weights of the path. The **distance** between two vertices s and t in a weighted graph is the length of a shortest s-t path.

Application 1.6.1:   *Shortest-Path Problem*   Suppose that each edge-weight in the digraph of Figure 1.6.1 represents the time it takes to traverse that edge. Find the shortest path (in time) from vertex s to vertex t (see Chapter 4).

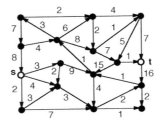

Figure 1.6.1   A weighted directed graph for a shortest-path problem.

Application 1.6.2:   *Minimum-Weight Spanning-Tree Problem*   Suppose that several computers in fixed locations are to be linked to form a computer network. The cost of creating a direct connection between a pair of computers is known for each possible pair and is represented by the edge weights given in Figure 1.6.2. Determine which direct links to construct so that the total networking cost is minimized (see Chapter 4).

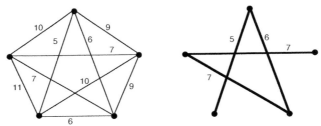

Figure 1.6.2   A weighted graph and its minimum spanning tree.

Application 1.6.3:   *Traveling Salesman Problem*   Suppose that a salesman must visit several cities during the next month. The edge-weights shown in Figure 1.6.3 represent the travel costs between each pair of cities. The problem is to sequence the visits so that the salesman's total travel cost is minimized. In other words, find a Hamiltonian cycle whose total edge-weight is a minimum (see Chapter 6).

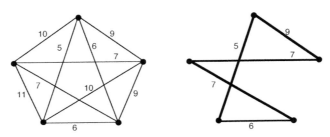

Figure 1.6.3   A weighted graph and its optimal traveling salesman tour.

Application 1.6.4:   *Maximum-Flow Problem*   Suppose that water is being pumped through a network of pipelines, from a *source s* to a *sink t*. Figure 1.6.4 represents the network, and each edge-weight is the upper bound on the flow through the corresponding pipeline. Find the maximum flow from *s* to *t* (see Chapter 10).

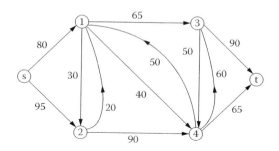

Figure 1.6.4   A maximum-flow problem.

It is interesting to note that there are *polynomial-time* algorithms for three of the problems, but the traveling salesman problem is notoriously difficult, belonging to the class *NP-Hard*. It has motivated a number of different approaches, including cutting-plane methods in integer programming, neural nets, simulated annealing, and the design of various heuristics.

Vertex-Weights and Labels

There are also many important problems requiring vertex attributes in the graph model. For example, a vertex-weight might represent production cost at an associated manufacturing site, and a vertex color might be a timeslot for a timetabling problem. Also, a vertex label might identify a state for a Markov process or a node in an interconnection network.

Application 1.6.5:   *Parallel Architectures*   *Shuffle-exchange graphs* serve as models for parallel architectures suitable for the execution of various distributed algorithms, including *card-shuffling* and the *Fast Fourier Transform*. The vertices of the shuffle-exchange graph SE_n are the bitstrings of length n.

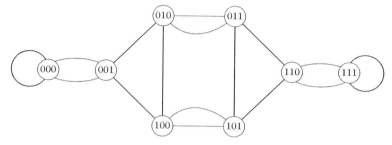

Figure 1.6.5   The shuffle-exchange graph SE_3.

Application 1.6.6: *Software Testing and the Chinese Postman Problem* During execution, a computer program's flow moves between various *states*, and the *transitions* from one state to another depend on the input. In testing software, one would like to generate input data that forces the program to test all possible transitions.

DEFINITION: A **directed postman tour** is a closed directed walk that uses each arc *at least* once.

Digraph Model: The program's execution flow is modeled as a digraph, where the states of the program are represented by vertices, the transitions are represented by arcs, and each of the arcs is assigned a label indicating the input that forces the corresponding transition. Then the problem of finding an input sequence for which the program invokes all transitions and minimizes the total number of transitions is equivalent to finding a directed postman tour of minimum length. This is a variation of the *Chinese Postman Problem*, which is discussed in §6.2.

Remark: Under certain reasonable assumptions, the flow digraph modeling a program's execution can be assumed to be strongly connected, which guarantees the existence of a postman tour.

EXERCISES for Section 1.6

1.6.1 Formulate the personnel-assignment problem (Application 1.3.1) as a maximum-flow problem. (Hint: Add an artificial source and sink to the bipartite graph.)

1.6.2^S Let x be a vertex of a complete graph K_n. How many different Hamiltonian cycles are there that begin at x?

In Exercises 1.6.3 through 1.6.8, describe an appropriate graph or digraph model, including vertex and/or edge attributes, for the given situation.

1.6.3^S Suppose that n final exams are to be scheduled in k non-overlapping timeslots. The undesirability of assigning certain pairs to the same timeslot varies from pair to pair. How should the assignments be made?

1.6.4 Suppose that a city owns ten snowplows. How should the snowplows be routed so that all the paved roads are plowed and the total distance that the snowplows travel is minimized?

1.6.5^S There are n employees and n tasks. Each employee-task pair has an efficiency rating, based on past performance. What is the most efficient matching of employees to tasks?

1.6.6 A manufacturer owns m warehouses, $w_1, w_2, \ldots, w_m$, from which she can ship units of a product to n customers. Warehouse w_i has s_i units of the product, and the jth customer requires d_j units. The cost of shipping from w_i to customer j is c_{ij}. How should the product be distributed so that demand is met while the total shipping cost is minimized?

1.6.7^S Suppose that an electric utility owns n power generators that serve a given region. The production cost for the ith generator is $\$p_i$ per KWH, and its maximum output per day is M KWH. Each generator has a separate transmission line to each of m holding stations. Each holding station requires d_j KWH each day. The percentage loss of power during transmission across the line from the ith generator to the jth station is l_{ij}. How much power should each generator produce, and how should that power be distributed?

1.6.8^S Suppose that a school owns m buses to transport n children to and from k bus stops. There are d_i children at the ith bus stop. The capacity of each bus is 50, and the distance from the ith stop to the jth stop is c_{ij}. How should the children be assigned to the buses, and how should the buses be routed through the stops, so that the total bus mileage is minimized?

1.7   SUPPLEMENTARY EXERCISES

1.7.1 A 20-vertex graph has 62 edges. Every vertex has degree 3 or 7. How many vertices have degree 3?

1.7.2 Either draw a 3-regular 7-vertex graph or prove that none exists.

1.7.3 Prove that no 5-vertex, 7-edge simple graph has diameter greater than 2.

1.7.4 Use the Havel-Hakimi algorithm to decide whether a simple graph can have the degree sequence $\langle 5, 3, 3, 3, 2, 2 \rangle$. If so, then construct the graph.

1.7.5 Draw a connected simple graph G whose line graph $L(G)$ has the degree sequence $\langle 3, 3, 2, 2, 2, 1, 1 \rangle$.

1.7.6 Draw a simple connected graph H whose line graph $L(H)$ has the degree sequence $\langle 3, 3, 2, 2, 2, 1, 1 \rangle$, such that H is *not* isomorphic to the graph G of Exercise 1.7.5.

1.7.7 Prove or disprove: Proposition 1.1.1 holds for digraphs.

1.7.8 How many edges are in the icosahedral graph?

1.7.9 How many edges are in the hypercube graph Q_4?

1.7.10 What is the diameter of the Petersen graph? Explain.

1.7.11 In the circulant graph $circ(24 : 1, 5)$, what vertices are in the open neighborhood $N(3)$?

1.7.12 In the circulant graph $circ(24 : 1, 5)$, what vertices are at distance 2 from vertex 3?

1.7.13 Is $circ(24 : 4, 9)$ connected? What is a general condition for the connectedness of a circulant graph? How would you calculate the number of components?

1.7.14 What vertex of $circ(13 : 1, 5)$ is most distant from 0?

1.7.15 Calculate the girth of $circ(13 : 1, 5)$.

1.7.16 Two opposite corners are removed from an 8-by-8 checkerboard. Prove that it is impossible to cover the remaining 62 squares with 31 dominoes, such that each domino covers two adjacent squares.

1.7.17 Prove that a graph G is bipartite if and only if it has the following property: every subgraph H contains an independent set of vertices whose cardinality is at least half the cardinality of V_H.

1.7.18 Prove that in a connected graph, any two longest paths must have a vertex in common.

1.7.**19** Count the number of different Eulerian tours in K_4.

1.7.**20** Let G be a simple n-vertex graph with an independent set of vertices of cardinality $c \leq n$. What is the maximum number of edges that G may have?

DEFINITION: The **edge-complement** of a simple graph G is the simple graph $\overline{G}$ on the same vertex set such that two vertices of $\overline{G}$ are adjacent if and only if they are *not* adjacent in G.

1.7.**21** Let G be a simple bipartite graph with at least 5 vertices. Prove that $\overline{G}$ is not bipartite. (See §2.4.)

1.7.**22** Let G be a graph whose vertices are the bitstrings of length 4 such that two bitstrings are adjacent if they differ either in exactly one bit or in all four bits.

   a. Write the vertex sequence of a minimum length cycle in the edge-complement $\overline{G}$.

   b. What is the diameter of the graph $\overline{G}$? Explain.

DEFINITION: A simple graph G is **self-complementary** if $G \cong \overline{G}$ (see §2.1).

1.7.**23** Prove that if G is an n-vertex self-complementary graph, then either $n \equiv 0 \,(\mathrm{mod}\ 4)$ or $n \equiv 1 \,(\mathrm{mod}\ 4)$.

1.7.**24** Can a self-complementary graph be non-connected?

1.7.**25** Suppose that a self-complementary graph has $4k + 1$ vertices. Prove that it must have at least one vertex of degree $2k$.

DEFINITION: An n-vertex, m-edge simple graph G is **graceful** if there is a one-to-one function $g : V_G \to \{0, 1, \ldots, m\}$ such that the function $g' : E_G \to \{0, 1, \ldots, m\}$ given by the rule $uv \mapsto |g(u) - g(v)|$ is one-to-one.

1.7.**26** Prove that the cycle C_5 is not graceful.

1.7.**27** Prove that the cycle C_8 is graceful.

GLOSSARY

acyclic: having no cycles.

adjacency table: a table with a row for each vertex, each containing the list of neighbors of that vertex.

adjacent edges: two edges that have an endpoint in common.

adjacent vertices: two vertices joined by an edge.

arc: an edge, one of whose endpoints is designated as the **tail**, and whose other endpoint is designated as the **head**; a synonym for *directed edge*.

arc multiplicity: the number of arcs within a multi-arc.

bigraphical pair of sequences of nonnegative integers $\langle a_1, a_2, \ldots, a_r \rangle$ and $\langle b_1, b_2, \ldots, b_t \rangle$: there exists a bipartite graph G with bipartition subsets $U = \{u_1, u_2, \ldots, u_r\}$ and $W = \{w_1, w_2, \ldots, w_t\}$ such that $\deg(u_i) = a_i$, $i = 1, 2, \ldots, r$, and $deg(w_i) = b_i$, $i = 1, 2, \ldots, t$.

bipartite graph: a graph whose vertex-set can be partitioned into two subsets (parts) such that every edge has one endpoint in one part and one endpoint in the other part.

bipartition of a bipartite graph G: the two subsets into which V_G is partitioned.

bouquet: a one-vertex graph with one or more self-loops.

central vertex of a graph G: a vertex v with minimum *eccentricity*, i.e., $ecc(v) = rad(G)$.

circulant graph $circ(n : S)$ with *connections* set $S \subseteq \{1, \dots, n-1\}$: the graph whose vertex-set is the group of integers $\mathbb{Z}_n = \{0, 1, \dots, n-1\}$ under addition modulo n, such that two vertices i and j are adjacent if and only if there is a number $s \in S$ such that $i + s = j \bmod n$ or $j + s = i \bmod n$.

circular ladder graph CL_n: a graph visualized as two concentric n-cycles in which each of the n pairs of corresponding vertices is joined by an edge.

clique: a subgraph that is a complete graph.

closed (directed) walk: a (directed) walk whose initial and final vertices are identical.

complete bipartite graph: a simple bipartite graph such that each pair of vertices in different sides of the partition is joined by an edge.

complete graph: a simple graph such that every pair of vertices is joined by an edge.

concatenation of walks W_1 and W_2: a walk whose traversal consists of a traversal of W_1 followed by a traversal of W_2 such that W_2 begins where W_1 ends.

connected graph: a graph in which every pair of distinct vertices has a walk between them.

connected digraph: a digraph whose underlying graph is connected.

cycle: a closed path with at least one edge.

cycle graph: a 1-vertex *bouquet* or a simple connected graph with n vertices and n edges that can be drawn so that all of its vertices and edges lie on a single circle.

decomposition of a closed trail T: a collection $C_1, C_2, \dots, C_m$ of edge-disjoint cycles, such that each cycle C_i is either a subwalk or a reduced walk of T, and the edge-sets of the cycles partition the edge-set of trail T.

degree of a vertex: the number of proper edges incident on that vertex plus twice the number of self-loops.

degree sequence: a list of the degrees of all the vertices in non-increasing order.

diameter of a graph G, denoted $diam(G)$: given by $diam(G) = \max\limits_{x,y \in V_G} \{d(x, y)\}$.

digraph: abbreviated name for *directed graph*.

digraphic pair of sequences $\langle a_1, a_2, \dots, a_n \rangle$ and $\langle b_1, b_2, \dots, b_n \rangle$: there exists a simple digraph with vertex-set $\{v_1, v_2, \dots, v_n\}$ such that $indegree(v_i) = b_i$ and $outdegree(v_i) = a_i$ for $i = 1, 2, \dots, n$.

dipole: a two-vertex graph with no self-loops.

directed distance from vertex u to v: the length of a shortest directed walk from u to v.

directed edge: a synonym for *arc*.

directed graph: a graph in which every edge is a directed edge.

directed walk from vertex u to vertex v: an alternating sequence of vertices and arcs representing a continuous traversal from u to v, where each arc is traversed from tail to head.

direction on an edge: an optional attribute that assigns the edge a one-way restriction or preference, said to be from *tail* to *head*; in a drawing the tail is the end behind the arrow and the head is the end in front.

distance between vertices u and v: the length of a shortest walk between u and v.

dominating set of vertices in a graph G: a vertex subset W such that every vertex of G is in W or is adjacent to at least one vertex in W.

eccentricity of a vertex v in a graph G: the distance from v to a vertex farthest from v; denoted $ecc(v)$.

edge: a connection between one or two vertices of a graph.

edge-complement of a simple graph G: the simple graph $\overline{G}$ on the same vertex set such that two vertices of $\overline{G}$ are adjacent if and only if they are *not* adjacent in G.

edge multiplicity: the number of edges within a *multi-edge*.

edge-weight: a number assigned to an edge (optional attribute).

endpoints of an edge: the one or two vertices that are associated with that edge.

Eulerian graph: a connected graph that has an Eulerian tour.

Eulerian tour: a closed Eulerian trail.

Eulerian trail in a graph: a trail that contains every edge of that graph.

formal specification of a graph or digraph: a triple (V, E, endpts) comprising a vertex list, an edge list, and an incidence table.

girth of a graph G: the length of a shortest cycle in G.

graceful graph: an n-*vertex*, m-edge simple graph G for which there is a one-to-one function $g : V_G \to \{0, 1, \ldots, m\}$ such that the function $g' : E_G \to \{0, 1, \ldots, m\}$ given by the rule $uv \mapsto |g(u) - g(v)|$ is one-to-one.

graph $G = (V, E)$: a mathematical structure consisting of two sets, V and E. The elements of V are called **vertices** (or **nodes**), and the elements of E are called **edges**. Each edge has a set of one or two vertices associated to it, which are called its **endpoints.**

graphic sequence: a non-increasing sequence $\langle d_1, d_2 \ldots d_n \rangle$ that is the degree sequence of a simple graph.

Hamiltonian cycle in a graph: a cycle that uses every vertex of that graph.

head of an arc: the endpoint to which that arc is directed.

hypercube graph Q_n: the n-regular graph whose vertex-set is the set of bitstrings of length n, and such that there is an edge between two vertices if and only if they differ in exactly one bit.

incidence: the relationship between an edge and its endpoints.

incidence table of a graph or digraph: a table that specifies the endpoints of every edge and, for a digraph, which endpoint is the head.

indegree of a vertex v: the number of arcs directed to v.

independent set of vertices: a set of mutually non-adjacent vertices.

intersection graph: a graph whose vertices are associated with a collection of sets, such that two vertices are adjacent if and only if their corresponding sets overlap.

interval graph: an intersection graph whose vertex-set corresponds to a collection of intervals on the real line.

length of a walk: the number of edge-steps in the walk sequence.

line graph $L(G)$ of a graph G: the graph that has a vertex for each edge of G, and two vertices in $L(G)$ are adjacent if and only if the corresponding edges in G have a vertex in common.

matching in a graph G: a set of mutually non-adjacent edges in G.

mixed graph: a graph that has undirected edges as well as directed edges.

multi-arc in a digraph: a set of two or more arcs such that each arc has the same head and the same tail.

multi-edge in a graph: a set of two or more edges such that each edge has the same set of endpoints.

mutual reachability of two vertices u and v in a digraph D: the existence of both a directed u-v walk and a directed v-u walk.

node: synonym for *vertex*.

null graph: a graph whose vertex- and edge-sets are empty.

open (directed) walk: a walk whose initial and final vertex are different.

outdegree of a vertex v: the number of arcs directed from v.

partially directed graph: synonym of *mixed graph*.

path: a walk with no repeated edges and no repeated vertices (except possibly the initial and final vertex); a trail with no repeated vertices.

path graph: a connected graph that can be drawn so that all of its vertices and edges lie on a single straight line.

Petersen graph: a special 3-regular, 10-vertex graph first described by Julius Petersen.

platonic graph: a vertex-and-edge configuration of a platonic solid.

platonic solid: any one of the five regular 3-dimensional polyhedrons — tetrahedron, cube, octahedron, dodecahedron, or icosahedron.

proper edge: an edge that is not a self-loop.

radius of a graph G, denoted $rad(G)$: the minimum of the vertex eccentricities.

reachability of a vertex v from a vertex u: the existence of a walk from u to v.

reduction of a walk W by a walk W': the walk that results when all of the vertices and edges of W', except for one of the vertices common to both walks, are deleted from W.

regular graph: a graph whose vertices all have equal degree.

self-complementary graph: a simple graph G such that $G \cong \overline{G}$ (see §2.1).

self-loop: an edge whose endpoints are the same vertex.

simple graph (digraph): a graph (digraph) with no multi-edges or self-loops.

1-skeleton of a polyhedron: the graph consisting of the vertices and edges of that polyhedron, retaining their incidence relation from the polyhedron.

strongly connected digraph: a digraph in which every two vertices are mutually reachable.

strong orientation of a graph: an assignment of directions to the edges of a graph so that the resulting digraph is strongly connected.

subwalk of a walk W: a walk consisting of a subsequence of consecutive entries of the walk sequence of W.

tail of an arc: the endpoint from which that arc is directed.

tournament: a digraph whose underlying graph is a complete graph.

trail: a walk with no edge that occurs more than once.

tree: a connected graph that has no cycles.

trivial walk, trail, path, or graph: a walk, trail, path, or graph consisting of one vertex and no edges.

underlying graph of a directed or mixed graph G: the graph that results from removing all the designations of *head* and *tail* from the directed edges of G (i.e., deleting all the edge directions).

valence: a synonym for *degree*.

vertex: an element of the first constituent set of a graph; a synonym for *node*.

vertex cover of a graph G: a vertex subset W such that every edge of G is incident on at least one vertex in W.

walk from vertex u to vertex v: an alternating sequence of vertices and edges, representing a continuous traversal from vertex u to vertex v.

weighted graph: a graph in which each edge is assigned a number, called an *edge-weight*.

Chapter 2

STRUCTURE AND REPRESENTATION

INTRODUCTION

The possible representations of a graph include drawings, incidence tables, formal specifications, and many other concrete descriptions of the abstract structure. Structure is what characterizes a graph itself and is independent of the way the graph is represented. The concepts and language needed for analyzing a graph's structure, and for understanding the distinction between the structure and its many forms of representation, are the main concern of this chapter.

Two different-seeming graph representations might actually be alternative descriptions of structurally equivalent graphs. Developing a universally applicable method for deciding structural equivalence is called the *graph isomorphism problem*. Although the analogous problem for vector spaces reduces simply to checking if the dimensions are equal, no such simple test is available in graph theory. Some strategies and tests work well in many instances, but no one has developed a practical method to handle all cases.

One prominent aspect of the structure is the system of smaller graphs inside a graph, which are called its *subgraphs*. Subgraphs arise explicitly or implicitly in almost any discussion of graphs. For instance, in the process of building a graph G from scratch, vertex by vertex, and edge by edge, the entire sequence of graphs formed along the way is a nested chain of subgraphs in G.

The basic operations of adding a vertex or an edge to a graph, together with the operations of deleting an edge or a vertex, provide a mechanism for transforming any graph, step by step, into any other graph. Computer scientists perceive of a graph as a variable, and they include these *primary operations* in the definition of a graph, in much the same way that mathematicians include the operations of vector addition and scalar multiplication in the definition of a vector space.

Upon the primary graph operations are built subsequent layers of graph algorithms, just as certain algorithms in elementary linear algebra (e.g., *Gaussian elimination*) are developed by combining the operations of vector addition and scalar multiplication. This layering perspective provides an organizational scheme that rescues graph theory from what might otherwise seem to be a grab bag of concepts and methods. Moreover, this layering approach is consistent with the established principles of software engineering.

Beyond pictures, adjacency lists, and incidence tables, the possible descriptions of a graph include various kinds of matrices. In view of the high level of computational success that matrix representations have achieved for vector spaces, it is not surprising that some forms of matrix representations were also introduced for graphs. The *incidence matrix* and the *adjacency matrix* are two classical matrix representations for graphs, introduced in §2.6, that allow us to establish certain graph properties using matrix-theoretic methods. Even though these matrix representations are in themselves rather inefficient for most graph computations, they do serve as a good start toward developing more efficient *information structures*.

2.1   GRAPH ISOMORPHISM

Deciding when two line drawings represent the same graph can be quite difficult for graphs containing more than a few vertices and edges. A related task is deciding when two graphs with different specifications are *structurally equivalent*, that is, whether they have the same pattern of connections. Designing a practical algorithm to make these decisions is a famous unsolved problem, called the *graph-isomorphism problem.*

Structurally Equivalent Graphs

Since the shape or length of an edge and its position in space are not part of a graph's specification, and neither are edge-crossings or other artifacts of a drawing, each graph has infinitely many spatial representations and drawings.

Example 2.1.1: The two line drawings in Figure 2.1.1 both depict the same graph.

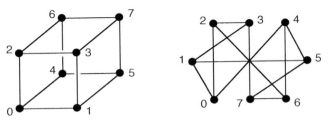

Figure 2.1.1   Two different drawings of the same graph.

The vertices and edges of the two drawings have matched labels, and they each have only eight vertices and twelve edges. It is easy to verify for every pair $\{i, j\}$ of vertices that they are adjacent in one graph if and only if they are adjacent in the other. We recognize, therefore, that these two drawings both represent the same graph.

If the graphs are unlabeled, or if the vertices of one drawing are labeled differently from the vertices of another, there are circumstances under which the graphs are nonetheless regarded as virtually the same.

Example 2.1.2: The names of the vertices and edges of graphs G and H in Figure 2.1.2 differ, but these two graphs are strikingly similar.

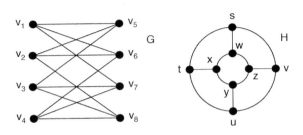

Figure 2.1.2   Two drawings of essentially the same graph.

In particular, the specification for graph G could be transformed into the specification for graph H by the following bijection f on the vertex names.

$$v_1 \rightarrow s \qquad v_5 \rightarrow v$$
$$v_2 \rightarrow z \qquad v_6 \rightarrow w$$
$$v_3 \rightarrow u \qquad v_7 \rightarrow t$$
$$v_4 \rightarrow x \qquad v_8 \rightarrow y$$

For instance, in graph G, we see that vertex v_1 is adjacent to vertices v_5, v_6 and v_7, but to no others. In graph H, we see that vertex $s = f(v_1)$ is adjacent to vertices $v = f(v_5)$, $w = f(v_6)$, and $t = f(v_7)$, but to no others. In this sense, the given vertex bijection $f : V_G \rightarrow V_H$ also acts bijectively on the adjacencies. Our immediate goal is to develop this notion.

Formalizing Structural Equivalence for Simple Graphs

DEFINITION: Let G and H be two simple graphs. A vertex function $f : V_G \rightarrow V_H$ **preserves adjacency** if for every pair of adjacent vertices u and v in graph G, the vertices $f(u)$ and $f(v)$ are adjacent in graph H. Similarly, f **preserves non-adjacency** if $f(u)$ and $f(v)$ are non-adjacent whenever u and v are non-adjacent.

DEFINITION: A vertex bijection $f : V_G \rightarrow V_H$ between the vertex-sets of two simple graphs G and H is **structure-preserving** if it preserves adjacency and non-adjacency. That is, for every pair of vertices in G,

$$u \text{ and } v \text{ are adjacent in } G \Leftrightarrow f(u) \text{ and } f(v) \text{ are adjacent in } H$$

DEFINITION: Two simple graphs G and H are **isomorphic**, denoted $G \cong H$, if there exists a structure-preserving vertex bijection $f : V_G \rightarrow V_H$. Such a function f between the vertex-sets of G and H is called an **isomorphism** from G to H.

NOTATION: When we think of a vertex function $f : V_G \rightarrow V_H$ as a mapping from one graph to another, we often write $f : G \rightarrow H$.

Example 2.1.3: Figure 2.1.3 specifies an isomorphism between the two simple graphs shown. Checking that the given vertex bijection is structure-preserving is left to the reader.

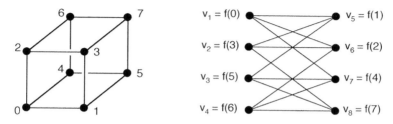

Figure 2.1.3   Specifying an isomorphism between two simple graphs.

Alternatively, one may simply relabel the vertices of the second graph with the names of the vertices in the first graph, as shown in Figure 2.1.4.

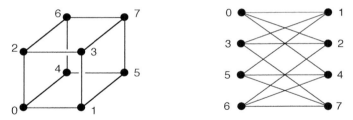

Figure 2.1.4   Another way of depicting an isomorphism.

Remark: A generalization of graph isomorphism, called a *linear graph mapping*, is an adjacency-preserving mapping between the vertex-sets that is not required to preserve non-adjacency and is not necessarily bijective. The reader familiar with *group theory* will notice this analogy: linear graph mapping is to graph isomorphism as group homomorphism is to group isomorphism. In fact, some authors use the term *graph homomorphism* instead of linear graph mapping.

Example 2.1.4: The vertex function $j \mapsto j + 4$ depicted in Figure 2.1.5 is bijective and adjacency-preserving, but it is not an isomorphism since it does not preserve non-adjacency. In particular, the non-adjacent pair $\{0, 2\}$ maps to the adjacent pair $\{4, 6\}$.

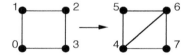

Figure 2.1.5   Bijective and adjacency-preserving, but not an isomorphism.

Example 2.1.5: The vertex function $j \mapsto j \mod 2$ depicted in Figure 2.1.6 preserves adjacency and non-adjacency, but it is not an isomorphism, since it is not bijective.

Figure 2.1.6   Preserves adjacency and non-adjacency, but is not bijective.

Extending the Definition of Isomorphism to General Graphs

As the following example illustrates, the concept of isomorphism for general graphs requires a more general definition of *structure-preserving* for a vertex bijection.

Example 2.1.6: The mapping $f : V_G \rightarrow V_H$ between the vertex-sets of the two graphs shown in Figure 2.1.7 given by $f(i) = i$, $i = 1, 2, 3$, preserves adjacency and non-adjacency, but the two graphs are clearly not structurally equivalent.

Figure 2.1.7   Two graphs that are not structurally equivalent.

DEFINITION: A vertex bijection $f : V_G \to V_H$ between the vertex-sets of two graphs G and H, simple or general, is **structure-preserving** if

(i) the number of edges (even if 0) between every pair of distinct vertices u and v in graph G equals the number of edges between their images $f(u)$ and $f(v)$ in graph H, and

(ii) the number of self-loops at each vertex x in G equals the number of self-loops at the vertex $f(x)$ in H.

This new definition, which reduces to our original definition of *structure-preserving* for simple graphs, allows us to extend the concept of isomorphism to general graphs.

DEFINITION: Two graphs G and H (simple or general) are **isomorphic graphs** if there exists a structure-preserving vertex bijection $f : V_G \to V_H$. This relationship is denoted $G \cong H$.

Specifying an Isomorphism between Graphs Having Multi-Edges

DEFINITION: Let G and H be two isomorphic graphs. A vertex bijection $f_V : V_G \to V_H$ and an edge bijection $f_E : E_G \to E_H$ are **consistent** if for every edge $e \in E_G$, the function f_V maps the endpoints of e to the endpoints of edge $f_E(e)$. A consistent mapping pair $(f_V : V_G \to V_H,\ f_E : E_G \to E_H)$ is often written shorthand as $f : G \to H$.

Proposition 2.1.1: *Let G and H be any two graphs. Then $G \cong H$ if and only if there is a vertex bijection $f_V : V_G \to V_H$ and an edge bijection $f_E : E_G \to E_H$ that are consistent.*

Proof: If vertex bijection f_V and edge bijection f_E are consistent, then f_V is structure-preserving. Conversely, if $G \cong H$, then, by definition, there is a structure-preserving vertex bijection $f_V : V_G \to V_H$. We may construct a consistent edge bijection as follows: for each pair of distinct vertices $u, v \in V_G$, map the set of edges between u and v bijectively to the sets of edges between vertices $f_V(u)$ and $f_V(v)$ (this is possible because these two edge sets have the same number of elements).   ◇

Remark: If G and H are isomorphic simple graphs, then every structure-preserving vertex bijection $f : V_G \to V_H$ induces a unique consistent edge bijection, implicitly given by the rule: $uv \mapsto f(u)f(v)$.

However, for a given structure-preserving bijection between the vertex-sets of two graphs with multi-edges, there is more than one *consistent* edge bijection.

DEFINITION: If G and H are graphs *with multi-edges*, then an **isomorphism** from G to H is specified by giving a vertex bijection $f_V : V_G \to V_H$ *and* an edge bijection $f_E : E_G \to E_H$ that are consistent.

Example 2.1.7: There are two structure-preserving vertex bijections for the isomorphic graphs G and H shown in Figure 2.1.8. Each of these vertex bijections has six consistent edge-bijections. Hence there are 12 distinct isomorphisms from G to H.

Figure 2.1.8   There are 12 distinct isomorphisms from G to H.

Isomorphic Graph Pairs

The following properties of isomorphisms provide some preliminary criteria to check when considering the existence or possible construction of a graph isomorphism. Although the examples given involve simple graphs, the properties apply to general graphs as well.

Theorem 2.1.2: *Let G and H be isomorphic graphs. Then they have the same number of vertices and the same number of edges.*

Proof: Let $f : G \to H$ be an isomorphism. The graphs G and H must have the same number of vertices and the same number of edges, because an isomorphism maps both sets bijectively.   ◇

Theorem 2.1.3: *Let $f : G \to H$ be a graph isomorphism and let $v \in V_G$. Then $\deg(f(v)) = \deg(v)$.*

Proof: Since f is structure-preserving, the number of proper edges and the number of self-loops incident on vertex v equal the corresponding numbers for vertex $f(v)$. Thus, $\deg(f(v)) = \deg(v)$.   ◇

Corollary 2.1.4: *Let G and H be isomorphic graphs. Then they have the same degree sequence.*

Corollary 2.1.5: *Let $f : G \to H$ be a graph isomorphism and let $e \in E_G$. Then the endpoints of edge $f(e)$ have the same degrees as the endpoints of e.*

Example 2.1.8: In Figure 2.1.9, we observe that both graphs have 8 vertices and 12 edges, and that both are 3-regular. The vertex labelings for the hypercube Q_3 and for the circular ladder CL_4 specify a vertex bijection. A careful examination reveals that this vertex bijection is structure-preserving. It follows that Q_3 and CL_4 are isomorphic graphs.

DEFINITION: The **Möbius ladder** ML_n is a graph obtained from the circular ladder CL_n by deleting from the circular ladder two of its parallel curved edges and replacing them with two edges that cross-match their endpoints.

Example 2.1.9: The complete bipartite graph $K_{3,3}$ is shown on the left in Figure 2.1.10, and the Möbius ladder ML_3 is shown on the right. Both graphs have 6 vertices and 9 edges, and both are 3-regular. The vertex labelings for the two drawings specify an isomorphism.

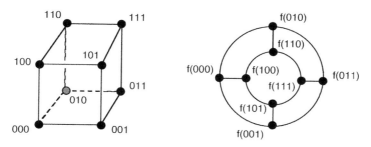

Figure 2.1.9   Hypercube graph Q_3 and circular ladder CL_4 are isomorphic.

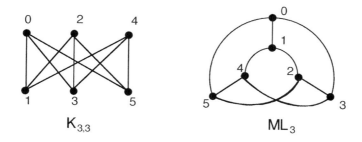

Figure 2.1.10   Bipartite graph $K_{3,3}$ and Möbius ladder ML_3 are isomorphic.

Isomorphism Type of a Graph

If $f = (f_V, f_E)$ is an isomorphism from G to H, then $f^{-1} = (f_V^{-1}, f_E^{-1})$ is an isomorphism from H to G. Thus, the relation $\cong$ ("isomorphic to") is symmetric. Reflexivity and transitivity are also easy to establish, and thus, the relation $\cong$ is an equivalence relation.

DEFINITION: Each equivalence class under $\cong$ is called an *isomorphism type*.

Isomorphic graphs are graphs of the same isomorphism type. The graphs in Figure 2.1.11 represent all four possible isomorphism types for a simple graph on three vertices. Counting isomorphism types of graphs generally involves the algebra of permutation groups.

Figure 2.1.11   The four isomorphism types for a simple 3-vertex graph.

Isomorphism of Digraphs

The definition of graph isomorphism is easily extended to digraphs.

DEFINITION: Two digraphs are *isomorphic* if there is an isomorphism f between their underlying graphs that preserves the direction of each edge. That is, e is directed from u to v if and only if $f(e)$ is directed from $f(u)$ to $f(v)$.

Example 2.1.10: The four different isomorphism types of a simple digraph with three vertices and three arcs are shown in Figure 2.1.12. Notice that non-isomorphic digraphs can have underlying graphs that are isomorphic.

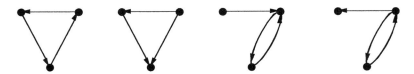

Figure 2.1.12   Four non-isomorphic digraphs.

The Isomorphism Problem

It would be very useful to have a reasonably fast general algorithm that accepts as input any two n-vertex graphs and that produces as output either an isomorphism between them or a report that the graphs are not isomorphic. Checking whether a given vertex bijection is an isomorphism would require an examination of all vertex pairs, which is not in itself overwhelming. However, since there are $n!$ vertex bijections to check, this brute-force approach is feasible only for very small graphs.

DEFINITION: The **graph-isomorphism problem** is to devise a practical general algorithm to decide graph isomorphism, or, alternatively, to prove that no such algorithm exists.

Application 2.1.1:   *Computer Chip Intellectual Property Rights*   Suppose that not long after ABC Corporation develops and markets a computer chip, it happens that the DEF Corporation markets a chip with striking operational similarities. If ABC could prove that DEF's circuitry is merely a rearrangement of the ABC circuitry (i.e., that the circuitries are isomorphic), they might have the basis for a patent-infringement suit. If ABC had to check structure preservation for each of the permutations of the nodes of the DEF chip, the task would take prohibitively long. However, knowledge of the organization of the chips might enable the ABC engineers to take a shortcut.

EXERCISES for Section 2.1

In Exercises 2.1.1 through 2.1.6, find all possible isomorphism types of the given kind of simple graph.

2.1.1[S]   A 4-vertex tree.

2.1.**2**   A 4-vertex connected graph.

2.1.**3**   A 5-vertex tree.

2.1.**4**   A 6-vertex tree.

2.1.**5**   A 5-vertex graph with exactly three edges.

2.1.**6**   A 6-vertex graph with exactly four edges.

In Exercises 2.1.7 through 2.1.10, find all possible isomorphism types of the given kind of simple digraph.

2.1.7[S] A simple 4-vertex digraph with exactly three arcs.

2.1.8 A simple 4-vertex digraph with exactly four arcs.

2.1.9 A simple 3-vertex strongly connected digraph.

2.1.10 A simple 3-vertex digraph with no directed cycles.

In Exercises 2.1.11 through 2.1.14, find all possible isomorphism types of the given kind of general graph.

2.1.11[S] A graph with two vertices and three edges.

2.1.12 A graph with three vertices and two edges.

2.1.13 A graph with three vertices and three edges.

2.1.14 A 4-vertex graph with exactly four edges including exactly one self-loop and a multi-edge of size 2.

In Exercises 2.1.15 through 2.1.18, find a vertex-bijection that specifies an isomorphism between the two graphs or digraphs shown.

2.1.15[S]

2.1.16

2.1.17

2.1.18

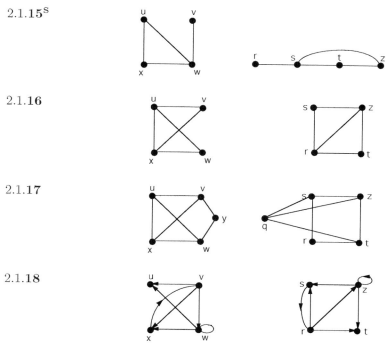

2.1.19 Give an example of two non-isomorphic 5-vertex digraphs whose underlying graphs are isomorphic.

2.1.20 Give an example of two non-isomorphic 9-vertex graphs with the same degree sequence.

2.2   AUTOMORPHISMS AND SYMMETRY

DEFINITION: An isomorphism from a graph G to itself is called an **automorphism**.

Thus, an automorphism π of graph G is a structure-preserving *permutation* π_V on the vertex-set of G, along with a (consistent) permutation π_E on the edge-set of G. We may write $\pi = (\pi_V, \pi_E)$.

NOTATION: When the context is clear, the subscripts that distinguish the vertex and edge actions of an automorphism are suppressed. Thus, we may simply write π in place of π_V or of π_E.

Remark: Any structure-preserving vertex-permutation is associated with one (if simple) or more (if there are any multi-edges) automorphisms of G. Two possible measures of symmetry of a graph are (1) simply the number of automorphisms or (2) the proportion of vertex permutations that are structure-preserving.

Permutations and Cycle Notation

The most convenient representation of a permutation, for our present purposes, is as a *product of disjoint cycles*.

NOTATION: In specifying a permutation, we use the notation (x) to show that the object x is fixed, i.e., permuted to itself; the notation $(x\ y)$ means that objects x and y are swapped; and the notation $(x_0\ x_1\ \cdots\ x_{n-1})$ means that the objects $x_0\ x_1\ \cdots\ x_{n-1}$ are cyclically permuted, so that $x_j \mapsto x_{j+1 (\mathrm{mod}\ n)}$, for $j = 0, 1, \ldots, n-1$. As explained in Appendix A4, every permutation can be written as a composition of disjoint cycles.

Example 2.2.1: The permutation

$$\pi = \begin{pmatrix} 1 & 2 & 3 & 4 & 5 & 6 & 7 & 8 & 9 \\ 7 & 4 & 1 & 8 & 5 & 2 & 9 & 6 & 3 \end{pmatrix}$$

which maps 1 to 7, 2 to 4, and so on, has the **disjoint cycle form**

$$\pi = (1\ 7\ 9\ 3)(2\ 4\ 8\ 6)(5)$$

Geometric Symmetry

A geometric symmetry on a graph drawing can be used to represent an automorphism on the graph.

Example 2.2.2: The graph $K_{1,3}$ has six automorphisms. Each of them is realizable by a rotation or reflection of the drawing in Figure 2.2.1. For instance, a 120° clockwise rotation of the figure corresponds to the graph automorphism with vertex-permutation $(x)(u\ v\ w)$ and edge-permutation $(a\ b\ c)$. Also, reflection through the vertical axis corresponds to the graph automorphism with vertex-permutation $(x)(u)(v\ w)$ and edge-permutation $(a)(b\ c)$. The following table lists all the automorphisms of $K_{1,3}$ and their corresponding vertex- and edge-permutations.

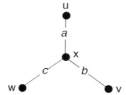

Figure 2.2.1   The graph $K_{1,3}$.

Symmetry	Vertex Permutation	Edge Permutation
identity	$(u)(v)(w)(x)$	$(a)(b)(c)$
120° rotation	$(x)(u\ v\ w)$	$(a\ b\ c)$
240° rotation	$(x)(u\ w\ v)$	$(a\ c\ b)$
refl. thru a	$(x)(u)(v\ w)$	$(a)(b\ c)$
refl. thru b	$(x)(v)(u\ w)$	$(b)(a\ c)$
refl. thru c	$(x)(w)(u\ w)$	$(c)(a\ b)$

Since the graph $K_{1,3}$ has four vertices, the total number of permutations on its vertex-set is 24. By Theorem 2.1.3, every automorphism on $K_{1,3}$ must fix the 3-valent vertex. Since there are only six permutations of four objects that fix one designated object, it follows that there can be no more than six automorphisms of $K_{1,3}$.

Remark: Except for a few exercises, the focus of this section is on simple graphs. Since each automorphism of a simple graph G is completely specified by a structure-preserving vertex-permutation, the automorphism and its corresponding vertex-permutation are often regarded as the same object.

Example 2.2.3: Figure 2.2.2 shows a graph with four vertex-permutations, each written in disjoint cyclic notation. It is easy to verify that these vertex-permutations are structure-preserving, so they are all graph automorphisms. We observe that the automorphisms λ_0, λ_1, λ_2, and λ_3 correspond, respectively, to the identity, vertical reflection, horizontal reflection, and 180° rotation. Proving that there are no other automorphisms is left as an exercise.

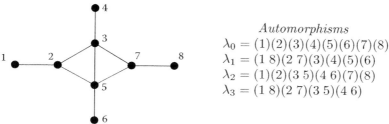

$Automorphisms$
$\lambda_0 = (1)(2)(3)(4)(5)(6)(7)(8)$
$\lambda_1 = (1\ 8)(2\ 7)(3)(4)(5)(6)$
$\lambda_2 = (1)(2)(3\ 5)(4\ 6)(7)(8)$
$\lambda_3 = (1\ 8)(2\ 7)(3\ 5)(4\ 6)$

Figure 2.2.2   A graph with four automorphisms.

Limitations of Geometric Symmetry

Although looking for geometric symmetry within a drawing may help in discovering automorphisms, there may be automorphisms that are not realizable as reflections or rotations of a particular drawing.

Example 2.2.4: Figure 2.2.3 shows three different drawings of the same-labeled Petersen graph. The leftmost drawing has 5-fold rotational symmetry that corresponds to the automorphism $(0\ 1\ 2\ 3\ 4)(5\ 6\ 7\ 8\ 9)$, but this automorphism does not correspond to any geometric symmetry of either of the other two drawings. The automorphism $(0\ 5)(1\ 8)(4\ 7)(2\ 3)(6)(9)$ is realized by 2-fold reflectional symmetry in the middle and rightmost drawings (about the axis through vertices 6 and 9) but is not realizable by any geometric symmetry in the leftmost drawing. There are several other automorphisms that are realizable by geometric symmetry in at least one of the drawings but not realizable in at least one of the other ones. (See Exercises.)

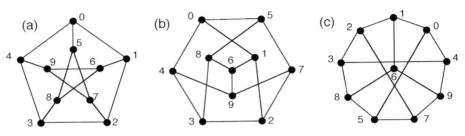

Figure 2.2.3   Three drawings of the Petersen graph.

Vertex- and Edge-Transitive Graphs

DEFINITION: A graph G is **vertex-transitive** if for every vertex pair $u, v \in V_G$, there is an automorphism that maps u to v.

DEFINITION: A graph G is **edge-transitive** if for every edge pair $d, e \in E_G$, there is an automorphism that maps d to e.

Example 2.2.5: The graph $K_{1,3}$, discussed in Example 2.2.2, is edge-transitive, but not vertex-transitive, since every automorphism must map the 3-valent vertex to itself.

Example 2.2.6: The complete graph K_n is vertex-transitive and edge-transitive for every n. (See Exercises.)

Example 2.2.7: The Petersen graph is vertex-transitive and edge-transitive. (See Exercises.)

Example 2.2.8: Every circulant graph $circ(n : S)$ is vertex-transitive. In particular, the vertex function $i \mapsto i + k \bmod n$ is an automorphism. In effect, it rotates a drawing of the circulant graph, as in Figure 2.2.4, $\frac{k}{n}$ of the way around. For instance, vertex 3 can be mapped to vertex 7 by rotating $\frac{4}{13}$ of the way around. Although $circ(13 : 1, 5)$ is edge-transitive, some circulant graphs are not. (See Exercises.)

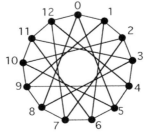

Figure 2.2.4   The circulant graph $circ(13 : 1, 5)$.

Vertex Orbits and Edge Orbits

Suppose that two vertices u and v of a graph G are to be considered related if there is an automorphism that maps u to v. This is clearly an *equivalence relation* on the vertices of G. There is a similar equivalence relation on the edges. (Equivalence relations are discussed in Appendix A.2.)

DEFINITION: The equivalence classes of the vertices of a graph under the action of the automorphisms are called **vertex orbits**. The equivalence classes of the edges are called **edge orbits**.

Example 2.2.9: The graph of Example 2.2.3 (repeated below in Figure 2.2.5) has the following vertex and edge orbits.

$$\text{vertex orbits:} \quad \{1,8\}, \quad \{4,6\}, \quad \{2,7\}, \quad \{3,5\}$$
$$\text{edge orbits:} \quad \{12,78\}, \quad \{34,56\}, \quad \{23,25,37,57\}, \quad \{35\}$$

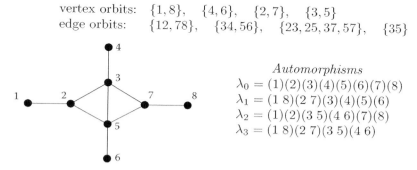

$$\textit{Automorphisms}$$
$$\lambda_0 = (1)(2)(3)(4)(5)(6)(7)(8)$$
$$\lambda_1 = (1\ 8)(2\ 7)(3)(4)(5)(6)$$
$$\lambda_2 = (1)(2)(3\ 5)(4\ 6)(7)(8)$$
$$\lambda_3 = (1\ 8)(2\ 7)(3\ 5)(4\ 6)$$

Figure 2.2.5   Graph of Example 2.2.3.

Theorem 2.2.1: *All vertices in the same orbit have the same degree.*

Proof: This follows immediately from Theorem 2.1.3.   ◇

Theorem 2.2.2: *All edges in the same orbit have the same pair of degrees at their endpoints.*

Proof: This follows immediately from Corollary 2.1.5.   ◇

Remark: A vertex-transitive graph is a graph with only one vertex orbit, and an edge-transitive graph is a graph with only one edge orbit.

Example 2.2.10: The complete graph K_n has one vertex orbit and one edge orbit.

Example 2.2.11: Each of the two partite sets of the complete bipartite graph $K_{m,n}$ is a vertex orbit. The graph is vertex-transitive if and only if $m = n$; otherwise it has two vertex orbits. However, $K_{m,n}$ is always edge-transitive (see Exercises).

Remark: Automorphism theory is one of several graph topics involving an interaction between graph theory and group theory.

Finding Vertex and Edge Orbits

We illustrate how to find orbits by consideration of two examples. (It is not known whether there exists a polynomial-time algorithm for finding orbits. Testing all $n!$ vertex-permutations for the adjacency preservation property is too tedious an approach.) In addition to using Theorems 2.2.1 and 2.2.2, we observe that if an automorphism maps vertex u to vertex v, then it maps the neighbors of u to the neighbors of v.

Example 2.2.12: In the graph of Figure 2.2.6, vertex 0 is the only vertex of degree 2, so it is in an orbit by itself. The vertical reflection $(0)(1\ 4)(2\ 3)$ establishes that vertices 1 and 4 are co-orbital and that 2 and 3 are co-orbital. Moreover, since vertices 1 and 4 have a 2-valent neighbor but vertices 2 and 3 do not, the orbit of vertices 1 and 4 must be different from that of 2 and 3. Thus, the vertex orbits are $\{0\}, \{1,4\}$, and $\{2,3\}$.

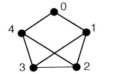

Vertex orbits: $\{0\}, \{1,4\}$, and $\{2,3\}$
Edge orbits: $\{23\}, \{01,04\}$, and $\{12,13,24,34\}$

Figure 2.2.6   **A graph and its orbits.**

Since edge 23 is the only edge, both of whose endpoints have three 3-valent neighbors, it is in an orbit by itself. The edges 01 and 04 are the only edges with a 2-valent endpoint and a 3-valent endpoint, so they could not be co-orbital with any edges except each other. The vertical reflection establishes that the pair $\{01,04\}$ is indeed co-orbital, as are the pair $\{12,34\}$ and the pair $\{13,24\}$. The latter two pairs combine into a single edge orbit, because of the automorphism $(0)(1)(4)(2\ 3)$, which swaps edges 12 and 13 as well as edges 24 and 34. Thus, the edge orbits are $\{23\}, \{01,04\}$, and $\{12,13,24,34\}$.

Example 2.2.13: Notice how the orbits in the previous example reflect the symmetry of the graph. In general, recognizing the symmetry of a graph expedites the determination of its orbits. Consider the 4-regular graph G in Figure 2.2.7.

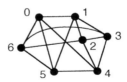

Figure 2.2.7   **A 4-regular graph G whose symmetry is less apparent.**

When we look at vertices 0, 2, 3, and 5, we discover that each of them has a set of 3 neighbors that are independent, while vertices 1, 4, and 6 each have two pairs of adjacent vertices. The redrawing of the graph G shown in Figure 2.2.8 reflects this symmetry.

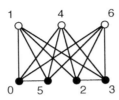

Vertex orbits: $\{0,2,3,5\}$ and $\{1,4,6\}$
Edge orbits: $\{05,23\}$ and $E_G - \{05,23\}$

Figure 2.2.8   **A redrawn graph and its orbits.**

In that form, we see immediately that there are two vertex orbits, namely $\{0,2,3,5\}$ and $\{1,4,6\}$. One of the two edge orbits is $\{05,23\}$, and the other contains all the other edges.

EXERCISES for Section 2.2

In Exercises 2.2.1 through 2.2.4, write down all the automorphisms of the given simple graph as vertex-permutations.

2.2.1^S 2.2.2

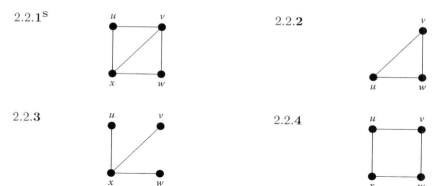

2.2.3   2.2.4

In Exercises 2.2.5 through 2.2.7, specify all the automorphisms of the given graph.

2.2.5^S 2.2.6   2.2.7

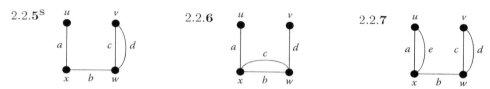

2.2.8 Determine the number of distinct automorphisms of the graph shown below.

In Exercises 2.2.9 through 2.2.12, prove that the specified graph is vertex-transitive.

2.2.9^S K_n 2.2.10 CL_n

2.2.11 $circ(n:S)$ 2.2.12 The Petersen graph

2.2.13 Prove that the complete bipartite graph $K_{m,n}$ is vertex-transitive if and only if $m = n$.

In Exercises 2.2.14 through 2.2.17, decide for which values of n (and/or m) the specified graph is edge-transitive.

2.2.14^S $circ(9:1,m)$ 2.2.15 $circ(13:1,m)$

2.2.16 $K_{m,n}$ 2.2.17 CL_n

In Exercises 2.2.18 through 2.2.27, find the vertex and edge orbits of the given graph.

2.2.18[S] The graph of Exercise 2.2.1.   2.2.19 The graph of Exercise 2.2.2.

2.2.20 The graph of Exercise 2.2.3.   2.2.21 The graph of Exercise 2.2.4.

2.2.22 The graph of Exercise 2.2.5.   2.2.23 The graph of Exercise 2.2.6.

2.2.24 The graph of Exercise 2.2.7.   2.2.25 The graph of Figure 2.2.9(a).

2.2.26 The graph of Figure 2.2.9(b).   2.2.27 The graph of Figure 2.2.9(c).

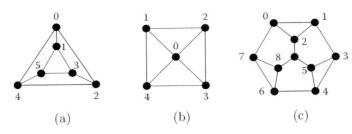

(a)   (b)   (c)

Figure 2.2.9

2.2.28 Each of the following refers to the three drawings of the Petersen graph that appear in Example 2.2.4.

(a) Specify an automorphism of the Petersen graph that is realizable by 2-fold reflectional symmetry in the leftmost drawing but not realizable by any geometric symmetry in the other two drawings.

(b) Specify an automorphism of the Petersen graph that is realizable by some geometric symmetry in the middle drawing but not realizable by any in the other two drawings.

(c) Specify an automorphism of the Petersen graph that is realizable by some geometric symmetry in the rightmost drawing but not realizable by any in the other two drawings.

2.2.29 Prove that the list of automorphisms in Example 2.2.3 is complete.

2.3 SUBGRAPHS

Properties of a given graph are often determined by the existence or non-existence of certain types of smaller graphs inside it. For instance, Theorem 1.5.4 asserts that a graph is bipartite if and only if it contains no odd cycle. Kuratowski's theorem, proved in Chapter 7, shows that a graph can be drawn in the plane without any edge-crossings if and only if the graph contains (in a more general sense) neither the complete graph K_5 nor the complete bipartite graph $K_{3,3}$. This section concentrates on basic kinds of structural containment.

DEFINITION: A **subgraph** of a graph G is a graph H whose vertices and edges are all in G. If H is a subgraph of G, we may also say that G is a **supergraph** of H.

DEFINITION: A **subdigraph** of a digraph G is a digraph H whose vertices and arcs are all in G.

Equivalently, one can say that a subgraph of a graph G comprises a subset V' of V_G and a subset E' of E_G such that the endpoints of every edge in E' are members of V'. Thus, the incidence table for a subgraph H can be obtained simply by deleting from the incidence table of G each column that does not correspond to an edge in H.

DEFINITION: A **proper subgraph** H of G is a subgraph such that V_H is a proper subset of V_G or E_H is a proper subset of E_G.

Example 2.3.1: Figure 2.3.1 shows the line drawings and corresponding incidence tables for two proper subgraphs of a graph.

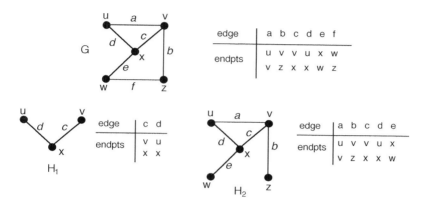

Figure 2.3.1   A graph G and two (proper) subgraphs H_1 and H_2.

A Broader Use of the Term "Subgraph"

The usual meaning of the phrase "H is a subgraph of G" is that H is merely isomorphic to a subgraph of G.

Example 2.3.2: The cycle graphs C_3, C_4, and C_5 are all subgraphs of the graph G of Figure 2.3.1.

Example 2.3.3: A graph with n vertices is Hamiltonian if and only if it contains a subgraph isomorphic to the n-cycle C_n.

Example 2.3.4: The graph G shown in Figure 2.3.2 has among its subgraphs a B_2(bouquet), a D_3 (dipole), and three copies of K_3.

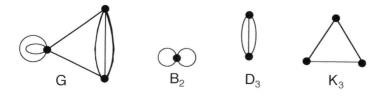

Figure 2.3.2   A graph and three of its subgraphs.

Spanning Subgraphs

DEFINITION: A subgraph H is said to **span** a graph G if $V_H = V_G$.

Example 2.3.5: In Figure 2.3.1, the subgraph H_2 spans G, but the subgraph H_1 does not.

DEFINITION: A **spanning tree** of a graph is a spanning subgraph that is a tree.

Example 2.3.6: The non-simple graph in Figure 2.3.3 has several different spanning trees. The edges of one of them are drawn in bold.

Figure 2.3.3   A spanning tree.

A number of important properties related to spanning trees are established in Chapters 3 and 4. Two of these properties are previewed here.

Proposition 2.3.1: *A graph is connected if and only if it contains a spanning tree.*
$\diamond$ *(See Exercises or §4.5.)*

Proposition 2.3.2: *Every acyclic subgraph of a connected graph G is contained in at least one spanning tree of G.*
$\diamond$ *(See Exercises or §4.5.)*

Since each of the components of an acyclic graph is a tree, the following terminology should not be surprising.

DEFINITION: An acyclic graph is called a **forest**.

Example 2.3.7: In Figure 2.3.4, the isolated vertex and the bold edges with their endpoints form a forest that spans a 12-vertex graph G. Each component of the forest is a spanning tree of the corresponding component of G.

Figure 2.3.4   A spanning forest H of graph G.

Cliques and Independent Sets

DEFINITION: A subset S of V_G is called a **clique** if every pair of vertices in S is joined by at least one edge, and no proper superset of S has this property.

Thus, a clique of a graph G is a maximal subset of mutually adjacent vertices in G.

TERMINOLOGY NOTE: The graph-theory community seems to be split on whether to require that a clique be maximal with respect to the pairwise-adjacency property. The prose sense of "clique" as an *exclusive* group of people motivates our decision to include maximality in the mathematical definition.

DEFINITION: The **clique number** of a graph G is the number $\omega(G)$ of vertices in a largest clique in G.

Example 2.3.8: There are three cliques in the graph shown in Figure 2.3.5. In particular, each of the vertex subsets, $\{u, v, y\}$, $\{u, x, y\}$, and $\{y, z\}$ corresponds to a complete subgraph, and no proper superset of any of them corresponds to a complete subgraph. The clique number is 3, the cardinality of the largest clique.

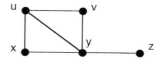

Figure 2.3.5   A graph with three cliques.

DEFINITION: A subset S of V_G is said to be an **independent set** if no pair of vertices in S is joined by an edge. That is, S is a subset of mutually non-adjacent vertices of G.

DEFINITION: The **independence number** of a graph G is the number $\alpha(G)$ of vertices in a largest independent set in G.

Remark: Thus, the clique number $\omega(G)$ and the independence number $\alpha(G)$ of a graph may be regarded as complementary concepts. In §2.4, a precise notion of complementation is introduced.

Induced Subgraphs

DEFINITION: For a given graph G, the **subgraph induced on a vertex subset** U of V_G, denoted $G(U)$, is the subgraph of G whose vertex-set is U and whose edge-set consists of all edges in G that have both endpoints in U. That is,

$$V_{G(U)} = U \quad \text{and} \quad E_{G(U)} = \{e \in E_G | \ endpts(e) \subseteq U\}$$

Example 2.3.9: If G is a simple graph, then it follows from the definition of a clique that the subgraph induced on a clique is a complete graph. Some authors refer to this complete graph as a clique.

Example 2.3.10: A 3-vertex graph is shown on the left in Figure 2.3.6, and the subgraph induced on the vertex subset $\{u, v\}$ is shown in bold on the right.

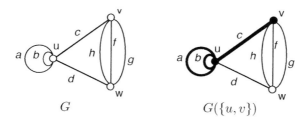

Figure 2.3.6   A subgraph induced on a subset of vertices.

DEFINITION: For a given graph G, the **subgraph induced on an edge subset** D of E_G, denoted $G(D)$, is the subgraph of G whose edge-set is D and whose vertex-set consists of all vertices that are incident with at least one edge in D. That is,

$$V_{G(D)} = \{v \in V_G \mid v \in endpts(e), \text{ for some } e \in D\} \text{ and } E_{G(D)} = D$$

Example 2.3.11: In Figure 2.3.7, the induced subgraph on the edge subset $\{a, c\}$ is shown on the right in bold.

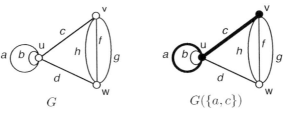

$$G \qquad\qquad G(\{a, c\})$$

Figure 2.3.7   A subgraph induced on a subset of edges.

The subdigraphs induced on a vertex subset or directed-edge subset are analogously defined.

DEFINITION: The **center of a graph** G, denoted $Z(G)$, is the subgraph induced on the set of central vertices of G (see §1.4).

Example 2.3.12: The vertices of the graph in Figure 2.3.8 are labeled by their eccentricities. Since the minimum eccentricity is 3, the vertices of eccentricity 3 lie in the center.

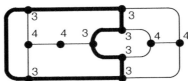

Figure 2.3.8   A graph whose center is a 7-cycle.

Remark: In §2.4, we show that any graph can be the center of a graph.

Local Subgraphs

DEFINITION: The (**open**) **local subgraph** (or (**open**) **neighborhood subgraph**) of a vertex v is the subgraph $L(v)$ induced on the neighbors of v.

Example 2.3.13: Figure 2.3.9 shows a graph and the local subgraphs of three of its vertices.

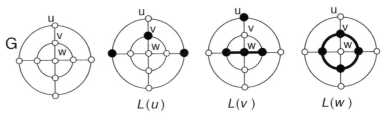

Figure 2.3.9   A graph G and three of its local subgraphs.

Theorem 2.3.3: *Let* $f : G \to H$ *be a graph isomorphism and* $u \in V_G$. *Then* f *maps the local subgraph* $L(u)$ *of* G *isomorphically to the local subgraph* $L(f(u))$ *of* H.

Proof: This follows since f maps the vertices of $L(u)$ bijectively to the vertices of $L(f(u))$, and since f is edge-multiplicity preserving.   ◇

Example 2.3.14: The local subgraphs for graph G of Figure 2.3.10 are all isomorphic to $4K_1$. The local subgraphs for graph H are all isomorphic to P_4. Thus, the two graphs are not isomorphic. Alternatively, we observe that $\alpha(G) = 4$, but $\alpha(H) = 2$, and also that $\omega(G) = 2$, but $\omega(H) = 3$.

Figure 2.3.10   Two 4-regular 8-vertex graphs.

Components

DEFINITION: A **component** of a graph G is a *maximal* connected subgraph of G. In other words, a connected subgraph H is a component of G if H is not a proper subgraph of any connected subgraph of G.

The only component of a connected graph is the entire graph. Intuitively, the components of a non-connected graph are the "whole pieces" it comprises.

Example 2.3.15: The 7-vertex graph in Figure 2.3.11 has four components.

Figure 2.3.11   A graph with four components.

NOTATION: The number of components of a graph G is denoted $c(G)$.

REVIEW FROM §1.4: A vertex v is said to be **reachable** from vertex u if there is a walk from u to v.

Notice that each vertex in a component is reachable from every other vertex in that component. This is because membership in the same component is an *equivalence relation* on V_G, called the *reachability relation*.[†]

DEFINITION: In a graph, the **component of a vertex** v, denoted $C(v)$, is the subgraph induced by the subset of all vertices reachable from v.

Observe that if w is any vertex in $C(v)$, then $C(w) = C(v)$. This suggests the following alternate definition of component.

[†] Equivalence relations are discussed in Appendix A.2.

ALTERNATIVE DEFINITION: A **component** of a graph G is a subgraph induced by an equivalence class of the reachability relation on V_G.

COMPUTATIONAL NOTE: Identifying the components of a small graph is trivial. But larger graphs that are specified by some computer representation require a computer algorithm. These "component-finding" algorithms are developed in Chapter 4 as straightforward applications of certain graph-traversal procedures.

Application to Parallel-Computer Computation

The following application shows how subgraph analysis might occur within a scaleddown model for porting a *distributed algorithm* from one kind of parallel architecture to another. A distributed algorithm is an algorithm that has been designed so that different steps are executed simultaneously on different processing units within a parallel computer.

Application 2.3.1: *Porting an Algorithm* Suppose that a distributed algorithm has been designed to run on an interconnection network whose architecture is an $8 \times 8 \times 32$ array. Suppose also that the only immediately available interconnection network is a 13-dimensional hypercube graph Q_{13} (see §1.2). If the 13-dimensional hypercube graph contains an $8 \times 8 \times 32$ mesh as a subgraph, then the algorithm could be directly ported to that hypercube.

Remark: In general, the ideal situation in which the available network (the *host*) is a supergraph of the required network (the *guest*) is not likely to occur. More realistically, one would like to associate the nodes of the guest with the nodes of the host in such a way as to minimize the decrease in performance of the algorithm.

EXERCISES for Section 2.3

In Exercises 2.3.1 and 2.3.2, determine whether each of the given vertex and edge subsets, W and D, respectively, form a subgraph of the graph at the right.

2.3.1^S a. $W = \{u, v, y\}$; $D = \{q, g\}$
b. $W = \{u, v, y\}$; $D = \{p, q\}$
c. $W = \{u, v, y\}$; $D = \{p, r, h\}$

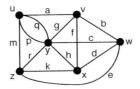

2.3.2 a. $W = \{u, w, y\}$; $D = \{a, b, c\}$
b. $W = \{u, w, y\}$; $D = \{c, q\}$
c. $W = \{u, w, y\}$; $D = \{g, r, h\}$

In Exercises 2.3.3 to 2.3.6, find the subgraphs $G(W)$ and $G(D)$ induced on the given vertex subset W and on the edge subset D, respectively, of the graph from Exercise 2.3.1.

2.3.3$^\text{S}$ $W = \{u, v, y\}$; $D = \{p, r, e\}$ **2.3.4** $W = \{u, v, x\}$; $D = \{g, c, h\}$

2.3.5 $W = \{u, w, y\}$; $D = \{b, c, d\}$ **2.3.6** $W = \{x, y, z, w\}$; $D = \{f, c, h\}$

In Exercises 2.3.7 to 2.3.12, find the subdigraph $G(U)$ induced on the given vertex subset U of the digraph at the right.

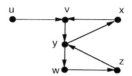

2.3.7$^\text{S}$ $U = \{u,\ v,\ x\}$ **2.3.8** $U = \{y\}$ **2.3.9** $U = \{u,\ v,\ y\}$

2.3.10 $U = \{x,\ y,\ z\}$ **2.3.11** $U = \{u,\ w,\ y,\ z\}$ **2.3.12** $U = \{u,\ x,\ z\}$

In Exercises 2.3.13 to 2.3.18, find the local subgraphs of the given vertex in the graph at the right.

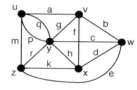

2.3.13$^\text{S}$ u **2.3.14** v **2.3.15** w

2.3.16 x **2.3.17** y **2.3.18** z

In Exercises 2.3.19 to 2.3.22, do all of the following.
(a) Find all of the cliques in the given graph. (b) Give the clique number. (c) Find all the maximal independent sets. (d) Give the independence number. (e) Find the center.

2.3.19$^\text{S}$

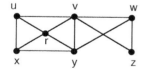

2.3.20

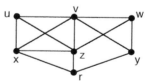

2.3.21

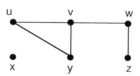

2.3.22

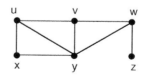

In Exercises 2.3.23 through 2.3.26, determine in the given graph the largest cardinality of a set of mutually adjacent vertices that is not a clique, and find one such set.

2.3.23$^\text{S}$ The graph of Exercise 2.3.19.   **2.3.24** The graph of Exercise 2.3.20.

2.3.25 The graph of Exercise 2.3.21.   **2.3.26** The graph of Exercise 2.3.22.

In Exercises 2.3.27 to 2.3.29, find a largest subset of mutually reachable vertices whose induced subgraph is <u>not</u> a component of the given graph.

2.3.**27**   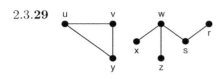   2.3.**28**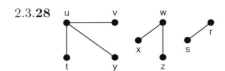

2.3.**29**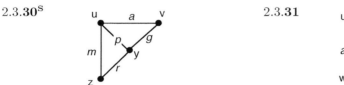

In Exercises 2.3.30 and 2.3.31, find the edge-sets of all spanning trees of the given graph.

2.3.**30**^{S} 2.3.**31**

In Exercises 2.3.32 to 2.3.35, find all possible isomorphism types of the given kind of graph.

2.3.**32** A 6-vertex forest with exactly two components.

2.3.**33** A simple 4-vertex graph with exactly two components.

2.3.**34** A simple 5-vertex graph with exactly three components.

2.3.**35** A simple 6-vertex graph with exactly three components.

2.3.**36** How many of the subgraphs of K_n are complete graphs?

2.3.**37** Prove that the reachability relation is an equivalence relation on V_G.

2.3.**38** Prove that the subgraph induced by an equivalence class of the vertex-reachability relation on a graph G is a component of G.

2.3.**39** Draw a forest having ten vertices, seven edges, and three components.

2.3.**40**^{S} Given six vertices in the plane, how many 2-component forests can be drawn having those six vertices?

2.3.**41** Prove or disprove: Every subgraph of a bipartite graph is a bipartite graph.

2.3.**42** Prove or disprove: For every subgraph H of any graph G, there exists a vertex subset W such that $H = G(W)$.

2.3.43 Prove or disprove: For every subgraph H of any graph G, there exists an edge subset D such that $H = G(D)$.

2.3.44 Let U and W be vertex subsets of a graph G. Under what conditions will $G(U) \cup G(W) = G(U \cup W)$?

2.3.45 Let D and F be edge subsets of a graph G. Under what conditions will $G(D) \cup G(F) = G(D \cup F)$?

2.3.46 Characterize those graphs with the property that the local subgraph of each vertex is a complete graph.

2.3.47 Characterize the class of graphs that have this property: every induced subgraph is connected.

2.3.48 Prove Proposition 2.3.1. (Hint: Prove that an edge-minimal connected spanning subgraph of G must be a spanning tree of G.)

2.3.49 Prove Proposition 2.3.2. (Hint: Let H be an acyclic subgraph of a connected graph G, and let S be the set of all supergraphs of H that are spanning subgraphs of G. Then show that an edge-minimal element of the set S must be a spanning tree of G that contains H.)

2.4 SOME GRAPH OPERATIONS

Computer scientists often regard a graph as a variable. Accordingly, the configuration that results when a vertex or edge is added to or deleted from a graph G is considered to be a new value of G. These primary operations are part of the datatype *graph*, just as the operations of addition and scalar multiplication are part of the definition of a vector space. In Chapter 4, other primary operations, intended for information retrieval, are constructed with the aid of spanning trees. In general, all non-primary operations can be constructed by combining and/or iterating the primary operations.

Deleting Vertices or Edges

DEFINITION: If v is a vertex of a graph G, then the **vertex-deletion subgraph** $G - v$ is the subgraph induced by the vertex-set $V_G - \{v\}$. That is,

$$V_{G-v} = V_G - \{v\} \text{ and } E_{G-v} = \{e \in E_G | v \notin \ endpts(e)\}$$

More generally, if $U \subseteq V_G$, then the result of iteratively deleting all the vertices in U is denoted $G - U$.

DEFINITION: If e is an edge of a graph G, then the **edge-deletion subgraph** $G - e$ is the subgraph induced by the edge-set $E_G - \{e\}$. That is,

$$V_{G-e} = V_G \text{ and } E_{G-e} = E_G - \{e\}$$

More generally, if $D \subseteq E_G$, then the result of iteratively deleting all the edges in D is denoted $G - D$.

The vertex-deletion and edge-deletion subdigraphs are analogously defined.

Example 2.4.1: The next two figures show one of the vertex-deletion subgraphs of a graph G and one of the edge-deletion subgraphs.

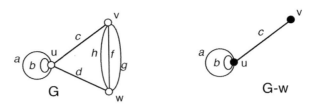

Figure 2.4.1   The result of deleting the vertex w from graph G.

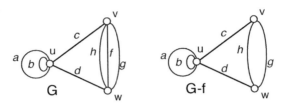

Figure 2.4.2   The result of deleting the edge f from graph G.

Network Vulnerability

The following definitions identify the most vulnerable parts of a network. These definitions lead to different characterizations of a graph's connectivity, which appear in Chapter 5.

DEFINITION: A **vertex-cut** in a graph G is a vertex-set U such that $G - U$ has more components than G.

DEFINITION: A vertex v is a **cut-vertex** (or **cutpoint**) in a graph G if $\{v\}$ is a vertex-cut.

Example 2.4.2: If the graph in Figure 2.4.3 represents a communications network, then a breakdown at any of the four cut-vertices w, x, y, or z would destroy the connectedness of the network. Also, $\{u, v\}$ and $\{w, x\}$ are both vertex-cuts, and $\{u, v\}$ is a *minimal* vertex-cut (i.e., no proper subset of $\{u, v\}$ is a vertex-cut).

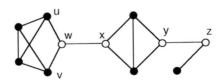

Figure 2.4.3   A graph with four cut-vertices.

DEFINITION: An **edge-cut** in a graph G is a set of edges D such that $G - D$ has more components than G.

DEFINITION: An edge e is a **cut-edge** (or **bridge**) in a graph G if $\{e\}$ is an edge-cut.

Example 2.4.3: For the graph shown in Figure 2.4.4, edges a, b, and c are cut-edges; $\{r, s, t\}$ is an edge-cut; and $\{r, s\}$ is a *minimal* edge-cut.

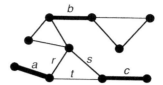

Figure 2.4.4   A graph with three cut-edges.

Observation: In a connected graph G, let e be any edge, with endpoints u and v. Then in the edge-deletion graph $G - e$, every vertex is reachable either from u or from v (or possibly from both). Thus, the only possible components of $G - e$ are the component $C_{G-e}(u)$ that contains the vertex u and the component $C_{G-e}(v)$. These two components coincide if and only if there is a path in G from u to v that *avoids* the edge e (in which case $G - e$ is connected). For example, in Figure 2.4.5, the graph $G - e$ has two components, $C_{G-e}(u)$ and $C_{G-e}(v)$. Notice that had there been an edge in the original graph joining the vertices w and z, then $G - e$ would have been connected.

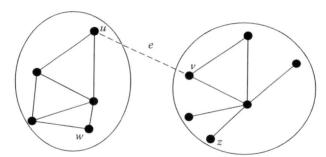

Figure 2.4.5   The components $C_{G-e}(u)$ and $C_{G-e}(v)$.

DEFINITION: An edge e of a graph is called a **cycle-edge** if e lies in some cycle of that graph.

Proposition 2.4.1: *Let e be an edge of a connected graph G. Then $G - e$ is connected if and only if e is a cycle-edge of G.*

Proof: Let u and v be the endpoints of the edge e. If e lies in some cycle of G, then the long way around that cycle is a path between vertices u and v, and by the observation above, the graph $G - e$ is connected. Conversely, if $G - e$ is connected, then there is a path P from u to v that avoids e. Thus, adding edge e to path P forms a cycle in G containing e.   ◇

The first of the following two corollaries restates Proposition 2.4.1 as a characterization of cut-edges. The second corollary combines the result of the proposition with the observation above to establish that deleting a cut-edge increases the number of components by at most one.

Corollary 2.4.2: *An edge of a graph is a cut-edge if and only if it is not a cycle-edge.*

Proof: Apply Proposition 2.4.1 to the component that contains that edge.   ◇

Corollary 2.4.3: *Let e be any edge of a graph G. Then*

$$c(G - e) = \begin{cases} c(G), & \text{if } e \text{ is a cycle-edge} \\ c(G) + 1, & \text{otherwise} \end{cases}$$

The Graph-Reconstruction Problem

DEFINITION: Let G be a graph with $V_G = \{v_1, v_2, \ldots, v_n\}$ The **vertex-deletion subgraph list** of G is the list of the n subgraphs $H_1, \ldots, H_n$, where $H_k = G - v_k$, $k = 1, \ldots, n$.

Remark: The word "list" is being used informally here to mean the unordered collection of the n vertex-deletion subgraphs, some of which may be isomorphic.

Example 2.4.4: A graph G and its labeled vertex-deletion subgraph list is shown in Figure 2.4.6.

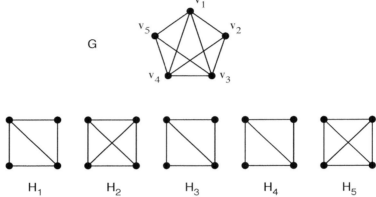

Figure 2.4.6   **A graph and its vertex-deletion subgraph list.**

DEFINITION: The **reconstruction deck** of a graph is its vertex-deletion subgraph list, with no labels on the vertices. We regard each individual vertex-deletion subgraph as being a **card** in the deck.

Example 2.4.4 continued: The reconstruction deck for the graph G is shown in Figure 2.4.7.

Figure 2.4.7   **The reconstruction deck for the graph G.**

DEFINITION: The **graph-reconstruction problem** is to decide whether two nonisomorphic simple graphs with three or more vertices can have the same reconstruction deck.

Remark: We observe that the 2-vertex graphs K_2 and $2K_1$ (two non-adjacent vertices) have the same deck.

Remark: The graph-reconstruction problem would be easy to solve if the vertex-deletion subgraphs included the vertex and edge names. However, the problem is concerned with a list of unlabeled graphs, and it is among the foremost unsolved problems in graph theory. The problem was originally formulated by P.J. Kelly and S.M. Ulam in 1941.

Example 2.4.5: In terms of Example 2.4.4, the graph-reconstruction problem asks whether the graph G in Figure 2.4.6 is the only graph (up to isomorphism) that has the deck shown in Figure 2.4.7.

Reconstruction Conjecture: Let G and G' be two graphs with at least three vertices and with $V_G = \{v_1, v_2, \ldots, v_n\}$ and $V_{G'} = \{w_1, w_2, \ldots, w_n\}$, such that $G - v_i \cong G' - w_i$, for each $i = 1 \ldots, n$. Then $G \cong G'$.

The following results indicate that some information about the original graph can be derived from its vertex-deletion subgraph list. This information is a big help in solving small reconstruction problems.

Theorem 2.4.4: *The number of vertices and the number of edges of a simple graph G can be calculated from its vertex-deletion subgraph list.*

Proof: If the length of the vertex-deletion subgraph list is n, then clearly $|V_G| = n$. Moreover, since each edge appears only in the $n - 2$ subgraphs that do not contain either of its endpoints, it follows that the sum $\sum_v |E_{G-v}|$ counts each edge $n - 2$ times. Thus,

$$|E_G| = \frac{1}{n - 2} \sum_v |E_{G-v}|$$

$\Diamond$

Corollary 2.4.5: *The degree sequence of a graph G can be calculated from its reconstruction deck.*

Proof: First calculate $|E_G|$. For each card, the degree of the missing vertex is the difference between $|E_G|$ and the number of edges on that card. $\Diamond$

The following example shows how Theorem 2.4.4 and its corollary can be used to find a graph G having a given reconstruction deck. Although the uniqueness is an open question, the example shows that if only one graph can be formed when v_k is joined to H_k, for some k, then G is unique and can be retrieved.

Example 2.4.6: Suppose G is a graph with the following reconstruction deck. Consider

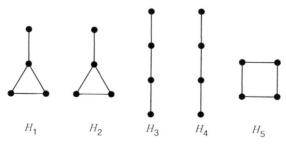

$$H_1 \qquad\qquad H_2 \qquad\qquad H_3 \qquad H_4 \qquad\quad H_5$$

H_5. By Theorem 2.4.4 and Corollary 2.4.5, $\deg(v_5) = 2$, and hence, there are two ways in which v_5 could be joined to H_5. Either v_5 is adjacent to two non-adjacent vertices of H_5 (case 1) or to two vertices that are adjacent (case 2). In the first case, three of the vertex-deletion subgraphs would be a 4-cycles, rather than just one. In the second case, the resulting graph would have the given reconstruction deck, and hence must be the graph G.

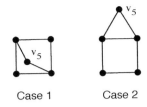

Case 1   Case 2

Corollary 2.4.6: *A regular graph can be reconstructed from its reconstruction deck.*

Proof: Suppose G is a d-regular graph, for some d. By Corollary 2.4.5, we can determine the value of d from the reconstruction deck. On any card in the deck, there would be d vertices of degree $d - 1$. To reconstruct the graph, join those vertices to the missing vertex.   ◇

Remark: B. McKay [Mc77] and A. Nijenhuis [Ni77] have shown, with the aid of computers, that a counterexample to the conjecture would have to have at least 10 vertices.

Adding Edges or Vertices

DEFINITION: **Adding an edge** between two vertices u and w of a graph G means creating a supergraph, denoted $G + e$, with vertex-set V_G and edge-set $E_G \cup \{e\}$, where e is a new edge with endpoints u and w.

Example 2.4.7: Figure 2.4.8 shows the effect of the operation of adding an edge.

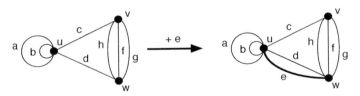

Figure 2.4.8   Adding an edge e with endpoints u and w.

DEFINITION: **Adding a vertex** v to a graph G, where v is a new vertex not already in V_G, means creating a supergraph, denoted $G \cup \{v\}$, with vertex-set $V_G \cup \{v\}$ and edge-set E_G.

DEFINITION: Any one of the operations of adding or deleting a vertex or adding or deleting an edge is called a **primary maintenance operation**.

Remark: *Secondary operations* on a graph are those that are realizable by a combination and/or repetition of one or more of the primary graph operations. Whether or not one is designing software, it is mathematically interesting to analyze or synthesize a graph construction as a sequence of primary operations.

Graph Union

The iterative application of adding vertices and edges results in the secondary graph operation called *graph union*.

DEFINITION: The **(graph) union** of two graphs $G = (V, E)$ and $G' = (V', E')$ is the graph $G \cup G'$ whose vertex-set and edge-set are the disjoint unions, respectively, of the vertex-sets and edge-sets of G and G'.

Example 2.4.8: The graph union $K_3 \cup K_3$ is two disjoint copies of K_3, as shown in Figure 2.4.9. This illustrates clearly that the graph union of two graphs is *not* given by the set-theoretic union of their vertex-sets and the set-theoretic union of their edge-sets, which in this example would be a single copy of K_3.

Figure 2.4.9   The graph union $K_3 \cup K_3$ of two copies of K_3.

DEFINITION: The *n-**fold self-union** nG* is the iterated union $G \cup \cdots \cup G$ of n disjoint copies of the graph G.

Example 2.4.9: The graph in Figure 2.4.9 is the 2-fold self-union $2K_3$.

Joining a Vertex to a Graph

Another secondary operation is the *join* of a new vertex to an existing graph. A more general operation joining two graphs is defined in §2.7.

DEFINITION: If a new vertex v is joined to each of the pre-existing vertices of a graph G, then the resulting graph is called the ***join of v to G*** or the ***suspension*** of G ***from*** v, and is denoted $G + v$.

Thus, to join a new vertex v to a graph G, using only primary operations, first add the vertex v to G. Then add one new edge between v and each pre-existing vertex of G.

DEFINITION: The *n-**wheel** W_n* is the join $C_n + v$ of a single vertex and an n-cycle. (The n-cycle forms the rim of the wheel, and the additional vertex is its hub.) If n is even, then W_n is called an ***even wheel***; if odd, then W_n is called an ***odd wheel***.

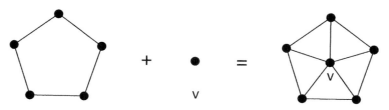

Figure 2.4.10   The 5-wheel $W_5 = C_5 + v$.

The following proposition hints at the general usefulness of the join construction.

Proposition 2.4.7: *Let H be a graph. Then there is a graph G of radius 2 of which H is the center.*

Proof: First construct a supergraph $\hat{H}$ by joining two new vertices u and v to every vertex of H, but not to each other; that is, $\hat{H} = ((H + u) + v) - uv$. Then form the supergraph G by attaching new vertices u_1 and v_1 to vertices u and v, respectively, with edges uu_1 and vv_1, as in Figure 2.4.11. Thus, $G = (\hat{H} \cup 2K_1) + uu_1 + vv_1)$, where the two copies of K_1 are the two trivial graphs consisting of vertices u_1 and v_1. It is easy to verify that, in graph G, each vertex of H has eccentricity 2, vertices u and v have eccentricity 3, and u_1 and v_1 have eccentricity 4. Thus, H is the center of G. $\diamond$

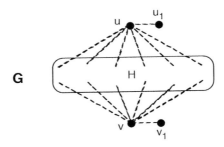

Figure 2.4.11   Making graph H the center of a graph G.

Edge-Complementation

DEFINITION: Let G be a simple graph. Its **edge-complement** (or **complement**) $\overline{G}$ is the graph on the same vertex-set, such that two vertices are adjacent in $\overline{G}$ if and only if they are not adjacent in G. More generally, if G is a subgraph of a graph X, the **relative complement** $X - G$ is the graph $X - E_G$. (Thus, if G has n vertices, then $\overline{G} \cong K_n - G$.)

Example 2.4.10: A graph and its edge-complement are shown in Figure 2.4.12.

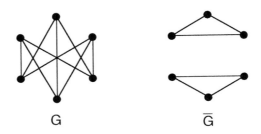

Figure 2.4.12   A graph and its complement.

To construct the edge-complement of the graph G from primary operations, initialize the graph variable H as the null graph (i.e., $V_H = E_H = 0$). Next, for each vertex of G, add a vertex to H. Then for each non-adjacent pair of vertices in G, add an edge to H. The final value of the graph variable H is then assigned to $\overline{G}$.

Remark: Observe that the edge-complement of the edge-complement always equals the original graph, i.e., $\overline{\overline{G}} = G$.

The following theorem formulates precisely the sense in which the independence number of a graph and the clique number are complementary.

Theorem 2.4.8: Let G be a simple graph. Then

$$\omega(G) = \alpha(\overline{G}) \text{ and } \alpha(G) = \omega(\overline{G})$$

◇ (*Exercises*)

Remark: Other graph operations are defined in §2.7.

EXERCISES for Section 2.4

In Exercises 2.4.1 through 2.4.5, find the indicated vertex-deletion subgraph and edge-deletion subgraph of the graph G.

2.4.1^S $G - u$; $G - q$.

2.4.2 $G - y$; $G - e$.

2.4.3 $G - \{w, z\}$; $G - \{m, p, q, a\}$.

2.4.4^S $G - \{w, v\}$; $G - \{c, a\}$.

2.4.5 $G - \{w, v, y\}$; $G - \{k, h, d\}$.

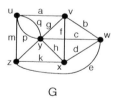

G

2.4.6 Draw a list of graphs that represents all possible isomorphism types of the vertex-deletion subgraphs of the Petersen graph (defined in §1.2).

2.4.7 Draw a list of graphs that represents all possible isomorphism types of the edge-deletion subgraphs of the Petersen graph.

In Exercises 2.4.8 through 2.4.11, find all the cut-vertices and cut-edges.

2.4.8^S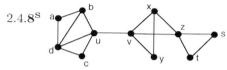

2.4.9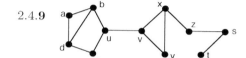

2.4.10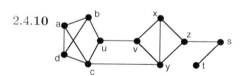

2.4.11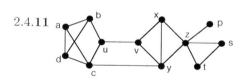

In Exercises 2.4.12 through 2.4.15, determine the largest number of vertices in a minimal vertex-cut for the specified graph, and find one such vertex-cut.

2.4.12^S The graph of Exercise 2.4.8.   2.4.13 The graph of Exercise 2.4.9.

2.4.14 The graph of Exercise 2.4.10.   2.4.15 The graph of Exercise 2.4.11.

In Exercises 2.4.16 through 2.4.19, determine the largest number of edges in a minimal edge-cut for the specified graph, and find one such edge-cut.

2.4.16^S The graph of Exercise 2.4.8.   2.4.17 The graph of Exercise 2.4.9.

2.4.18 The graph of Exercise 2.4.10.   2.4.19 The graph of Exercise 2.4.11.

In Exercises 2.4.20 and 2.4.21, find all possible isomorphism types of the given kind of graph.

2.4.20^S A 4-vertex connected simple graph with no cut-vertices.

2.4.21 A 4-vertex connected simple graph with no cut-edges.

In Exercises 2.4.22 through 2.4.25, find a graph with the given vertex-deletion subgraph list.

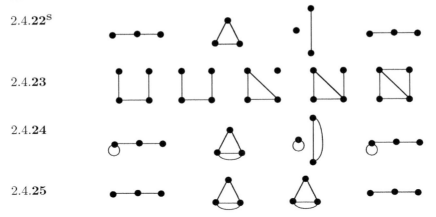

2.4.22[S]

2.4.23

2.4.24

2.4.25

2.4.26 How would you recognize from the reconstruction deck of a graph whether it is bipartite? (Give a proof that your method works.)

2.4.27 How would you recognize from the reconstruction deck of a graph whether it is connected? (Give a proof that your method works.)

2.4.28 Characterize those graphs G that satisfy the equality $(G - v) + v = G$ for every vertex v in G.

2.4.29 Prove that an edge is a cut-edge of a graph if and only if it is not part of any cycle in that graph.

2.4.30 Let e be a cut-edge of a connected graph G. Prove that $G - e$ has exactly two components.

2.4.31 Let v be one of the n vertices of a connected graph G. Find an upper bound for the number of components of $G - v$, and give an example that achieves the upper bound.

2.4.32 Draw a 5-vertex connected graph G that has no cut-vertices, and then verify that G satisfies each of the following properties.

    a. Given any two vertices, there exists a cycle containing both.
    b. For any vertex v and any edge e of G, there exists a cycle containing v and e.
    c. Given any two vertices x and y, and any edge e, there exists a path from x to y that contains e.
    d. Given any two edges, there exists a cycle containing both.
    e. Given any three distinct vertices u, v, and w, there exists a $u - v$ path that contains w.
    f. Given any three distinct vertices u, v, and w, there exists a $u - v$ path that does not contain w.

2.4.33 Draw a 5-vertex graph G that has no cut-edges and such that the suspension $G + v$ has exactly five cut-edges.

2.4.34 Let G be an n-vertex graph with no cut-edges. Prove or disprove: The suspension $G + v$ has exactly n cut-edges.

2.4.**35**[S] Prove or disprove: For every graph G, the suspension $G + v$ has no cut-vertices.

2.4.**36** State a necessary and sufficient condition for an n-vertex graph G to be a suspension of some subgraph H from a vertex in $V_G - V_H$.

2.4.**37** Give a recursive definition for the complete graph K_n, using the join operation.

2.4.**38** Explain why Theorem 2.4.8 follows immediately from the definitions.

2.4.**39** Prove that for any simple graph G on 6 vertices, a copy of K_3 appears as a subgraph of either G or $\overline{G}$ (or both).

2.4.**40** Prove that no vertex of a graph G can be a cut-vertex of both G and $\overline{G}$.

2.4.**41**[S] Let G be a simple graph. Prove that at least one of the graphs G and $\overline{G}$ is connected.

2.5   TESTS FOR NON-ISOMORPHISM

In §2.1, we derived the most basic necessary conditions for two graphs to be isomorphic: same number of vertices, same number of edges, and same degree sequence. We noticed also that two isomorphic graphs could differ in various artifacts of their representations, including vertex names and some kinds of differences in drawings. And we observed that a brute-force approach of considering all $n!$ possible vertex bijections for two n-vertex graphs would be too labor-intensive.

The fact is, there is no known short list of easily applied tests to decide for any possible pair of graphs whether they are isomorphic. Thus, the isomorphism problem is formidable, but not all is bleak. We will establish a collection of graph properties that are preserved under isomorphism. These will be used to show that various pairs of graphs are not isomorphic.

DEFINITION: A **graph invariant** (or **digraph invariant**) is a property of graphs (digraphs) that is preserved by isomorphisms.

Example 2.5.1: We established in §2.1 that the number of vertices, the number of edges, and the degree sequence are the same for any two isomorphic graphs, so they are graph invariants.

In this section, we establish several other graph invariants that are useful for isomorphism testing and in the construction of isomorphisms.

A Local Invariant

The following theorem provides additional necessary criteria for isomorphism.

Theorem 2.5.1: *Let $f : G \to H$ be a graph isomorphism, and let $v \in V_G$. Then the multiset of degrees of the neighbors of v equals the multiset of degrees of the neighbors of $f(v)$.*

Proof: This is an immediate consequence of Corollary 2.1.5.   ◇

Example 2.5.2: Figure 2.5.1 shows two non-isomorphic graphs with the same number of vertices, the same number of edges, and the same degree sequence: $(3,2,2,1,1,1)$. By Theorem 2.1.2, an isomorphism would have to map vertex v in graph G to vertex w in graph H since they are the only vertices of degree 3 in their respective graphs. However, the three neighbors of v have degrees 1, 1, and 2, and the structure-preservation property implies that they would have to be mapped bijectively to the three neighbors of w (i.e., *a forced match*), with degrees 1, 2, and 2. This would violate Theorem 2.5.1.

Figure 2.5.1   Non-isomorphic graphs with the same degree sequence.

Example 2.5.3: Although the indegree and outdegree sequences of the digraphs in Figure 2.5.2 are identical, these digraphs are not isomorphic. To see this, first observe that vertices u, v, x, and y are the only vertices in their respective digraphs that have indegree 2. Since indegree is a local isomorphism invariant (by a digraph analogy to Theorem 2.5.1), u and v must map to y and x, respectively. But the directed path of length 3 that ends at u would have to map to a directed path of length 3 that ends at y. Since there is no such path, the digraphs are not isomorphic.

Figure 2.5.2   Non-isomorphic digraphs with identical degree sequences.

Distance Invariants

DEFINITION: Let $W = \langle v_0, e_1, v_1, \ldots, e_n, v_n \rangle$ be a walk in the domain G of a graph isomorphism (or more generally, under any graph mapping) $f : G \to H$. Then the ***image of the walk*** W is the walk $f(W) = \langle f(v_0), f(e_1), f(v_1), \ldots, f(e_n), f(v_n) \rangle$ in graph H.

Theorem 2.5.2: *The isomorphic image of a graph walk is a walk of the same length.*

Proof: This follows directly from the edge-multiplicity-preserving property of an isomorphism.   ◇

Corollary 2.5.3: *The isomorphic image of a trail, path, or cycle is a trail, path, or cycle, respectively, of the same length.*

Proof: This is an immediate consequence of Theorem 2.5.2 and the bijectivity of the isomorphism.   ◇

Corollary 2.5.4: *For each integer l, two isomorphic graphs must have the same number of trails (paths) (cycles) of length l.*   ◇

Corollary 2.5.5: *The diameter, the radius, and the girth are graph invariants.*

Proof: This is an immediate consequence of Corollary 2.5.3 and the bijectivity of the isomorphism. ◇

Example 2.5.4: Figure 2.5.3 shows the circular ladder CL_4 and the Möbius ladder ML_4, which are 8-vertex graphs. Since both graphs are 3-regular, it is pointless to apply isomorphism tests based on vertex degree.

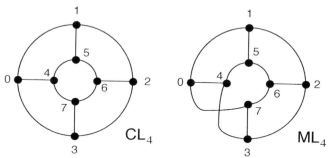

Figure 2.5.3   Non-isomorphic graphs with the same degree sequence.

To reduce the calculation of diameter from checking all $\binom{8}{2} = 28$ vertex pairs to checking the maximum distance from any one vertex, we first establish that both graphs are vertex-transitive. For the circular ladder we observe the symmetries of rotation and the isomorphism that swaps the inner cycle with the outer cycle. For the Möbius ladder, we observe that $j \mapsto j + 1 \mod 8$ is an automorphism, whose iteration establishes vertex-transitivity.

The maximum distance from vertex 0 in CL_4 is 3, to vertex 6. The maximum distance from vertex 0 in ML_4 is 2. Thus, they are not isomorphic. We may also observe that $\alpha(CL_4) = 4$, but $\alpha(ML_4) = 3$.

Subgraph Presence

Theorem 2.5.6: *For each graph-isomorphism type, the number of distinct subgraphs in a graph having that isomorphism type is a graph invariant.*

Proof: Let $f : G_1 \to G_2$ be a graph isomorphism and H a subgraph of G_1. Then $f(H)$ is a subgraph of G_2, of the same isomorphism type as H. The bijectivity of f establishes the invariant. ◇

Example 2.5.5: The five graphs in Figure 2.5.4 are mutually non-isomorphic, even though they have the same degree sequence. Whereas A and C have no K_3 subgraphs, B has two, D has four, and E has one. Thus, Theorem 2.5.6 implies that the only possible isomorphic pair is A and C. However, graph C has a 5-cycle, but graph A does not, because it is bipartite.

Alternatively, we may observe that graphs A and C are the only pair with the same multiset of local subgraphs. We could distinguish this pair by observing that $\alpha(A) = 4$ and $\alpha(C) = 3$.

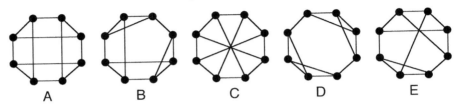

Figure 2.5.4   Five mutually non-isomorphic, 8-vertex, 3-regular graphs.

Edge-Complementation

The next invariant is particularly useful when analyzing simple graphs that are *dense* (i.e., most of the vertex pairs are adjacent).

Theorem 2.5.7: *Let G and H both be simple graphs. They are isomorphic if and only if their edge-complements are isomorphic.*

Proof: By definition, a graph isomorphism necessarily preserves non-adjacency as well as adjacency.   ◇

Example 2.5.6: The two graphs in Figure 2.5.5 are relatively dense, simple graphs (both with 20 out of 28 possible edges). The edge-complement of the left graph consists of two disjoint 4-cycles, and the edge-complement of the right graph is an 8-cycle. Since these edge-complements are non-isomorphic, the original graphs must be non-isomorphic.

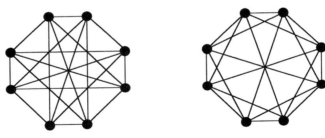

Figure 2.5.5   Two relatively dense, non-isomorphic 5-regular graphs.

Summary

Table 2.5.1 summarizes these results on graph-isomorphism invariants. It is straight-forward to establish digraph versions (see Exercises).

Table 2.5.1: Some graph invariants

1.	The number of vertices
2.	The number of edges
3.	The degree sequence
4.	The multiset of degrees of the neighbors of a *forced match* (see Example 2.5.2)
5.	The multiset of local subgraphs of the neighbors of a *forced match*
6.	Diameter, radius, girth
7.	Independence number, clique number
8.	For any possible subgraph, the number of distinct copies
9.	For a simple graph, the edge-complement

Using Invariants to Construct an Isomorphism

The last example in this section illustrates how invariants can guide the construction of an isomorphism.

Example 2.5.7: The two graphs in Figure 2.5.6 both have open paths of length 9, indicated by consecutive vertex numbering $1, \ldots, 10$. By Corollary 2.5.3, any isomorphism must map the length-9 path in the left graph to some length-9 path in the right graph. The label-preserving bijection that maps the left length-9 path to the right length-9 path is a reasonable candidate for such an isomorphism. Adjacency is clearly preserved for the nine pairs of consecutively numbered vertices, $(i, i + 1)$, since they represent edges along the paths. Also, it is easy to see that adjacency is preserved for the pairs of vertices that are not consecutively numbered, since each of the six remaining edges in the left graph has a corresponding edge in the right graph with matching endpoints. Thus, the label-preserving bijection is an isomorphism.

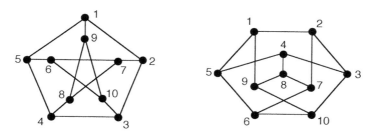

Figure 2.5.6   Two copies of the Petersen graph.

EXERCISES for Section 2.5

2.5.1^{S} For each of the following graphs, either show that it is isomorphic to one of the five regular graphs of Example 2.5.5, or argue why it is not.

a.   b.

In Exercises 2.5.2 and 2.5.3, explain why the graphs in the given graph pair are not isomorphic.

2.5.2^{S}

2.5.3

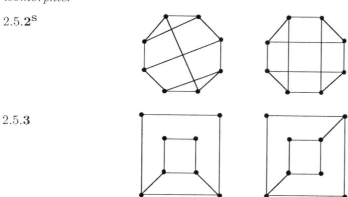

In Exercises 2.5.4 through 2.5.11, determine whether the graphs in the given pair are isomorphic.

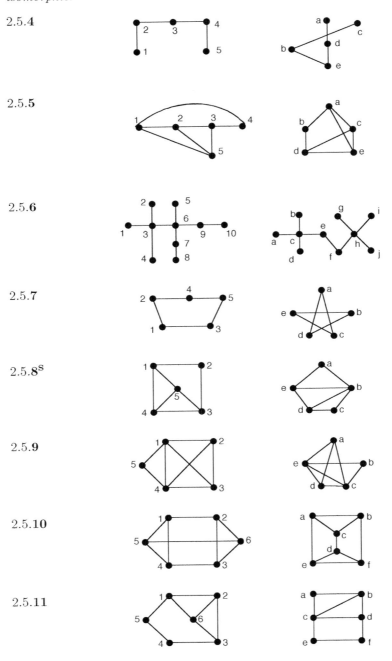

2.5.4

2.5.5

2.5.6

2.5.7

2.5.8^S

2.5.9

2.5.10

2.5.11

2.5.12 Suppose that $f : G \to H$ is a graph isomorphism. Establish each of the following isomorphism properties.

    a. For each $v \in V_G$, the local subgraph of v is isomorphic to the local subgraph of $f(v)$.

    b. For each $v \in V_G$, $G - v$ is isomorphic to $H - f(v)$.

In Exercises 2.5.13 through 2.5.18, state and prove digraph versions of the specified assertion.

2.5.13^S Theorem 2.5.1.

2.5.14 Theorem 2.5.2.

2.5.15 Corollary 2.5.4.

2.5.16 Corollary 2.5.5.

2.5.17 Theorem 2.5.6.

2.5.18 Theorem 2.5.7.

2.5.19 Compile a table of digraph-isomorphism invariants analogous to Table 2.5.1.

In Exercises 2.5.20 through 2.5.25, determine whether the digraphs in the given pair are isomorphic.

2.5.20^S

2.5.21

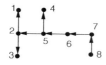 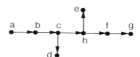

2.5.22

 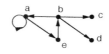

2.5.23

2.5.24

2.5.25

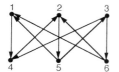 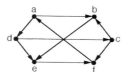

2.6   MATRIX REPRESENTATIONS

Representing graphs by matrices has conceptual and theoretical importance. It helps bring the power of linear algebra to graph theory.

Adjacency Matrices

DEFINITION: The **adjacency matrix of a simple graph** G, denoted A_G, is the symmetric matrix whose rows and columns are both indexed by identical orderings of V_G, such that

$$A_G[u, v] = \begin{cases} 1 & \text{if } u \text{ and } v \text{ are adjacent} \\ 0 & \text{otherwise} \end{cases}$$

Example 2.6.1: Figure 2.6.1 shows the adjacency matrix of a graph G, with respect to the vertex ordering u, v, w, x.

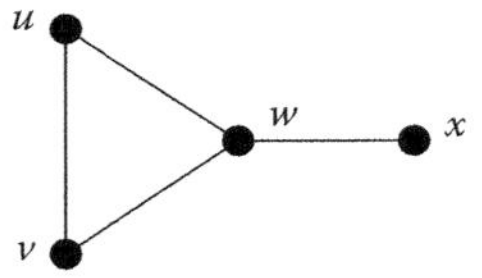

$$A_G = \begin{array}{c} \\ u \\ v \\ w \\ x \end{array} \begin{array}{c} \begin{matrix} u & v & w & x \end{matrix} \\ \begin{pmatrix} 0 & 1 & 1 & 0 \\ 1 & 0 & 1 & 0 \\ 1 & 1 & 0 & 1 \\ 0 & 0 & 1 & 0 \end{pmatrix} \end{array}$$

Figure 2.6.1   A graph and its adjacency matrix.

Usually the vertex order is implicit from context, in which case the adjacency matrix A_G can be written as a matrix without explicit row or column labels.

Proposition 2.6.1: *Let G be a graph with adjacency matrix A_G. Then the value of element $A_G^r[u, v]$ of the rth power of matrix A_G equals the number of u-v walks of length r.*

Proof: The assertion holds for $r = 1$, by the definition of the adjacency matrix and the fact that walks of length 1 are the edges of the graph. The inductive step follows from the definition of matrix multiplication (see Exercises).   ◇

The adjacency matrix of a simple digraph and the relationship between its powers and the number of directed walks in the digraph is analogous to the result for graphs.

DEFINITION: The **adjacency matrix of a simple digraph** D, denoted A_D, is the matrix whose rows and columns are both indexed by identical orderings of V_G, such that

$$A_D[u, v] = \begin{cases} 1 & \text{if there is a edge from } u \text{ to } v \\ 0 & \text{otherwise} \end{cases}$$

Example 2.6.2: The adjacency matrix of the digraph D in Figure 2.6.2 uses the vertex ordering u, v, w, x.

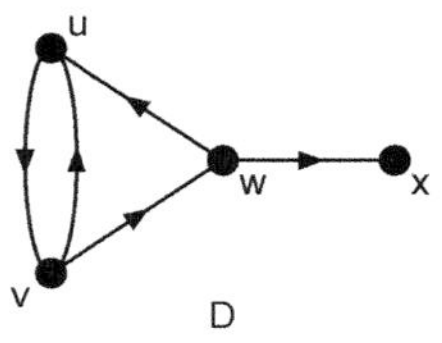

$$A_D = \begin{array}{c} \\ u \\ v \\ w \\ x \end{array} \begin{array}{c} \begin{matrix} u & v & w & x \end{matrix} \\ \begin{pmatrix} 0 & 1 & 0 & 0 \\ 1 & 0 & 1 & 0 \\ 1 & 0 & 0 & 1 \\ 0 & 0 & 0 & 0 \end{pmatrix} \end{array}$$

Figure 2.6.2   A digraph and its adjacency matrix.

Proposition 2.6.2: *Let D be a digraph with $V_D = v_1, v_2, \ldots, v_n$. Then the sum of the elements of row i of the adjacency matrix A_D equals the outdegree of vertex v_i, and the sum of the elements of column j equals the indegree of vertex v_j.* $\diamond$ *(Exercises)*

Proposition 2.6.3: *Let D be a digraph with adjacency matrix A_D. Then the value of the entry $A_D^r[u, v]$ of the rth power of matrix A_D equals the number of directed u-v walks of length r.* $\diamond$ *(Exercises)*

Remark: To extend the definition of adjacency matrix to a general graph (or digraph), for distinct vertices u and v, set $A_G[u, v]$ equal to the number of edges between vertex u and vertex v (or from vertex u to vertex v), and set $A_G[u, v]$ equal to twice the number of self-loops at u.

Brute-Force Graph-Isomorphism Testing

The definition of graph isomorphism implies that two graphs are isomorphic if it is possible to order their respective vertex-sets so that their adjacency matrices are identical.

Example 2.6.3: The graphs in Figure 2.6.3 are isomorphic, because under the orderings u, v, w, x and a, d, c, b, they have identical adjacency matrices.

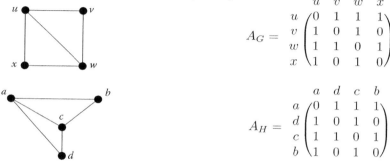

$$A_G = \begin{array}{c} u \\ v \\ w \\ x \end{array}\begin{pmatrix} 0 & 1 & 1 & 1 \\ 1 & 0 & 1 & 0 \\ 1 & 1 & 0 & 1 \\ 1 & 0 & 1 & 0 \end{pmatrix}$$

$$A_H = \begin{array}{c} a \\ d \\ c \\ b \end{array}\begin{pmatrix} 0 & 1 & 1 & 1 \\ 1 & 0 & 1 & 0 \\ 1 & 1 & 0 & 1 \\ 1 & 0 & 1 & 0 \end{pmatrix}$$

Figure 2.6.3   Establishing isomorphism using adjacency matrices.

Algorithm 2.6.1:   Brute-Force Test for Graph Isomorphism

Input: graphs G and H.

Output: Return YES or NO, according to whether G is isomorphic to H.

   If $|V_G| \neq |V_H|$

      Return NO.

   If degree sequences are not equal

      Return NO.

   Fix a vertex ordering for graph G.

   Write the adjacency matrix A_G with respect to that ordering.

   For each vertex ordering τ of graph H

      Write A_H with respect to ordering τ

      If $A_H(\text{w.r.t. } \tau) = A_G$

         Return YES.

   Return NO.

COMPUTATIONAL NOTE: For all but the smallest graphs, this exhaustive method is hopelessly inefficient. This should not be surprising in light of the discussion of the *isomorphism problem* in §2.1.

Incidence Matrices for Undirected Graphs

DEFINITION: The **incidence matrix** of a graph G is the matrix I_G whose rows and columns are indexed by some orderings of V_G and E_G, respectively, such that

$$I_G[v, e] = \begin{cases} 0 \text{ if } v \text{ is not an endpoint of } e \\ 1 \text{ if } v \text{ is an endpoint of } e \\ 0 \text{ if } e \text{ is self-loop at } v^\dagger \end{cases}$$

Example 2.6.4: Figure 2.6.4 shows a graph and its incidence matrix.

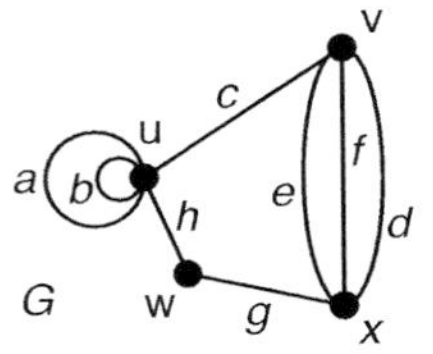

$$I_G = \begin{array}{c} \\ u \\ v \\ w \\ x \end{array} \begin{array}{cccccccc} a & b & c & d & e & f & g & h \\ \begin{pmatrix} 2 & 2 & 1 & 0 & 0 & 0 & 0 & 1 \\ 0 & 0 & 1 & 1 & 1 & 1 & 0 & 0 \\ 0 & 0 & 0 & 0 & 0 & 0 & 1 & 1 \\ 0 & 0 & 0 & 1 & 1 & 1 & 1 & 0 \end{pmatrix} \end{array}$$

Figure 2.6.4   A graph and its incidence matrix.

Observe that changing the orderings of V_G and E_G permutes the rows and columns of I_G. The next two propositions follow immediately from the definition of the incidence matrix.

Proposition 2.6.4: *The sum of the entries in any row of an incidence matrix is the degree of the corresponding vertex.*   ◇ *(Exercises)*

Proposition 2.6.5: *The sum of the entries in any column of an incidence matrix is equal to 2.*   ◇ *(Exercises)*

These two propositions lead to another simple proof of Euler's theorem that the sum of the degrees equals twice the number of edges. In particular, the sum of the degrees is simply the sum of the row-sums of the incidence matrix, and the sum of the column-sums equals twice the number of edges. The result follows since these two sums both equal the sum of all the entries of the matrix.

Incidence Matrices for Digraphs

The incidence matrices for digraphs are analogously defined.

DEFINITION: The **incidence matrix** of a digraph D is the matrix whose rows and columns are indexed by some orderings of V_D and E_D, respectively, such that

$$I_D[v, e] = \begin{cases} 0 & \text{if } v \text{ is not an endpoint of } e \\ 1 & \text{if } v \text{ is the head of } e \\ -1 & \text{if } v \text{ is the tail of } e \\ 2 & \text{if } e \text{ is a self - loop at } v \end{cases}$$

---

†An alternative definition seen elsewhere, which uses 1 for a self-loop instead of 2, takes more effort to find the self-loops. It also has the theoretical inconsistency that row-sums are not necessarily equal to vertex-degrees and column-sums are not necessarily equal to 2.

Example 2.6.5: Figure 2.6.5 shows a digraph and its incidence matrix.

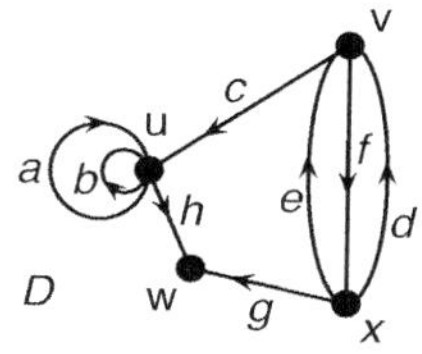

$$I_G = \begin{array}{c} \\ u \\ v \\ w \\ x \end{array} \begin{array}{cccccccc} a & b & c & d & e & f & g & h \\ \begin{pmatrix} 2 & 2 & 1 & 0 & 0 & 0 & 0 & -1 \\ 0 & 0 & -1 & 1 & 1 & -1 & 0 & 0 \\ 0 & 0 & 0 & 0 & 0 & 0 & 1 & 1 \\ 0 & 0 & 0 & -1 & -1 & 1 & -1 & 0 \end{pmatrix} \end{array}$$

Figure 2.6.5   A digraph and its incidence matrix.

Using Incidence Tables to Save Space

The incidence matrix and the adjacency matrix of most graphs have the undesirable feature of consisting mostly of zeros. The following definition gives an alternative representation of a graph that is more space-efficient.

DEFINITION: The **table of incident edges** for a graph G is an incidence table that lists, for each vertex v, all the edges incident on v. This table is denoted $I_{V:E}(G)$.

Example 2.6.6: The table of incident edges for the graph of Example 2.6.4 (repeated below) is as follows:

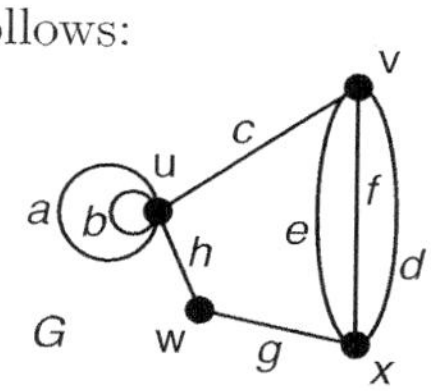

$$I_{V:E}(G) = \begin{array}{ll} \underline{u}: & a\ b\ c\ h \\ \underline{v}: & c\ d\ e\ f \\ \underline{w}: & g\ h \\ \underline{x}: & d\ e\ f\ g \end{array}$$

COMPUTATIONAL NOTE: A "big O" comparison of the space required by the three types of graph representation reveals the advantage that a table of incident edges holds over the two matrix representations.[†] In particular, the total space requirements of the incidence matrix, the adjacency matrix, and the table of incident-edges are, respectively, $O(|V| \cdot |E|)$, $O(|V|^2)$, and $O(|E|)$.

DEFINITION: For a digraph D, the **table of outgoing arcs**, denoted $out_{V:E}(D)$, lists, for each vertex v, all arcs that are directed from v. The **table of incoming arcs**, denoted $in_{V:E}(D)$, is defined similarly.

Example 2.6.7: Here are $in_{V:E}(D)$ and $out_{V:E}(D)$ for the digraph in Figure 2.6.5.

$$in_{V:E}(D) = \begin{array}{ll} \underline{u}: & a\ b\ c \\ \underline{v}: & d\ e \\ \underline{w}: & g\ h \\ \underline{x}: & f \end{array} \qquad out_{V:E}(D) = \begin{array}{ll} \underline{u}: & a\ b\ h \\ \underline{v}: & c\ f \\ \underline{w}: & \\ \underline{x}: & d\ e\ g \end{array}$$

Remark: Since software requirements vary from application to application, it is not possible to prescribe a universally optimal representation of the incidence structure of a graph. The objective here is to introduce software-performance criteria along with the environment required for achieving good performance.

[†]A brief discussion of algorithmic complexity and "big O" notation appears in Appendix A.5.

EXERCISES for Section 2.6

In Exercises 2.6.1 through 2.6.3, determine the matrices A_G, I_G, and the table $I_{V:E}(G)$ for the given graph G.

2.6.1^S 2.6.2   2.6.3

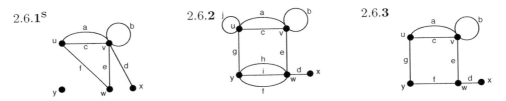

In Exercises 2.6.4 through 2.6.6, determine the matrices A_D and I_D, and the tables $out_{V:E}(D)$ and $in_{V:E}(D)$ for the given digraph D.

2.6.4^S 2.6.5   2.6.6

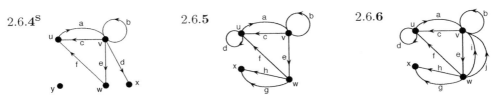

In Exercises 2.6.7 and 2.6.8, draw a graph that has the given adjacency matrix.

2.6.7

$$A_G = \begin{array}{c} \\ a \\ b \\ c \\ d \end{array} \begin{array}{cccc} a & b & c & d \\ \left(\begin{array}{cccc} 0 & 1 & 1 & 1 \\ 1 & 0 & 1 & 0 \\ 1 & 1 & 0 & 1 \\ 1 & 0 & 1 & 0 \end{array} \right) \end{array}$$

2.6.8

$$A_G = \begin{array}{c} \\ a \\ b \\ c \\ d \end{array} \begin{array}{cccc} a & b & c & d \\ \left(\begin{array}{cccc} 2 & 1 & 1 & 1 \\ 1 & 0 & 1 & 0 \\ 1 & 1 & 0 & 1 \\ 1 & 0 & 1 & 0 \end{array} \right) \end{array}$$

In Exercises 2.6.9 through 2.6.13, describe the adjacency matrix of each of the following graph families.

2.6.9^S $K_{m,n}$ 2.6.10 K_n 2.6.11 P_n 2.6.12 C_n 2.6.13 Q_n

In Exercises 2.6.14 and 2.6.15, draw a digraph that has the given adjacency matrix.

2.6.14^S

$$A_D = \begin{array}{c} \\ a \\ b \\ c \\ d \end{array} \begin{array}{cccc} a & b & c & d \\ \left(\begin{array}{cccc} 0 & 1 & 1 & 1 \\ 0 & 0 & 0 & 0 \\ 0 & 1 & 0 & 0 \\ 0 & 0 & 1 & 0 \end{array} \right) \end{array}$$

2.6.15

$$A_D = \begin{array}{c} \\ a \\ b \\ c \\ d \end{array} \begin{array}{cccc} a & b & c & d \\ \left(\begin{array}{cccc} 0 & 1 & 1 & 1 \\ 0 & 0 & 1 & 0 \\ 1 & 1 & 0 & 1 \\ 1 & 0 & 1 & 0 \end{array} \right) \end{array}$$

In Exercises 2.6.16 and 2.6.17, verify that Proposition 2.6.1 holds for the entries $A_G^2[a,a]$ and $A_G^2[d,a]$ for the indicated graph and adjacency matrix.

2.6.16^S The graph and adjacency matrix of Exercise 2.6.7.

2.6.17 The graph and adjacency matrix of Exercise 2.6.8.

In Exercises 2.6.18 and 2.6.19, verify that Proposition 2.6.3 holds for the entries $A_D^2[a,b]$ and $A_D^2[d,a]$ for the indicated digraph and adjacency matrix.

2.6.18^S The digraph and adjacency matrix of Exercise 2.6.14.

2.6.19 The digraph and adjacency matrix of Exercise 2.6.15.

In Exercises 2.6.20 and 2.6.21, draw a graph that has the given incidence matrix.

2.6.20^S

$$I_G = \begin{array}{c} \\ u \\ v \\ w \\ x \end{array} \begin{array}{ccccc} a & b & c & d & e \\ \begin{pmatrix} 0 & 1 & 0 & 0 & 0 \\ 1 & 0 & 2 & 1 & 1 \\ 1 & 0 & 0 & 0 & 0 \\ 0 & 1 & 0 & 1 & 1 \end{pmatrix} \end{array}$$

2.6.21

$$I_G = \begin{array}{c} \\ u \\ v \\ w \\ x \end{array} \begin{array}{ccccc} a & b & c & d & e \\ \begin{pmatrix} 1 & 2 & 1 & 0 & 0 \\ 0 & 0 & 1 & 1 & 1 \\ 0 & 0 & 0 & 0 & 0 \\ 1 & 0 & 0 & 1 & 1 \end{pmatrix} \end{array}$$

In Exercises 2.6.22 and 2.6.23, draw a digraph that has the given incidence matrix.

2.6.22^S

$$I_D = \begin{array}{c} \\ u \\ v \\ w \\ x \end{array} \begin{array}{ccccc} a & b & c & d & e \\ \begin{pmatrix} 1 & 2 & 1 & 0 & 0 \\ 0 & 0 & -1 & 1 & -1 \\ 0 & 0 & 0 & 0 & 0 \\ -1 & 0 & 0 & -1 & 1 \end{pmatrix} \end{array}$$

2.6.23

$$I_D = \begin{array}{c} \\ u \\ v \\ w \\ x \end{array} \begin{array}{ccccc} a & b & c & d & e \\ \begin{pmatrix} -1 & 0 & 1 & 0 & 0 \\ 0 & 0 & 0 & 0 & 1 \\ 0 & 0 & -1 & 1 & 0 \\ 1 & 2 & 0 & -1 & -1 \end{pmatrix} \end{array}$$

2.6.24 Complete the proof of Proposition 2.6.1 by establishing the inductive step.

2.6.25 Prove Proposition 2.6.2.

2.6.26 Prove Proposition 2.6.3.

2.6.27 Prove Proposition 2.6.4.

2.6.28 Prove Proposition 2.6.5.

2.6.29 [*Computer Project*] Implement Algorithm 2.6.1 and test it on at least four of the graph pairs in Exercises 2.5.4 through 2.5.11.

2.7   MORE GRAPH OPERATIONS

The graph-theoretic binary operations of *Cartesian product* and *join* are constructed by iteratively applying the primary operations of adding or deleting vertices and edges. Thus, in a computer-science sense, both these operations may be regarded as secondary operations. The operation of Cartesian product is adapted from set theory, but the join operation has no set-theoretic counterpart. Another binary operation of interest is *amalgamation*, which means pasting two graphs together.

Cartesian Product

DEFINITION: The **Cartesian product** $G \times H$ of the graphs G and H has as its vertex-set the Cartesian product

$$V_{G \times H} = V_G \times V_H$$

and its edge-set is the union of two products:

$$E_{G \times H} = (V_G \times E_H) \cup (E_G \times V_H)$$

The endpoints of the edge (u, d) are the vertices (u, x) and (u, y), where x and y are the endpoints of edge d in graph H. The endpoints of the edge (e, w) are (u, w) and (v, w), where u and v are the endpoints of edge e in graph G.

Example 2.7.1: Figure 2.7.1 illustrates the Cartesian product of the 4-cycle graph C_4 and the complete graph K_2. Observe that $C_4 \times K_2$ is isomorphic to the hypercube graph Q_3 defined in §1.2.

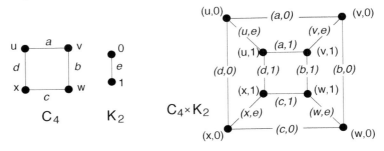

Figure 2.7.1   The labeled Cartesian product $C_4 \times K_2$.

Example 2.7.2: Figure 2.7.2 illustrates the Cartesian product of the complete graph K_3 and the path graph P_3.

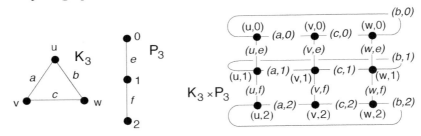

Figure 2.7.2   The labeled Cartesian product $K_3 \times P_3$.

Often there are many ways to visualize a spatial model of the product of two graphs. For instance, two alternative ways to draw the product $K_3 \times P_3$ would be as three concentric triangles, or as a row of triangles, both illustrated in Figure 2.7.3.

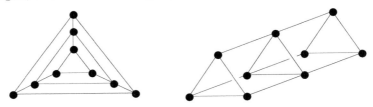

Figure 2.7.3   Two alternative views of $K_3 \times P_3$.

DEFINITION: The product $K_2 \times C_n$ is called a **circular ladder with n rungs** and often denoted CL_n.

Remark: This formal definition is consistent with the informal definition of circular ladders in §1.2.

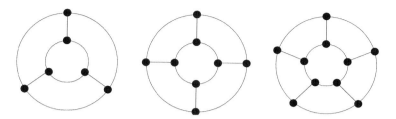

Figure 2.7.4   The circular ladders CL_3, CL_4, and CL_5.

REVIEW FROM §1.2: The **hypercube graph** Q_n is the n-regular graph whose vertex-set is the set of bitstrings of length n, such that there is an edge between two vertices if and only if they differ in exactly one bit.

The labeling in Figure 2.7.5 suggests how Q_n may be defined recursively.

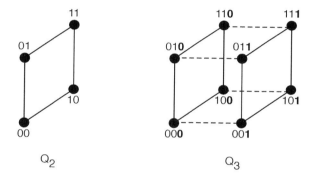

Figure 2.7.5   Bitstring labels showing $Q_3 = Q_2 \times K_2$.

RECURSIVE DEFINITION OF Q_n:

$$Q_1 = K_2 \text{ and } Q_n = Q_{n-1} \times K_2, \text{ for } n \geq 2$$

Figure 2.7.6 illustrates the recursion for Q_4.

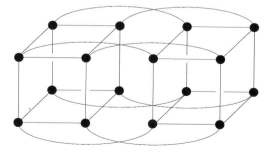

Figure 2.7.6 $Q_4 = Q_3 \times K_2$.

Observation: This recursive definition shows that Q_n is the iterated product of n copies of K_2. That is,

$$Q_n = \underbrace{K_2 \times \cdots \times K_2}_{n \text{ copies}}$$

DEFINITION: The $m_1 \times m_2 \times \cdots \times m_n$ **-mesh** is the iterated product $P_{m_1} \times P_{m_2} \times \cdots \times P_{m_n}$ of paths.

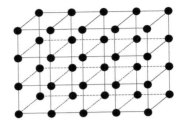

Figure 2.7.7   A 5 $\times$ 4 $\times$ 2 mesh.

DEFINITION: The **wraparound** $m_1 \times m_2 \times \cdots \times m_n$**-mesh** is the iterated product $C_{m_1} \times C_{m_2} \times \cdots \times C_{m_n}$ of cycles.

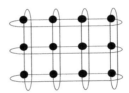

Figure 2.7.8   A wraparound 4 $\times$ 3 mesh.

Remark: The product operation is both commutative and associative.

Join

DEFINITION: The **join** $G + H$ of the graphs G and H is obtained from the graph union $G \cup H$ by adding an edge between each vertex of G and each vertex of H.

Figure 2.7.9   The join $P_2 + P_3$.

Example 2.7.3: The join $K_m + K_n$ is isomorphic to the complete graph K_{m+n}. To see this, consider any two vertices in $K_m + K_n$. If both are from Km or if both are from K_n, then they are obviously still adjacent in $K_m + K_n$. Moreover, if one is from K_m and the other from K_n, then the join construction places an edge between them.

Example 2.7.4: The join $mK_1 + nK_1$ is isomorphic to the complete bipartite graph $K_{m,n}$. One part of the bipartition is the vertices from mK_1 and the other part is the vertices from nK_1.

DEFINITION: The *n-dimensional octahedral graph* $\mathcal{O}_n$, is defined recursively, using the join operation.

$$\mathcal{O}_n = \begin{cases} 2K_1 & \text{if } n = 1 \\ 2K_1 + \mathcal{O}_{n-1} & \text{if } n \geq 2 \end{cases}$$

It is also called the *n-octahedron graph* or, when $n = 3$, the **octahedral graph**, because it is the 1-skeleton of the octahedron, a platonic solid.

Example 2.7.5: Figure 2.7.10 illustrates the first four octahedral graphs.

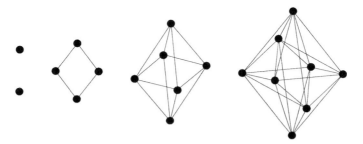

Figure 2.7.10   The octahedral graphs $\mathcal{O}_1$, $\mathcal{O}_2$, $\mathcal{O}_3$, and $\mathcal{O}_4$.

Example 2.7.6: Figure 2.7.11 illustrates that the edge-complement of the graph nK_2 (in K_{2n}) is isomorphic to $\mathcal{O}_n$. This follows recursively, because the two non-adjacent vertices from $2K_1$ and the $n - 1$ pairs of non-adjacent vertices from $\mathcal{O}_{n-1}$ remain nonadjacent in the join $2K_1 + \mathcal{O}_{n-1}$, yielding n pairs of non-adjacent vertices in $2K_1 + \mathcal{O}_{n-1}$. The joining construction guarantees that apart from these n non-adjacencies, all other vertices are adjacent.

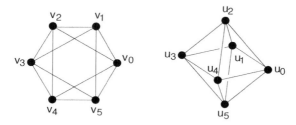

Figure 2.7.11   The graph $K_6 - 3K_2$ is isomorphic to the octahedral graph $\mathcal{O}_3$.

Amalgamations

One way to construct new graphs is to paste together a few standard recognizable "graph parts," which become subgraphs of the resulting big new graph. Pasting on vertices leads directly to pasting on arbitrary subgraphs, via an isomorphism between them.

DEFINITION: Let G and H be disjoint graphs, with $u \in V_G$ and $v \in V_H$. The **vertex amalgamation** $(G \cup H)/\{u = v\}$ is the graph obtained from the union $G \cup H$ by merging (or amalgamating) vertex u of graph G and vertex v of graph H into a single vertex, called uv.[†] The vertex-set of this new graph is $(V_G - \{u\}) \cup (V_H - \{v\}) \cup \{uv\}$, and the edge-set is $E_G \cup E_H$, except that any edge that had u or v as an endpoint now has the amalgamated vertex uv as an endpoint instead.

Example 2.7.7: Figure 2.7.12 illustrates an amalgamation of a 3-cycle and a 4-cycle, in which vertex u of the 3-cycle is identified with vertex v of the 4-cycle.

Figure 2.7.12   A vertex amalgamation of a 3-cycle and a 4-cycle.

In this example, no matter which vertices are chosen in the 3-cycle and the 4-cycle, respectively, the isomorphism type of the amalgamated graph is the same. This is due to the symmetry of the two cycles.

Example 2.7.8: Figure 2.7.13 illustrates the four different isomorphism types of graphs that can be obtained by amalgamating P_3 and P_4 at a vertex.

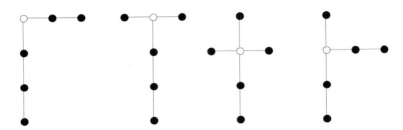

Figure 2.7.13   Four different vertex amalgamations of P_3 and P_4.

When pasting on an arbitrary pair of isomorphic subgraphs, the isomorphism type of the resulting graph may depend on exactly how the vertices and edges of the two subgraphs are matched together. The matching of subgraph to subgraph is achieved by an isomorphism.

DEFINITION: Let G and H be disjoint graphs, with X a subgraph of G and Y a subgraph of H. Let $f : X \to Y$ be an isomorphism between these subgraphs. The **amalgamation of G and H modulo the isomorphism** $f : X \to Y$ is the graph obtained from the union $G \cup H$ by merging each vertex u and each edge e of subgraph X with their images $f(u)$ and $f(e)$ in subgraph Y. The amalgamated vertex is generically denoted $uf(u)$, and the amalgamated edge is generically denoted $ef(e)$. The vertex-set of this new graph is $(V_G - V_X) \cup (V_H - V_Y) \cup \{uf(u)|u \in V_X\}$, and the edge-set is $(E_G - E_X) \cup (E_H - E_Y) \cup \{ef(e)|e \in V_X\}$, except that any edge that had $u \in V_X$ or $f(u) \in V_Y$ as an endpoint now has the amalgamated vertex $uf(u)$ as an endpoint instead. This general amalgamation is denoted $(G \cup H)/f : X \to Y$.

DEFINITION: In an amalgamated graph $(G \cup H)/f : X \to Y$, the image of the pasted subgraphs X and Y is called the **subgraph of amalgamation**.

[†]In other contexts, the juxtaposition notation uv often denotes an edge with endpoints u and v. In this section, we use juxtaposition only for amalgamation.

Example 2.7.9: Each of the six different isomorphisms from the 3-cycle in graph G to the 3-cycle in graph H in Figure 2.7.14 leads to a different amalgamated graph.

Figure 2.7.14   Two graphs that each contain a 3-cycle.

The six different possible results are illustrated in Figure 2.7.15.

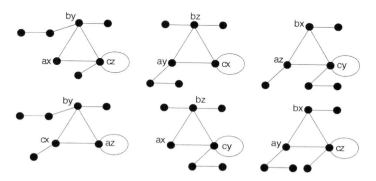

Figure 2.7.15   Six different possible amalgamated graphs.

Remark: The operation of amalgamating graphs G and H by pasting vertex $u \in V_G$ to vertex $v \in V_H$ is equivalent to amalgamating the graphs G and H modulo an isomorphism from the one-vertex subgraph u to the one-vertex subgraph v.

EXERCISES for Section 2.7

For Exercises 2.7.1 through 2.7.4, draw the indicated Cartesian product.

2.7.1^{S} $K_2 \times C_5$.

2.7.3 $W_5 \times P_3$.

2.7.2 $P_3 \times C_5$.

2.7.4 $P_3 \times 2K_3$.

2.7.5 Prove that the Cartesian product of two bipartite graphs is always a bipartite graph.

2.7.6 Prove that the Cartesian product $C_4 \times C_4$ of two 4-cycles is isomorphic to the hypercube graph Q_4.

For Exercises 2.7.7 through 2.7.10, draw the indicated join.

2.7.7^{S} $K_3 + K_3$.

2.7.9 $W_6 + P_2$.

2.7.8 $B_2 + K_4$.

2.7.10 $P_3 + K_3 + C_2$.

2.7.11 Give a necessary and sufficient condition that the join of two bipartite graphs be non-bipartite.

For Exercises 2.7.12 through 2.7.15, draw all the different isomorphism types of graph that can be obtained by a vertex amalgamation of the indicated graphs.

2.7.12^S P_3 and W_4.

2.7.14 $K_4 - K_2$ and P_3.

2.7.13 P_3 and P_5.

2.7.15 $K_4 - P_3$ and C_3.

For Exercises 2.7.16 through 2.7.19, draw all the different isomorphism types of graph that can be obtained by an amalgamation across an edge (i.e., formally, across a K_2-subgraph) of the indicated graphs.

2.7.16^S P_3 and W_4.

2.7.18 $K_4 - K_2$ and P_3.

2.7.17 P_3 and P_5.

2.7.19 $K_4 - P_3$ and C_3.

For Exercises 2.7.20 through 2.7.23, draw all the different isomorphism types of graph that can be obtained by an amalgamation across a pair of vertices of the indicated graphs.

2.7.20^S P_3 and W_4.

2.7.22 $K_4 - K_2$ and P_3.

2.7.21 P_3 and P_5.

2.7.23 $K_4 - P_3$ and C_3.

2.7.24 It is possible to amalgamate $2C_4$ to $2C_4$ modulo an isomorphism into the circular ladder CL_4. Describe the subgraph of amalgamation.

2.7.25 It is possible to amalgamate the join $K_1 + P_r$ to the join $2K_1 + P_s$ modulo an isomorphism into the wheel W_{r+s-2}. Draw a representation of this amalgamation for $r = 3$ and $s = 4$, clearly identifying all parts of the drawing.

2.7.26 Prove that $K_{2n} - 2K_n$ is isomorphic to $K_{n,n}$.

2.7.27 As an entertaining way to obtain a physical model of Q_4, construct from flexible wires (such as soft paper clips) a frame in the shape of the cube graph Q_3, with an attached "wand," as illustrated.

Then dip the frame into a solution of water and liquid dish soap. A cubic bubble will cling to the frame, and inside there will appear a smaller cube, attached only by sheets of soap film surface to the outer cube. The edges within this total configuration form a physical model of Q_4.

[Computer Projects] *For Exercises 2.7.28 through 2.7.31, write an algorithm to construct the indicated graph operation, using only the primary graph operations of additions and deletions of vertices and edges. Test your algorithm on the pair (P_4, W_5) and on the pair $(K_4 - K_2, C_4)$.*

2.7.28 Cartesian product of two graphs.

2.7.29 Join of two graphs.

2.7.30 Vertex amalgamation of two graphs.

2.7.31 Amalgamation of two graphs modulo an isomorphism.

2.8   SUPPLEMENTARY EXERCISES

2.8.1 Draw all isomorphism types of general graphs with

     a. 2 edges and no isolated vertices.   b. 2 vertices and 3 edges.

2.8.2 Draw all isomorphism types of digraphs with

     a. 2 edges and no isolated vertices.   b. 2 vertices and 3 arcs.

2.8.3 Draw all the isomorphism types of simple graphs with

     a. 6 vertices and 3 edges.   b. 6 vertices and 4 edges.

     c. 5 vertices and 5 edges.   d. 7 vertices and 4 edges.

2.8.4 Draw the isomorphism types of simple graphs with degree sequence

     a. 433222.   b. 333331.

2.8.5 Draw two non-isomorphic 4-regular, 7-vertex simple graphs, and prove that every such graph is isomorphic to one of them. Hint: Consider edge complements.

2.8.6 Either draw two non-isomorphic 10-vertex, 4-regular, bipartite graphs, or prove that there is only one such graph.

2.8.7 Show that the graphs $circ(13:1,4)$ and $circ(13:1,5)$ are not isomorphic.

2.8.8 Prove that the complete bipartite graph $K_{4,4}$ is not isomorphic to the Cartesian product graph $K_4 \times K_2$.

2.8.9 Indicate the isomorphic pairs of graphs in this set:

    $\{circ(8:1,2), circ(8:1,3), circ(8:1,4), circ(8:2,3), circ(8:2,4), circ(8:3,4)\}$

2.8.10 Show that the following two graphs are not isomorphic.

2.8.11 Decide which pairs of these three graphs are isomorphic.

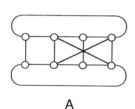 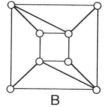 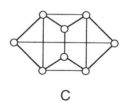

          A   B   C

2.8.**12** Decide which pairs of these three graphs are isomorphic.

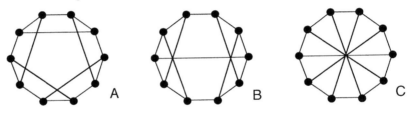

2.8.**13** Decide which pairs of these three graphs are isomorphic.

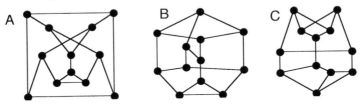

2.8.**14** List the vertex orbits and the edge orbits of the graph of Figure 2.8.1.

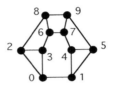

Figure 2.8.1

2.8.**15** Calculate the independence number of the graph in Figure 2.8.1.

2.8.**16** List the vertex orbits and the edge orbits of the following graph.

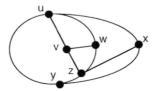

2.8.**17** Some of the 4-vertex, simple graphs have exactly two vertex orbits. Draw an illustration of each such isomorphism type.

DEFINITION: A *rigid graph* is a graph whose only automorphism is the identity automorphism.

2.8.**18** Draw a non-trivial rigid graph.

2.8.**19** Reconstruct the graph with the following deck.

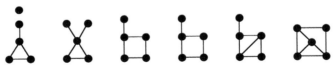

2.8.**20** Reconstruct the graph with the following deck.

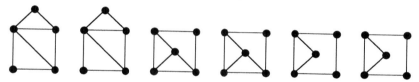

2.8.**21** Suppose that v is a cut-vertex in an n-vertex graph G. Determine the maximum possible number of components of the vertex-deletion subgraph $G - v$, and then describe a graph that achieves that maximum.

2.8.**22** Suppose that a graph G is obtained from the hypercube Q_5 by joining each vertex to the vertex that differs from it in all five bits. (For instance, 01101 is joined to 10010.)

a. Prove that the diameter of the graph G equals 3.

b. Prove that the graph G does not contain a 3-cycle.

2.8.**23** Characterize those graphs with the property that every connected subgraph is an induced subgraph.

2.8.**24** Prove that the cube-graph Q_3 is a subgraph of the product graph $K_4 \times K_2$.

DEFINITION: A graph G is **edge-critical** for a property P if G has property P, but for every edge e, the graph $G - e$ does *not* have property P.

2.8.**25** We consider diameter-related properties.

a. The graphs $K_{m,n}$ clearly have diameter 2 (for $m, n \geq 2$). Prove they are edge-critical graphs for this property.

b. What is the maximum increase in diameter that can be caused by deleting an edge from a 2-connected bipartite graph that is edge-critical with respect to its diameter? Explain.

2.8.**26** Find the diameter of the Cartesian product $C_m \times C_n$ of two cycle graphs.

2.8.**27** Calculate the diameter of the Möbius ladder ML_n.

2.8.**28** Prove that the hypercube Q_n is vertex transitive.

GLOSSARY

adding an edge between two vertices u and v of a graph G: creating a supergraph with vertex-set V_G and edge-set $E_G \cup \{e\}$.

adding a vertex v to a graph G: creating a supergraph, denoted $G \cup \{v\}$, with vertex-set $V_G \cup \{v\}$ and edge-set E_G.

adjacency matrix of a simple digraph D, denoted A_D: the matrix whose rows and columns are both indexed by identical orderings of V_G, such that

$$A_D [u, v] = \begin{cases} 1 & \text{if there is a edge from } u \text{ to } v \\ 0 & \text{otherwise} \end{cases}$$

adjacency matrix of a simple graph G, denoted A_G: the matrix whose rows and columns are both indexed by identical orderings of V_G, such that

$$A_G\,[u,v] = \begin{cases} 1 & \text{if } u \text{ to } v \text{ are adjacent} \\ 0 & \text{otherwise} \end{cases}$$

adjacency-preserving vertex function f $V_G \to V_H$ between two simple graphs G and H: for every pair of adjacent vertices u and v in graph G, the vertices $f(u)$ and $f(v)$ are adjacent in graph H.

——, **non-** : $f(u)$ and $f(v)$ are non-adjacent in H whenever u and v are non-adjacent in G.

amalgamation, vertex of two disjoint, graphs G and H: the graph $(G \cup H)/\{u = v\}$ obtained from the union $G \cup H$ by merging (or amalgamating) a vertex u of graph G and a vertex v of graph H into a single vertex called uv. The vertex-set of this new graph is $(V_G - \{u\}) \cup (V_H - \{v\}) \cup \{uv\}$ and the edge-set is $E_G \cup E_H$, except that any edge that had u or v as an endpoint now has the amalgamated vertex uv as an endpoint instead.

amalgamation modulo an isomorphism $f\colon X \to Y$ of disjoint graphs G and H with X a subgraph of G and Y a subgraph of H: the graph $(G \cup H)/f : X \to Y$ obtained from the union $G \cup H$ by merging each vertex u and each edge e of subgraph X with their images $f(u)$ and $f(e)$ in subgraph Y. See §2.7.

——, **subgraph of:** the image of the pasted subgraphs X and Y in the amalgamated graph $(G \cup H)/f : X \to Y$.

automorphism of a graph G: an isomorphism from the graph to itself, that is, a structure-preserving permutation π_V on V_G and a consistent permutation π_E on E_G; often written as $\pi = (\pi_V, \pi_E)$.

bridge: a synonym for *cut-edge*.

card in a reconstruction deck: one of the vertex-deletion subgraphs in the reconstruction deck.

Cartesian product $G \times H$ of graphs G and H: the graph whose vertex-set is the Cartesian product $V_{G \times H} = V_G \times V_H$ and whose edge-set is the union $E_{G \times H} = (V_G \times E_H) \cup (E_G \times V_H)$. The endpoints of edge (u, d) are the vertices (u, x) and (u, y), where x and y are the endpoints of edge d in graph H. The endpoints of the edge (e, w) are (u, w) and (v, w), where u and v are endpoints of edge e in graph G.

center of a graph G: the subgraph induced on the set of *central vertices* of G (see §1.4); denoted $Z(G)$.

circular ladder with n rungs: the product $K_2 \times C_n$; denoted CL_n (defined informally in §1.2).

clique in a graph G: a maximal subset of mutually adjacent vertices in G.

clique number of a graph G: the number of vertices in a largest clique in G; denoted $\omega(G)$.

complement of a graph: shortened term for *edge-complement*.

component of a graph: a maximal connected subgraph; that is, a connected subgraph which is not contained in any larger connected subgraph.

component of a vertex v: a subgraph induced on the subset of all vertices reachable from v.

consistent vertex and edge bijections: see *isomorphism between general graphs*.

cut-edge of a graph: an edge whose removal increases the number of components.

cutpoint: a synonym for *cut-vertex*.

cut-vertex of a graph: a vertex whose removal increases the number of components.

digraph invariant: a property of a digraph that is preserved by isomorphisms.

disjoint cycle form of a permutation: a notation for specifying a permutation; see Example 2.2.1.

edge-complement of a simple graph G: a graph $\overline{G}$ with the same vertex-set as G, such that two vertices are in adjacent in $\overline{G}$ if and only if they are *not* adjacent in G.

edge-critical graph G for a property P: G has property P, but for every edge e, the graph $G - e$ does not have property P.

edge-cut of a graph: a subset of edges whose removal increases the number of components.

edge-deletion subgraph $G - e$: the subgraph of G induced on the edge subset $E_G - \{e\}$.

edge-multiplicity: the number of edges joining a given pair of vertices or the number of self-loops at a given vertex.

edge orbit of a graph G: an edge subset $F \subseteq E_G$ such that for every pair of edges d, $e \in F$, there is an automorphism of G that maps d to e. Thus, an edge orbit is an equivalence class of E_G under the action of the automorphisms of G.

edge-transitive graph: a graph G such that for every edge pair d, $e \in E_G$, there is an automorphism of G that maps d to e.

forest: an acyclic graph.

graph invariant: a property of a graph that is preserved by isomorphisms.

(n-) hypercube graph Q_n of dimension n: the iterated product $K_2 \times \cdots \times K_2$ of n copies of K_2. Equivalently, the graph whose vertices correspond to the 2^n bitstrings of length n and whose edges correspond to the pairs of bitstrings that differ in exactly one coordinate.

incidence matrix I_D **of a digraph** D: a matrix whose rows and columns are indexed by V_D and E_D, respectively, such that

$$I_D[v, e] = \begin{cases} 0 & \text{if } v \text{ is not an endpoint of } e \\ 1 & \text{if } v \text{ is the head of } e \\ -1 & \text{if } v \text{ is the tail of } e \\ 2 & \text{if } e \text{ is a self-loop at } v \end{cases}$$

incidence matrix I_G **of a graph** G**:** a matrix whose rows and columns are indexed by V_G and E_G, respectively, such that

$$I_G[v,e] = \begin{cases} 0 & \text{if } v \text{ is not an endpoint of } e \\ 1 & \text{if } v \text{ is an endpoint of } e \\ 2 & \text{if } e \text{ is a self-loop at } v \end{cases}$$

independence number of a graph G: the number of vertices in a largest independent set in G; denoted $\alpha(G)$.

independent set of vertices in a graph G: a vertex subset $W \subseteq V_G$ such that no pair of vertices in S is joined by an edge, i. e., S is a subset of mutually non-adjacent. vertices of G.

induced subgraph on an edge set D**:** the subgraph with edge-set D and with vertex-set consisting of the endpoints of all edges in D.

induced subgraph on a vertex set W**:** the subgraph with vertex-set W and edge-set consisting of all edges whose endpoints are in W.

invariant: a shortened term for graph (or digraph) invariant.

isomorphic digraphs: two digraphs that have an isomorphism between their underlying graphs that preserves the direction of each edge.

isomorphic graphs: two graphs G and H that have a structure-preserving vertex bijection between them; denoted $G \cong H$.

isomorphism between two **general** graphs G and H: a structure-preserving vertex bijection $f_V : V_G \rightarrow V_H$ and an edge bijection $f_E : E_G \rightarrow E_H$ such that for every edge $e \in E_G$, the function f_V maps the endpoints of e to the endpoints of the edge $f_E(e)$. Such a mapping pair $(f_V : V_G \rightarrow V_H, f_E : E_G \rightarrow E_H)$ is often written shorthand as $f : G \rightarrow H$.

isomorphism between two **simple** graphs G and H: a structure-preserving vertex bijection $f : V_G \rightarrow V_H$.

isomorphism problem for graphs: the unsolved problem of devising a practical general algorithm to decide graph isomorphism, or, alternatively, to prove that no such algorithm exists.

isomorphism type of a graph G: the class of all graphs H isomorphic to G, i.e., such that there is an isomorphism of G and H.

join of a new vertex v to a graph G: the graph that results when each of the preexisting vertices of G is joined to vertex v; denoted $G + v$.

join of two graphs G and H: the graph obtained from the graph union $G \cup H$ by adding an edge between each vertex of G and each vertex of H; denoted $G + H$.

local subgraph of a vertex v: synonym of *neighborhood subgraph* of v; denoted $L(v)$. mesh, $m_1 \times m_2 \times \cdots \times m_n$—: the iterated product $P_{m_1} \times P_{m_2} \times \cdots \times P_{m_n}$ of paths.

—, **wraparound:** the iterated product $C_{m_1} \times C_{m_2} \times \cdots \times C_{m_n}$ of paths.

Möbius ladder graph ML_n: a graph obtained from the circular ladder CL_n by deleting from the circular ladder two of its parallel curved edges and replacing them with two edges that cross; analogous to the relationship between a Möbius band and a cylindrical band.

neighbor of a vertex v: any vertex adjacent to v.

(open) neighborhood subgraph of a vertex v: the subgraph induced on the neighbors of v; denoted $L(v)$. Also called *local subgraph*.

octahedral graph, n-dimensional: the graph $\mathcal{O}_n$ defined recursively, using the join operation, as

$$\mathcal{O}_n = \begin{cases} 2K_1 & \text{if } n = 0 \\ 2K_1 + \mathcal{O}_{n-1} & \text{if } n \geq 1 \end{cases}$$

preserves adjacency: see *adjacency-preserving*.

preserves non-adjacency: see *adjacency-preserving, non-*.

primary maintenance operation on a graph: adding or deleting a vertex or an edge.

product $G \times H$ of graphs G and H: shortened term for *Cartesian product*.

proper subgraph of a graph G: a subgraph of G that is neither G nor an isomorphic copy of G.

reachable from a vertex v: said of a vertex for which there is a walk from v.

reachability relation for a graph G: the equivalence relation on V_G, where two vertices are related if one is reachable from the other.

reconstruction deck of a graph G: the vertex-deletion subgraph list of G, with no labels on the vertices.

reconstruction problem for graphs: the unsolved problem of deciding whether two non-isomorphic graphs with three or more vertices can have the same vertex-deletion subgraph list.

rigid graph: a graph whose only automorphism is the identity automorphism.

spanning subgraph of a graph G: a subgraph H of G with $V_H = V_G$.

spanning tree: a spanning subgraph that is a tree.

structure-preserving vertex bijection between two general graphs G and H: a vertex bijection $f : V_G \to V_H$ such that

(i) the number of edges (even if 0) joining each pair of distinct vertices u and v in G equals the number of edges joining their images $f(u)$ and $f(v)$ in H, and

(ii) the number of self-loops at each vertex x in G equals the number of self-loops at the vertex $f(x)$ in H.

structure-preserving vertex bijection between two simple graphs G and H: a vertex bijection $f : V_G \to V_H$ that preserves adjacency and non-adjacency, i.e., for every pair of vertices in G,

u and v are adjacent in $G \Leftrightarrow f(u)$ and $f(v)$ are adjacent in H.

subdigraph of a digraph D: a digraph whose underlying graph is a subgraph of the underlying graph of D, and whose edge directions are inherited from D.

subgraph of a graph G: a graph H whose vertices and edges are all in G, or any graph isomorphic to such a graph.

——, **proper:** a subgraph that is neither G nor an isomorphic copy of G.

——, **spanning:** a subgraph H with $V_H = V_G$.

supergraph: the "opposite" of subgraph; that is, H is a supergraph of G if and only if G is a subgraph of H.

suspension of a graph G from a new vertex v: a synonym for *join*.

table of incident edges for a graph G: a table $I_{V:E}(G)$ that lists, for each vertex v, all the edges incident on v.

table of incoming arcs for a digraph D: a table $in_{V:E}(D)$ that lists, for each vertex v, all arcs that are directed to v.

table of outgoing arcs for a digraph D: a table $out_{V:E}(D)$ that lists, for each vertex v, all arcs that are directed from v.

(graph) union of two graphs $G = (V, E)$ and $G' = (V', E')$: the graph $G \cup G'$ whose vertex-set and edge-set are the disjoint unions, respectively, of the vertex-sets and edge-sets of G and G'.

——, n-**fold self- :** the iterated disjoint union $G \cup \cdots \cup G$ of n copies of the graph G; denoted nG.

vertex-cut of a graph: a subset of vertices whose removal increases the number of components.

vertex-deletion subgraph $G - v$: the subgraph of G induced on the vertex subset $V_G - \{v\}$.

vertex-deletion subgraph list: a list of the isomorphism types of the collection of vertex-deletion subgraphs.

vertex orbit of a graph G: a vertex subset $W \subseteq V_G$ such that for every pair of vertices $u, v \in W$, there is an automorphism of G that maps u to v. Thus, a vertex orbit is an equivalence class of V_G under the action of the automorphisms of G.

vertex-transitive graph: a graph G such that for every vertex pair $u, v \in V_G$, there is an automorphism of G that maps u to v.

n-**wheel** W_n: the join $K_1 + C_n$ of a single vertex and an n-cycle.

——, **even:** a wheel for n even.

——, **odd:** a wheel for n odd.

Chapter 3

TREES

INTRODUCTION

Trees are important to the structural understanding of graphs and to the algorithmics of information processing, and they play a central role in the design and analysis of connected networks. In fact, trees are the backbone of optimally connected networks.

A main task in information management is deciding how to store data in space-efficient ways that also allow their retrieval and modification to be time-efficient. Tree-based structures are often the best way of balancing these competing goals. For example, the binary-search-tree structure introduced in §3.4 leads to an optimally efficient search algorithm. Several other applications of binary trees are given in §3.3 through §3.6.

The final two sections include a brief excursion into *enumerative combinatorics*. In §3.7, Cayley's formula for the number of labeled n-vertex trees is derived using *Prüfer Encoding*. In §3.8, a recurrence relation for the number of different binary trees is established and then solved to obtain a closed formula.

3.1   CHARACTERIZATIONS AND PROPERTIES OF TREES

REVIEW FROM §1.5: A *tree* is a connected graph with no cycles.

Characterizing trees in a variety of ways provides flexibility for their application. The first part of §3.1 establishes some basic properties of trees that culminate in Theorem 3.1.8, where six different but equivalent characterizations of a tree are given.

Basic Properties of Trees

DEFINITION: In an undirected tree, a *leaf* is a vertex of degree 1.

If a leaf is deleted from a tree, then the resulting graph is a tree having one vertex fewer. Thus, induction is a natural approach to proving tree properties, provided one can always find a leaf. The following proposition guarantees the existence of such a vertex.

Proposition 3.1.1: *Every tree with at least one edge has at least two leaves.*

Proof: Let $P = (v_1, v_2, \cdots, v_m)$ be a path of maximum length in a tree T. Suppose one of its endpoints, say v_1, has degree greater than 1. Then v_1 is adjacent to vertex v_2 on path P and also to some other vertex w (see Figure 3.1.1). If w is different from all of the vertices v_i, then P could be extended, contradicting its maximality. On the other hand, if w is one of the vertices v_i on the path, then the acyclic property of T would be contradicted. Thus, both endpoints of path P must be leaves in tree T.   ◇

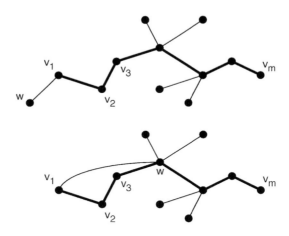

Figure 3.1.1   The two cases in the proof of Proposition 3.1.1.

Corollary 3.1.2: *If the degree of every vertex of a graph is at least 2, then that graph must contain a cycle.*

Proof: Apply Proposition 3.1.1 to any one of the components of that graph.   ◇

The next proposition establishes a fundamental property of trees. Its proof is the first of several instances that demonstrate the effectiveness of the inductive approach to proving assertions about trees.

Proposition 3.1.3: *Every tree on n vertices contains exactly $n - 1$ edges.*

Proof: A tree on one vertex is the trivial tree, which has no edges.

Assume for some number $k \geq 1$, as an induction hypothesis, that every tree on k vertices has exactly $k - 1$ edges. Next consider any tree T on $k + 1$ vertices. By Proposition 3.1.1, T contains a leaf, say v. Then the graph $T - v$ is acyclic, since deleting pieces from an acyclic graph cannot create a cycle. Moreover, $T - v$ is connected, since the vertex v has degree 1 in T. Thus, $T - v$ is a tree on k vertices, and, hence, $T - v$ has $k - 1$ edges, by the induction hypothesis. But since $deg(v) = 1$, it follows that $T - v$ has one edge fewer than T. Therefore, T has k edges, which completes the proof.   ◇

REVIEW FROM §2.3: An acyclic graph is called a **forest**.

REVIEW FROM §2.4: The number of components of a graph G is denoted $c(G)$.

Corollary 3.1.4: *A forest G on n vertices has $n - c(G)$ edges.*

Proof: Apply Proposition 3.1.3 to each of the components of G.   ◇

Corollary 3.1.5: *Any graph G on n vertices has at least $n - c(G)$ edges.*

Proof: If G has cycle-edges, then remove them one at a time until the resulting graph $\widehat{G}$ is acyclic. Then $\widehat{G}$ has $n - c(\widehat{G})$ edges, by Corollary 3.1.4; and $c(\widehat{G}) = c(G)$, by Corollary 2.4.3.   ◇

Corollary 3.1.5 provides a lower bound for the number of edges in a graph. The next two results establish an upper bound for certain simple graphs. This kind of result is typically found in *extremal graph theory* .

Proposition 3.1.6: *Let G be a simple graph with n vertices and k components. If G has the maximum number of edges among all such graphs, then*

$$|E_G| = \binom{n - k}{2}$$

Proof: Since the number of edges is maximum, each component of G is a complete graph. If $n = k$, then G consists of k isolated vertices, and the result is trivially true. If $n > k$, then G has at least one non-trivial component. We show that G has exactly one nontrivial component. Suppose, to the contrary, that $C_1 = K_r$ and $C_2 = K_s$, where $r \geq s \geq 2$. Then the total number of edges contained in these two components is $\binom{r}{2} + \binom{s}{2}$. However, the graph that results from replacing C_1 and C_2 by K_{r+1} and K_{s-1}, respectively, has $\binom{r+1}{2} + \binom{s-1}{2}$ edges in those two components. By expanding these formulas, it is easy to show that the second graph has $r - s + 1$ more edges than the first, contradicting the maximality of the first graph. Thus, G consists of $k - 1$ isolated vertices and a complete graph on $n - k + 1$ vertices, which shows that $|E_G| = \binom{n-k+1}{2}$ and completes the proof.   ◇

Corollary 3.1.7: *A simple n-vertex graph with more than $\binom{n-1}{2}$ edges must be connected.*   ◇

Six Different Characterizations of a Tree

Trees have many possible characterizations, and each contributes to the structural understanding of graphs in a different way. The following theorem establishes some of the most useful characterizations.

Theorem 3.1.8: *Let T be a graph with n vertices. Then the following statements are equivalent.*

1. *T is a tree.*
2. *T contains no cycles and has $n-1$ edges.*
3. *T is connected and has $n-1$ edges.*
4. *T is connected, and every edge is a cut-edge.*
5. *Any two vertices of T are connected by exactly one path.*
6. *T contains no cycles, and for any new edge e, the graph $T+e$ has exactly one cycle.*

Proof: If $n = 1$, then all six statements are trivially true. So assume $n \geq 2$.

$(1 \Rightarrow 2)$ By Proposition 3.1.3.

$(2 \Rightarrow 3)$ Suppose that T has k components. Then, by Corollary 3.1.4, the forest T has $n - k$ edges. Hence, $k = 1$.

$(3 \Rightarrow 4)$ Let e be an edge of T. Since $T - e$ has $n - 2$ edges, Corollary 3.1.5 implies that $n - 2 \geq n - c(T - e)$. So $T - e$ has at least two components.

$(4 \Rightarrow 5)$ By way of contradiction, suppose that Statement 4 is true, and let x and y be two vertices that have two different paths between them, say P_1 and P_2. Let u be the first vertex from which the two paths diverge (this vertex might be x), and let v be the first vertex at which the paths meet again (see Figure 3.1.2). Then these two $u - v$ paths taken together form a cycle, and any edge on this cycle is not a cut-edge, which contradicts Statement 4.

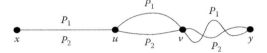

Figure 3.1.2

$(5 \Rightarrow 6)$ T contains no cycles since any two vertices on a cycle have two different paths between them, consisting of the opposite routes around the cycle. Furthermore, the addition of any new edge e to T will create a cycle, since the endpoints of e, say u and v, are already connected by a path in T. To show that this cycle is unique, suppose two cycles were created. They both would contain edge e, and the long way around each of these cycles would then be two different $u - v$ paths in T, contradicting Statement 5.

$(6 \Rightarrow 1)$ By way of contradiction, assume that T is not connected. Then the addition of an edge joining a vertex in one component to a vertex in a different component would not create a cycle, which would contradict Statement 6. $\diamond$

The fundamental properties of a tree are summarized in Table 3.1.1 for easy reference.

Table 3.1.1: Summary of Basic Properties of a Tree T on n vertices

1.	T is connected.
2.	T contains no cycles.
3.	Given any two vertices u and v of T, there is a unique $u - v$ path.
4.	Every edge in T is a cut-edge.
5.	T contains $n - 1$ edges.
6.	T contains at least two vertices of degree 1 if $n \geq 2$.
7.	Adding an edge between two vertices of T yields a graph with exactly one cycle.

The Center of a Tree

REVIEW FROM §1.4 AND §2.3:

- The **eccentricity** of a vertex v in a graph G, denoted $ecc(v)$, is the distance from v to a vertex farthest from v. That is, $ecc(v) = \max\limits_{x \in V_G}\{d(v, x)\}$.

- A **central vertex** of a graph is a vertex with minimum eccentricity.

- The **center of a graph** G, denoted $Z(G)$, is the subgraph induced on the set of central vertices of G.

In an arbitrary graph G, the center $Z(G)$ can be anything from a single vertex to all of G. However, C. Jordan showed in 1869 that the center of a tree has only two possible cases. We begin with some preliminary results concerning the eccentricity of vertices in a tree.

Lemma 3.1.9: *Let T be a tree with at least three vertices.*

(a) *If v is a leaf of T and w is its neighbor, then $ecc(v) = ecc(w) + 1$.*

(b) *If v is a central vertex of T, then $deg(v) \geq 2$.*

Proof: (a) Since T has at least three vertices, $deg(w) \geq 2$. Then there exists a vertex $z \neq v$ such that $d(w, z) = ecc(w)$. But z is also a vertex farthest from v, and hence $ecc(v) = d(v, z) = d(w, z) + 1 = ecc(w) + 1$.

(b) By Part (a), a vertex of degree 1 cannot have minimum eccentricity in T and hence, cannot be a central vertex of T.   ◇

Lemma 3.1.10: *Let v and w be two vertices in a tree T such that w is of maximum distance from v (i.e., $ecc(v) = d(v, w)$). Then w is a leaf.*

Proof: Let P be the unique $v - w$ path in tree T. If $deg(w) \geq 2$, then w would have a neighbor whose distance from v would equal $d(v, w) + 1$, contradicting the premise that w is at maximum distance.   ◇

Lemma 3.1.11: *Let T be a tree with at least three vertices, and let $T*$ be the subtree of T obtained by deleting from T all its leaves. If v is a vertex of $T*$, then $ecc_T(v) = ecc_{T*}(v) + 1$.*

Proof: Let w be a vertex of T such that $ecc_T(v) = d(v, w)$. By Lemma 3.1.10, w is a leaf of T and hence, $w \notin V_{T*}$ (as illustrated in Figure 3.1.3). It follows that the neighbor of w, say z, is a vertex of T^* that is farthest from v among all vertices in T^*, that is, $ecc_{T*}(v) = d(v, z)$. Thus,

$$ecc_T(v) = d(v, w) = d(v, z) + 1 = ecc_{T*}(v) + 1 \qquad\qquad ◇$$

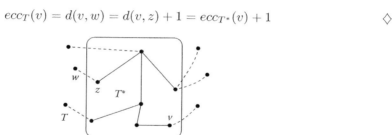

Figure 3.1.3

Proposition 3.1.12: *Let T be a tree with at least three vertices, and let T^* be the subtree of T obtained by deleting from T all its leaves. Then*

$$Z(T) = Z(T^*)$$

Proof: From the two preceding lemmas, we see that deleting all the leaves decreases all the remaining vertex eccentricities by 1. It follows that the resulting tree has the same center. ◇

Corollary 3.1.13: [Jordan, 1869]. *Let T be an n-vertex tree. Then the center $Z(T)$ is either a single vertex or a single edge.*

Proof: The assertion is trivially true for $n = 1$ and $n = 2$. The result follows by induction, using Proposition 3.1.12. ◇

Tree Isomorphisms and Automorphisms

As we saw in §2.5, graph invariants are helpful both in isomorphism testing and in the construction of isomorphisms. For instance, the center of a graph must map to the center of the image graph, and a leaf and its image leaf must be the same distance from their respective centers. The next two examples illustrate these ideas for trees.

Example 3.1.1: The two graphs in Figure 3.1.4 have the same degree sequence, but they can be readily seen to be non-isomorphic in several ways. For instance, the center of the left graph is a single vertex, but the center of the right graph is a single edge. Another way is to observe that the two graphs have unequal diameters.

Figure 3.1.4   Why are these trees non-isomorphic?

Example 3.1.2: The graph shown in Figure 3.1.5 below does not have a non-trivial automorphism because the three leaves are all different distances from the 3-valent vertex, and hence, an automorphism must map each of them to itself.

Figure 3.1.5   A tree that has no non-trivial automorphisms.

Remark: There is a linear-time algorithm for testing the isomorphism of two trees (see [AhHoUl83]).

Tree-Graphic Sequences

The next result characterizes those sequences that are degree sequences of trees.

DEFINITION: A sequence $\langle d_1, d_2, \ldots, d_n \rangle$ is said to be **tree-graphic** if there is a permutation of it that is the degree sequence of some n-vertex tree.

Theorem 3.1.14: *A sequence $\langle d_1, d_2, \ldots, d_n \rangle$ of $n \geq 2$ positive integers is tree-graphic if and only if $\sum_{i=1}^{n} d_i = 2n - 2$.*

Proof: If some permutation of the sequence $\langle d_1, d_2, \cdots, d_n \rangle$ is the degree sequence of an n-vertex tree T, then, by Euler's Degree-Sum Theorem (§1.1) and Proposition 3.1.3, we have $\sum_{i=1}^{n} d_i = 2|E_T| = 2n - 2$.

To prove the sufficiency assertion, we use induction on n. The assertion is trivially true for $n = 2$. Assume that the assertion is true for some $n \geq 2$ and let $\langle d_1, d_2, \cdots, d_{n+1} \rangle$ be a sequence of positive integers satisfying $\sum_{i=1}^{n+1} d_i = 2(n + 1) - 2 = 2n$. We may assume without loss of generality that $d_1 \geq d_2 \geq \cdots \geq d_{n+1}$. It follows from a simple counting argument that $2 \leq d_1 \leq n$ and $d_n = d_{n+1} = 1$. Thus, the terms of the sequence $\langle d_1 - 1, d_2, d_3, \cdots, d_n \rangle$ are positive and sum to $2n - 2$, and hence, by the induction hypothesis, there is an n-vertex tree T whose degree sequence is a permutation of $\langle d_1 - 1, d_2, d_3, \cdots, d_n \rangle$. Let $\hat{T}$ be the tree obtained by joining a new vertex to a vertex of T of degree $d_1 - 1$. Then the degree sequence of $\hat{T}$ is a permutation of the sequence $\langle d_1, d_2, \cdots, d_{n+1} \rangle$. $\diamond$

Trees as Subgraphs

An arbitrary graph is likely to contain a number of different trees as subgraphs. The following theorem shows that if a simple n-vertex graph is sufficiently *dense* (i.e., it has sufficiently many edges), then it will contain every type of tree up to a certain order.

REVIEW FROM §1.1: The minimum degree of the vertices of a graph G is denoted $\delta_{\min}(G)$.

Theorem 3.1.15: *Let T be any tree on n vertices, and let G be a simple graph such that $\delta_{\min}(G) \geq n - 1$. Then T is a subgraph of G.*

Proof: The result is clearly true if $n = 1$ or $n = 2$, since K_1 and K_2 are subgraphs of every graph having at least one edge.

Assume that the result is true for some $n \geq 2$. Let T be a tree on $n + 1$ vertices, and let G be a graph with $\delta_{\min}(G) \geq n$. We show that T is a subgraph of G.

Let v be a leaf of T, and let vertex u be its only neighbor in T. The vertex-deletion subgraph $T - v$ is a tree on n vertices, so, by the induction hypothesis, $T - v$ is a subgraph of G. Since $deg_G(u) \geq n$, there is some vertex w in G but not in $T - v$ that is adjacent to u. Let e be the edge joining vertices u and w (see Figure 3.1.6). Then $T - v$ together with vertex w and edge e form a tree that is isomorphic to T and is a subgraph of G. $\diamond$

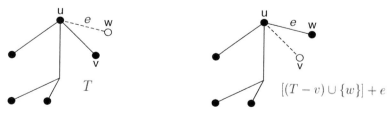

Figure 3.1.6 T **is isomorphic to** $[(T - v) \cup \{w\}] + e$.

EXERCISES for Section 3.1

3.1.1 Draw a 6-vertex connected graph that has exactly seven edges and exactly three cycles.

3.1.2 Draw a l2-vertex forest that has exactly 10 edges.

In Exercises 3.1.3 through 3.1.12, either draw the required graph or explain why no such graph exists.

3.1.3^S A 7-vertex, 3-component, simple graph with 10 edges.

3.1.4 A 6-vertex, 2-component, simple graph with 11 edges.

3.1.5 An 8-vertex, 2-component, simple graph with exactly nine edges and three cycles.

3.1.6 An 8-vertex, 2-component, simple graph with exactly 10 edges and three cycles.

3.1.7 A 9-vertex, 2-component, simple graph with exactly 10 edges and two cycles.

3.1.8 A 9-vertex, simple, connected graph with exactly 10 edges and three cycles.

3.1.9^S A 9-vertex, simple, connected graph with exactly 12 edges and three cycles.

3.1.10 A 10-vertex, 2-component, forest with exactly nine edges.

3.1.11 A 10-vertex, 3-component, forest with exactly nine edges.

3.1.12 An 11-vertex, simple, connected graph with exactly 14 edges that contains five edge-disjoint cycles.

3.1.13^S Let G be a connected simple graph on n vertices. Determine a lower bound on the average degree of a vertex, and characterize those graphs that achieve the lower bound.

3.1.14 Prove that if G is a tree having an even number of edges, then G must contain at least one vertex having even degree.

3.1.15 Suppose the average degree of the vertices of a connected graph is exactly 2. How many cycles does G have?

3.1.16 Prove or disprove: If a simple graph G has no cut-edges, then every vertex of G has even degree.

3.1.17^S Prove or disprove: A connected n-vertex simple graph with n edges must contain exactly one cycle.

3.1.18 Prove or disprove: There exists a connected n-vertex simple graph with $n+2$ edges that contains exactly two cycles.

3.1.19 Prove or disprove: There does not exist a connected n-vertex simple graph with $n+2$ edges that contains four edge-disjoint cycles.

3.1.20^S Prove that H is a maximal acyclic subgraph of a connected graph G if and only if H is a spanning tree of G. What analogous statement can be made for graphs that are not necessarily connected?

3.1.21 Prove that if a graph has exactly two vertices of odd degree, then there must be a path between them.

3.1.22 Show that any nontrivial simple graph contains at least two vertices that are not cut-vertices.

3.1.23 Prove that if two distinct closed paths have an edge e in common, then there must be a closed path that does not contain e.

3.1.24^S Let G be a simple graph on n vertices. Prove that if $\delta_{\min}(G) \geq \frac{n-1}{2}$, then G is connected.

3.1.25 Prove that if any single edge is added to a connected graph G, then at least one cycle is created.

3.1.26 Prove that a graph G is a forest if and only if every induced subgraph of G contains a vertex of degree 0 or 1.

3.1.27^S Characterize those graphs with the property that every connected subgraph is an induced subgraph.

3.1.28 Let T be a tree with at least three vertices, and let T^* be the subtree of T obtained by deleting from T all its vertices of degree 1. Prove that $diam(T^*) = diam(T) - 2$ and $rad(T^*) = rad(T) - 1$.

3.1.29 Let T be a tree with at least two vertices. Prove that the center $Z(T)$ is a single edge if and only $diam(T) = 2rad(T) - 1$.

3.1.30 Determine the smallest integer $n \geq 2$ for which there exists an n-vertex tree whose only automorphism is the trivial one.

3.1.31 Determine the smallest integer $n \geq 2$ for which there exists an n-vertex tree whose center is a single edge and whose only automorphism is the trivial one.

DEFINITION: A tree T with $m + 1$ vertices and m edges is **graceful** if it is possible to label the vertices of T with distinct elements from the set $\{0, 1, \cdots, m\}$ in such a way that the induced edge-labeling, which assigns the integer $|i - j|$ to the edge joining the vertices labeled i and j, assigns the labels $1, 2, \ldots, m$ to the m edges of G.

3.1.32 a. Prove that every path is graceful.

   b. Prove that $K_{1,n}$ (also called a *star*) is graceful for all $n \geq 2$.

   c. Show that every tree on 6 vertices is graceful.

   d. Prove or disprove: Every tree is graceful. (This is an open problem.)

DEFINITION: The **distance sum** (or **Wiener index**[†]) of a graph G, denoted $ds(G)$, is the sum of the distances between all pairs of vertices in G; that is, $ds(G) = \sum\limits_{u,v \in V_G} d(u,v)$.

3.1.33 Determine the distance sum for the n-vertex star $K_{1,n-1}$.

3.1.34 Let T be an n-vertex tree different from the star $K_{1,n}$. Prove that $ds(T) > ds(K_{1,n-1})$

3.1.35 Determine the distance sum for the n-vertex path graph P_n.

3.1.36 Let T be an n-vertex tree different from the path graph P_n. Prove that $ds(T) < ds(P_n)$.

[†]Named for the chemist H. Wiener, who introduced this measure in 1947.

3.2 ROOTED TREES, ORDERED TREES, and BINARY TREES

Rooted Trees

DEFINITION: A *directed tree* is a directed graph whose underlying graph is a tree.

DEFINITION: A *rooted tree* is a tree with a designated vertex called the **root**. Each edge is considered to be directed away from the root.

Thus, a rooted tree is a directed tree such that the root has indegree 0 and all other vertices have indegree 1.

In drawing a rooted tree, the arrows are usually omitted from the arcs, since their direction is always away from the root. Figure 3.2.1 shows the two most common ways of drawing a rooted tree.

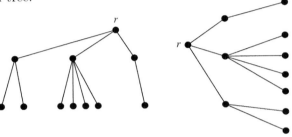

Figure 3.2.1   Two common ways of drawing a rooted tree.

Some Applications of Rooted Trees

Application 3.2.1:   *Decision Trees*   The organization of large computer programs that perform complex decision-making strategies is often based on rooted trees. For example, Figure 3.2.2 below shows some of the sequences for the first three moves of a game of tic-tac-toe, starting with an x in the center. In fact, if we take into account symmetry, then the figure actually shows *all* the possible sequences.

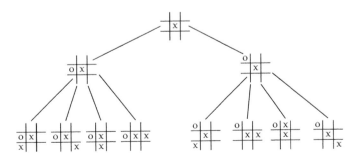

Figure 3.2.2   The first three moves of tic-tac-toe.

Application 3.2.2:   *Data Organization*   Rooted trees are used to store information that is organized into categories and subcategories. The *Dewey Decimal Classification System* for libraries is one such example. The system starts with ten broad subject areas, each assigned a range of numbers. More specialized areas within a given area are assigned subranges. Eventually single numbers are assigned to the specific subjects within an area, and these subjects are further broken down by using decimal fractions. The rooted tree shown in Figure 3.2.3 shows how books in the fine arts are classified in the Dewey decimal system.

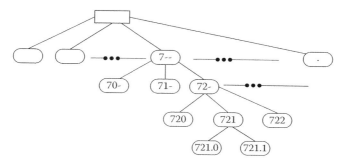

Figure 3.2.3   Fine arts subtree under Dewey decimal system.

Application 3.2.3:   *Sentence Parsing*   A rooted tree can be used to parse a sentence in a natural language, such as English. The tree in Figure 3.2.4 diagrams the underlying syntactic structure of the words and phrases in the sentence "Sammy has two ears."

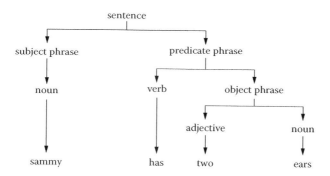

Figure 3.2.4   Parsing a sentence.

Application 3.2.4:   *Shortest-Path Trees*   For a given connected graph G with a vertex v, a rooted tree provides a compact way of exhibiting for each $w \in V_G$, a shortest path from v to w.

DEFINITION: Let v be a vertex in a connected graph G. A **shortest-path tree** for G from v is a rooted tree T with vertex-set V_G and root v such that the unique path in T from v to each vertex w is a shortest path in G from v to w.

Example 3.2.1:   Figure 3.2.5 shows an 8-vertex graph and a shortest-path tree from vertex a.

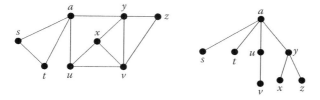

Figure 3.2.5   A graph and a shortest-path tree.

Remark: The *breadth-first search* produces a shortest-path tree for an unweighted graph, and *Dijkstra's algorithm* produces one for a weighted graph. These and other *tree-growing* algorithms are discussed in Chapter 4.

Rooted Tree Terminology

Designating a root imposes a hierarchy on the vertices of a rooted tree, according to their distance from that root.

DEFINITION: In a rooted tree, the **depth** or **level** of a vertex v is its distance from the root, i.e., the length of the unique path from the root to v. Thus, the root has depth 0.

Remark: Typically, a rooted tree is drawn so that the root is at the top, and the vertices at each level are horizontally aligned.

DEFINITION: The **height** of a rooted tree is the length of a longest path from the root (or the greatest depth in the tree).

DEFINITION: If vertex v immediately precedes vertex w on the path from the root to w, then v is the **parent** of w and w is the **child** of v.

DEFINITION: Vertices having the same parent are called **siblings**.

DEFINITION: A vertex w is called a **descendant** of a vertex v (and v is called an **ancestor** of w), if v is on the unique path from the root to w. If, in addition, $w \neq v$, then w is a **proper** descendant of v (and v is a **proper** ancestor of w).

DEFINITION: A **leaf** in a rooted tree is any vertex having no children.

DEFINITION: An **internal vertex** in a rooted tree is any vertex that has at least one child. The root is internal, unless the tree is trivial (i.e., a single vertex).

Example 3.2.2: A rooted tree is shown in Figure 3.2.6. The height of this tree is 3, and vertex j is the only vertex whose depth achieves this value. Also, r, a, b, c, and d are the internal vertices; vertices e, f, g, h, i, and j are the leaves; vertices g, h, and i are siblings; vertex a is an ancestor of j, and j is a descendant of a.

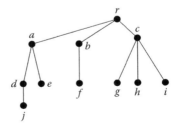

Figure 3.2.6   A rooted tree of height 3.

Many applications impose an upper bound on the number of children that a given vertex can have.

DEFINITION: An *m-ary tree* $(m \geq 2)$ is a rooted tree in which every vertex has at most m children, and in which at least one vertex has exactly m children.

DEFINITION: A **complete** *m-ary tree* is an m-ary tree in which every internal vertex has exactly m children and all leaves have the same depth.

Example 3.2.3: Figure 3.2.7 shows two *ternary* (3-ary) trees; the one on the left is complete; the other one is not.

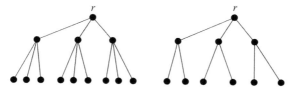

Figure 3.2.7   Two 3-ary trees, one complete and the other not complete.

Proposition 3.2.1: *A complete m-ary tree of height h has m^k vertices at level k, for $k \le h$.*

Proof: The statement is trivially true for $k = 1$.
Assume as an induction hypothesis that there are m^l vertices at level l, for some $l \ge 1$. Since each of these vertices has m children, there are $m \cdot m^l = m^{l+1}$ children at level $l + 1$.   ◇

Corollary 3.2.2: *An m-ary tree has at most m^k vertices at level k.*

Theorem 3.2.3: *Let T be an n-vertex m-ary tree of height h. Then*

$$h + 1 \le n \le \frac{m^{h+1} - 1}{m - 1}$$

Proof: Let n_k be the number of vertices at level k, so that $1 \le n_k \le m^k$, by Corollary 3.2.2. Thus,

$$h + 1 = \sum_{k=0}^{h} 1 \le \sum_{k=0}^{h} n_k \le \sum_{k=0}^{h} m^k = \frac{m^{h+1} - 1}{m - 1}$$

The result follows since $\sum_{k=0}^{h} n_k = n$.   ◇

Corollary 3.2.4: *The complete m-ary tree of height h has $\frac{m^{h+1}-1}{m-1}$ vertices.*   ◇

Isomorphism of Rooted Trees

DEFINITION: Two rooted trees are said to be ***isomorphic as rooted trees*** if there is a graph isomorphism between them that maps root to root.

Example 3.2.4: Figure 3.2.8 illustrates two pairs of rooted trees. The two on the left are isomorphic as rooted trees; the two on the right are isomorphic trees, but they are not isomorphic as rooted trees. This shows that there are more isomorphism types of rooted trees than there are of trees.

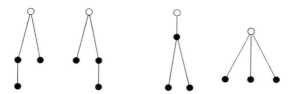

Figure 3.2.8   Isomorphic trees need not be isomorphic as rooted trees.

COMPUTATIONAL NOTE: *Representing a Rooted Tree:* If the n vertices of a rooted tree are labeled $0, 1, \ldots, n - 1$, with the root assigned label 0, then a list whose kth entry is the parent of vertex k is called an **array-of-parents** representation. It fully specifies the rooted tree.

Example 3.2.5: Figure 3.2.9 shows the array-of-parents representation for a rooted tree with seven vertices.

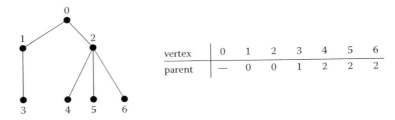

vertex	0	1	2	3	4	5	6
parent	—	0	0	1	2	2	2

Figure 3.2.9   Representing a rooted tree with an array of parents.

Ordered Trees

DEFINITION: An **ordered tree** is a rooted tree in which the children of each vertex are assigned a fixed ordering.

DEFINITION: In a **standard plane drawing** of an ordered tree, the root is at the top, the vertices at each level are horizontally aligned, and the left-to-right order of the vertices agrees with their prescribed order.

Remark: In an ordered tree, the prescribed *local ordering* of the children of each vertex extends to several possible *global orderings* of the vertices of the tree. One of them, the *level order,* is equivalent to reading the vertex names top to bottom, left to right in a standard plane drawing. Level order and three other global orderings, *pre-order, post-order,* and *in-order,* are explored in §3.3.

Example 3.2.6: Standard plane drawings of two different ordered trees are shown in Figure 3.2.10 below. As rooted trees, they are isomorphic, and one of the rooted-tree isomorphisms preserves the vertex labeling. The ordered tree on the left stores the expression $a * b - c$, whereas the one on the right stores $c - a * b$.

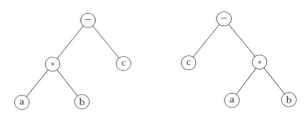

Figure 3.2.10   Two different ordered trees that are isomorphic rooted trees.

Remark: The two ordered trees in Example 3.2.6 are a special type of 2-ary ordered tree known as a *binary tree*. The example previews an application that appears in §3.3, in which binary trees are used to store and produce arithmetic expressions.

Binary Trees

Binary trees are among the most frequently used information structures in computer science. A variety of applications are given in §3.3 through §3.6.

DEFINITION: A **binary tree** is an ordered 2-ary tree in which each child is designated either a **left-child** or a **right-child**.

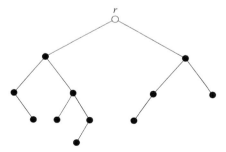

Figure 3.2.11   A binary tree of height 4.

DEFINITION: The **left (right)subtree** of a vertex v in a binary tree is the binary subtree spanning the left (right)-child of v and all of its descendants.

The mandatory designation of left-child or right-child means that two different binary trees may be indistinguishable when regarded as ordered trees.

Example 3.2.7: In Figure 3.2.12, the four different binary trees shown on the right are all realizations of the ordered tree on the left. A formula for the number of different binary trees with a given number of vertices is developed in §3.8.

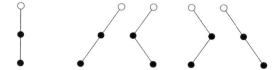

Figure 3.2.12   Four different binary trees that are the same ordered tree.

A binary tree consists of a root and its left and right subtrees, which are themselves smaller binary trees. This *recursive* view makes mathematical induction the method of choice for proving many important facts about binary trees. Typically, the induction is on the height of the binary tree, because of the following recursive property.

RECURSIVE PROPERTY OF A BINARY TREE: If T is a binary tree of height h, then its left and right subtrees both have height less than or equal to $h - 1$, and equality holds for at least one of them.

Example 3.2.8: The binary tree shown in Figure 3.2.13. has height 4, and its left and right subtrees have heights 3 and 2, respectively.

The next two results provide an upper bound on the number of vertices in a binary tree of fixed height. The upper bound is fundamental to certain results in algorithmic analysis. Although Theorem 3.2.5 is an immediate consequence of Corollary 3.2.4, the inductive proof given here is instructive in its use of the recursive property given above.

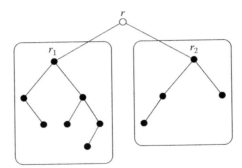

Figure 3.2.13   Recursive view of a binary tree.

Theorem 3.2.5: *The complete binary tree of height h has $2^{h+1} - 1$ vertices.*

Proof: The assertion is trivially true if $h = 0$. Assume for some $k \geq 0$, that a complete binary tree of height k has $2^{k+1} - 1$ vertices, and let T be a complete binary tree of height $k + 1$. Since T is complete, its left and right subtrees, say T_1 and T_2, must also be complete. Furthermore, trees T_1 and T_2 are both of height k, so by the induction hypothesis, they each contain $2^{k+1} - 1$ vertices. Thus, the number of vertices of T is

$$1 + 2^{k+1} - 1 + 2^{k+1} - 1 = 2^{k+2} - 1 \qquad\qquad \Diamond$$

Corollary 3.2.6: *Every binary tree of height h has at most $2^{h+1} - 1$ vertices.*

Example 3.2.9: Figure 3.2.14 illustrates Theorem 3.2.5 for the case $h = 3$.

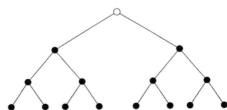

Figure 3.2.14   The complete binary tree of height 3 has 15 vertices.

EXERCISES for Section 3.2

3.2.1^S Draw all the rooted tree types with 3 or 4 vertices. How many different graph isomorphism types do they represent?

3.2.2 Draw all the rooted tree types with 5 vertices. How many different graph isomorphism types do they represent?

3.2.3^S Draw all possible binary trees of height 2, and group these into classes of isomorphic rooted trees.

3.2.4 Draw all possible binary trees of height 3 whose internal vertices have exactly two children. Group these into classes of isomorphic rooted trees.

In Exercises 3.2.5 through 3.2.13, draw the specified tree(s) or explain why no such tree(s) can exist.

3.2.5[S] A 12-vertex binary tree of height 5 that has exactly five leaf vertices.

3.2.6 A 12-vertex binary tree of height 4.

3.2.7 A 14-vertex binary tree of height 3.

3.2.8[S] A 14-vertex tree such that every leaf vertex has exactly three proper ancestors.

3.2.9 Three different 6-vertex rooted trees whose underlying graphs are isomorphic.

3.2.10 A ternary tree of height 3 with exactly 10 vertices.

3.2.11 A ternary tree of height 3 with exactly 28 leaf vertices.

3.2.12 A ternary tree of height 3 with exactly four vertices.

3.2.13[S] A ternary tree of height 3 with exactly 13 vertices, such that every internal vertex has exactly three children.

3.2.14 Prove that a directed tree that has more than one vertex with indegree 0 cannot be a rooted tree.

3.2.15 Represent the folder and subfolder organization in your computer as a rooted tree.

3.2.16 What is the relationship between the depth of a vertex in a rooted tree and the number of ancestors of v? Prove your answer.

3.2.17[S] How many different non-complete ternary trees (i.e., non-isomorphic as rooted trees) are there of height 3 such that every internal vertex has exactly three children?

3.2.18 Demonstrate the semantic power of the comma by using a rooted tree to parse the sentence "The Republicans, say the Democrats, are crooks." in two ways, one with the commas and one without.

In Exercises 3.2.19 and 3.2.20, draw the rooted tree specified by the given array of parents.

3.2.19 –,0,1,1,1,1,1,2,2,2.   **3.2.20** –,0,1,1,2,2,3,4,4.

In Exercises 3.2.21 and 3.2.22, specify the given rooted tree with an array of parents.

3.2.21[S] 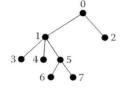   **3.2.22**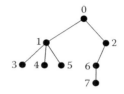

3.2.23[S] Let $f(h)$ be the number of nodes in a complete binary tree of height h. Express $f(h)$ in terms of $f(h-1)$ without appealing to the result of Theorem 3.2.5. Then use this *recurrence relation* to simplify the induction step in the proof of Theorem 3.2.5.

3.2.24 Instead of using Theorem 3.2.5, prove Corollary 3.2.6 directly with the "strong form" of induction, using the following weaker induction hypothesis: For some $h > 0$ and all $k \leq h - 1$, a binary tree of height k has at most $2^{k+1} - 1$ vertices.

3.3   BINARY-TREE TRAVERSALS

A systematic visit of each vertex of a tree is called a **tree traversal**. Thus, a traversal of a binary tree is a global ordering of its vertices. Four kinds of traversals are presented in this section: *level-order, pre-order, post-order,* and *in-order.*

Level-Order Traversal

DEFINITION: The **level-order** of an ordered tree is a listing of the vertices in the top-to-bottom, left-to-right order of a standard plane drawing of that tree.

Thus, the level-order is consistent with the prescribed local ordering of the children of each vertex in the ordered tree.

Example 3.3.1: The level-order of the labeled binary tree in Figure 3.3.1 is $r, a, b, c, d, e, f, h, i, j, k, l$.

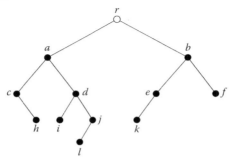

Figure 3.3.1   Level-order: $r, a, b, c, d, e, f, h, i, j, k, l$

Pre-Order, Post-Order, and In-Order Traversals

The pre-order, post-order, and in-order traversals are defined using the recursive view of a binary tree introduced in §3.2. Then a geometric description is given for each one, and at the end of the section, algorithmic descriptions using stacks are presented. The vertex sequence given for each of the traversals refers to the binary tree in Figure 3.3.1.

DEFINITION: The **left pre-order traversal** of a binary tree T is defined recursively as follows:

   i. List (process) the root of T.
  ii. Perform a left pre-order traversal of the left subtree of T.
 iii. Perform a left pre-order traversal of the right subtree of T.

Example 3.3.1 continued: Left Pre-Order: $r, a, c, h, d, i, j, l, b, e, k, f$

DEFINITION: The **post-order traversal** of a binary tree T is defined recursively as follows:

   i. Perform a post-order traversal of the left subtree of T.
  ii. Perform a post-order traversal of the right subtree of T.
 iii. List (process) the root of T.

Example 3.3.1 continued: Post-Order: $h, c, i, l, j, d, a, k, e, f, b, r$

DEFINITION: The **in-order traversal** of a binary tree T is defined recursively as follows:

 i. Perform an in-order traversal of the left subtree of T.
 ii. List (process) the root of T.
 iii. Perform an in-order traversal of the right subtree of T.

Example 3.3.1 continued: In-Order: $c, h, a, i, d, l, j, r, k, e, b, f$

Processing Arithmetic Expressions

An expression tree is the most natural information structure for storing and processing arithmetic expressions.

DEFINITION: An **expression tree** for an arithmetic expression is either a single vertex labeled by an identifier, or is described by the following recursive rule:

 • The operator to be executed during an evaluation is at the root.

 • The left subtree of the root is an expression tree representing the left operand.

 • The right subtree is an expression tree representing the right operand.

Example 3.3.2: The expression tree in Figure 3.3.2 stores the expression $((a + b) * (c - d))/e$.

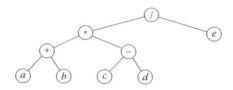

Figure 3.3.2   An expression tree for $((a + b) * (c - d))/e$.

Application 3.3.1:   *Prefix, Postfix, and Infix Notation*   When an expression tree is traversed, the resulting expression will appear in either *prefix*, *postfix*, or *infix* notation, according to which of the traversals is used.

Example 3.3.3: For the expression tree in Figure 3.3.2., the pre-order, post-order, and in-order traversals yield, respectively, the following.

prefix: $/ * + ab - cde$
postfix: $ab + cd - *e/$
infix: $((a + b) * (c - d))/e$

The infix expression requires parentheses, which can be incorporated into the recursive in-order traversal by producing a left parenthesis at the beginning of each subtree traversal and producing a right parenthesis at the end of each subtree traversal.

Geometric Descriptions of Pre-Order, Post-Order, and In-Order Traversals

The three traversals can be described geometrically in terms of a *left-first walk* around a standard plane drawing of the binary tree. A left-first walk may be imagined by starting at the root, and walking alongside the edges of the tree so that the edge you are passing along is always on your left (see Figure 3.3.3).

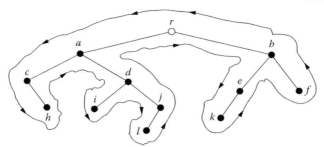

Figure 3.3.3 A left-first walk around the tree.

Pre-Order Traversal: $r, a, c, h, d, i, j, l, b, e, k, f$

As you take a left-first walk around the tree, list each vertex the first time you see it, but not again thereafter.

Post-Order Traversal: $h, c, i, l, j, d, a, k, e, f, b, r$

While taking a left-first walk around the tree, list each vertex the last time you see it and not beforehand.

In-Order Traversal: $c, h, a, i, d, l, j, r, k, e, b, f$

While taking a left-first walk around the tree, list each node with a left-child the second time you see it, and list each other node the first time you see it.

Stack Implementations of Pre-Order, Post-Order, and In-Order Traversals

DEFINITION: A **stack** is a sequence of elements such that each new element is added (or **pushed**) onto one end, called the *top*, and an element is removed (or **popped**) from that same end.

These iterative versions of the three traversals use the functions *pop*(*stack*) and *top*(*stack*). Both functions return the value at the top of the stack, but *pop* removes that value from the stack, whereas *top* does not.

Algorithm 3.3.1:   **Left Pre-Order**
   Input: a binary tree.
   Output: a list of the vertices in left pre-order.
      Push *root* onto stack.
      While stack is not empty
         Pop a vertex off stack, and write it on the output list.
         Push its children right-to-left onto stack.

Algorithm 3.3.2:   **Post-Order**
   Input: a binary tree.
   Output: a list of the vertices in post-order.
      Push *root* onto stack.
      While stack is not empty
         If *top*(*stack*) is unmarked
            Mark it, and push its children right-to-left onto stack.
         Else
            $v := pop(stack)$.
            List v

Algorithm 3.3.3:   **In-Order**

Input: a binary tree.

Output: a list of the vertices in in-order.

Push *root* onto stack.

While stack is not empty

$\quad v := top(stack)$.

$\quad$ While v has a left-child

$\qquad$ Push *leftchild*(v) onto stack.

$\qquad v := leftchild(v)$.

$\quad v := pop(stack)$.

$\quad$ List v.

$\quad$ If v has a right-child,

$\qquad$ Push *rightchild*(v) onto stack.

$\qquad v := rightchild(v)$.

$\quad$ Else

$\qquad$ While stack not empty and v has no right-child

$\qquad\quad v := pop(stack)$.

$\qquad\quad$ List v.

$\quad$ If v has a right-child,

$\qquad$ Push *rightchild*(v) onto stack.

$\qquad v := rightchild(v)$.

Queue Implementation of Level-Order Traversal

DEFINITION: A **queue** is a sequence of elements such that each new element is added (**enqueued**) to one end, called the *back* of the queue, and an element is removed (**dequeued**) from the other end, called the *front*.

Algorithm 3.3.4:   **Level-Order: Top-to-Bottom, Left-to-Right**

Input: A binary tree.

Output: A list of the vertices in level-order.

Enqueue *root*.

While queue is not empty

$\quad$ Dequeue a vertex and write it on the output list.

$\quad$ Enqueue its children left-to-right.

EXERCISES for Section 3.3

In Exercises 3.3.1 through 3.3.3, give the level-order, pre-order, in-order, and post-order traversals of the given binary tree.

3.3.1^{S}

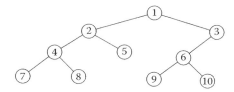

3.3.**2**

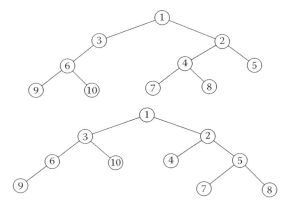

3.3.**3**

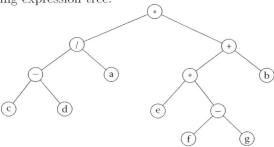

3.3.**4** Give the prefix, infix, and postfix notations for the arithmetic expression represented by the following expression tree.

3.3.4-tree-image

In Exercises 3.3.5 through 3.3.9, represent the given arithmetic expression by an expression tree. Then give the prefix and postfix notations for the arithmetic expression by performing the appropriate tree traversal.

3.3.**5**^{S} $(a + b) * (c - (d + e))$

3.3.**6** $((a + (b * c)) - d)/g$

3.3.**7** $((a + b) * (c - (d + e))) * (((a + (b * c)) - d)/g)$

3.3.**8** $((b * (c + a) * (g - d))/(b + d)) * ((g + c) * h)$

3.3.**9** $(((a + b) * c) - (d + e)) * (((a + (b * c)) - d)/g)$

3.3.**10** Give an example of two 4-vertex binary trees whose level-order and pre-order traversals are a, b, c, d but whose post-order is not.

3.3.**11** Prove that two binary trees that have the same level-order and the same post-order must be equivalent as 2-ary trees. (Hint: Show that the information obtained from a post-order traversal uniquely determines the parent of each vertex.)

3.3.**12** Show that a post-order traversal is equivalent to taking a *right pre-order* traversal (i.e., taking a right-first walk around the tree) and reversing the order of the output list.

3.3.**13** Reconcile the symmetry suggested by Exercise 3.3.12 with the asymmetry suggested by Exercises 3.3.10 and 3.3.11.

3.3.**14** [*Computer Project*] Implement Algorithm 3.3.1 and test it on each of the binary trees in Exercises 3.3.1 through 3.3.3.

3.3.15 [*Computer Project*] Implement Algorithm 3.3.2 and test it on each of the binary trees in Exercises 3.3.1 through 3.3.3.

3.3.16 [*Computer Project*] Implement Algorithm 3.3.3 and test it on each of the binary trees in Exercises 3.3.1 through 3.3.3.

3.3.17 [*Computer Project*] Implement recursive versions of the pre-order, post-order, and in-order traversal algorithms based on the definitions given at the beginning of the section. Test them on each of the binary trees in Exercises 3.3.1 through 3.3.3.

3.4   BINARY-SEARCH TREES

DEFINITION: A **random-access table** is a database whose domain is a sequence of entries, where each entry consists of two fields. One field is for the actual data element, and the other one is for the **key**, whose value determines that entry's position in the database.

Thus, an entry is located in a random-access table by searching for its key. The most generally useful implementation of a random-access table uses the following information structure.

DEFINITION: A **binary-search tree** (BST) is a binary tree, each of whose vertices is assigned a key, such that the key assigned to any vertex v is greater than the key at each vertex in the left subtree of v, and is less than the key at each vertex in the right subtree of v.

Example 3.4.1: Both of the binary-search trees in Figure 3.4.1 store the keys:

$$3, 8, 9, 12, 14, 21, 22, 23, 28, 35, 40, 46$$

Notice that the smallest key in either BST can be found by starting at the root and proceeding to the left until you reach a vertex with no left-child. Similarly, the largest key can be found by proceeding from the root iteratively to the right as long as possible. A straightforward inductive proof can be used to show that these two properties hold for an arbitrary binary-search tree (see Exercises).

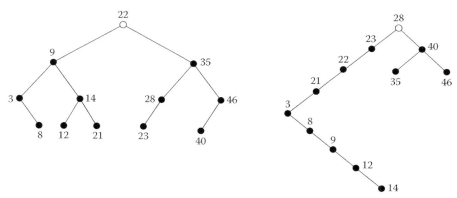

Figure 3.4.1   Two binary-search trees storing the same set of keys.

Iterative and recursive versions of the binary-search algorithm are shown below. The algorithm is based on the following simple strategy.

Strategy: In each iteration, exclude either the left or right subtree from the rest of the search, depending on whether the target key is less than or greater than the key at the current vertex.

Remark: An apology is due to the computer scientists because the versions below blur the distinction between a variable and a pointer to that variable. For instance, the variable T is typically a pointer to the root of the binary tree that is called T. Here, instead, $root(T)$ is used to refer to the root of the binary tree T. Of course, any implementation requires a careful treatment of the pointer datatype, but for the purposes here, pointers are avoided for fear of obscuring the essence of a fundamentally important graph algorithm.

Algorithm 3.4.1:   **Binary-Search-Tree Search**
    Input: a binary-search tree T and a target key t.
    Output: a vertex v of T such that $key(v) = t$ if t is found,
            or NULL if t is not found.
    $v := root(T)$
    While $(v \neq \text{NULL})$ and $(t \neq key(v))$
        If $t > key(v)$
            $v := rightchild(v)$
        Else $v := leftchild(v)$
    Return v.

Algorithm 3.4.2:   **Recursive BinarySearch** (T, t)
    Input: a binary-search tree T and a target key t.
    Output: a vertex v of T such that $key(v) = t$ if t is found,
            or NULL if t is not found.
    If $root(T) = \text{NULL}$
        Return NULL
    Else If $t = key(root(T))$
        Return $root$
    Else If $t > key(root(T))$
        Return BinarySearch($rightsubtree(T), t$)
    Else Return BinarySearch($leftsubtree(T), t$)

Balanced Binary-Search Trees

If each of the vertices of a binary-search tree has left and right subtrees of approximately equal size, then each comparison in the algorithm will rule out roughly half of the vertices remaining to be searched. This property motivates the following definition, which is a generalization of a complete binary tree.

DEFINITION: A binary tree is **balanced** if for every vertex, the number of vertices in its left and right subtrees differ by at most one.

Example 3.4.2: In Figure 3.4.1, the binary tree on the left is balanced; the one on the right is quite unbalanced. Observe that the binary-search algorithm generally performs faster on the balanced one. For instance, it takes four comparisons to determine that 20 is not one of the keys stored in the tree on the left, but it takes nine comparisons for the tree on the right.

COMPUTATIONAL NOTE: The **search** operation is one of the *primary operations* of the *random-access table* datatype. Two others are the **insert** and **delete** operations. An evaluation of a particular implementation of the random-access table should include an analysis of the computation required for each of these three primary operations.

Three different software implementations of the random-access table of n elements are as follows:

- Sorted array of records
- Sorted linked list
- Binary-search tree

Searching a sorted array using a binary search requires at worst $O(\log_2 n)$ computations, since each comparison eliminates half of the elements. A search through a linked list must be sequential and therefore is, in the worst case, $O(n)$.

Since each comparison of a binary search performed on a binary-search tree moves the search down to the next level, the number of comparisons is at most the height of the tree plus one. If the tree is balanced, then it is not hard to show that the number of vertices n is between 2^h and 2^{h+1}. Hence, the worst-case performance of the binary search on a perfectly balanced binary-search tree is $O(\log_2 n)$. The other extreme occurs when each internal vertex of the binary tree has only one child. Such a binary tree is actually an ordinary linked list, and therefore the performance of the search degenerates to $O(n)$.

If the random-access table is relatively *static*, that is, if insertions or deletions are infrequent, then the array implementation is the clear-cut winner. However, if insertions and deletions occur regularly, then the computational cost of these two operations must be taken into account.

For the array implementation, it will require $O(\log_2 n)$ comparisons to find where the insertion or deletion must occur, but then $O(n)$ movements of data are needed either to make room for the new element or to close the gap for the deleted element. The linked-list implementation requires $O(n)$ comparisons to find where the insertion or deletion must occur, but then a reassignment of a few pointer values ($O(1)$) is all that is needed to perform the actual insertion or deletion. Thus, both the array and the linked-list implementations perform the insertion and deletion in $O(n)$ instruction executions.

The binary-search tree combines the best characteristics of the other two implementations to achieve $O(\log_2 n)$ execution time for the search.

Insertions and Deletions in a Binary-Search Tree

A recursive description of the insert operation is shown. From an iterative view, a binary search is performed until it terminates at a leaf w. Then the new key k is assigned to a new vertex v, which becomes the left-child or right-child of w, depending on the comparison of k with $key(w)$.

Algorithm 3.4.3:   Binary-Search-Tree Insert (T, x)
    Input: a binary-search tree T and an entry x
    Output: tree T with a new leaf v such that $entry(v) = x$
            If $root(T) = $ NULL
                    Create a vertex v with $entry(v) = x$.
                    $root(T) := v$
            Else If $key(x) < key(root)$
                    Insert $(leftsubtree(T), x)$
            Else $Insert(rightsubtree(T), x)$

The deletion operation requires a bit more care. If an internal vertex v is deleted, then the way in which the two resulting pieces are pasted back together falls into one of three cases.

Case 1: If v has no left-child, then v is replaced by the right subtree of v.

Case 2: If v has no right-child, then v is replaced by the left subtree of v.

Case 3: If v has both a left-child and a right-child, then v is replaced by the leftmost (minimum) element l of the right subtree of v, and l is replaced by its right subtree (it cannot have a left subtree). It is straightforward to show that the resulting tree still satisfies the binary-search-tree property.

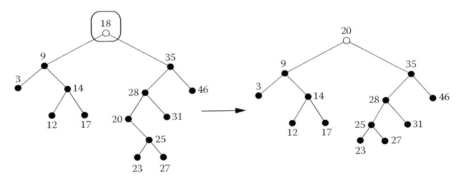

Figure 3.4.2   Deleting a vertex having two children.

EXERCISES for Section 3.4

In Exercises 3.4.1 through 3.4.4, draw three different binary-search trees for the given list of keys, two that are balanced and one that is not.

3.4.1^S 1, 4, 9, 19, 41, 49.

3.4.2 2, 3, 5, 7, 15, 20, 30, 40, 45, 50.

3.4.3 12, 23, 35, 37, 45, 50, 54, 60, 66.

3.4.4 1, 2, 3, 5, 8, 13, 21, 34, 55, 89, 144.

In Exercises 3.4.5 through 3.4.10, perform the specified sequence of insertions and deletions on the following search tree. After each individual insertion or deletion, draw the resulting search tree.

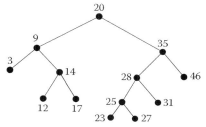

3.4.5^{S} Insert 10, insert 15, delete 20.

3.4.6 Insert 10, delete 20, insert 15.

3.4.7 Delete 35, delete 28, insert 35

3.4.8 Delete 28, insert 35, delete 9.

3.4.9 Insert 7, delete 9, insert 5.

3.4.10 Delete 9, insert 37, insert 5.

3.4.11^{S} Find a sequence of the keys $1, 2, 3, 5, 8, 13, 21, 34, 55, 89, 144$ such that their insertion by Algorithm 3.4.3, in that order, results in a balanced binary-search tree. Begin the sequence of insertions with the null tree.

3.4.12 How many different insertion sequences of the keys 1,4,9,19,41,49 are there that result in a balanced search tree?

3.4.13^{S} When will an insertion sequence for a set of n keys result in the worst possible search tree?

3.4.14 Give an iterative description of the insert operation.

3.4.15 Show that Case 3 of the deletion operation can also be accomplished by replacing v by the rightmost element r of the left subtree of v, and by then replacing r by the left subtree of r.

3.4.16 Use induction to prove that the smallest key in a binary-search tree can be found by starting at the root and proceeding to the left until you reach a vertex with no left-child.

3.4.17 [*Computer Project*] Implement Algorithm 3.4.2, and test it on the balanced search tree in Figure 3.4.1 by searching for 23 and then for 25.

3.5   HUFFMAN TREES AND OPTIMAL PREFIX CODES

Binary Codes

The existence of fast two-state materials makes sequences of 0's and 1's (*bitstrings*) the most natural way of encoding information for computers. Every upper- and lowercase English letter, punctuation mark, and mathematical symbol that can occur (as well as blank, $, etc.) is encoded as a different bitstring. The most common such encoding is known as ASCII code. Each of the letters and other symbols is regarded as the *meaning* of its ASCII code.

DEFINITION: A **binary code** is an assignment of symbols or other meanings to a set of bitstrings. Each bitstring is referred to as a **codeword**.

For some applications, it may be desired to allow codewords to vary in length. Confusion might result, in scanning a string from left to right, if one codeword were also part of another codeword. The following property, referred to as the *prefix property*, avoids this kind of ambiguity.

DEFINITION: A **prefix code** is a binary code with the property that no codeword is an initial substring of any other codeword.

Application 3.5.1:   *Constructing Prefix Codes*   A binary tree can be used to construct a prefix code for a given set of symbols. First, draw an arbitrary binary tree whose leaves are bijectively labeled by the symbols (i.e., each leaf gets a different symbol). Next, label every edge that goes to a left-child with a 0, and label every edge to a right-child with a 1. Then each symbol corresponds to the codeword formed by the sequence of edge labels on the path from the root to the leaf labeled by that symbol. It is not hard to show that the resulting set of codewords satisfies the prefix property.

Example 3.5.1: Suppose that each relevant message can be expressed as a string of letters (repetitions allowed) drawn from the restricted alphabet $\{a, b, c, d, e, f, g\}$. The binary tree shown in Figure 3.5.1 represents the prefix code whose seven codewords correspond to the unique paths from the root to each of the seven leaves. The resulting encoding scheme is shown below.

letter	a	b	c	d	e	f	g
codeword	000	0010	0011	0101	011	100	101

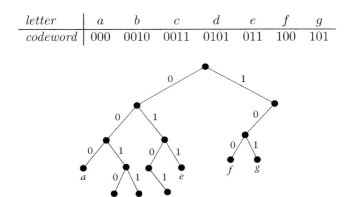

Figure 3.5.1   A binary tree used to construct a prefix code.

Huffman Codes

In a prefix code that uses its shorter codewords to encode the more frequently occurring symbols, the messages will tend to require fewer bits than in a code that does not. For instance, in a prefix code for ordinary English prose, it would make sense to use a short codeword to represent the letter "e" and a longer codeword to represent "x."

This suggests that one measure of a code's efficiency be the *average weighted length* of its codewords, where the length of each codeword is multiplied by the frequency of the symbol it encodes.

Example 3.5.2: Suppose that the frequency for each letter of the restricted alphabet of the previous example is given by the following table.

letter	a	b	c	d	e	f	g
frequency	.2	.05	.1	.1	.25	.15	.15

Then the average weighted length of a codeword for the prefix code in Figure 3.5.1 is

$$3 \times .2 + 4 \times .05 + 4 \times .1 + 4 \times .1 + 3 \times .25 + 3 \times .15 + 3 \times .15 = 3.25$$

For a prefix code constructed from a binary tree, as described above, the length of each codeword is simply the depth of its corresponding leaf. This motivates the next definition, which is used in the discussion that follows.

DEFINITION: Let T be a binary tree with leaves $s_1, s_2, \ldots, s_l$, such that each leaf s_i is assigned a weight w_i. Then the **average weighted depth** of the binary tree T, denoted $wt(T)$, is given by

$$wt(T) = \sum_{i=1}^{l} depth(s_i) \cdot w_i$$

Thus, if the weight assigned to a leaf is the frequency of the symbol that labels that leaf, then the average weighted length of a codeword equals the average weighted depth of the binary tree.

Application 3.5.2:   *Constructing Efficient Codes — the Huffman Algorithm*   The following algorithm, developed by David Huffman [Hu52], constructs a prefix code whose codewords have the smallest possible average weighted length.

Algorithm 3.5.1:   **Huffman Prefix Code**

Input: a set $\{s_1, \cdots, s_l\}$ of symbols; a list $\{w_1, \cdots, w_l\}$ of weights, where w_i is the weight associated with symbol S_i.

Output: a binary tree representing a prefix code for a set of symbols whose codewords have minimum average weighted length.

Initialize F to be a forest of isolated vertices, labeled $s_1, \cdots, s_l$, with respective weights $w_1, \ldots, w_l$.

For $i = 1$ to $l - 1$

Choose from forest F two trees, T and T', of smallest weights in F.

Create a new binary tree whose root has T and T' as its left and right subtrees, respectively.

Label the edge to T with a 0 and the edge to T' with a 1.

Assign to the new tree the weight $w(T) + w(T')$

Replace trees T and T' in forest F by the new tree.

Return F.

COMPUTATIONAL NOTE: In choosing the two trees of smallest weights, ties are resolved by some default ordering of the trees in the forest. In Example 3.5.3 , we use a left-to-right ordering.

DEFINITION: The binary tree produced from Algorithm 3.5.1 is called the **Huffman tree** for the list of symbols, and its corresponding prefix code is called the **Huffman code**.

Example 3.5.3: The Huffman algorithm is applied to the symbols and frequencies given in Example 3.5.2 to obtain an optimal prefix code. For each iteration, the current set of binary trees is displayed, and the root of each binary tree is labeled with the weight assigned to that tree.

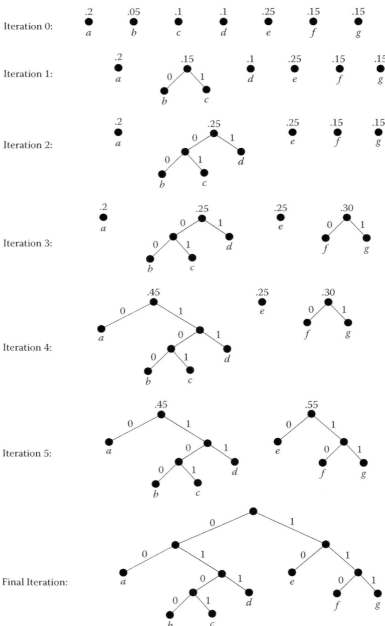

The resulting Huffman code has the following encoding scheme.

letter	a	b	c	d	e	f	g
codeword	00	0100	0101	011	10	110	111

The average length of a codeword for this prefix code is

$$2 \times .2 + 4 \times .05 + 4 \times .1 + 3 \times .1 + 2 \times .25 + 3 \times .15 + 3 \times .15 = 2.7$$

The Huffman tree also provides an efficient decoding scheme. A given codeword determines the unique path from the root to the leaf that stores the corresponding symbol. As the codeword is scanned from left to right, the path is traced downward from the root by traversing 0-edges or 1-edges, according to each bit. When a leaf is reached, its corresponding symbol is recorded. The next bit begins a new path from the root to the next symbol. The process continues until all the bits in the codeword have been read.

Observation: Notice that each internal vertex of a Huffman Tree has both a left and a right child.

The next assertion, which can be proved by induction, is used in showing that the Huffman algorithm produces an optimal prefix code.

Lemma 3.5.1: *If the leaves of a binary tree are assigned weights, and if each internal vertex is assigned a weight equal to the sum of its children's weights, then the tree's average weighted depth equals the sum of the weights of its internal vertices.*

$\diamond$ *(Exercises)*

Theorem 3.5.2: *For a given list of weights $w_1, w_2, \cdots, w_l$, a Huffman tree has the smallest possible average weighted depth among all binary trees whose leaves are assigned those weights.*

Proof: The proof uses induction on the number l of weights. If $l = 2$, then the Huffman tree consists only of a root with weight $w_1 + w_2$ and a left-child and a right-child. Any other binary tree with two leaves has at least two internal vertices, one with weight $w_1 + w_2$, and by Lemma 3.5.1, has greater average weighted depth.

Assume for some $l \geq 2$ that the Huffman algorithm produces a Huffman tree of minimum average weighted depth for any list of l weights. Let $w_1, w_2, \cdots, w_{l+1}$ be any list of $l + 1$ weights, and assume that w_1 and w_2 are two of the smallest ones, chosen first by the algorithm. Then the Huffman tree H that results consists of a Huffman tree $\widehat{H}$ for the weights $w_1 + w_2, w_3, \cdots, w_{l+1}$, where the leaf of weight $w_1 + w_2$ is replaced by a vertex with leaves assigned weights w_1 and w_2 (see Figure 3.5.2). Lemma 3.5.1 implies $wt(H) = wt(\widehat{H}) + w_1 + w_2$. By the induction hypothesis, $\widehat{H}$ is optimal among all binary trees whose leaves are assigned weights $w_1 + w_2, w_3, \ldots, w_{l+1}$.

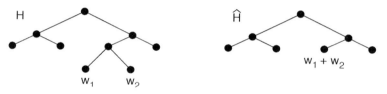

Figure 3.5.2   Huffman trees H and $\widehat{H}$ in proof.

Now suppose T^* is an optimal binary tree for the weights $w_1, w_2, \ldots, w_{l+1}$. Let x be an internal vertex of T^* of greatest depth and suppose that y and z are its left-child and right-child, respectively. Without loss of generality, we can assume that y and z have weight w_1 and w_2, since otherwise we could swap their weights with w_1 and w_2 to produce a tree with smaller average weighted depth. Consider the tree $\widehat{T}$ formed by deleting the leaves y and z from T^*. Then $wt(T^*) = wt(\widehat{T}) + w_1 + w_2$. But $\widehat{T}$ is a binary tree with leaves of weight $w_1 + w_2, w_3, \cdots, w_{l+1}$, and, hence, $wt(\widehat{T}) \geq wt(\widehat{H})$. Thus, $wt(T^*) \geq wt(H)$, proving H is optimal. $\diamond$

EXERCISES for Section 3.5

In Exercises 3.5.1 through 3.5.3, use the Huffman tree constructed in Example 3.5.3 to decode the given string.

3.5.1^S 1100001010001110.

3.5.2   0100001111110011110.

3.5.3   0111011000010110011.

In Exercises 3.5.4 through 3.5.6, construct a Huffman code for the given list of symbols and weights, calculate its average weighted length, and encode the strings "defaced" and "baggage." Use the left-to-right ordering to break ties.

3.5.4^S

letter	a	b	c	d	e	f	g	h
frequency	.2	.05	.1	.1	.18	.15	.15	.07

3.5.5

letter	a	b	c	d	e	f	g	h
frequency	.1	.15	.2	.17	.13	.15	.05	.05

3.5.6

letter	a	b	c	d	e	f	g	h
frequency	.15	.1	.15	.12	.08	.25	.05	.1

3.5.7 Use the Huffman algorithm to construct a prefix code so that the following line is encoded using the shortest possible bitstring. Remember to encode the blank.

meandering with a mazy motion

3.5.8 Explain why the output forest F in Algorithm 3.5.1 is actually a tree.

3.5.9 Prove Lemma 3.5.1.

3.5.10 [*Computer Project*] Implement the Huffman Code algorithm and test the computer program on Exercises 3.5.4 through 3.5.6.

3.6   PRIORITY TREES

A *priority tree* is a special type of binary tree used in implementing the abstract datatype called a *priority queue*, an important structure for information processing.

DEFINITION: A binary tree of height h is called **left-complete** if the bottom level has no gaps as one traverses from left to right. More precisely, it must satisfy the following three conditions.

- Every vertex of depth $h - 2$ or less has two children.
- There is at most one vertex v at depth $h - 1$ that has only one child (a left one).
- No vertex at depth $h - 1$ has fewer children than another vertex at depth $h - 1$ to its right.

Figure 3.6.1 shows a left-complete binary tree of height 3.

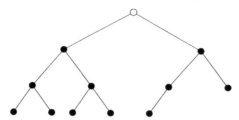

Figure 3.6.1   A left-complete binary tree of height 3.

DEFINITION: A **priority tree** is a left-complete binary tree whose vertices have labels (called *priorities*) from an ordered set (or sometimes, a *partially ordered set*), such that no vertex has higher priority than its parent.

Although no vertex can have higher priority than any of its ancestors, a vertex can have higher priority than a sibling of an ancestor.

Example 3.6.1: In the priority tree shown in Figure 3.6.2, all of the vertices in the right subtree of the vertex labeled 21 have higher priority than the sibling of that vertex.

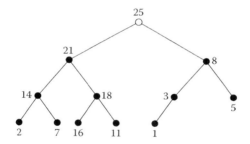

Figure 3.6.2   A priority tree of height 3.

DEFINITION: A **priority queue** is a set of entries, each of which is assigned a *priority*. When an entry is to be removed, or *dequeued*, from the queue, an entry with the highest priority is selected.

The stack and queue structures are extreme cases of the priority queue, and are opposites to each other, because the newest entry to a stack gets the highest priority (*Last-In-First-Out*), whereas the newest entry to a queue gets the lowest priority (*First-In-First-Out*).

The most obvious implementation of a priority queue is as a linked list, sorted by descending priority. Each insertion must be "bubbled" to its correct position, and, hence, its worst-case behavior is $O(n)$, where n is the number of items. The deletion operation is $O(1)$, since deletion from a priority queue occurs at the beginning of the sorted list, where the highest priority item is stored.

A less obvious and generally more efficient implementation of the priority queue uses a priority tree.

Inserting an Entry into a Priority Tree

The enqueue operation for a priority queue is implemented by the following algorithm, which inserts an entry x into a priority tree T representing that priority queue.

The idea is to start by appending x at the first vacant spot in T, and then to bubble x upward toward the root until the priority of x is less than or equal to the priority of its parent.

Algorithm 3.6.1:   PriorityTreeInsert
Input: a priority tree T and a new entry x.
Output: tree T with x inserted so that T is still a priority tree.
    Append entry x to the first vacant spot v in the left-complete tree T.
    While $v \neq root(T)$ AND $priority(v) > priority(parent(v))$
        Swap v with $parent(v)$.

Example 3.6.2: Figure 3.6.3 shows the sequence of exchanges that result when an entry with priority 13 is inserted into the priority tree from Figure 3.6.2.

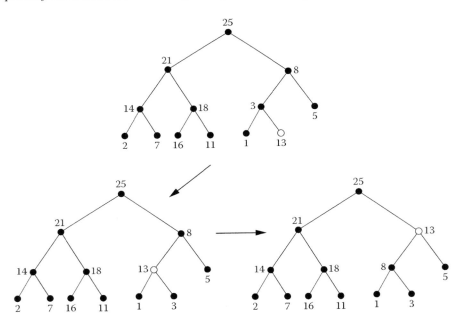

Figure 3.6.3   Inserting the entry with priority 13 into the priority tree.

COMPUTATIONAL NOTE: The worst case occurs if entry x must be bubbled all the way up to the root. The number of exchanges this would require is the height of priority tree T, and thus, the worst-case running time of PriorityTreeInsert is $O(\log_2 n)$.

Deleting an Entry from a Priority Tree

The dequeue operation on a priority queue is implemented by the following algorithm, which deletes an entry x from a priority tree T. First, entry x is replaced by the entry y that occupies the rightmost spot at the bottom level of the priority tree. Then y is trickled iteratively downward, each time swapped with the larger of its two children, until it exceeds the values of both its children.

Algorithm 3.6.2:   PriorityTreeDelete

Input: a priority tree T and an entry x in T.

Output: tree T with x deleted so that it remains a priority tree.

      Replace x by the entry y that occupies the rightmost spot at the bottom level of T.

      While y is not a leaf AND $[priority(y) \leq priority\ (leftchild\ (y))$.

         OR $priority\ (y) \leq priority\ (rightchild(y))]$

      If $priority\ leftchild\ (y) > priority\ (rightchild\ (y))$

         Swap y with $leftchild\ (y)$.

      Else swap y with $rightchild(y)$

Example 3.6.3: Figure 3.6.4 illustrates this process when the entry with priority 21 is deleted from the given priority tree.

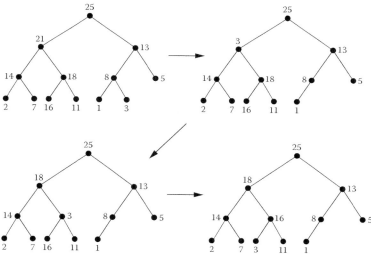

Figure 3.6.4   Deleting the entry 21 from the priority tree.

Heaps

DEFINITION: A **heap** is a representation of a priority tree as an array, having the following address pattern.

- $index(root) = 0$
- $index(leftchild(v)) = 2 \times index(v) + 1$
- $index(rightchild(v)) = 2 \times index(v) + 2$
- $index(parent(v))\ \left\lfloor \frac{index(v)-1}{2} \right\rfloor$

A heap saves space because the address pattern eliminates the need to store pointers to children and to parents.

Example 3.6.4: Figure 3.6.5 shows a priority tree and the heap that represents it.

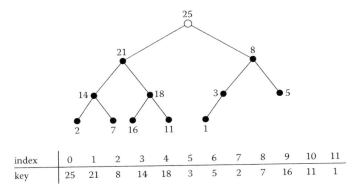

index	0	1	2	3	4	5	6	7	8	9	10	11
key	25	21	8	14	18	3	5	2	7	16	11	1

Figure 3.6.5   Representing a priority tree using a heap.

EXERCISES for Section 3.6

3.6.1 Draw all possible left-complete binary trees of height 2.

3.6.2[S] How many different left-complete binary trees of height h are there?

In Exercises 3.6.3 through 3.6.8, perform the specified sequence of insertions and deletions on the following priority tree. After each individual insertion or deletion, draw the resulting priority tree.

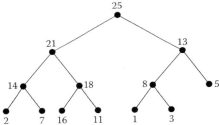

3.6.3 Insert 10, insert 15, delete 21.

3.6.5 Delete 14, delete 3, insert 14.

3.6.7 Insert 17, delete 8, insert 15.

3.6.4 Insert 10, delete 21, insert 15.

3.6.6[S] Delete 18, insert 35, delete 25.

3.6.8 Delete 8, insert 17, insert 22.

In Exercises 3.6.9 through 3.6.12, either draw the priority tree represented by the heap whose entries are as shown, or explain why the given sequence does not represent a priority tree.

3.6.9[S] $80, 60, 40, 50, 25, 10, 5, 40, 25, 10, 15$.

3.6.10 $45, 30, 32, 20, 28, 25, 16, 15, 18$.

3.6.11[S] $45, 32, 37, 20, 25, 16, 29, 10, 22$.

3.6.12 $50, 20, 42, 15, 16, 25, 37, 11, 5, 8, 3$.

3.6.13 Prove or disprove: The level-order of a priority tree is the same as the order of the heap that represents it.

3.7   COUNTING LABELED TREES: PRÜFER ENCODING

In 1875, Arthur Cayley presented a paper to the British Association describing a method for counting certain hydrocarbons containing a given number of carbon atoms. In the same paper, Cayley also counted the number of n-vertex trees with the standard vertex labels $1, 2, \ldots, n$. Two labeled trees are considered the same if their respective edge-sets are identical. For example, in Figure 3.7.1, the two labeled 4-vertex trees are different, even though their underlying unlabeled trees are both isomorphic to the path graph P_4.

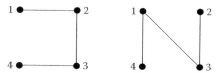

Figure 3.7.1   Two different labeled trees.

The number of n-vertex labeled trees is n^{n-2}, for $n \geq 2$, and is known as **Cayley's Formula**. A number of different proofs have been given for this result, and the one presented here, due to H. Prüfer, is considered among the most elegant. The strategy of the proof is to establish a one-to-one correspondence between the set of standard-labeled trees with n vertices and certain finite sequences of numbers.

Prüfer Encoding

DEFINITION: A **Prüfer sequence** of length $n - 2$, for $n \geq 2$, is any sequence of integers between 1 and n, with repetitions allowed.

The following encoding procedure constructs a Prüfer sequence from a given standard labeled tree, and thus, defines a function $f_e : \mathcal{T}_n \to \mathcal{P}_{n-2}$ from the set $\mathcal{T}_n$ of trees on n labeled vertices to the set $\mathcal{P}_{n-2}$ of Prüfer sequences of length $n - 2$.

Algorithm 3.7.1:   Prüfer Encode
  Input: an n-vertex tree with a standard 1-based vertex-labeling.
  Output: a Prüfer sequence of length $n - 2$.
      Initialize T to be the given tree.
      For $i = 1$ to $n - 2$
          Let v be the 1-valent vertex with the smallest label.
          Let s_i be the label of the neighbor of v.
          $T := T - v$.
      Return sequence $\langle s_1, s_2 \ldots, s_{n-2} \rangle$.

Example 3.7.1: The encoding procedure for the tree shown in Figure 3.7.2 is illustrated with the two figures that follow. The first figure shows the first two iterations of the construction, and the second figure shows iterations 3, 4, and 5. The portion of the Prüfer sequence constructed after each iteration is also shown.

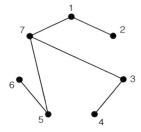

Figure 3.7.2   A labeled tree to be encoded into a Prüfer sequence S**.**

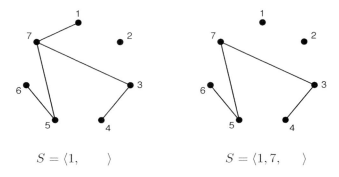

$$S = \langle 1, \quad \rangle \qquad\qquad S = \langle 1, 7, \quad \rangle$$

Figure 3.7.3   First two iterations of the Prüfer encoding.

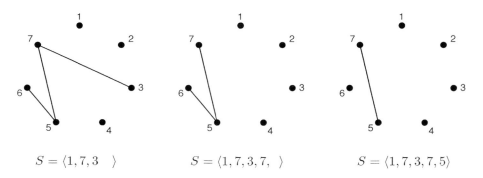

$$S = \langle 1, 7, 3 \quad \rangle \qquad S = \langle 1, 7, 3, 7, \quad \rangle \qquad S = \langle 1, 7, 3, 7, 5 \rangle$$

Figure 3.7.4   Iterations 3, 4, and 5 of the Prüfer encoding.

Notice that if we allow the label-set for the n vertices of the tree to be *any* set of positive integers (not necessarily consecutive integers starting at 1), then the encoding proceeds exactly as before. Working with this more general label-set enables us to write the encoding algorithm recursively, shown on the next page, and leads more naturally to the inductive arguments that establish the one-to-one correspondence between labeled trees and Prüfer sequences.

The resulting Prüfer sequence will now need to indicate the label-set.

DEFINITION: A **Prüfer sequence of length** $n-2$ **on a label-set** L of n positive integers is any sequence of integers from L with repetitions allowed.

Algorithm 3.7.2:   **Prüfer Encode (recursive) (T, L)**

   Input: an n-vertex labeled tree T and its label-set L of n positive integers.

   Output: a Prüfer sequence of length $n - 2$ of integers from L.

     If T is a 2-vertex tree

       Return the empty sequence

     Else

         Let v be the leaf vertex having the smallest label t.

         Let s be the label of the neighbor of v

         $P :=$ Prüfer Encode$(T - v, L - \{t\})$

         Return $\langle s, P \rangle$

Proposition 3.7.1: *Let d_k be the number of occurrences of the number k in a Prüfer encoding sequence for a labeled tree T on a set L. Then the degree of the vertex with label k in T equals $d_k + 1$.*

Proof: The assertion is true for any tree on 2 vertices, because the Prüfer sequence for such a tree is the empty sequence and both vertices in T have degree 1.

Assume that the assertion is true for every n-vertex labeled tree, for some $n \geq 2$ and suppose T is an $(n+1)$-vertex labeled tree. Let v be the leaf vertex with the smallest label, let w be the neighbor of v, and let $l(w)$ be the label of w. Then the Prüfer sequence S for T consists of the label $l(w)$ followed by the Prüfer sequence S^* of the n-vertex labeled tree $T^* = T - v$.

By the inductive hypothesis, for every vertex u of the tree T^*, $deg_{T^*}(u)$ is one more than the number of occurrences of its label $l(u)$ in S^*. But for all $u \neq w$, the number of occurrences of the label $l(u)$ in S^* is the same as in S, and $\deg_T(u) = \deg_{T^*}(u)$. Furthermore, $\deg_T(w) = \deg_{T^*}(w) + 1$, and $l(w)$ has one more occurrence in S than in S^*. Thus, the condition is true for every vertex in T.   ◇

Corollary 3.7.2: *If T is an n-vertex labeled tree with label-set L, then a label $k \in L$ occurs in the Prüfer sequence $f_e(T)$ if and only if the vertex in T with label k is not a leaf vertex.*

Prüfer Decoding

   The following decoding procedure maps a given Prüfer sequence to a standard labeled tree.

Algorithm 3.7.3:   **Prüfer Decode (recursive) (P, L)**

   Input: a Prüfer sequence $P = \langle p_1 p_2 \ldots p_{n-2} \rangle$ on a label-set L of n positive

      integers.

   Output: a labeled n-vertex tree T on label-set L.

     If P is the empty sequence (i.e., $n = 2$)

       Return a 2-vertex tree on label-set L.

     Else

         Let k be the smallest number in L that is not in P.

         Let P^* be the Prüfer sequence $\langle p_2 \ldots p_{n-2} \rangle$ on label-set $L^* = L - \{k\}$.

         Let $T^* = $ Prüfer Decode(P^*, L^*).

         Let T be the tree obtained by adding a new vertex with label k and

            an edge joining it to the vertex in T^* labeled p_1.

         Return T.

Example 3.7.2: To illustrate the decoding procedure, start with the Prüfer sequence $P = \langle 1, 7, 3, 7, 5 \rangle$ and label-set $L = \{1, 2, 3, 4, 5, 6, 7\}$. Proposition 3.7.1 implies that the corresponding tree has: $\deg(2) = \deg(4) = \deg(6) = 1$; $\deg(1) = \deg(3) = \deg(5) = 2$; and $\deg(7) = 3$. Among the leaf vertices, vertex 2 has the smallest label, and its neighbor must be vertex 1. Thus, an edge is drawn joining vertices 1 and 2. The number 2 is removed from the list, and the first occurrence of label 1 is removed from the sequence. The sequence of figures that follows shows each iteration of the decoding procedure. Shown in each figure are: the edges to be inserted up to that point, the label-set, and the remaining part of the Prüfer sequence.

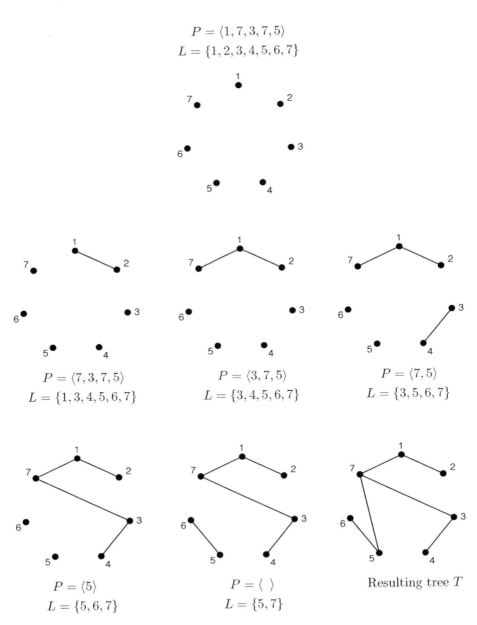

Proposition 3.7.3: *For any label-set L of n positive integers, the decoding procedure defines a function $f_d : \mathcal{P}_{n-2} \to \mathcal{T}_n$ from the set of Prüfer sequences on L to the set of n-vertex labeled trees with label-set L.*

Proof: First observe that at each step of the procedure, there is never any choice as to which edge must be drawn. Thus, the procedure defines a function from the Prüfer sequences on L to the set of labeled graphs with label-set L. Therefore, proving the following assertion will complete the proof.

Assertion: When the decoding procedure is applied to a Prüfer sequence of length $n - 2$, the graph produced is an n-vertex tree.

The assertion is trivially true for $n = 2$, since the procedure produces a single edge. Assume that the assertion is true for some $n \geq 2$, and consider a label-set L of $n + 1$ positive integers and a Prüfer sequence $\langle p_1, p_2, \ldots, p_{n-1} \rangle$ on L. Let k be the smallest number in L that does not appear in $\langle p_1, p_2, \ldots, p_{n-1} \rangle$.

The first call of the procedure creates a new vertex v with label k and joins v to the vertex with label p_1. By the inductive hypothesis, Prüfer Decode($\langle p_2, \ldots, p_{n-1} \rangle, L - \{k\}$) produces an n-vertex tree. Since k is not in the label-set $L - \{k\}$, this tree has no vertex with label k. Therefore, adding the edge from v to the vertex with label p_1 does not create a cycle and the resulting graph is a tree. $\diamond$

Notice that the tree obtained in Example 3.7.2 by the Prüfer decoding of the sequence $\langle 1, 7, 3, 7, 5 \rangle$ is the same as the tree in Example 3.7.1 that was Prüfer-encoded as $\langle 1, 7, 3, 7, 5 \rangle$. This inverse relationship between the encoding and decoding functions holds in general, as the following proposition asserts.

Proposition 3.7.4: *The decoding function $f_d : \mathcal{P}_{n-2} \to \mathcal{T}_n$ is the inverse of the encoding function $f_e : \mathcal{T}_n \to \mathcal{P}_{n-2}$.*

Proof: We show that for any list L of n positive integers, $f_d \circ f_e$ is the identity function on the set of n-vertex labeled trees with n distinct labels from L. We use induction on n.

Let T be a labeled tree on 2 vertices labeled s and t. Since $f_e(T)$ is the empty sequence, $f_d(f_e(T)) = T$.

Assume for some $n \geq 2$, that for n-vertex labeled tree T, $(f_d \circ f_e)(T) = T$. Let T be an $(n + 1)$-vertex labeled tree, and suppose v is the leaf vertex with the smallest label t. If s is the label of the neighbor of v, then $f_e(T) = \langle s, f_e(T - v) \rangle$. It remains to show that $f_d(\langle s, f_e(T - v) \rangle) = T$.

By Corollary 3.7.2, the label of every non-leaf vertex appears in $\langle s, f_e(T - v) \rangle$, and since t is the smallest label among the leaf vertices, t is the smallest label that does not appear in $\langle s, f_e(T - v) \rangle$. Therefore, $f_d(\langle s, f_e(T - v) \rangle)$ consists of a new vertex labeled t and an edge joining it to the vertex labeled s in $f_d(f_e(T - v))$ (see figure below).

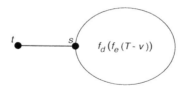

By the inductive hypothesis, $f_d(f_e(T - v)) = T - v$. Thus, $(f_d \circ f_e)(T)$ consists of the tree $T - v$, a new vertex labeled t and an edge joining that vertex to the vertex labeled s in $T - v$. That is, $(f_d \circ f_e)(T) = T$.

A similar argument shows that $f_e(f_d(P)) = P$, where P is a Prüfer sequence of length $n - 2$ on a label-set L of n positive integers. (See exercises.) $\diamond$

Theorem 3.7.5: [*Cayley's Tree Formula*]. *The number of different trees on n labeled vertices is* n^{n-2}.

Proof: By Proposition 3.7.4, $f_e \circ f_d : \mathcal{P}_{n-2} \to \mathcal{P}_{n-2}$ and $f_d \circ f_e : \mathcal{T}_n \to \mathcal{T}_n$ are both identity functions, and hence, f_d and f_e are both bijections. This establishes a one-to-one correspondence between the trees in $\mathcal{T}_n$ and the sequences in $\mathcal{P}_{n-2}$, and, by the Rule of Product, there are n^{n-2} such sequences. $\diamond$

Remark: A slightly different view of Cayley's Tree Formula gives us the number of different spanning trees of the complete graph K_n. The next chapter is devoted to spanning trees.

EXERCISES for Section 3.7

In Exercises 3.7.1 through 3.7.6, encode the given labeled tree as a Prüfer sequence. Then decode the resulting sequence, to demonstrate that Proposition 3.7.4 holds.

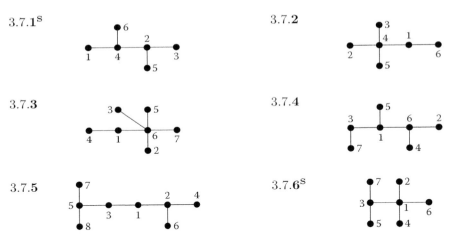

3.7.1^S

3.7.2

3.7.3

3.7.4

3.7.5

3.7.6^S

In Exercises 3.7.7 through 3.7.12, construct the labeled tree corresponding to the given Prüfer sequence.

3.7.7 $(6, 7, 4, 4, 4, 2)$.

3.7.8 $(2, 1, 1, 3, 5, 5)$.

3.7.9 $(1, 3, 7, 2, 1)$.

3.7.10^S $(1, 3, 2, 3, 5)$.

3.7.11 $(1, 1, 5, 1, 5)$.

3.7.12 $(1, 1, 5, 2, 5)$.

3.7.13 Draw the 4^{4-2} labeled trees on four vertices.

3.7.14^{S} How many spanning trees are there of the labeled complete bipartite graph $K_{m,2}$?

3.7.15 Prove Proposition 3.7.4, by showing that the Prüfer encoding of an arbitrary n-vertex labeled tree, followed by the Prüfer decoding of the resulting Prüfer sequence, recaptures the original tree.

3.8   COUNTING BINARY TREES: CATALAN RECURSION

In this section, we derive a formula for the number of different binary trees on n vertices. As is frequently the case in *enumerative combinatorics*, the derivation begins by establishing a recursive formula.

Let b_n denote the number of binary trees on n vertices. The recursion can be expressed more conveniently by artificially defining $b_0 = 1$. Since the only binary tree on one vertex is a root with no children, it follows that $b_1 = 1$. For $n > 1$, a binary tree T on n vertices has a left subtree with j vertices and a right subtree with $n - 1 - j$ vertices, for some j between 0 and $n - 1$. Pairing the b_j possible left subtrees with the b_{n-1-j} possible right subtrees results in $b_j \times b_{n-1-j}$ different combinations of left and right subtrees for T. Hence,

$$b_n = b_0 b_{n-1} + b_1 b_{n-2} + \cdots + b_{n-1} b_0$$

This *recurrence relation* is known as the **Catalan recursion**, and the quantity b_n is called the *nth Catalan number*. The Catalan numbers occurs in many different applications besides tree-counting (see Exercises).

Example 3.8.1: Applying the Catalan recursion to the cases $n = 2$ and $n = 3$ yields $b_2 = 2$ and $b_3 = 5$. Figure 3.8.1 shows the five different binary trees on 3 vertices.

Figure 3.8.1   The five different binary trees on 3 vertices.

With the aid of generating functions, it is possible to derive the following *closed formula* for b_n.

Theorem 3.8.1: *The number b_n of different binary trees on n vertices is given by*

$$b_n = \frac{1}{n+1} \binom{2n}{n}$$

Outline of Proof:

(a) Let $g(x) = \sum_{k=0}^{\infty} b_k x^k$ be the generating function for the number b_n of different binary trees on n vertices. Then $1 + x[g(x)]^2 = g(x)$.

(b) Applying the quadratic formula to step (a) implies that $g(x) = \frac{1-\sqrt{1-4x}}{2x}$.

(c) The *Generalized Binomial Theorem* states that $(1+z)^r = \sum_{k=0}^{\infty} \binom{r}{k} z^k$, for *any* real number r, where the generalized binomial coefficient $\binom{r}{k}$ is defined by

$$\binom{r}{k} = \begin{cases} 1, & \text{if } k = 0; \\ \frac{r(r-1)(r-2)\cdots(r-k+1)}{k!}, & \text{if } k \geq 1. \end{cases}$$

Using the Generalized Binomial Theorem and the result from part (b) we obtain

$$g(x) = \frac{1}{2x} \left(1 - \sum_{k=0}^{\infty} \binom{1/2}{k} (-4x)^k \right).$$

(d) A change of variable for the index of summation in part (c) yields

$$g(x) = \sum_{n=0}^{\infty} \binom{1/2}{n+1} (-1)^n 2^{2n+1} x^n,$$

from which we conclude that $b_n = \binom{1/2}{n+1} (-1)^n 2^{2n+1}$.

(e) The definition of $\binom{1/2}{n+1}$ implies that

$$\binom{1/2}{n+1} = \frac{(-1)^n}{2^{n+1}} \cdot \frac{(1)(3)(5)\cdots(2n-1)}{(1)(2)(3)\cdots(n+1)},$$

and, hence,

$$b_n = \frac{1}{n+1} \binom{2n}{n}.$$

EXERCISES for Section 3.8

3.8.1^S Use the Catalan recursion to determine the number of different binary trees on four vertices. Compare your answer with that of the closed formula.

3.8.2 Use the Catalan recursion to determine the number of different binary trees on five vertices. Compare your answer with that of the closed formula.

3.8.3 Illustrate the Catalan recursion for $n = 4$ by drawing the different binary trees on four vertices and grouping them according to the size of their left and right subtrees.

3.8.4^S Let p_n be the number of ways to place parentheses to multiply the n numbers $a_1 \times a_2 \times \cdots \times a_n$ on a calculator. For instance. $p_3 = 2$, because there are two ways to multiply $a_1 \times a_2 \times a_3$, namely, $(a_1 \times a_2) \times a_3$ and $a_1 \times (a_2 \times a_3)$. Show that p_n equals the $(n-1)$st Catalan number b_{n-1}.

3.8.5 Fill in the details of steps (a) and (b) of the proof outline for Theorem 3.8.1.

3.8.6 Fill in the details of steps (c) and (d) of the proof outline for Theorem 3.8.1.

3.8.7^S Fill in the details of step (e) of the proof outline for Theorem 3.8.1.

3.8.8 [*Computer Project*] Write a computer program that uses a recursive function *Catalan(n)* whose value is the nth Catalan number, and write a second program that uses an iterative version of *Catalan(n)*, based on the closed formula established in Theorem 3.8.1. Then use both programs to produce the first 15 Catalan numbers, and if your computing environment allows it, compare the execution time of both programs.

3.9   SUPPLEMENTARY EXERCISES

3.9.1 Draw every tree T such that the edge-complement $\bar{T}$ is a tree.

3.9.2 What are the minimum and maximum independence numbers of an n-vertex tree?

3.9.3 What are the minimum and maximum number of vertex orbits in an n-vertex tree?

3.9.4 What are the minimum and maximum number of rooted n-vertex trees that have the same underlying tree?

3.9.5 Reconstruct the tree whose deck appears in Figure 3.9.1.

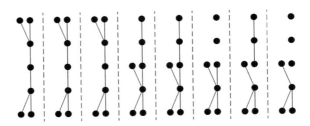

Figure 3.9.1

3.9.6 Characterize the class of connected graphs G such that for every edge $e \in E_G$, the graph $G - e$ is a tree.

3.9.7 List all tree-graphic sequences for a 7-vertex tree.

3.9.8 Draw all 11 isomorphism types of trees with 7 vertices.

3.9.9 List all the possible degree sequences for an 8-vertex tree with maximum degree 3.

3.9.10 For each of the possible degree sequences of Exercise 3.9.9, draw all the different isomorphism types of tree with that degree sequence.

3.9.11 List all tree-graphic sequences for an 8-vertex tree.

3.9.12 Draw all five isomorphism types of trees with 9 vertices, such that no vertex has degree two.

3.9.13 Draw all the isomorphism types (without repetition) of trees with degree sequence 33221111.

3.9.14 Prove or disprove: For a fixed n, all n-vertex trees have the same average degree.

DEFINITION: A *rigid tree* is a tree whose only automorphism is the identity automorphism.

3.9.15 Prove that for all $n \geq 7$, there exists an n-vertex rigid tree.

3.9.16 Draw two non-isomorphic rigid 9-vertex trees.

3.9.17 Prove that there is no 5-vertex rigid tree.

3.9.18 Prove that there is no 6-vertex rigid tree.

3.9.**19** Write the post-order for the following plane tree. Hint: Be careful.

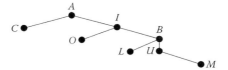

3.9.**20** Draw the BST that results from deleting node 41 from the BST in Figure 3.9.2.

3.9.**21** Draw the BST that results from inserting 44 into the BST in Figure 3.9.2.

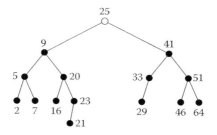

Figure 3.9.2

3.9.**22** Draw the result of dequeuing highest priority node 25 from the heap in Figure 3.9.3 below. Then draw the result of reinserting 25.

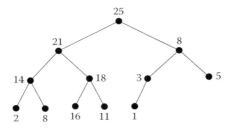

Figure 3.9.3

3.9.**23** Write the Prüfer code for the following labeled tree.

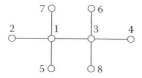

Figure 3.9.4

In Exercises 3.9.24 through 3.9.29, draw the tree with the given Prüfer code.

3.9.**24** 112050   3.9.**25** 316313   3.9.**26** 512256

3.9.**27** 61661   3.9.**28** 713328   3.9.**29** 01201201

GLOSSARY

ancestor of a vertex w in a rooted tree: a vertex on the path from the root to vertex w; vertex v is an ancestor of vertex w if and only if w is a descendant of v.

array-of-parents representation of a rooted tree with vertices labeled $0, 1, \ldots, n-1$: a list whose kth entry is the parent of vertex k.

average weighted depth of a binary tree T with leaves $s_1, s_2, \ldots, s_l$: the quantity, denoted $wt(T)$, given by $wt(T) = \sum_{i=1}^{l} depth(s_i) \cdot w_i$, where w_i is the weight assigned to s_i, $1 \leq i \leq l$.

balanced binary tree: a binary tree such that for every vertex, the number of vertices in its left and right subtrees differ by at most one.

binary code: an assignment of symbols or other meanings to a set of bitstrings.

binary-search tree (BST): a binary tree, each of whose vertices is assigned a key, such that the key assigned to any vertex v is greater than the key at each vertex in the left subtree of v, and is less than the key at each vertex in the right subtree of v.

binary tree: a rooted tree in which each vertex has no more than two children, each designated as either a *left-child* or *right-child;* an ordered 2-ary tree.

Catalan number, nth the quantity b_n in the Catalan recursion.

Catalan recursion: the recurrence relation for the number b_n of different binary trees on n vertices, given by $b_n = b_0 b_{n-1} + b_1 b_{n-2} + \cdots + b_{n-1} b_0$.

center of a graph G, denoted $Z(G)$: the subgraph induced on the set of central vertices of G.

central vertex of a graph: a vertex of minimum eccentricity.

child of a vertex v in a rooted tree: a vertex w that immediately succeeds vertex v on the path from the root to w; w is the child of v if and only if v is the parent of w.

codeword: one of the bitstrings in a binary code.

complete n-ary tree: an n-ary tree in which every internal vertex has exactly n children, and all leaves have the same depth.

cycle-edge: an edge that lies on a cycle.

depth of a vertex v in a rooted tree: the length of the unique path from the root to v.

descendant of a vertex v in a rooted tree: a vertex w whose path from the root contains vertex v;

directed tree: a digraph whose underlying graph is a tree.

distance sum of a graph G, denoted $ds(G)$: the sum of the distances between all pairs of vertices in G; that is, $ds(G) = \sum_{u,v \in V_G} d(u,v)$.

eccentricity of a vertex v in a graph G, denoted $ecc(v)$: the distance from v to a vertex farthest from v; that is, $ecc(v) = \max_{x \in V_G} \{d(v,x)\}$.

expression tree for an arithmetic expression: either a single vertex labeled by an identifier, or a binary tree described by the following recursive rule:

- The operator to be executed during an evaluation is at the root.
- The left subtree of the root is an expression tree representing the left operand.
- The right subtree is an expression tree representing the right operand.

forest: an acyclic graph.

graceful tree: a tree T with $m+1$ vertices and m edges such that it is possible to label the vertices *of* T with distinct elements from the set $\{0, 1, \ldots, m\}$ in such a way that the induced edge-labeling, which assigns the integer $|i - j|$ to the edge joining the vertices labeled i and j, assigns the labels $1, 2, \ldots, m$ to the m edges of G.

heap: a representation of a priority tree as an array, having the following address pattern:

- $index(root) = 0$
- $index(leftchild(v)) = 2 \times index(v) + 1$
- $index(rightchild(v)) = 2 \times index(v) + 2$
- $index(parent(v)) = \left\lfloor \frac{index(v)-1}{2} \right\rfloor$

height of a rooted tree: the length of a longest path from the root; the greatest depth in the tree.

Huffman code: the prefix code corresponding to a Huffman tree.

Huffman tree: the binary tree that results from an application of the Huffman algorithm (Algorithm 3.5.1).

internal vertex in a rooted tree: a vertex that is not a leaf.

isomorphic rooted trees: two rooted trees that have a graph isomorphism between them that maps root to root.

leaf in a rooted tree: a vertex having no children.

leaf in an undirected tree: a vertex of degree 1.

left-child of a vertex in a binary tree: one of two possible designations for a child.

left-complete binary tree: a binary tree such that the bottom level has no gaps as one traverses from left to right.

level of a vertex in a rooted tree: synonym for *depth*.

level-order of an ordered tree: a listing of the vertices in the top-to-bottom, left-to-right order of a standard plane drawing of that tree.

m-ary tree: a rooted tree in which every vertex has m or fewer children.

ordered tree: a rooted tree in which the children of each vertex are assigned a fixed ordering.

parent of a vertex w in a rooted tree: the vertex that immediately precedes vertex w on the path from the root to w.

prefix code: a binary code with the property that no codeword is an initial substring of any other codeword.

priority queue: a set of entries, each of which is assigned a *priority*. When an entry is to be removed, an entry with the highest priority is selected.

priority tree: a left-complete binary tree whose vertices have labels (called *priorities*) from an ordered set, such that no vertex has higher priority than its parent.

proper descendant of a vertex v: a descendant of v that is different from v.

Prüfer sequence of length $n-2$, for $n \geq 2$: any sequence of integers between 1 and n, with repetitions allowed; used to encode trees (see Algorithm 3.7.1).

queue: a sequence of elements such that each new element is added (*enqueued*) to one end, called the *back* of the queue, and an element is removed (*dequeued*) from the other end, called the *front*.

random-access table: a database whose domain is a sequence of entries, where each entry consists of two fields. One field is for the actual data element, and the other one is for the key, whose value determines that entry's position in the database.

right-child: one of two possible designations for a child of a vertex in a binary tree.

rigid tree: a tree whose only automorphism is the identity automorphism.

root of a rooted tree: the distinguished vertex.

rooted tree: a directed tree having a distinguished vertex r called the *root* such that for every other vertex v, there is a directed path from r to v.

shortest-path tree for a connected graph G from a vertex v: a rooted tree T with vertex-set V_G and root v such that the unique path in T from v to each vertex w is a shortest path in G from v to w.

siblings in a rooted tree: vertices that have the same parent.

stack: a sequence of elements such that each new element is added (*pushed*) onto one end, called the *top*, and an element is removed (*popped*) from that same end.

standard plane drawing of a rooted tree: a plane drawing of the rooted tree such that the root is at the top, and the vertices at each level are horizontally aligned.

standard plane representation of an ordered tree: a standard plane drawing of the ordered tree such that at each level, the left-to-right order of the vertices agrees with their prescribed order.

ternary tree: synonym for *3-ary tree*.

tree-graphic sequence: a sequence for which there is a permutation that is the degree sequence of some tree.

tree traversal of a binary tree: a systematic visit of each vertex of the tree.

—, **in-order:** defined recursively as follows:

    i. Perform an in-order traversal of the left subtree of T.

    ii. List (process) the root of T.

    iii. Perform an in-order traversal of the right subtree of T.

—, **left pre-order:** defined recursively as follows:

    i. List (process) the root of T.

    ii. Perform a left pre-order traversal of the left subtree of T.

    iii. Perform a left pre-order traversal of the right subtree of T.

—, **level-order:** a visit of the vertices in the top-to-bottom, left-to-right order of a standard plane drawing of that binary tree.

—, **post-order:** defined recursively as follows:

    i. Perform a post-order traversal of the left subtree of T.

    ii. Perform a post-order traversal of the right subtree of T.

    iii. List (process) the root of T.

Wiener index: synonym for *distance sum*.

Chapter 4

SPANNING TREES

INTRODUCTION

Systematic processing of the vertices and edges of a graph often involves the strategy of growing a spanning tree. This chapter demonstrates that several classical algorithms, including depth-first and breadth-first searches, Prim's minimum-spanning-tree method, and Dijkstra's shortest-path method, are special cases or extensions of this strategy. Sketches of these different algorithms highlight their similarities and provide an overview from which a deeper study can be made. That there is potentially an exponentially large number of spanning trees (established by Cayley's Formula in §3.7) makes the low-order-polynomial-time algorithms that find a spanning tree of minimum weight all the more impressive.

Spanning trees also provide a foundation for a systematic analysis of the cycle structure of a graph, as described in the last three sections of this chapter. Mathematicians regard the algebraic structures underlying the collection of cycles and edge-cuts of a graph as beautiful in their own right. Their use in various applications here offers further evidence of the merger of mathematical elegance with techniques of practical importance.

4.1   TREE GROWING

Several different problem-solving algorithms involve growing a spanning tree, one edge and one vertex at a time. All these techniques are refinements and extensions of the same basic tree-growing scheme given in this section.

TERMINOLOGY: For a given tree T in a graph G, the edges and vertices of T are called **tree edges** and **tree vertices**, and the edges and vertices of G that are not in T are called **non-tree edges** and **non-tree vertices**.

Frontier Edges

DEFINITION: A **frontier edge** for a given tree T in a graph is a non-tree edge with one endpoint in T, called its **tree endpoint**, and one endpoint not in T, its **non-tree endpoint**.

Example 4.1.1: For the graph in Figure 4.1.1, the tree edges of a tree T are drawn in bold. The tree vertices are black, and the non-tree vertices are white. The frontier edges for T, appearing as dashed lines, are edges a, b, c, and d. The plain edges are the non-tree edges that are not frontier edges for T.

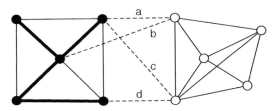

Figure 4.1.1   A tree with frontier edges a, b, c, and d.

Observe that when any one of the frontier edges in Figure 4.1.1 is added to the tree T, the resulting subgraph is still a tree. This property holds in general, as we see in the next proposition, and its iterative application forms the core of the tree-growing scheme.

Proposition 4.1.1: *Let T be a tree in a graph G, and let e be a frontier edge for T. Then the subgraph of G formed by adding edge e to tree T is a tree.*

Proof: The addition of the frontier edge e to the tree T cannot create a cycle, since one of its endpoints is outside of T. Moreover, the vertex that was added to the tree is clearly reachable from any other vertex in the resulting tree.   ◇

Remark: Formally, adding a frontier edge to a tree involves adding a new vertex to the tree, as well as the primary operation of *adding an edge*, defined in §2.4.

Choosing a Frontier Edge

An essential component of our tree-growing scheme is a function called *nextEdge*, which chooses a frontier edge to add to the current tree.

DEFINITION: Let T be a tree subgraph of a graph G, and let S be the set of frontier edges for T. The function **nextEdge(G, S)** chooses and returns as its value the frontier edge in S that is to be added to tree T.

Remark: Ordinarily, the function *nextEdge* is deterministic; however, it may also be randomized. In either case, the full specification of *nextEdge* must ensure that it always returns a frontier edge.

DEFINITION: After a frontier edge is added to the current tree, the procedure **update-Frontier(G,S)** removes from S those edges that are no longer frontier edges and adds to S those that have become frontier edges.

Example 4.1.1 continued: The current set S of frontier edges shown in the top half of Figure 4.1.2 is $S = \{a, b, c, d\}$. Suppose that the value of $nextEdge(G, S)$ is edge c. The bottom half of Figure 4.1.2 shows the result of adding edge c to tree T and applying *updateFrontier*(G, S). Notice that *updateFrontier*(G, S) added four new frontier edges to S and removed edges c and d, which are no longer frontier edges.

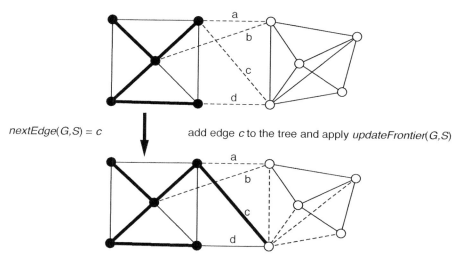

Figure 4.1.2   Result after adding edge c to the tree.

Algorithm 4.1.1:   Tree-Growing
   Input: a connected graph G, a starting vertex $v \in V_G$, and a function *nextEdge*.
   Output: an ordered spanning tree T of G with root v.
      Initialize tree T as vertex v.
      Initialize S as the set of proper edges incident on v.
      While $S \neq \emptyset$
         Let $e = nextEdge(G, S)$.
         Let w be the non-tree endpoint of edge e.
         Add edge e and vertex w to tree T.
         updateFrontier(G, S).
      Return tree T.

Remark: Each different version of *nextEdge*, based on how its selection rule is defined, creates a different instance of Tree-Growing.[†] In §4.2 and §4.3, we will see that the four classical algorithms, *depth-first search*, *breadth-first search*, *Prim's spanning-tree*, and *Dijkstra's shortest-path*, are four different instances of Tree-Growing.

[†] That is, *nextEdge* is regarded as a function-valued variable whose instantiation turns Tree-Growing into a specific algorithm.

Discovery Order of the Vertices

DEFINITION: Let T be the spanning tree produced by any instance of Tree-Growing in a graph G. The **discovery order** is a listing of the vertices of G in the order in which they are added (discovered) as tree T is grown. The position of a vertex in this list, starting with 0 for the start vertex, is called the **discovery number** of that vertex.

Proposition 4.1.2: *Let T be the output tree produced by any instance of Tree-Growing in a graph. Then T is an ordered tree with respect to the discovery order of its vertices.*
$\diamond$ *(Exercises)*

Example 4.1.2: Figure 4.1.3 shows the two output trees that result when two different instances of Tree-Growing are applied to the graph shown, with starting vertex v. For the first instance, the function *next Edge* selects the frontier edge whose name is lexicographically (alphabetically) first, and for the second instance, *nextEdge* uses reverse lexicographic order.

Each output tree is depicted in two ways: as a spanning tree whose edges are drawn in bold, and as an ordered tree with root v. The discovery number appears next to each vertex. Notice that the left-to-right order of the children of each vertex is consistent with the discovery order, as asserted by Proposition 4.1.2.

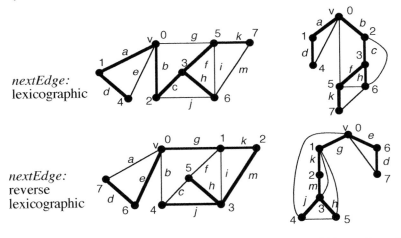

Figure 4.1.3   Output produced by two instances of Tree-Growing.

Example 4.1.3: *Cycle Detection in a Connected Graph.* Observe that a connected graph G is acyclic if and only if the output tree produced by Tree-Growing for the input graph G contains all the edges of G. Moreover, if graph G does have cycles, then Tree-Growing can be used to find a special subcollection of them that forms a basis in a certain vector subspace associated with graph (see §4.5 and §4.6). In particular, as we saw in §3.1 and as illustrated in Figure 4.1.3 above, each non-tree edge creates a unique cycle when that edge alone is added to the tree. In §4.5 and §4.6, we see how these *fundamental cycles* generate *all* the cycles of the graph.

Two Kinds of Non-Tree Edges

The non-tree edges that appear in both output trees in Figure 4.1.3 above, or any other output tree, fall into two categories.

DEFINITION: Two vertices in a rooted tree are **related** if one is a descendant of the other.

DEFINITION: For a given output tree grown by Tree-Growing, a **skip-edge** is a non-tree edge whose endpoints are related; a **cross-edge** is a non-tree edge whose endpoints are not related.

Example 4.1.2 continued: For the output tree shown in the top half of Figure 4.1.3, there are three skip-edges and two cross-edges. For the other output tree, there are four skip-edges and one cross-edge.

Depth-First and Breadth-First Search as Tree-Growing

We preview the two classical search algorithms presented in the next section by specifying two different selection rules for the function *nextEdge*. In fact, the *depth-first* and *breadth-first* searches use opposite versions of *nextEdge*.

- **depth-first**: *nextEdge* selects a frontier edge whose tree endpoint was most recently discovered.

- **breadth-first**: *nextEdge* selects a frontier edge whose tree endpoint was discovered earliest.

Resolving Ties when Choosing the Next Frontier Edge

It must be emphasized that these last two selection criteria do *not* fully specify the function *nextEdge*, because, in general, they will produce ties. Typically, ties are resolved by some *default priority* that is likely to be part of the implementation of the data structures involved. For instance, if the graph is stored using a *table of incident edges* (§2.6), the table imposes a global ordering of the vertices and a local ordering of the incident edges on each vertex. These orderings can then be used to complete the specification of the function *nextEdge* by incorporating a tie-breaking rule.

To avoid getting bogged down in implementation details when we illustrate the different instances of Tree-Growing in this and the next two sections, we will often rely on the somewhat artificial lexicographic (alphabetical) order of the edge names and/or vertex names to resolve ties in choosing the next frontier edge. In practice, it is highly unlikely that lexicographic order would be used.

Tree-Growing Applied to a Non-Connected Graph

REVIEW FROM §2.3: The **component of a vertex** v in a graph G, denoted $C_G(v)$, is the subgraph of G induced on the set of vertices that are reachable from v.

Proposition 4.1.3: *Let* T *be the output tree produced when an instance of Tree-Growing (Algorithm 4.1.1) is applied to a graph* G *(not necessarily connected), starting at a vertex* $v \in V_G$. *Then* T *is a spanning tree of the component* $C_G(v)$.

Proof: The result follows by induction, using Proposition 4.1.1.   ◇

Corollary 4.1.4: *A graph* G *is connected if and only if the output tree produced when any instance of Tree-Growing is applied to* G *is a spanning tree of* G.   ◇

Tree-Growing in a Digraph

DEFINITION: A ***frontier arc*** for a rooted tree T in a digraph is an arc whose tail is in T and whose head is not in T.

Tree-growing for digraphs looks the same as for undirected graphs: in each iteration, a frontier arc is selected and added to the growing tree. However, there are notable differences. For instance, Corollary 4.1.4 provides an easy way of determining whether a given undirected graph is connected, but determining whether a digraph is strongly connected is more involved. In contrast with undirected graphs, the number of vertices in the output tree for a digraph depends on the choice of a starting vertex, as the next example shows.

Example 4.1.4: If Tree-Growing is applied to the 5-vertex digraph in Figure 4.1.4, then the number of vertices in the output tree ranges between 1 and 5, depending on the starting vertex. For instance, the output tree spans the digraph if the tree is grown starting at vertex u, but the output tree contains only one vertex if the starting vertex is v.

Figure 4.1.4   The output tree for a digraph depends on the starting vertex.

DEFINITION: For tree-growing in digraphs, the non-tree edges (arcs) fall into three categories:

- A ***back-arc*** is directed from a vertex to one of its ancestors.

- A ***forward-arc*** is directed from a vertex to one of its descendants.

- A ***cross-arc*** is directed from a vertex to another vertex that is unrelated.

TERMINOLOGY: There are two kinds of cross-arcs: a ***left-to-right cross-arc*** is directed from smaller discovery number to larger one; a ***right-to-left cross-arc*** is the opposite.

Remark: In Chapter 9, a digraph version of tree-growing is adapted to determine whether a given digraph is strongly connected and to finding the strong components if it is not. As the last example suggests, the solution requires more than just checking to see whether all the vertices are discovered.

Forest-Growing

The forest-growing scheme shown on the next page is simply the repeated application of Tree-Growing. Each iteration grows a spanning tree for a new component by restarting the tree-growing at a vertex that had not been discovered previously. Forest-Growing uses a function ***nextVertex***, which selects (deterministically) a restart vertex from the set of undiscovered vertices.

DEFINITION: A ***full spanning forest*** of a graph G is a spanning forest consisting of a collection of trees, such that each tree is a spanning tree of a different component of G.

Algorithm 4.1.2:   **Forest-Growing and Component-Counting**
 Input: a graph G
 Output: a full spanning forest F of G and the number of components $c(G)$.
  Initialize forest F as the empty graph.
  Initialize component counter $t := 1$
  While forest F does not yet span graph G
   Let $v = nextVertex(V_G - V_F)$.
   Use Tree-Growing to produce a spanning tree T_t of $C_G(v)$.
   Add tree T_t to forest F.
   $t := t + 1$
  Return forest F and component count $c(G) = t$.

Example 4.1.5: Figure 4.1.5 shows the forest (with edges in bold) that results when the function *nextVertex* uses lexicographic order of vertex names and *nextEdge* uses lexicographic order of edge names. The bracketed numbers give the discovery order of the vertices. Notice that the forest consists of two trees that are spanning trees of their respective components.

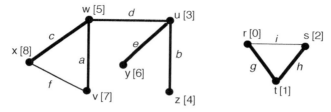

Figure 4.1.5   Growing a 2-component spanning forest of a graph.

Some Implementation Considerations

COMPUTATIONAL NOTE 1: Any implementation of Tree-Growing (Algorithm 4.1.1) requires concrete data structures and operations. One of the options for the implementation of the procedure *updateFrontier* is to have it immediately discard at each iteration all old frontier edges that end at the newly discovered vertex w. However, certain applications may require some additional bookkeeping involving the old frontier edges, making their immediate removal premature.

Deciding how best to manage the frontier set is just one of several implementation details that depend on how the algorithm is to be used. But the goal here is to unify several classical algorithms in computer science and operations research, and sidestepping these issues makes it easier to view each of these algorithms as essentially the same tree-growing scheme, except for the rules by which the function *nextEdge* is defined. The reader who wants to learn more about the data structures necessary for a full implementation is encouraged to consult a standard computer science text on algorithms and data structures (e.g., [AhHoU183], [Ba83], [Se88]).

COMPUTATIONAL NOTE 2: **Algorithmic Analysis of Tree-Growing**
A complete analysis of the computational requirements of Tree-Growing (Algorithm 4.1.1) is not possible without fully specifying data structures and other implementation details. Instead, we provide a rough estimate.

Suppose that the input graph has n vertices and m edges. Even if the adjacency-matrix representation (§2.6) is used to represent incidence (a possibility for simple graphs), Tree-Growing is implement able in $O(n^2)$ running time. However, if a table of incident edges is used, then Tree-Growing is essentially $O(m)$, since each edge is examined once. Thus, for *sparse* graphs (graphs with relatively few edges, i.e., m much less than $\binom{n}{2}$), the tree-growing runs faster when the incidence-table representation is used.

One possibility is to store the frontier edges in a *priority queue*, leading to an algorithm having $O(m \log m)$ running time.

EXERCISES for Section 4.1

In Exercises 4.1.1 through 4.1.3, indicate the set of frontier edges for the given tree.

4.1.1^S

4.1.2

4.1.3

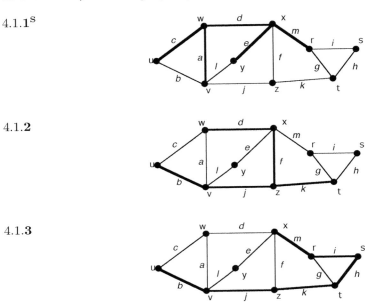

In Exercises 4.1.4 through 4.1.8, draw the output tree, including the discovery numbers, that results when Algorithm 4.1.1 is applied to the graph in Exercise 4.1.1. Start at the specified vertex and use: (a) lexicographic order for nextEdge; (b) reverse lexicographic order for nextEdge.

4.1.4^S Vertex w 4.1.5 Vertex u 4.1.6 Vertex x

4.1.7 Vertex y 4.1.8^S Vertex t

4.1.9 Apply Algorithm 4.1.2 to the graph in Figure 4.1.5, using *reverse* lexicographic order as the default priority for vertices and for edges.

In Exercises 4.1.10 through 4.1.13, draw the output tree that results when the digraph version of Algorithm 4.1.1 is applied to the digraph shown, starting at the specified vertex and using lexicographic order for nextEdge. Include the discovery numbers.

4.1.10^S Vertex u.

4.1.11 Vertex x.

4.1.12 Vertex w.

4.1.13 Vertex y.

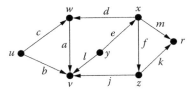

4.1.14 Add an edge q to the graph in Figure 4.1.3 so that the outputs produced by two instances of Algorithm 4.1.1, one using lexicographic order for $nextEdge$, and the other using reverse lexicographic order, both contain at least one cross-edge whose endpoints are at different levels of the output (rooted) tree.

4.1.15 Prove Proposition 4.1.2.

4.1.16^S Characterize those connected graphs for which there is a starting vertex such that the discovery order produced by Algorithm 4.1.1 is independent of the default priority of the edges.

4.1.17 Use induction to prove that for any application of Algorithm 4.1.1, the discovered vertices are all reachable from the starting vertex.

4.1.18 [*Computer Project*] Implement Algorithm 4.1.1 and compare its output with that obtained by hand for Exercises 4.1.4 through 4.1.8.

4.1.19 [*Computer Project*] Implement Algorithm 4.1.2, and test the program on a 10-vertex 2-component graph and a 10-vertex 3-component graph.

4.2 DEPTH-FIRST AND BREADTH-FIRST SEARCH

Depth-first search and *breadth-first search* are integral parts of numerous algorithms used in computer science, operations research, and other engineering disciplines. To mention just a few of these, depth-first search is central to algorithms for finding the cut-vertices and blocks of a graph (§4.4 and §5.4), for finding the strong components of a digraph (§9.5), and for determining whether a graph can be drawn in the plane without edge-crossings (§7.6); breadth-first search is used for finding flow-augmenting paths as part of a flow-maximization algorithm for capacitated networks (§10.2).

TERMINOLOGY: The output trees produced by the depth-first and breadth-first searches of a graph are called the **depth-first tree** (or **dfs-tree**) and the **breadth-first tree** (or **bfs-tree**).

TERMINOLOGY NOTE: The use of the word "search" here is traditional. Although one does occasionally use these algorithms to search for something, they are tree-growing algorithms with much wider use.

As previewed in §4.1, depth-first search and breadth-first search use two opposite priority rules for the function $nextEdge$.

Depth-First Search

The idea of depth-first search is to select a frontier edge incident on the most recently discovered vertex of the tree grown so far. When that is not possible, the algorithm *backtracks* to the next most recently discovered vertex and tries again. Thus, the search moves deeper into the graph (hence, the name "depth-first"). In implementing Tree-Growing, depth-first search uses the following version of the function $nextEdge$.

DEFINITION: Let S be the current set of frontier edges. The function **dfs-nextEdge** is defined as follows: $dfs\text{-}nextEdge(G, S)$ selects and returns as its value the frontier edge whose tree-endpoint has the largest discovery number. If there is more than one such edge, then $dfs\text{-}nextEdge(G, S)$ selects the one determined by the default priority.

Algorithm 4.2.1:   **Depth-First Search**

  Input: a connected graph G, a starting vertex $v \in V_G$.

  Output: an ordered spanning tree T of G with root v.

   Initialize tree T as vertex v.

   Initialize S as the set of proper edges incident on v.

   While $S \neq \emptyset$

      Let $e = \textit{dfs-nextEdge}(G, S)$.

      Let w be the non-tree endpoint of edge e.

      Add edge e and vertex w to tree T.

      $\textit{updateFrontier}(G, S)$.

   Return tree T.

TERMINOLOGY: The discovery number of each vertex w for a depth-first search is called the **dfnumber** of w and is denoted $\textit{dfnumber}(w)$.

Remark: The term *dfnumber* is used instead of discovery number to underscore its special role in a number of applications of depth-first search. For instance, in §4.4 and §5.4, *dfnumbers* are used to determine the cut-vertices and blocks of a graph. In Chapter 9, they are used to find the strong components of a digraph.

Example 4.2.1: Figure 4.2.1 shows the output when the depth-first search is applied to the connected graph on the left, starting at vertex v. The function *dfs-nextEdge* resolves ties using lexicographic order of the non-tree endpoints of the frontier edges. The *dfnumbers* are shown in parentheses, and the tree edges are drawn in bold. Notice that there are no cross-edges. More specifically, for each non-tree edge, the endpoint with the smaller *dfnumber* is an ancestor of the other endpoint. The next proposition shows that this is always the case for a depth-first search of an *undirected* graph. In Chapter 9, it is shown that this property does not hold for digraphs.

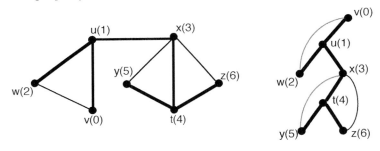

Figure 4.2.1   Depth-first search using lexicographic order for ties.

Proposition 4.2.1: *Depth-first search trees have no cross-edges.*

Proof: Let T be the output tree produced by a depth-first search, and let e be a non-tree edge whose endpoints x and y satisfy $\textit{dfnumber}(x) < \textit{dfnumber}(y)$. At the point when the depth-first search discovered vertex x, edge e become a frontier edge. Since e never became a tree edge, the search had to have discovered vertex y before backtracking to vertex x for the last time. Thus, y is in the subtree rooted at x and, hence, is a descendant of x. $\diamondsuit$

Remark: A pre-order traversal (§3.3) of the depth-first tree reproduces the discovery order generated by the depth-first search (see Exercises).

Recursive Depth-First Search

A depth-first search, starting at a vertex v, has the following informal recursive description.

DFS(v):
    While v has an undiscovered neighbor,
        Let w be an undiscovered neighbor of v.
        DFS(w)
    Return

COMPUTATIONAL NOTE: We mentioned in §4.1 that for tree-growing, the natural way to store the frontier edges is to use a *priority queue*. For the particular case of depth-first search, the priority queue emulates the simpler *stack* data structure (Last-In-First-Out), because the newest frontier edges are given the highest priority by being pushed onto the stack (in increasing default priority order, if there is more than one new frontier edge). The recursive aspect of depth-first search also suggests the feasibility of implementation as a stack.

Depth-First Search in a Digraph

The depth-first search in a digraph is Algorithm 4.2.1 with the function *dfs-nextArc* replacing *dfs-nextEdge*. Also, we do not assume that the input digraph is strongly connected, so the dfs-tree produced will not necessarily be a spanning tree.

DEFINITION: The function **dfs-nextArc** selects and returns as its value the frontier arc whose tree-endpoint has the largest *dfnumber*.

Proposition 4.2.2: *When a depth-first search is executed on a digraph, the only kind of non-tree arc that cannot occur is a left-to-right cross arc.*   ◇ *(Exercises)*

Remark: Applications of depth-first search in a digraph appear in §4.4 and in §9.5.

Breadth-First Search

In the breadth-first search, the search "fans out" from the starting vertex and grows the tree by selecting frontier edges incident on vertices as close to the starting vertex as possible.

Example 4.2.2: Figure 4.2.2 below compares the results of depth-first and breadth-first searches by displaying typical possibilities for their partial output trees after 11 iterations, starting at vertex v. Of course, the exact partial trees depend on the exact rule for choosing a frontier edge. Assume that the respective *nextEdge* functions resolve ties by the same default priority.

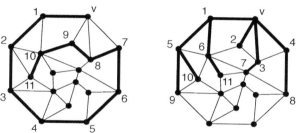

Figure 4.2.2   Depth-first and breadth-first search after 11 iterations.

Algorithm 4.2.2 below uses the following version of the function *nextEdge*.

DEFINITION: Let S be the current set of frontier edges. The function **bfs-nextEdge** is defined as follows: *bfs-nextEdge*(G, S) selects and returns as its value the frontier edge whose tree-endpoint has the smallest discovery number. If there is more than one such edge, then *bfs-nextEdge*(G, S) selects the one determined by the default priority.

TERMINOLOGY: The discovery number of each vertex w for a breadth-first search is called the **bfnumber** of w and is denoted *bfnumber*(w).

Algorithm 4.2.2:   **Breadth-First Search**
  Input: a connected graph G, a starting vertex $v \in V_G$.
  Output: an ordered spanning tree T of G with root v.
        Initialize tree T as vertex v.
        Initialize S as the set of proper edges incident on v.
        While $S \neq \emptyset$
              Let $e = $ *bfs-nextEdge*(G, S).
              Let w be the non-tree (undiscovered) endpoint of edge e.
              Add edge e and vertex w to tree T.
              update Frontier(G, S).
        Return tree T.

Example 4.2.3: Figure 4.2.3 shows the result of a breadth-first search applied to the same graph as in Example 4.2.1, again starting at vertex v. The function *bfs-nextEdge* resolves ties using lexicographic order of the non-tree endpoints of the frontier edges. Observe that the non-tree edges are all cross-edges, which is opposite from the result of depth-first search. Also notice that the breadth-first tree is a shortest-path tree (§3.2). Both these properties hold in general, as asserted by the next two propositions.

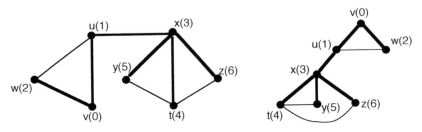

Figure 4.2.3   Breadth-first search using lexicographic order for ties.

The following result shows that the breadth-first search proceeds level by level, in contrast to the depth-first search.

Proposition 4.2.3: If x and y are vertices in a bfs-tree and $level(y) > level(x)$, then $bfnumber(y) > bfnumber(x)$.

Proof: We use induction on $level(x)$. The base case ($level(x) = 0$) is trivially true, since the root has *bfnumber* 0 and is the only vertex at level 0. Assume the result holds when $level(x) = k$, for some $k \geq 0$, and suppose $level(y) > level(x) = k+1$. Let v be the parent of x, and let w be the parent of y in the bfs-tree. Then $level(w) > level(v) = k$, and thus, by the inductive hypothesis, $bfnumber(w) > bfnumber(v)$. This implies that the breadth-first search chose frontier edge vx before edge wy, and hence, $bfnumber(y) > bfnumber(x)$.
$\diamondsuit$

Proposition 4.2.4: *When breadth-first search is applied to an undirected graph, the endpoints of each non-tree edge are either at the same level or at consecutive levels.*

Proof: Let xy be a non-tree edge in a bfs-tree with $bfnumber(x) < bfnumber(y)$. By Proposition 4.2.3, $level(x) \le level(y)$. Let z be the parent of y in the tree. Since the edge zy was chosen from the set of frontier edges, rather than xy, $bfnumber(z) < bfnumber(x)$. Therefore, by Proposition 4.2.3, $level(y) = level(z) + 1 \le level(x) + 1$. Since, $level(x) \le level(y) \le level(x) + 1$, the result follows.   ◇

Proposition 4.2.5: *The breadth-first tree produced by any application of Algorithm 4.2.2 is a shortest-path tree for the input graph.*   ◇ (Exercises)

COMPUTATIONAL NOTE: Whereas the stack (last-In-First-Out) is the appropriate data structure to store the frontier edges in a depth-first. search, the *queue* (First-In-First-Out) is most appropriate for the breadth-first search, since the frontier edges that are oldest have the highest priority.

EXERCISES for Section 4.2

In Exercises 4.2.1 through 4.2.7, draw the depth-first tree that results when Algorithm 4.2.1 is applied to the graph shown below, starting at the specified vertex. Include the dfnumbers and use: (a) lexicographic order as the default priority; (b) reverse lexicographic order as the default priority.

4.2.1^{S} Vertex w.

4.2.2 Vertex u.

4.2.3 Vertex x.

4.2.4 Vertex y.

4.2.5^{S} Vertex t.

4.2.6 Vertex z.

4.2.7 Vertex s.

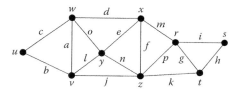

In Exercises 4.2.8 through 4.2.14, draw the breadth-first tree, including the discovery numbers, that results when Algorithm 4.2.2 is applied to the following graph. Start at the specified vertex and use: (a) lexicographic order as the default priority; (b) reverse lexicographic order as the default priority.

4.2.8 Vertex w.

4.2.9^{S} Vertex u.

4.2.10 Vertex x.

4.2.11 Vertex y.

4.2.12 Vertex t.

4.2.13^{S} Vertex z.

4.2.14 Vertex s.

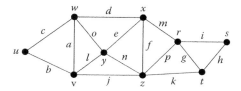

4.2.15 Characterize those graphs for which the depth-first and breadth-first trees that result are identical, no matter what the default priority and starting vertex are.

4.2.16^{S} What kinds of graphs have a unique depth-first spanning tree that is independent of the starting vertex?

4.2.**17** Verify that a pre-order traversal of a depth-first search tree re-creates the discovery order.

4.2.**18** Is there a traversal from Chapter 3 that can be applied to a breadth-first tree to re-create the discovery order of the breadth-first search? Justify your answer.

4.2.**19** Prove Proposition 4.2.2.

4.2.**20** Prove Proposition 4.2.4. Hint: Use Proposition 4.2.3 and an argument analogous to the one given in the proof of Proposition 4.2.1.

4.2.**21** Prove Proposition 4.2.5.

4.2.**22** [*Computer Project*] Implement a recursive depth-first search and use an adjacency matrix (§2.6) to store the input graph. Compare the output from the computer program with the results obtained by hand in Exercises 4.2.1 through 4.2.7.

4.2.**23** [*Computer Project*] Implement a recursive depth-first search and use an incidence matrix (§2.6) to store the input graph. Compare the output from the computer program with the results obtained by hand in Exercises 4.2.8 through 4.2.14.

4.2.**24** [*Computer Project*] Implement Algorithm 4.2.1 using stacks, and use an adjacency matrix (§2.6) to store the input graph. Compare the output from the computer program with the results obtained by hand in Exercises 4.2.1 through 4.2.7.

4.2.**25** [*Computer Project*] Implement Algorithm 4.2.1 using stacks, and use an incidence matrix (§2.6) to store the input graph. Compare the output from the computer program with the results obtained by hand in Exercises 4.2.8 through 4.2.14.

4.2.**26** [*Computer Project*] Implement Algorithm 4.2.2 using queues, and use an adjacency matrix to store the input graph. Compare the output from the computer program with the results obtained by hand in Exercises 4.2.1 through 4.2.7.

4.2.**27** [*Computer Project*] Implement Algorithm 4.2.2 using queues, and use an incidence matrix to store the input graph. Compare the output from the computer program with the results obtained by hand in Exercises 4.2.8 through 4.2.14.

4.2.**28** [*Computer Project*] Modify Algorithm 4.2.1 so that it halts as soon as a cycle is detected. Then implement the algorithm and test the program on at least two acyclic graphs and at least two with cycles.

4.3   MINIMUM SPANNING TREES AND SHORTEST PATHS

The two classical algorithms presented in this section are instances of the generic tree-growing algorithm given in §4.1, applied on a weighted graph.

REVIEW FROM §1.6: A **weighted graph** is a graph in which each edge is assigned a number, called its **edge-weight**.

Finding the Minimum Spanning Tree: Prim's Algorithm

Minimum-Spanning-Tree Problem: Let G be a connected weighted graph. Find a spanning tree of G whose total edge-weight is a minimum.

We introduced this problem in §1.6. By Cayley's formula (§3.7) the number of different spanning trees of an n-vertex graph could be as many as n^{n-2}. Thus, an exhaustive examination of all the spanning trees is clearly impractical for large graphs.

The following algorithm, attributed to R. C. Prim [Pr57] finds a minimum spanning tree without explicitly examining all of the spanning trees. The basic idea is to start at any vertex and "grow" a **Prim tree** by adding the frontier edge of smallest edge-weight at each iteration. Thus, Prim's algorithm is an instance of Tree-Growing, that uses the following version of the function *nextEdge*. Since *Prim-nextEdge*(G, S) and its associated procedure *update Frontier* run in low-order polynomial-time, so does Prim's algorithm.

DEFINITION: Let S be the current set of frontier edges. The function **Prim-nextEdge** is defined as follows: *Prim-nextEdge*(G, S) selects and returns as its value the frontier edge with smallest edge-weight. If there is more than one such edge, then *PrimnextEdge*(G, S) selects the one determined by the default priority.

Algorithm 4.3.1:   Prim's Minimum Spanning Tree
   Input: a weighted connected graph G and starting vertex v.
   Output: a minimum spanning tree T.
      Initialize tree T as vertex v.
      Initialize S as the set of proper edges incident on v.
      While $S \neq \emptyset$
         Let $e = Prim\text{-}nextEdge(G, S)$.
         Let w be the non-tree endpoint of edge e.
         Add edge e and vertex w to tree T.
         update Frontier(G, S).
      Return tree T.

Example 4.3.1: Figure 4.3.1 shows the minimum spanning tree produced by Prim's algorithm for the input graph shown, starting at vertex v. The function *Prim-nextEdge* uses lexicographic order of the names of the non-tree endpoints of frontier edges as a first tie-breaker and of the tree-endpoints as a second tie-breaker. The discovery numbers appear in parentheses.

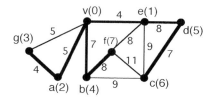

Figure 4.3.1   Minimum spanning tree produced by Prim's algorithm.

The correctness of Prim's algorithm is an immediate consequence of the next proposition.

Proposition 4.3.1: *Let T_k be the Prim tree after k iterations of Prim's algorithm on a connected graph G, for $0 \leq k \leq |V_G| - 1$. Then T_k is a subtree of a minimum spanning tree of G.*

Proof: The assertion is trivially true for $k = 0$, since T_0 is the trivial tree consisting of the starting vertex.

By way of induction, assume for some $k, 0 \leq k \leq |V_G| - 2$, that T_k is a subtree of a minimum spanning tree T of G, and consider tree T_{k+1}. The algorithm grew T_{k+1} during the $(k + 1)^{\text{st}}$ iteration by adding to n a frontier edge e of smallest weight. Let u and v be the endpoints of edge e, such that u is in tree T_k and v is not.

If spanning tree T contains edge e, then T_{k+1} is a subtree of T. If e is not an edge in tree T, then e is part of the unique cycle contained in $T + e$. Consider the path in T from u to v consisting of the "long way around the cycle." On this path, let f be the first edge that joins a vertex in T_k to a vertex not in T_k. The situation is illustrated in Figure 4.3.2; the black vertices and bold edges make up Prim tree T_k, the spanning tree T consists of everything except edge e, and Prim tree $T_{k+1} = (T_k \cup v) + e$.

Since f was a frontier edge at the beginning of the $(k+1)$st iteration, $w(e) \leq w(f)$ (since the algorithm chose e). Tree $\hat{T} = T + e - f$ clearly spans G, and T_{k+1} is a subtree of T (since f was not part of T_k). Finally, $w\left(\hat{T}\right) = w(T) + w(e) - w(f) \leq w(T)$, which shows that $\hat{T}$ is a minimum spanning tree of G. $\diamond$

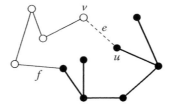

Figure 4.3.2

Corollary 4.3.2: *When Prim's algorithm is applied to a connected graph, it produces a minimum spanning tree.*

Proof: Proposition 4.3.1 implies that the Prim tree resulting from the final iteration is a minimum spanning tree. $\diamond$

At the end of this chapter, a different (i.e., not an instance of Tree-Growing) but comparably fast algorithm for finding a minimum spanning tree, known as *Kruskal's algorithm*, is presented.

The Steiner-Tree Problem

If instead of requiring that the minimum-weight tree *span* the given graph, we require that it simply include a prescribed *subset* of the vertices, we get a generalization of the minimum-spanning-tree problem, called the *Steiner-tree problem*.

DEFINITION: Let U be a subset of vertices in a connected edge-weighted graph G. The **Steiner-tree problem** is to find a minimum-weight tree subgraph of G that contains all the vertices of U.

Observe that if $U = V_G$, then the Steiner-tree problem reduces to the minimum-spanning tree problem.

Example 4.3.2: Figure 4.3.3 shows a weighted graph and a Steiner tree (with bold edges) for the vertex subset $U = \{x, y, z\}$.

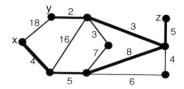

Figure 4.3.3   Steiner tree for $U = \{x, y, z\}$.

Remark: The Steiner-tree problem often arises in network-design and wiring-layout problems. It does not lend itself to the tree-growing strategy that we used for the minimum-spanning tree problem. The case in which all the edges have equal weight (or are unweighted) gives rise to a generalization of distance, the *Steiner distance* of a subset of vertices.

Finding the Shortest Path: Dijkstra's Algorithm

REVIEW FROM §1.6: The **length** of a path in a weighted graph is the sum of the edge-weights of the path. The **distance** between two vertices s and t in a weighted graph is the length of a shortest s-t path.

As mentioned in Section §1.6, the word *distance* is used loosely and can represent various other measurable quantities (e.g., time or cost).

Shortest-Path Problem: Let s and t be two vertices of a connected weighted graph. Find a shortest path from s to t.

Remark: If the edge-weights are all equal, then the problem reduces to one that can be solved by breadth-first search (Algorithm 4.2.2).

The algorithm presented here is another instance of generic tree-growing and is attributed to E. Dijkstra [Di59]. The algorithm does slightly more than the problem requires: it finds a shortest path from vertex s to *each* of the vertices of the graph. That is, it produces a shortest-path tree for the input (weighted) graph. We assume that all edge-weights are nonnegative. Dijkstra's algorithm is not guaranteed to work for graphs that have edges with negative weight, but a good alternative in this case is Floyd's algorithm [F162].

The strategy for Dijkstra's algorithm is to grow a **Dijkstra tree**, starting at vertex s, by adding, at each iteration, a frontier edge whose non-tree endpoint is as close as possible to s. Thus, Dijkstra's algorithm is an instance of Tree-Growing that uses the following version of the function *nextEdge*.

DEFINITION: The function **Dijkstra-nextEdge** is defined as follows: Let S be the current set of frontier edges. *Dijkstra-nextEdge*(G, S) selects and returns as its value the frontier edge whose non-tree endpoint is closest to the start vertex s. If there is more than one such edge, then *Dijkstra-nextEdge*(G, S) selects the one determined by the default priority.

Algorithm 4.3.2:   **Dijkstra's Shortest Path**

> *Input*: a weighted connected graph G and starting vertex s.
> *Output*: a shortest-path tree T with root s.
>   Initialize tree T as vertex s.
>   Initialize S as the set of proper edges incident on s.
>   While $S \neq \emptyset$
>   Let $e := Dijkstra\text{-}nextEdge(G, S)$.
>   Let w be the non-tree endpoint of edge e.
>   Add edge e and vertex w to tree T.
>   *update Frontier*(G, S).
>   Return tree T.

NOTATION: For each tree vertex x , let $dist[x]$ denote the distance from vertex s to x.

Example 4.3.3: Figure 4.3.4 shows the shortest-path tree produced by Dijkstra's algorithm for the input graph shown, starting at vertex s. The function $Dijkstra\text{-}nextEdge$ uses lexicographic order: of the names of the non-tree endpoints of frontier edges as a first tie-breaker and of the tree-endpoints as a second tie-breaker. In the parentheses at each vertex v, the discovery number appears first and $dist[v]$ appears second.

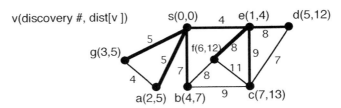

Figure 4.3.4   The shortest-path tree produced by Dijkstra's algorithm.

Calculating Distances as the Dijkstra Tree Grows

NOTATION: For each frontier edge e in the weighted graph, $w(e)$ denotes its edge-weight.

Notice that each tree edge q in Figure 4.3.4 has the following property: if x is the endpoint with the smaller discovery number and y is the other endpoint, then $dist[y] = dist[x] + w(q)$. Thus, when q was the frontier edge selected by $Dijkstra\text{-}nextEdge$, the value of $dist[x] + w(q)$ must have been a minimum over all frontier edges in that iteration.

This suggests the following definition, which enables the function $Dijkstra\text{-}nextEdge$ to be efficiently calculated. It is also used in the proof of correctness of Dijkstra's algorithm (Theorem 4.3.3).

DEFINITION: Let e be a frontier edge of the Dijkstra tree grown so far, and let x be the tree endpoint of c. The *P*-**value** of edge c, denoted $P(e)$, is given by

$$P(e) = dist[x] + w(e)$$

Thus, $Dijkstra\text{-}nextEdge(G, S)$ selects and returns as its value the edge e^* such that $P(e^*) = \min_{e \in S}\{P(e)\}$. (As usual, if there is more than one such edge, then $Dijkstra\text{-}nextEdge(G, S)$ selects the one determined by the default priority.)

Correctness of Dijkstra's Algorithm

The following theorem establishes the correctness of Dijkstra's algorithm. Its proof is similar to the one used to show the correctness of Prim's algorithm.

Theorem 4.3.3: *Let T_j be the Dijkstra tree after j iterations of Dijkstra's algorithm on a connected graph G, for $0 \leq j \leq |V_G| - 1$. Then for each v in T_j, the unique s-v path in T_j is a shortest s-v path in G.*

Proof: The assertion is trivially true for T_0. By way of induction, assume for some j, $0 \leq j \leq |V_G| - 2$, that T_j satisfies the property. Let e, with labeled endpoint x and unlabeled endpoint y, be the frontier edge added to T_j in the $(j+1)^{st}$ iteration. Since y is the only new vertex in T_{j+1}, it suffices to show that the $s - y$ path Q in T_{j+1} is a shortest $s - y$ path in G. The algorithm formed Q by concatenating the $s - x$ path in T_j with edge e, and $length(Q) = P(e)$.

Now let R be any $s - y$ path in G. We must show that $length(R) \geq length(Q)$. Suppose that edge f, with endpoints v and z, is the first edge of path R that is not in T_j, and let K be the portion of R from z to y (see Figure 4.3.5 below).

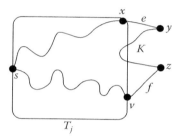

Figure 4.3.5   Proof of correctness of Dijkstra's algorithm.

Since the edge e was the frontier edge selected in the $(j+1)^{st}$ iteration, it follows that $P(e) \leq P(f)$. Thus,

$$length(R) = dist[v] + w(f) + length(K) = P(f) + length(K)$$
$$\geq P(e) + length(K) \geq P(e) = length(Q)$$

where the rightmost inequality follows by the nonnegativity of the edge-weights.   ◇

COMPUTATIONAL NOTE: Each time a new vertex y is added to the Dijkstra tree, the set of frontier edges can be updated without recomputing any P-values. In particular, delete each previous frontier edge incident on vertex y (because both its endpoints are now in the tree), and compute the priorities of the new frontier edges only (i.e., those that are incident on vertex y). This modification significantly decreases the running time.

Also, if there are two frontier edges e and f having the same undiscovered endpoint, then the edge with the larger P-value will never be selected and therefore need never be added to the set of frontier edges. A full-blown implementation of Dijkstra's algorithm might include these strategies for maintaining the set of frontier edges.

EXERCISES for Section 4.3

In Exercises 4.3.1 through 4.3.4, apply Prim's algorithm (Algorithm 4.3.1) to the given weighted graph, starting at vertex s and resolving ties as in Example 4.3.1. Draw the minimum spanning tree that results, and indicate its total weight. Also give the discovery number at each vertex.

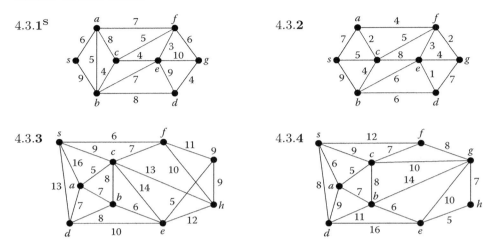

4.3.1$^\text{S}$

4.3.2

4.3.3

4.3.4

In Exercises 4.3.5 through 4.3.8, apply Dijkstra's algorithm (Algorithm 4.3.2) to the given weighted graph, starting at vertex s and resolving ties as in Example 4.3.3. Draw the shortest-path tree that results, and indicate for each vertex v, the discovery number and the distance from vertex s to v.

4.3.5$^\text{S}$ The graph of Exercise 4.3.1.   4.3.6 The graph of Exercise 4.3.2.

4.3.7 The graph of Exercise 4.3.3.   4.3.8 The graph of Exercise 4.3.4.

4.3.9 Suppose that the weighted graph shown below represents a communication network, where the weight p_{ij} on arc ij is the probability that the link from i to j does not fail. If the link failures are independent of one another, then the probability that a path does not fail is the product of the link probabilities for that path. Under this assumption, find the most reliable path from s to t. (Hint: Consider $-\log_{10} p_{ij}$.)

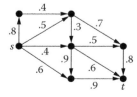

4.3.10$^\text{S}$ Give an example of a weighted graph that has a unique minimum spanning tree, even though it contains a cycle having two edges of equal weight.

4.3.11 State a sufficient condition to guarantee that a given weighted graph does not have a unique minimum spanning tree. Is the condition also necessary?

4.3.12 Suppose that a breadth-first search is executed starting at each vertex of the input graph G. How would you identify the central vertices of G?

4.3.13 Draw a graph G and a shortest-path tree for G that cannot be a breadth-first tree regardless of how ties are resolved.

4.3.14 [*Computer Project*] Implement Prim's spanning-tree algorithm (Algorithm 4.3.1), and run the program on each of the instances of the problem given in Exercises 4.3.1 through 4.3.4.

4.3.15 [*Computer Project*] Implement Dijkstra's shortest-path algorithm (Algorithm 4.3.2), and run the program on each of the instances of the problem given in Exercises 4.3.1 through 4.3.4.

4.4   APPLICATIONS OF DEPTH-FIRST SEARCH

In this section, depth-first search is applied to a connected graph. The results can easily be extended to non-connected graphs by considering each component separately.

The Finish Order of a Depth-First Search

DEFINITION: During a depth-first search, a discovered vertex v becomes **finished** when all of its neighbors have been discovered and all those with higher discovery number are finished.

Example 4.4.1: The table below shows the *finish order* when the depth-first search is applied to the following graph, starting at vertex x. The function *nextEdge* breaks ties by choosing the frontier edge whose non-tree endpoint comes first alphabetically.

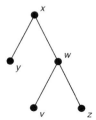

	nextEdge	vertex(discovery #)	frontier-edge set	vertex(finish #)
Initialization:		$x(0)$	$\{xy, xw\}$	
Iteration 1:	xw	$w(1)$	$\{xy, wv, wz\}$	
Iteration 2:	wv	$v(2)$	$\{xy, wz\}$	$v(1)$
Iteration 3:	wz	$z(3)$	$\{xy\}$	$z(2), w(3)$
Iteration 4:	xy	$y(4)$	$\emptyset$	$y(4), x(5)$

Remark: In §4.2, we observed that a pre-order traversal of a dfs-tree reproduces the discovery order (Exercise 4.2.17). Regarding finish order, observe that the root is the last vertex finished in a depth-first search. In fact, a post-order traversal (§3.3) of the dfs-tree reproduces the finish order generated by the depth-first search (see Exercises).

Growing a DFS-Path

A key strategy in many applications of depth-first search is to execute the search only until you have to backtrack for the first time.

DEFINITION: A *dfs-path* is a path in the depth-first tree produced by executing a depth-first search and stopping before you backtrack for the first time.

Proposition 4.4.1: *Let G be a graph, and let P be the dfs-path produced by executing a depth-first search until the first vertex t becomes finished. Then all of the neighbors of t lie on path P.*

Proof: If a neighbor of t does not lie on the path P, then there is a frontier edge incident on t, contradicting the premise that t is finished.   ◇

Corollary 4.4.2: *Let G be a connected graph with at least two vertices. If a dfs-path is grown until the first vertex t is finished, then either $deg(t) = 1$ or t and all its neighbors lie on a cycle of G.*

Proof: Let w be the neighbor of t that has the smallest *dfnumber*. By Proposition 4.4.1, the subpath of P from w to t contains at least $deg(t)$ vertices. If $deg(t) > 1$, then vertices w and t are joined by a back edge, which completes a cycle that contains t and all of its neighbors.   ◇

Corollary 4.4.3: *Let G be a connected simple graph with minimum degree δ. Then G contains a cycle of length greater than δ.*

Finding the Cut-Edges of a Connected Graph

REVIEW FROM §2.4:

- A *cut-edge* (or *bridge*) of a connected graph G is an edge e such that the edge-deletion subgraph $G - e$ has two components.

- An edge in a connected graph is a cut-edge if and only if it is not a cycle edge.

DEFINITION: Let B be the set of cut-edges (bridges) of a connected graph G. A *bridge component* (*BC*) of G is a component of the subgraph $G - B$ (§2.4).

DEFINITION: Let H be a subgraph of a graph G. The *contraction of H to a vertex* is the replacement of H by a single vertex k. Each edge that joined a vertex $v \in V_G - V_H$ to a vertex in H is replaced by an edge with endpoints v and k.

Example 4.4.2: The graph on the left in Figure 4.4.1 below has a subgraph H shown with bold edges. The graph on the right is the result of contracting H to a vertex.

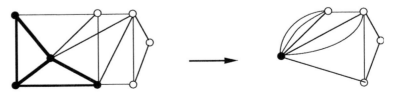

Figure 4.4.1   Contracting H to a vertex.

Proposition 4.4.4: *Let G be a connected graph. All vertices on a cycle are in the same bridge component of G.*   ◇ (*Exercises*)

Proposition 4.4.5: *Let G be a connected graph. The graph that results from contracting each bridge component of G to a vertex is a tree.* ◇
(Exercises)

Example 4.4.3: The graph on the left in Figure 4.4.2, with its three cut-edges drawn with dashed lines, has four bridge components (two of them are single vertices). The contraction of each of them to a vertex results in the tree on the right.

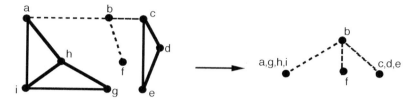

Figure 4.4.2   Contracting each bridge component to a vertex.

Proposition 4.4.6: *Let v be a vertex of a connected graph G with deg(v) ≤ 1. Then the bridge components of G are the trivial subgraph v and the bridge components of G − v.* ◇ *(Exercises)*

The last three propositions together with Corollary 4.4.2 justify the following high-level algorithm for finding the cut-edges of a graph.

Algorithm 4.4.1:   **Finding Cut-Edges**
    Input: a connected graph G.
    Output: the cut-edges of G.
        Initialize graph H as graph G.
        While $|V_H| > 1$
            Grow a dfs-path to the first vertex t that becomes finished.
            If $deg(t) = 1$
                Mark the edge incident on t as a bridge.
                $H := H - t$.
            Else /* vertex t and all its neighbors lie on a cycle C* /
                Let H be the result of contracting cycle C to a vertex.
        Return the edges of H

COMPUTATIONAL NOTE: For efficiency, the dfs-path grown in each iteration should start at the same vertex and use as much of the previous dfs-path as possible.

Topological Sorting by Depth-First Search

As we saw in §4.2, depth-first search extends easily to digraphs. Moreover, the strategy of growing a dfs-path is applicable to digraphs and is adapted here to perform a *topological sort* on a directed acyclic graph.

DEFINITION: A **dag** is a directed acyclic graph, i.e., it has no directed cycles.

DEFINITION: A **topological sort** (or **topological order**) of an n-vertex dag G is a bijection ts from the set $\{1, 2, \ldots, n\}$ to the vertex-set V_G such that every arc e is directed from a smaller number to a larger one, i.e., $ts(tail(e)) < ts(head(e))$.

DEFINITION: A **source** in a dag is a vertex with indegree 0, and a **sink** is a vertex with outdegree 0.

Proposition 4.4.7: *Every dag has at least one source and at least one sink.*

$\diamond$ *(Exercises)*

Remark: Simply using the discovery order (i.e., the *dfnumbers*) produced by a depth-first search will not necessarily be a topological sort because right-to-left cross arcs may occur (Proposition 4.2.2). However, the reverse finish order can be used.

Proposition 4.4.8: *Let G be a dag, and let t be the first vertex that becomes finished during an execution of a depth-first search. Then t is a sink.*

Proof: If vertex t were not a sink, then G would contain a directed cycle. $\diamond$

This last proposition leads to the following high-level algorithm for a topological sort. The algorithm grows a dfs-path, assigns n to the vertex that finishes first, deletes that vertex, and grows another dfs-path on the digraph that remains, etc.

Algorithm 4.4.2:   **Topological Sort**

   Input: an n-vertex dag G.

   Output: a topological sort ts of the vertices of G.

      Initialize graph H as graph G.

      Initialize $k := n$.

      While $V_H \neq \emptyset$

         Start at a source and grow a dfs-path until the first vertex t is

            finished.

         Set $ts(t) = k$

         $H := H - t$.

         $k := k - 1$.

      Return the function ts.

COMPUTATIONAL NOTE: Again, as in Algorithm 4.4.1, efficiency dictates that the dfs-path grown in each iteration should start at the same vertex and use as much of the previous dfs-path as possible.

Remark: Topological sorting is discussed in the context of partial orders in Chapter 9.

Finding the Cut-Vertices of a Connected Graph

REVIEW FROM §2.4: A **cut-vertex** of a connected graph G is an vertex v such that the vertex-deletion subgraph $G - v$ has more components than G.

The next two characterizations of a cut-vertex lead to a method for finding cut-vertices.

Proposition 4.4.9: *A vertex v in a connected graph is a cut-vertex if and only if there exist two distinct vertices u and w, both different from v, such that v is on every u-w path in the graph.*

Proof: *Necessity* ($\Rightarrow$) If v is a cut-vertex of a connected graph G, then there are vertices, say u and w, in separate components of $G - v$. It follows that every u-w path in G must contain v.

Sufficiency ($\Leftarrow$) If u and w are vertices of a connected graph G, such that every u-w path in G contains v, then $G - v$ contains no u-w path. Thus, u and w are in different components of $G - v$.   ◇

Example 4.4.4: In graph G of Figure 4.4.3, vertices v and x are the only cut-vertices. The rooted tree on the left represents one of the possible depth-first trees (depending on how ties are resolved) that could result when the search starts at cut-vertex v. The rooted tree on the right is generated by starting the search at the non-cut-vertex w. Observe that vertex w as the root has only one child, whereas v as the root has two children. Also notice that v as a non-root in the tree on the right has a child, none of whose descendants (including itself) is joined to any proper ancestors of v. The only other non-root in either search tree that has this property is the other cut-vertex x.

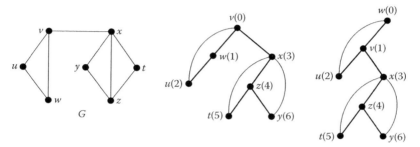

Figure 4.4.3   Depth-first trees with two different roots.

These observations lead to the next two propositions, which show how a depth-first tree and its non-tree edges can be used to determine whether a given vertex is a cut-vertex.

Proposition 4.4.10: *Let tree T be the result of applying depth-first search to a connected graph G. Then the root r of T is a cut-vertex of G if and only if r has more than one child in T.*

Proof: *Necessity* ($\Rightarrow$) By way of contrapositive, suppose that the root r has only one child w in T. Then all other vertices are descendants of w. Thus, the subtree rooted at w is a spanning tree of $G - r$, and, hence, $G - r$ is connected.

Sufficiency ($\Leftarrow$) Suppose that x and y are two of the children of r. Neither x nor y is an ancestor of the other; furthermore, no descendant of x is an ancestor or descendant of any descendant of y (see Figure 4.4.4). Thus, by Proposition 4.2.1, there is no non-tree edge joining any vertex in the subtree rooted at x to any vertex in the subtree rooted at y. This implies that every x-y path in G must go through r, and, hence, by Proposition 4.4.9, r is a cut-vertex.   ◇

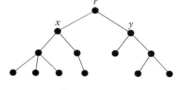

Figure 4.4.4

Proposition 4.4.11: *Let tree T be the result of applying depth-first search to a connected graph G. Then a non-root v of T is a cut-vertex of G if and only if v has a child w such that no descendant of w is joined to a proper ancestor of v by a non-tree edge.*

Proof: *Necessity* ($\Rightarrow$) By way of contrapositive, suppose that every child of v has a descendant joined to a proper ancestor of v (see Figure 4.4.5). First suppose that x and y are any two proper descendants of v. We claim that there exists an x-y path that does not contain v. If x and y are descendants of the same child of v, then the claim is trivially true. Otherwise, x and y are descendants of two different children of v, say w_1 and w_2, respectively. Vertex w_1 has a descendant d_1 joined by a non-tree edge to a proper ancestor of v, say a_1. Similarly, w_2 has a descendant d_2 joined by a non-tree edge to a proper ancestor of v, say a_2 (see Figure 4.4.5). Then the concatenation of the x-w_1 path, the w_1-d_1 path, edge $d_1 a_1$, the a_1-a_2 path, edge $a_2 d_2$, the d_2-w_2 path, and the w_2-y path forms an x-y path that does not contain v.

A similar argument shows that for any proper ancestor of v and any proper descendant of v, there is a path between them that avoids v (see Exercises). It follows that there is a path from any proper descendant of v to any vertex in a different subtree of r. Hence, by Proposition 4.4.9, v is not a cut-vertex.

Sufficiency ($\Leftarrow$) Suppose that v has a child w such that no descendant of w is joined to an ancestor of v by a non-tree edge. Then every path in G from w to r must pass through v, and, hence, v is a cut-vertex. $\Diamond$

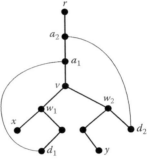

Figure 4.4.5

Characterizing Cut-Vertices in Terms of the *dfnumbers*

By Propositions 4.2.1 and 4.4.11, it follows that a non-root v of a depth-first search tree is a cut-vertex if and only if v has a child w such that no descendant of w is joined by a non-tree edge to a vertex whose *dfnumber* is smaller than *dfnumber*(v).

NOTATION: Let *low*(w) denote the smaller of *dfnumber*(w) and the smallest *dfnumber* among all vertices joined by a non-tree edge to some descendant of w.

Example 4.4.5: The *dfnumber* and *low* values are computed for the example graph in Figure 4.4.6 and are shown in parentheses with the *dfnumber* listed first.

Proposition 4.4.11 may now be restated as the following corollary, on which a scheme for finding the cut-vertices (Algorithm 4.4.3) is based.

Corollary 4.4.12: *Let tree T be the result of applying depth-first search to a connected graph. Then a non-root v of T is a cut-vertex if and only if v has a child w such that* *low*$(w) \geq$ *dfnumber*(v). $\Diamond$

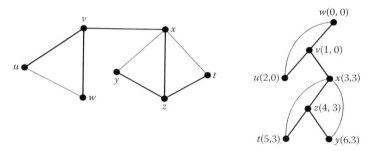

Figure 4.4.6   The *dfnumber* and *low* values for a depth-first search.

Algorithm 4.4.3:   Finding Cut-Vertices

Input: a connected graph G.
Output: a set K of the cut-vertices of graph G.
      Initialize the set K of cut-vertices as empty.
      Choose an arbitrary vertex r of graph G.
      Execute a depth-first search of graph G, starting at vertex r.
      Let T be the output tree.
      If root r has more than one child in T
            Add vertex r to set K.
      For each vertex w
            Compute $low(w)$.
      For each non-root v
            If there is a child w of v such that $low(w) \geq dfnumber(v)$
                  Add vertex v to set K.
      Return set K.

COMPUTATIONAL NOTE: It is not hard to show that for any vertex v, $low(v)$ is the minimum of the following values:

a. $dfnumber(v)$,

b. $dfnumber(z)$ for any z joined to v by a non-tree edge, and

c. $low(y)$ for any child y of v.

This relationship allows the computation of the *low* values to take place *during* the depth-first search. In particular, when the vertex v has been backtracked to for the last time, the *low* values of each of the children of v will have already been computed, allowing $low(v)$ then to be computed. (For further details, see [AhHoU183].)

We close this section with an example illustrating how depth-first search can arise in unexpected ways.

Escaping from a Maze: Tarry's Algorithm

Suppose you wake up inside a room in a maze consisting of a network of tunnels and rooms. At the end of each tunnel is an open doorway through which you can enter or exit a room. Your goal is to find your way to a *freedom* room from which you can exit the maze, if such a room exists, or determine that no such room exists. Each room has a piece of marking chalk and nothing else.

Gaston Tarry published a maze-tracing algorithm in 1895, now known as **Tarry's algorithm**, that avoids the danger of cycling forever. The algorithm is based on the following strategy: *never backtrack through a tunnel unless there's no alternative, and never go through a tunnel a second time in the same direction.* Thus, you must always traverse an unexplored tunnel before you resort to backtracking, which makes the strategy a variation of a depth-first search.

The chief difficulty here is in keeping track of which tunnels have already been traversed in a given direction. For an ordinary computer implementation of the corresponding graph, this problem is easily solved with the kind of routine bookkeeping that we have alluded to earlier. But without a map, there is no graph to store, and hence, all you can do is keep *local information* at each room. Tarry's solution involves marking the doorways according to the following rules:

a. If you are in a room that is not the freedom room, and it has at least one unmarked doorway, then pass through one of the unmarked doorways to a tunnel, mark that doorway OUT, traverse the tunnel, pass through the doorway at the other end, and enter the room there.

b. If you pass through a doorway and enter a room that is not the freedom room, and it has all doorways unmarked, mark the doorway you passed through IN.

c. If you pass through a doorway and enter a room that is not the freedom room, and it has all doorways marked, pass through a doorway marked IN, if one exists, traverse the tunnel, pass through the doorway at the other end, and enter the room there.

d. If you pass through a doorway and enter a room with all doorways marked OUT, stop.

Proposition 4.4.13: *If there is no freedom room in the maze, Tarry's algorithm will stop after each tunnel is traversed exactly twice, once in each direction.*   ◇ *(Exercises)*

EXERCISES for Section 4.4

In Exercises 4.4.1 through 4.4.7, (a) give the dfnumber and low values for each vertex that result from a depth-first search on the graph shown, starting at the specified vertex. Use the alphabetical order of the vertex, names as the default priority; (b) verify the characterization given by Corollary 4.4.12 for the calculations of part (a).

4.4.1[S] Vertex a.

4.4.2 Vertex b.

4.4.3 Vertex c.

4.4.4 Vertex e.

4.4.5[S] Vertex g.

4.4.6 Vertex i.

4.4.7 Vertex f.

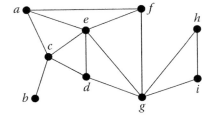

4.4.8 Prove that a post-order traversal of a depth-first tree reproduces the finish order of the depth-first search.

4.4.9 Show that the assertion in Corollary 4.4.3 is not true for all non-simple graphs.

4.4.10 Prove Proposition 4.4.4.

4.4.11 Prove Proposition 4.4.5.

4.4.12 Prove Proposition 4.4.6.

4.4.13 Prove Proposition 4.4.7.

4.4.14 Complete the *Necessity* part of the proof of Proposition 4.4.11 by showing that for any ancestor of vertex v and any descendant of v, there is a path between them that avoids v.

4.4.15 Prove Proposition 4.4.13.

4.4.16 [*Computer Project*] Implement Algorithm 4.4.1, and test it on the graph used for Exercise 4.4.1.

4.4.17 [*Computer Project*] Implement Algorithm 4.4.3, and test it on the graph used for Exercises 4.4.1 through 4.4.7. Execute the program seven times, one for each of the starting vertices specified in those exercises.

4.5   CYCLES, EDGE-CUTS, AND SPANNING TREES

This section lays the foundation for the chapter's final two sections, where an algebraic structure for the cycles, edge-cuts, and spanning trees of a graph is revealed. Toward that end, we establish here a number of properties that highlight the relationship and a certain duality between the edge-cuts and cycles of a graph.

The first two propositions can be proved using properties of tree-growing (e.g., Corollary 4.1.4). However, the proofs presented here are more in keeping with the focus of these final sections.

Proposition 4.5.1: *A graph G is connected if and only if it has a spanning tree.*

Proof: If G is connected, then among the connected, spanning subgraphs of G, there is at least one, say T, with the least number of edges. If subgraph T contained a cycle, then the deletion of one of its cycle-edges would create a smaller connected, spanning subgraph, contradicting the minimality of T. Thus, T is acyclic and therefore, a spanning tree of graph G.

Conversely, if G contains a spanning tree, then every pair of vertices is connected by a path in the tree, and, hence, G is connected.   ◇

Proposition 4.5.2: *A subgraph H of a connected graph G is a subgraph of some spanning tree if and only if H is acyclic.*

Proof: *Necessity* ($\Rightarrow$) A subgraph of a tree is acyclic by definition.

Sufficiency ($\Leftarrow$) Let H be an acyclic subgraph of G, and let T be any spanning tree of G. Consider the connected, spanning subgraph G_1, where $V_{G_1} = V_T \cup V_H$ and $E_{G_1} = E_T \cup E_H$. If G_1 is acyclic, then every edge of H must already be an edge of the spanning tree T, since otherwise a cycle would have been created (by the characterization theorem in §3.1). Thus, subgraph H is contained in tree T, and we are done. So suppose alternatively that G_1 has a cycle C_1. Since H is acyclic, it follows that some edge e_1 of cycle C_1 is not in H. Then the graph $G_2 = G_1 - e_1$ is still a connected, spanning subgraph of G and still contains H as a subgraph. If G_2 is acyclic, then it is the required spanning tree. Otherwise, repeat the process of removing a cycle-edge not in H until a spanning tree is obtained.   ◇

Partition-Cuts and Minimal Edge-Cuts

REVIEW FROM §2.4: An **edge-cut** S in a graph G is a set of edges such that the edge-deletion subgraph $G - S$ has more components than G.

The following definition is closely linked to the concept of a minimal edge-cut. It is used in this and the next section, and it also plays a central role in the study of *network flows*, which appears in Chapter 10.

DEFINITION: Let G be a graph, and let X_1 and X_2 form a partition of V_G. The set of all edges of G having one endpoint in X_1 and the other endpoint in X_2 is called a **partition-cut** of G and is denoted $\langle X_1, X_2 \rangle$.

The next two propositions make explicit the relationship between partition-cuts and minimal edge-cuts.

Proposition 4.5.3: *Let $\langle X_1, X_2 \rangle$ be a partition-cut of a connected graph G. If the subgraphs of G induced by the vertex sets X_1 and X_2 are connected, then $\langle X_1, X_2 \rangle$ is a minimal edge-cut.*

Proof: The partition-cut $\langle X_1, X_2 \rangle$ is an edge-cut of G, since X_1 and X_2 lie in different components of $G - \langle X_1, X_2 \rangle$. Let S be a proper subset of $\langle X_1, X_2 \rangle$, and let edge $e \in \langle X_1, X_2 \rangle - S$. By definition of $\langle X_1, X_2 \rangle$, one endpoint of e is in X_1 and the other endpoint is in X_2. Thus, if the subgraphs induced by the vertex sets X_1 and X_2 are connected, then $G - S$ is connected. Therefore, S is not an edge-cut of G, which implies that $\langle X_1, X_2 \rangle$ is a minimal edge-cut.   ◇

Proposition 4.5.4: *Let S be a minimal edge-cut of a connected graph G, and let X_1 and X_2 be the vertex-sets of the two components of $G - S$. Then $S = \langle X_1, X_2 \rangle$.*

Proof: By minimality, $S \subseteq \langle X_1, X_2 \rangle$. If $e \in \langle X_1, X_2 \rangle - S$, then its endpoints would lie in different components of $G - \langle X_1, X_2 \rangle$, contradicting the definition of S as an edge-cut. ◇

Remark: The premise of Proposition 4.5.4 assumes that the removal of a minimal edge-cut from a connected graph creates *exactly* two components. This is a generalization of Corollary 2.4.3 and can be argued similarly, using the minimality condition (see Exercises).

Proposition 4.5.5: *A partition-cut $\langle X_1, X_2 \rangle$ in a connected graph G is a minimal edge-cut of G or a union of edge-disjoint minimal edge-cuts.*

Proof: Since $\langle X_1, X_2 \rangle$ is an edge-cut of G, it must contain a minimal edge-cut, say S_1. If $\langle X_1, X_2 \rangle \neq S_1$, then let $e \in \langle X_1, X_2 \rangle - S_1$, where the endpoints v_1 and v_2 of e lie in X_1 and X_2, respectively. Since S_1 is a minimal edge-cut, the X_1-endpoints of S_1 are in one of the components of $G - S_1$, and the X_2-endpoints of S_1 are in the other component. Furthermore, v_1 and v_2 are in the same component of $G - S_1$ (since $e \in G - S_1$). Suppose, without loss of generality, that v_1 and v_2 are in the same component as the X_1-endpoints of S_1 (see Figure 4.5.1).

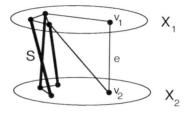

Figure 4.5.1

Then every path in G from v_1 to v_2 must use at least one edge of $\langle X_1, X_2 \rangle - S_1$. Thus, $\langle X_1, X_2 \rangle - S_1$ is an edge-cut of G and, hence, contains a minimal edge-cut S_2. Applying the same argument, $\langle X_1, X_2 \rangle - (S_1 \cup S_2)$ either is empty or is an edge-cut of G. Eventually, the process ends with $\langle X_1, X_2 \rangle - (S_1 \cup S_2 \cup \ldots S_r) = \emptyset$, where the S_i are edge-disjoint minimal edge-cuts of G. $\diamondsuit$

Fundamental Cycles and Fundamental Edge-Cuts

DEFINITION: Let G be a graph with $c(G)$ components. The **edge-cut rank** of G is the number of edges in a full spanning forest of G. Thus, by Corollary 3.1.4, the edge-cut rank equals $|V_G| - c(G)$.

REVIEW FROM §2.4: Let H be a fixed subgraph of a graph G. The **relative complement** of H (**in** G), denoted $G - H$, is the edge-deletion subgraph $G - E(H)$.

DEFINITION: Let G be a graph with $c(G)$ components. The **cycle rank** (or **Betti number**) of G, denoted $\beta(G)$, is the number of edges in the relative complement of a full spanning forest of G. Thus, the cycle rank is $\beta(G) = |E_G| - |V_G| + c(G)$.

Remark: Observe that *all* of the edges in the relative complement of a spanning forest could be removed without increasing the number of components. Thus, the cycle rank $\beta(G)$ equals the maximum number of edges that can be removed from G without increasing the number of components. Therefore, $\beta(G)$ is a measure of the *edge redundancy* with respect to the graph's connectedness.

REVIEW FROM §4.1: A **full spanning forest** of a graph G is a spanning forest consisting of a collection of trees, such that each tree is a spanning tree of a different component of G.

DEFINITION: Let F be a full spanning forest of a graph G, and let e be any edge in the relative complement of forest F. The cycle in the subgraph $F+e$ (existence and uniqueness guaranteed by the characterization theorem in §3.1) is called a **fundamental cycle of** G (**associated with the spanning forest** F).

Remark: Each of the edges in the relative complement of a full spanning forest F gives rise to a *different* fundamental cycle.

DEFINITION: The **fundamental system of cycles** associated with a full spanning forest F of a graph G is the set of all fundamental cycles of G associated with F.

By the remark above, the cardinality of the fundamental system of cycles of G associated with a given full spanning forest of G is the cycle rank $\beta(G)$.

DEFINITION: Let F be a full spanning forest of a graph G, and let e be any edge of F. Let V_1 and V_2 be the vertex-sets of the two new components of the edge-deletion subgraph

$F - e$. Then the partition-cut $\langle V_1, V_2 \rangle$ which is a minimal edge-cut of G by Proposition 4.5.3, is called a **fundamental edge-cut** (*associated with* F).

Remark: For each edge of F, its deletion gives rise to a different fundamental edge-cut.

DEFINITION: The **fundamental system of edge-cuts** associated with a full spanning forest F is the set of all fundamental edge-cuts associated with F.

Thus, the cardinality of the fundamental system of edge-cuts associated with a given full spanning forest of G is the edge-cut rank of G.

Remark: If F is a full spanning forest of a graph G, then each of the components of F is a spanning tree of the corresponding component of G. Since the removal or addition of an edge in a general graph affects only one of its components, the definitions of fundamental cycle and fundamental edge-cut are sometimes given in terms of a spanning tree of a connected graph. All of the remaining assertions in this section are stated in terms of a connected graph but can easily be restated for graphs having two or more components.

Example 4.5.1: Figure 4.5.2 below shows a fundamental system of cycles and a fundamental system of edge-cuts for a graph G. Both systems are associated with the spanning tree whose edges are drawn in bold. Notice that the fundamental system of edge-cuts does not contain every minimal edge-cut of graph G. For instance, the edge-cut consisting of the three edges incident on vertex v is a minimal one but is not in the fundamental system.

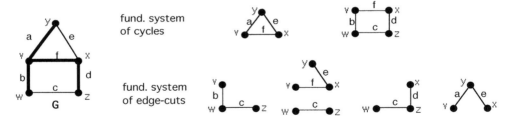

Figure 4.5.2 **Fundamental system of cycles and of edge-cuts.**

Relationship between Cycles and Edge-Cuts

The next series of propositions reveals a dual relationship between the cycles and minimal edge-cuts of a graph. This duality is an instance of a more general property that holds for certain abstract structures called *matroids*, which are introduced in §4.7.

Proposition 4.5.6: *Let S be a set of edges in a connected graph G. Then S is an edge-cut of G if and only if every spanning tree of G has at least one edge in common with S.*

Proof: By Proposition 4.5.1, S is an edge-cut if and only if $G - S$ contains no spanning tree of G, which means that every spanning tree of G has at least one edge in common with S.   ◇

Proposition 4.5.7: *Let C be a set of edges in a connected graph G. Then C contains a cycle if and only if the relative complement of every spanning tree of G has at least one edge in common with C.*

Proof: By Proposition 4.5.2, edge set C contains a cycle if and only if C is not contained in any spanning tree of G, which means that the relative complement of every spanning tree of G has at least one edge in common with C. $\diamond$

Proposition 4.5.8: *A cycle and a minimal edge-cut of a connected graph have an even number of edges in common.*

Proof: Let C be a cycle and S be a minimal edge-cut of a connected graph G. Let V_1 and V_2 be the vertex-sets of the two components G_1 and G_2 of $G - S$. Then each edge of S joins a vertex in V_1 to a vertex in V_2. Now consider a traversal of the edges of the cycle C. Without loss of generality, assume that the traversal begins at some vertex in V_1. Then each time the traversal uses an edge in S in moving from V_1 to V_2, it will have to return to V_2 by traversing another edge of S. This is possible only if C and S have an even number of edges in common. $\diamond$

Example 4.5.1 continued: It is easy but tedious to check that each of the three cycles in graph G of Figure 4.5.2 has either zero or two edges in common with each minimal edge-cut of G.

Proposition 4.5.9: *Let T be a spanning tree of a connected graph, and let C be a fundamental cycle with respect to an edge e^* in the relative complement of T. Then the edge-set of cycle C consists of edge e^* and those edges of tree T whose fundamental edge-cuts contain e^*.*

Proof: Let e_1, e_2, $\ldots$, e_k be the edges of T that, with e^*, make up the cycle C, and let S_i be the fundamental edge-cut with respect to e_i, $1 \le i \le k$.

Edge e_i is the only edge of T common to both C and S_i (by the definitions of C and S_i). By Proposition 4.5.8, C and S_i must have an even number of edges in common, and, hence, there must be an edge in the relative complement of T that is also common to both C and S_i. But e^* is the only edge in the complement of T that is in C. Thus, the fundamental edge-cut S_i must contain e^*, $1 \le i \le k$.

To complete the proof, we must show that no other fundamental edge-cuts associated with T contain e^*. So let S be the fundamental edge-cut with respect to some edge b of T, different from e_1, e_2, $\ldots$, e_k. Then S does not contain any of the edges e_1, e_2, $\ldots$, e_k (by definition of S). The only other edge of cycle C is e^*, so by Proposition 4.5.8, edge-cut S cannot contain e^*. $\diamond$

Example 4.5.1 continued: The 3-cycle in Figure 4.5.2 is the fundamental cycle obtained by adding edge e to the given spanning tree. Of the four fundamental edge-cuts associated with that spanning tree, only the second and fourth ones contain edge e. The tree edges in these two edge-cuts, namely f and a, are the other two edges of the 3-cycle.

The proof of the next proposition uses an argument similar to the one just given and is left as an exercise.

Proposition 4.5.10: *The fundamental edge-cut with respect to an edge e of a spanning tree T consists of e and exactly those edges in the relative complement of T whose fundamental cycles contain e.* $\diamond$ *(Exercises)*

This section closes with a characterization of Eulerian graphs that dates back to as early as 1736, when Euler solved and generalized the Königsberg Bridge Problem (§6.1). The result is used in the next section, but it is important in its own right and is essential for parts of Chapter 6.

REVIEW FROM §1.5: An **Eulerian tour** in a graph is a closed trail that contains every edge of that graph. An **Eulerian graph** is a graph that has an Eulerian tour.

Theorem 4.5.11: [**Eulerian-Graph Characterization**] *The following statements are equivalent for a connected graph G.*

 1. *G is Eulerian.*
 2. *The degree of every vertex in G is even.*
 3. *E_G is the union of the edge-sets of a set of edge-disjoint cycles of G.*

Proof:

$(1 \Rightarrow 2)$ Let C be an Eulerian tour of G, and let v be the starting point of some traversal of C. The initial edge and final edge of the traversal contribute 2 toward the degree of v, and each time the traversal passes through a vertex, a contribution of 2 toward that vertex's degree also results. Thus, there is an even number of traversed edges incident with each vertex, and since each edge of G is traversed exactly once, this number is the degree of that vertex.

$(2 \Rightarrow 3)$ Since G is connected and every vertex has even degree, G cannot be a tree and therefore contains a cycle, say C_1. If C_1 includes all the edges of G, then statement 3 is established. Otherwise, the graph $G_1 = G - E_{C_1}$ has at least one nontrivial component. Furthermore, since the edges that were deleted from G form a cycle, the degree in G_1 of each vertex on C_1 is reduced by 2, and, hence, every vertex of G_1 still has even degree. It follows that G_1 contains a cycle C_2. If all of the edges have been exhausted, then $E_G = E_{C_1} \cup E_{C_2}$. Otherwise, consider $G_2 = G - E_{C_1} - E_{C_2}$, and continue the procedure until all the edges are exhausted. If C_n is the cycle obtained at the last step, then $E_G = E_{C_1} \cup E_{C_2} \cup \ldots \cup E_{C_n}$, which completes the proof of the implication $2 \Rightarrow 3$.

$(3 \Rightarrow 1)$ Assume that E_G is the union of the edge-sets of m edge-disjoint cycles of G. Start at any vertex v_1 on one of these cycles, say C_1, and consider $T_1 = C_1$ as our first closed trail. There must be some vertex of T_1, say v_2, that is also a vertex on some other cycle, say C_2. Form a closed trail T_2 by *splicing* cycle C_2 into T_1 at vertex v_2. That is, trail T_2 is formed by starting at vertex v_1, traversing the edges of T_1 until v_2 is reached, traversing all the edges of cycle C_2, and completing the closed trail T_2 by traversing the remaining edges of trail T_1. The process continues until all the cycles have been spliced in, at which point T_m is an Eulerian tour of G.

$\diamondsuit$

EXERCISES for Section 4.5

In Exercises 4.5.1 through 4.5.5, consider the spanning tree T with the specified edge subset of the following graph G. (a) Find the fundamental system of cycles associated with T. (b) Find the fundamental system of edge-cuts associated with T.

4.5.1^S $\{a, e, c, d\}$

4.5.2 $\{a, e, g, d\}$

4.5.3 $\{e, f, c, d\}$

4.5.4 $\{a, b, g, c\}$

4.5.5 $\{e, f, g, d\}$

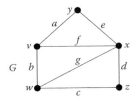

In Exercises 4.5.6 through 4.5.9, consider the fundamental systems of cycles and edge-cuts obtained in the specified exercise. (a) Show that Proposition 4.5.6 holds for one of the fundamental edge-cuts. (b) Show that Proposition 4.5.7 holds for one of the fundamental cycles. (c) Show that the 5-cycle in graph G has an even number of edges in common with each of the fundamental edge-cuts.

4.5.6[S] Exercise 4.5.1.   **4.5.7** Exercise 4.5.2.

4.5.8 Exercise 4.5.3.   **4.5.9** Exercise 4.5.4.

In Exercises 4.5.10 through 4.5.13, consider the fundamental systems of cycles and edge-cuts obtained in the specified exercise. (a) Show that Proposition 4.5.9 holds for one of the fundamental cycles. (b) Show that Proposition 4.5.10 holds for one of the fundamental edge-cuts.

4.5.10[S] Exercise 4.5.1.   **4.5.11** Exercise 4.5.2.

4.5.12 Exercise 4.5.3.   **4.5.13** Exercise 4.5.4.

4.5.14 Prove that each edge of a connected graph G lies in a spanning tree of G.

4.5.15 Give an alternative proof of Proposition 4.5.2 that avoids the "repeat the process" phrase by considering a connected spanning subgraph of G that contains H, and that has the least number of edges among all such subgraphs.

4.5.16[S] Prove that in a tree, every vertex of degree greater than 1 is a cut-vertex.

4.5.17 Prove that every nontrivial connected graph contains a minimal edge-cut.

4.5.18 Prove that the removal of a minimal edge-cut from any graph increases the number of components by exactly 1.

4.5.19 Let T_1 and T_2 be two different spanning trees of a graph. Show that if e is any edge in tree T_1, then there exists an edge f in tree T_2 such that $T_1 - e + f$ is also a spanning tree. (Hint: Apply Proposition 4.5.6 to the fundamental edge-cut associated with T_1 and e.)

4.5.20 Prove that a subgraph of a connected graph G is a subgraph of the relative complement of some spanning tree if and only if it contains no edge-cuts of G.

4.5.21 Prove Proposition 4.5.10.

4.5.22[S] Give a counterexample to show that the assertion of Proposition 4.5.8 no longer holds if the minimality condition is removed.

4.6   GRAPHS AND VECTOR SPACES

The results of the previous section are used here to define a vector space associated with a graph. Two important subspaces of this vector space, the *edge-cut space* and the *cycle space*, are also identified and studied in detail. Understanding the algebraic structure underlying a graph's cycles and edge-cuts makes it possible to apply powerful and elegant results from linear algebra that lead to a deeper understanding of graphs and of certain graph algorithms. Supporting this claim is an application, appearing later this section, to a problem concerning electrical circuits.

Vector Space of Edge Subsets

NOTATION: For a graph G, let $W_E(G)$ denote the set of all subsets of E_G.

DEFINITION: The **ring sum** of two elements of $W_E(G)$, say E_1 and E_2, is defined by

$$E_1 \oplus E_2 = (E_1 - E_2) \cup (E_2 - E_1)$$

The next proposition asserts that $W_E(G)$ under the ring-sum operation forms a vector space over the field of scalars $GF(2)$, where the scalar multiplication $*$ is defined by $1 * S$ = S and $0 * S = \emptyset$ for any S in $W_E(G)$. Its proof is a straightforward verification of each of the defining properties of a vector space and is left as an exercise. (The basic definitions and properties of a vector space and of the finite field $GF(2)$ appear in Appendix A.4.)

Proposition 4.6.1: $W_E(G)$ *is a vector space over* $GF(2)$. $\diamond$ *(Exercises)*

TERMINOLOGY: The vector space $W_E(G)$ is called the **edge space** of G.

Proposition 4.6.2: *Let* $E_G = \{e_1, e_2, \ldots, e_m\}$ *be the edge-set of a graph* G. *Then the subsets* $\{e_1\}, \{e_2\}, \ldots, \{e_m\}$ *form a basis for the edge space* $W_E(G)$. *Thus,* $W_E(G)$ *is an* m-*dimensional vector space over* $GF(2)$.

Proof: If $H = \{e_{i_1}, e_{i_2}, \ldots, e_{i_r}\}$ is any vector in $W_E(G)$, then $H = \{e_{i_1}\} \oplus \{e_{i_2}\} \oplus \cdots \oplus \{e_{i_r}\}$. Clearly, the elements $\{e_1\}, \{e_2\}, \ldots, \{e_m\}$ are also linearly independent. $\diamond$

DEFINITION: Let $s_1, s_2 \ldots, s_n$ be any sequence of objects, and let A be a subset of $S = \{s_1, s_2 \ldots, s_n\}$. The **characteristic vector** of the subset A is the n-tuple whose jth component is 1 if $s_j \in A$ and 0 otherwise.

A general result from linear algebra states that every m-dimensional vector space over a given field F is isomorphic to the vector space of m-tuples over F. For the vector space $W_E(G)$, this result may be realized in the following way. If $E_G = \{e_1, e_2, \ldots, e_m\}$, then the mapping *charvec* that assigns to each subset of E_G its *characteristic vector* is an isomorphism from $W_E(G)$ to the vector space of m-tuples over $GF(2)$. Proving this assertion first requires showing that the mapping *preserves the vector-space operations*. If E_1 and E_2 are two subsets of E_G, then the definitions of the ring-sum operator $\oplus$ and mod 2 component-wise addition $+_2$ imply

$$charvec\,(E_1 \oplus E_2) = charvec\,(E_1) +_2 charvec\,(E_2)$$

The remaining details are left to the reader.

Remark: Each subset E_i of E_G uniquely determines a subgraph of G, namely, the edge-induced subgraph $G_i = G(E_i)$. Thus, the vectors of $W_E(G)$ may be viewed as the edge-induced subgraphs G_i instead of as the edge subsets E_i. Accordingly, $1 * G_i = G_i$ and $0 * G_i = \emptyset$, where G_i is any edge-induced subgraph of G and where $\emptyset$ now refers to the *null graph* (with no vertices and no edges). This vector space will still be denoted $W_E(G)$.

Furthermore, the only subgraphs of a graph that are *not* edge-induced subgraphs are those that contain isolated vertices. Since isolated vertices play no role in the discussions in this section and in the section that follows, the adjective "edge-induced" will no longer be used when referring to the elements of the edge space $W_E(G)$.

The Cycle Space of a Graph

DEFINITION: The **cycle space** of a graph G, denoted $W_C(G)$, is the subset of the edge space $W_E(G)$ consisting of the null set (graph) $\emptyset$, all cycles in G, and all unions of edge-disjoint cycles of G.

Example 4.6.1: Figure 4.6.1 shows a graph G and the seven non-null elements of the cycle space $W_C(G)$.

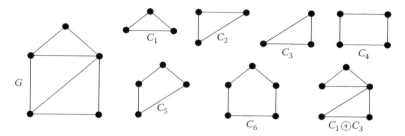

Figure 4.6.1   A graph G and the non-null elements of cycle space $W_C(G)$.

Notice that each vector of $W_C(G)$ is a subgraph having no vertices of odd degree, and that the sum of any two of the vectors of $W_C(G)$ is again a vector of $W_C(G)$. The first of these observations follows directly from the characterization of Eulerian graphs from the previous section (Theorem 4.5.11). The next result shows that the second property of $W_C(G)$ is also true in general.

Proposition 4.6.3: Given a graph G, the cycle space $W_C(G)$ is a subspace of the edge space $W_E(G)$.

Proof: It suffices to show that the elements of $W_C(G)$ are closed under $\oplus$. So consider any two distinct members C_1 and C_2 of $W_C(G)$ and let $C_3 = C_1 \oplus C_2$. By Theorem 4.5.11, it must be shown that $deg_{C_3}(v)$ is even for each vertex v in C_3.

Consider any vertex v in C_3, and let X_i denote the set of edges incident with v in C_i for $i = 1, 2, 3$. Since $|X_i|$ is the degree of v in C_i, $|X_1|$ and $|X_2|$ are both even, and $|X_3|$ is nonzero. Since $C_3 = C_1 \oplus C_2$, $X_3 = X_1 \oplus X_2$. But this implies that $|X_3| = |X_1| + |X_2| - 2|X_1 \cap X_2|$, which shows that $|X_3|$ must be even. $\diamondsuit$

The Edge-Cut Subspace of a Graph

DEFINITION: The **edge-cut space** of a graph G, denoted $W_S(G)$, is the subset of the edge space $W_E(G)$ consisting of the null graph $\emptyset$, all minimal edge-cuts in G, and all unions of edge-disjoint minimal edge-cuts of G.

Example 4.6.2: Figure 4.6.2 shows a graph G and three of its edge-cuts. Since R and T are edge-disjoint minimal ones, all three edge-cuts are in $W_S(G)$.

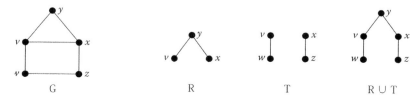

Figure 4.6.2   A graph G and three edge-cuts in $W_S(G)$.

Example 4.6.2 continued: Figure 4.6.3 shows graph G and the 15 non-null elements of $W_S(G)$. The non-minimal edge-cuts appear in the bottom row.

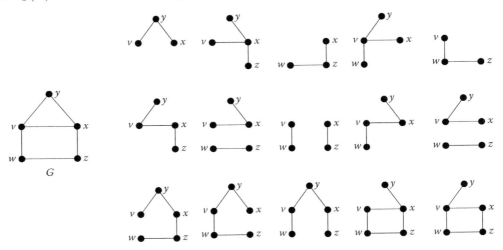

Figure 4.6.3   Graph G and the 15 non-null elements of $W_S(G)$.

Proposition 4.6.4: *Given a graph G, the edge-cut space $W_S(G)$ is closed under the ring-sum operation $\oplus$ and, hence, is a subspace of the edge space $W_E(G)$.*

Proof: By Proposition 4.5.3, it suffices to show that the ring sum of any two partition-cuts in a graph G is also a partition-cut in G. So let $S_1 = \langle X_1, X_2 \rangle$ and $S_2 = \langle X_3, X_4 \rangle$ be any two partition-cuts in G, and let $V_{ij} = X_i \cap X_j$, for $i = 1, 2$ and $j = 3, 4$. Then the V_{ij} are mutually disjoint, with

$$S_1 = \langle V_{13} \cup V_{14}, V_{23} \cup V_{24} \rangle = \langle V_{13}, V_{23} \rangle \cup \langle V_{13}, V_{24} \rangle \cup \langle V_{14}, V_{23} \rangle \cup \langle V_{14}, V_{24} \rangle \quad \text{and}$$
$$S_2 = \langle V_{13} \cup V_{23}, V_{14} \cup V_{24} \rangle = \langle V_{13}, V_{14} \rangle \cup \langle V_{13}, V_{24} \rangle \cup \langle V_{23}, V_{14} \rangle \cup \langle V_{23}, V_{24} \rangle$$

Hence,
$$S_1 \oplus S_2 = \langle V_{13}, V_{23} \rangle \cup \langle V_{14}, V_{24} \rangle \cup \langle V_{13}, V_{14} \rangle \cup \langle V_{23}, V_{24} \rangle$$

But
$$\langle V_{13}, V_{23} \rangle \cup \langle V_{14}, V_{24} \rangle \cup \langle V_{13}, V_{14} \rangle \cup \langle V_{23}, V_{24} \rangle = \langle V_{13} \cup V_{24}, V_{14} \cup V_{23} \rangle$$

which is a partition-cut in G, and the proof is complete. $\diamondsuit$

Example 4.6.3: To illustrate Proposition 4.6.4, it is easy to check that the elements of $W_S(G)$ in Figure 4.6.3 are closed under ring sum $\oplus$ and that each element is a partition-cut. For instance, if S_1 denotes the last element in the second row of five elements, and S_2 and S_3 are the first and last elements of the third row, then $S_1 = \langle \{v, w\}, \{x, y, z\} \rangle$; $S_2 = \langle \{v, x, y\}, \{w, z\} \rangle$; and $S_3 = S_1 \oplus S_2 = \langle \{x, y, w\}, \{v, z\} \rangle$.

Bases for the Cycle and Edge-Cut Spaces

Theorem 4.6.5: *Let T be a spanning tree of a connected graph G. Then the fundamental system of cycles associated with T is a basis for the cycle space $W_C(G)$.*

Proof: By the construction, each fundamental cycle associated with T contains exactly one non-tree edge that is not part of any other fundamental cycle associated with T.

Thus, no fundamental cycle is a ring sum of some or all of the other fundamental cycles. Hence, the fundamental system of cycles is a linearly independent set.

To show the fundamental system of cycles spans $W_C(G)$, suppose H is any element of $W_C(G)$. Now let $e_1, e_2, \ldots, e_r$ be the non-tree edges of H, and let C_i be the fundamental cycle in $T + e_i$, $i = 1, \ldots, r$. The completion of the proof requires showing that $H = C_1 \oplus C_2 \oplus \cdots \oplus C_r$, or equivalently, that $B = H \oplus C_1 \oplus C_2 \oplus \cdots \oplus C_r$. is the null graph.

Since each e_i appears only in C_i, and e_i is the only non-tree edge in C_i, B contains no non-tree edges. Thus, B is a subgraph of T and is therefore acyclic. But B is an element of $W_C(G)$ (since $W_C(G)$ is closed under ring sum), and the only element of $W_C(G)$ that is acyclic is the null graph.   ◇

Example 4.6.1 continued: Figure 4.6.4 shows a spanning tree and the associated fundamental system of cycles $\{C_3, C_4, C_6\}$ for the graph of Figure 4.6.1. Each of the other four non-null elements of cycle space $W_C(G)$ can be expressed as the ring sum of some or all of the fundamental cycles. In particular,

$$
\begin{aligned}
C_1 &= C_4 \oplus C_6 \\
C_2 &= C_3 \oplus C_4 \\
C_5 &= C_3 \oplus C_6 \\
C_1 \oplus C_3 &= C_3 \oplus C_4 \oplus C_6
\end{aligned}
$$

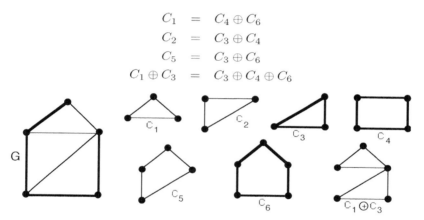

Figure 4.6.4   Fundamental system $\{C_3, C_4, C_6\}$ is a basis for $W_C(G)$.

The next theorem and its proof are analogous to Theorem 4.6.5 and its proof.

Theorem 4.6.6: *Let T be a spanning tree of a connected graph G. Then the fundamental system of edge-cuts associated with T is a basis for the edge-cut space $W_S(G)$.*
                                                                        ◇ *(Exercises)*

Example 4.6.2 continued: Figure 4.6.5 shows a spanning tree and the associated fundamental system of edge-cuts $\{S_1, S_2, S_3, S_4\}$ for the graph of Figure 4.6.3. It is easy to verify that each of the 15 non-null elements of the edge-cut space $W_S(G)$ can be expressed as the ring sum of some or all of the fundamental edge-cuts S_1, S_2, S_3, S_4 (see Exercises). For instance, the edge-cut appearing on the lower right in Figure 4.6.3 is equal to $S_1 \oplus S_2 \oplus S_3 \oplus S_4$.

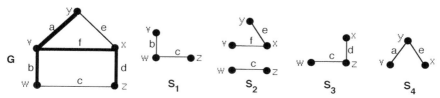

Figure 4.6.5   Fundamental system $\{S_1, S_2, S_3, S_4\}$ is a basis for $W_S(G)$.

Application 4.6.1:   *Applying Ohm's and Kirchhoff's Laws*   Suppose that the graph shown in Figure 4.6.6 represents an electrical network with a given voltage E on wire e_1, oriented as shown. Also let R_j be the resistance on e_j.

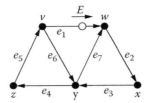

Figure 4.6.6   An electrical network.

The problem is to determine the electric current i_j for wire e_j, using Ohm's Law, Kirchhoff's current law (KCL), and Kirchhoff's voltage law (KVL), given by:

- Ohm's Law   For a current i flowing through a resistance r, the voltage drop v across the resistance satisfies $v = ir$.
- KCL   The algebraic sum of the currents at each vertex is zero.
- KVL   The algebraic sum of voltage drops around any cycle is zero.

To apply these laws, a direction must be assigned to the current in each wire. These directions are arbitrary and do not affect the final solution, in the sense that a negative value for an i_j simply means that the direction of flow is opposite to the direction assigned for e_j.

Illustration: Figure 4.6.7 shows an arbitrary assignment of directions for the wires in the example network.

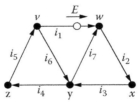

Figure 4.6.7   An electrical network.

The five (KCL)-equations corresponding to the five vertices are

$$-i_1 + i_5 - i_6 = 0$$
$$i_2 - i_3 = 0$$
$$i_4 - i_5 = 0$$
$$i_3 - i_4 + i_6 - i_7 = 0$$
$$i_1 - i_2 + i_7 = 0$$

Notice that the sum of these equations is the equation $0 = 0$, which indicates that one of them is redundant. Furthermore, if one circuit is the sum of other circuits, then its (KVL)-equation is redundant. For instance, the (KVL)-equations for the circuits $\langle v, w, y, v \rangle$, $\langle w, x, y, w \rangle$, and $\langle v, w, x, y, v \rangle$ are, respectively,

$$i_1 R_1 - i_7 R_7 - i_6 R_6 - E = 0$$
$$i_2 R_2 + i_3 R_3 + i_7 R_7 = 0$$
$$i_1 R_1 + i_2 R_2 + i_3 R_3 - i_6 R_6 - E = 0$$

The third equation is the sum of the first two, since the third circuit is the sum of the first two circuits.

A large network is likely to have a huge number of these redundancies. The objective for an efficient solution strategy is to find a minimal set of circuits whose corresponding equations, together with the equations from Kirchhoff's circuit law, are just enough to solve for the i_j's. Since a fundamental system of cycles is a basis for the cycle space, their corresponding equations will constitute a full set of linearly independent equations and will meet the objective.

Illustration: One of the spanning trees of the example network is shown in Figure 4.6.8 below. The fundamental system of cycles and their corresponding equations for this spanning tree are as follows:

$$\langle z, v, w, y, z \rangle : \quad i_5 R_5 + i_1 R_1 - i_7 R_7 + i_4 R_4 - E = 0$$
$$\langle v, w, y, v \rangle : \quad i_1 R_1 - i_7 R_7 - i_6 R_6 - E = 0$$
$$\langle w, x, y, w \rangle : \quad i_2 R_2 + i_3 R_3 + i_7 R_7 = 0$$

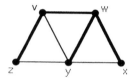

Figure 4.6.8   A spanning tree for the electrical network.

These three equations, together with any four of the five (KCL)-equations will determine the i_j's. If, for example, the voltage $E = 28$ and all the R_j's are taken to be 1, then it is easy to verify that the solution for the currents is

$$i_1 = 12; \quad i_2 = i_3 = 4; \quad i_4 = i_5 = 4; \quad i_6 = i_7 = -8$$

Dimension of the Cycle Space and of the Edge-Cut Space

In light of the discussion and remark preceding Example 4.5.1 in §4.5, Theorems 4.6.5 and 4.6.6 extend easily to non-connected graphs. For such graphs, the fundamental systems are associated with a full spanning forest instead of a spanning tree. The resulting corollary establishes the dimensions of the cycle space and the edge-cut space, thereby justifying the use of the terms *cycle rank* and *edge-cut rank*.

Corollary 4.6.7: *Let G be a graph with $c(G)$ components. Then the dimension of the cycle space $W_C(G)$ is the cycle rank $\beta(G) = |E_G| - |V_G| + c(G)$, and the dimension of the edge-cut space $W_S(G)$ is the edge-cut rank $|V_G| - c(G)$.*

Proof: This follows directly from Theorems 4.6.5 and 4.6.6 and the definitions of cycle rank and edge-cut rank. $\diamond$

Relationship between the Cycle and Edge-Cut Spaces

The next two propositions characterize the elements of the cycle space and the edge-cut space of a graph in terms of each other. The proof of the second characterization is similar to that of the first and is left as an exercise.

Proposition 4.6.8: *A subgraph H of a graph G is in the cycle space $W_C(G)$ if and only if it has an even number of edges in common with every subgraph in the edge-cut space $W_S(G)$.*

Proof: *Necessity* ($\Rightarrow$) Each subgraph in the cycle space is a union of edge-disjoint cycles, and each subgraph in the edge-cut space is a union of edge-disjoint edge-cuts. Thus, necessity follows from Proposition 4.5.8.
Sufficiency ($\Leftarrow$) It may be assumed without loss of generality that G is connected, since the argument that follows may be applied separately to each of the components of G if G is not connected.

Suppose H has an even number of edges in common with each subgraph in the edge-cut space of G, and let T be a spanning tree of G. Let $e_1, e_2, \ldots, e_r$ be the non-tree edges of H, and consider $C = C_1 \oplus C_2 \oplus \ldots \oplus C_r$, where each C_i is the fundamental cycle in $T + e_i$, $i = 1, \ldots, r$. Arguing as in the proof of Theorem 4.6.5, $H \oplus C$ has no non-tree edges. Thus, the only possible edges in $H \oplus C$ are edges of T. So suppose b is an edge in both T and $H \oplus C$. Now if S is the fundamental edge-cut associated with b, then b is the only edge in $H \oplus C \oplus S$. But, since $C \in W_C(G), C$ has an even number of edges in common with each subgraph in the edge-cut space of G, as does H. This implies that $H \oplus C$ has an even number of edges in common with each subgraph in the edge-cut space. In particular, $H \oplus C \oplus S$ must have an even number of edges. This contradiction shows that $H \oplus C$ must be the null graph, that is, $H = C$. Hence, H is a subgraph in the cycle space of G. $\diamond$

Proposition 4.6.9: *A subgraph H of a graph G is an element of the edge-cut space of G if and only if it has an even number of edges in common with every subgraph in the cycle space of G.* $\diamond$ *(Exercises)*

Orthogonality of the Cycle and Edge-Cut Spaces

DEFINITION: Two vectors in the edge space $W_E(G)$ are said to be **orthogonal** if the dot product of their characteristic vectors equals 0 in $GF(2)$. Thus, two subsets of edges are orthogonal if they have an even number of edges in common.

Theorem 4.6.10: *Given a graph G, the cycle space $W_C(G)$ and the edge-cut space $W_S(G)$ are orthogonal subspaces of $W_E(G)$.*

Proof: Let H be a subgraph in the cycle space, and let K be a subgraph in the edge-cut space. By either Proposition 4.6.8 or Proposition 4.6.9, H and K have an even number of edges in common and, hence, are orthogonal. $\diamond$

Theorem 4.6.11: *Given a graph G, the cycle space $W_C(G)$ and the edge-cut space $W_S(G)$ are orthogonal complements in $W_E(G)$ if and only if $W_C(G) \cap W_S(G) = \{\emptyset\}$.*

Proof: If $W_C(G) \oplus W_S(G)$ denotes the direct sum of the subspaces $W_C(G)$ and $W_S(G)$, then

$$dim\,(W_C(G) \oplus W_S(G)) = dim\,(W_C(G)) + dim\,(W_S(G)) - dim\,(W_C(G) \cap W_S(G))$$

By Corollary 4.6.7, $dim\,(W_C(G)) + dim\,(W_S(G)) = |E_G|$. Thus, $dim\,(W_C(G) \oplus W_S(G)) = |E_G| = dim\,(W_E(G))$ if and only if $dim\,(W_C(G) \cap W_S(G)) = 0$. $\diamond$

Vector Space of Vertex Subsets

This section concludes by showing that a spanning tree of a simple graph G corresponds to a basis of the column space of the incidence matrix of G.

FROM LINEAR ALGEBRA: The column space of a matrix M is the set of column vectors that are linear combinations of the columns of M. When the entries of M come from $GF(2)$, linear combinations are simply mod 2 sums. (Also see Appendix A.4.)

Let G be a simple graph with n vertices and m edges, and let $\langle e_1, e_2, \ldots, e_m \rangle$ and $\langle v_1, v_2, \ldots, v_n \rangle$ be fixed orderings of E_G and V_G, respectively.

REVIEW FROM §2.6: The incidence matrix I_G of G is the matrix whose (i, j)th entry is given by

$$I_G[i, j] = \begin{cases} 0, & \text{if } v_i \text{ is not an endpoint of } e_j \\ 1, & \text{otherwise} \end{cases}$$

Analogous to the edge space $W_E(G)$, the collection of vertex subsets of V_G under ring sum forms a vector space over $GF(2)$, which is called the **vertex space of G** and is denoted $W_V(G)$. Each element of the vertex space $W_V(G)$ may be viewed as an n-tuple over $GF(2)$.

In this setting, I_G represents a linear transformation from edge space $W_E(G)$ to vertex space $W_V(G)$, mapping the characteristic vectors of edge subsets to characteristic vectors of vertex subsets.

Example 4.6.4: Consider the graph G and its corresponding incidence matrix I_G shown in Figure 4.6.9.

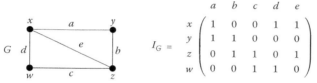

Figure 4.6.9   A graph G and its incidence matrix I_G.

The characteristic vector of the image of the subset $E_1 = \{a, c, e\}$ under the mapping is obtained by multiplying the characteristic vector of E_1 by I_G (mod 2), as follows:

$$\begin{pmatrix} 1 & 0 & 0 & 1 & 1 \\ 1 & 1 & 0 & 0 & 0 \\ 0 & 1 & 1 & 0 & 1 \\ 0 & 0 & 1 & 1 & 0 \end{pmatrix} \begin{pmatrix} 1 \\ 0 \\ 1 \\ 0 \\ 1 \end{pmatrix} = \begin{pmatrix} 0 \\ 1 \\ 0 \\ 1 \end{pmatrix}$$

Thus, E_1 is mapped to the vertex subset $V_1 = \{y, w\}$. Notice that V_1 consists of the endpoints of the path formed by the edges of E_1. It is not hard to show that every open path in G is mapped to the vertex subset consisting of the path's initial and terminal vertices. For this reason, I_G is sometimes called a *boundary operator*.

TERMINOLOGY NOTE: Algebraists and topologists sometimes refer to the edge subsets of $W_E(G)$ and the vertex subsets of $W_V(G)$ as *1-chains* and *0-chains*, respectively, and denote these vector spaces $C_1(G)$ and $C_0(G)$. In this context, the chains are typically represented as sums of edges or vertices. For example, the expression $e_1 + e_2 + e_3$ would represent the edge subset $\{e_1, e_2, e_3\}$.

In general, the image under I_G of any edge subset is obtained by simply computing the mod 2 sum of the corresponding set of columns of I_G. For instance, in Example 4.6.4, I_G maps the edge subset $E_2 = \{c, d, e\}$ to $\emptyset$ (which is the zero element of the vertex space

$W_V(G)$), since the mod 2 sum of the third, fourth, and fifth columns of I_G is the column vector of all zeros. It is not a coincidence that E_2 comprises the edges of a cycle, as the next proposition confirms.

NOTATION: For any edge subset D of a simple graph G, let C_D denote the corresponding set of columns of I_G.

Proposition 4.6.12: *Let D be the edge-set of a cycle in a simple graph G. Then the mod 2 sum of the columns in C_D is the zero vector.*

Proof: For a suitable ordering of the edges and vertices of G, a cycle of length k corresponds to the following submatrix of I_G.   ◇

$$\begin{pmatrix} 1 & 0 & \cdots & 0 & 1 \\ 1 & 1 & \ddots & \vdots & 0 \\ 0 & 1 & \ddots & 0 & \vdots \\ \vdots & \ddots & \ddots & 1 & 0 \\ 0 & \cdots & 0 & 1 & 1 \end{pmatrix}$$

Corollary 4.6.13: *A set D of edges of a simple graph G forms a cycle or a union of edge-disjoint cycles if and only if the mod 2 sum of the columns in C_D is the zero vector.*
◇ *(Exercises)*

Each column of I_G may be regarded as an element of the vector space of all n-tuples over $GF(2)$. Therefore, a subset S of the columns of I_G is *linearly independent* over $GF(2)$ if no nonempty subset of S sums to the zero vector. Using this terminology, the following characterization of acyclic subgraphs is an immediate consequence of Corollary 4.6.13.

Proposition 4.6.14: *An edge subset D of a simple graph G forms an acyclic subgraph of G if and only if the column set C_D is linearly independent over $GF(2)$.*

The next proposition is needed to complete the characterization of spanning trees promised earlier.

Proposition 4.6.15: *Let D be a subset of edges of a connected, simple graph G. Then D forms (induces) a connected spanning subgraph of G if and only if column set C_D spans the column space of I_G.*

Proof: The column set C_D spans the column space of I_G if and only if each column of I_G can be expressed as the sum of the columns in a subset of C_D. But each column of I_G corresponds to some edge $e = xy$ of G, and that column is a sum of the columns in some subset of C_D if and only if there is a path from x to y whose edges are in D.   ◇

Proposition 4.6.16: *Let G be a connected, simple graph. Then an edge subset D induces a spanning tree of G if and only if the columns corresponding to the edges of D form a basis for the column space of I_G over $GF(2)$.*

Proof: This follows directly from Propositions 4.6.14 and 4.6.15.   ◇

EXERCISES for Section 4.6

In Exercises 4.6.1 through 4.6.3, find (a) the non-null elements of the cycle space $W_C(G)$ for the given graph G; and (b) the non-null elements of the edge-cut space $W_S(G)$.

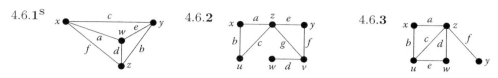

4.6.1^S 4.6.2   4.6.3

In Exercises 4.6.4 through 4.6.6, consider the spanning tree T defined by the specified edge subset of the graph G in the indicated exercise. (a) Show that the fundamental system of cycles associated with T is a basis for the cycle-space $W_C(G)$. (b) Show that the fundamental system of edge-cuts associated with T is a basis for the edge-cut space $W_S(G)$.

4.6.4^S $\{a, e, d\}$ in Exercise 4.6.1.

4.6.5 $\{a, c, g, e, d\}$ in Exercise 4.6.2.

4.6.6 $\{a, b, e, f\}$ in Exercise 4.6.3.

4.6.7 a. Show that the collection $\{\{a, c, d, f\}, \{b, c, e, g\}, \{a, b, h\}\}$ of edge subsets of E_G forms a basis for the cycle space $W_C(G)$ of the graph G shown. b. Find a different basis by choosing some spanning tree and using the associated fundamental system of cycles.

4.6.8 Express each of the 15 non-null elements of $W_S(G)$ in Figure 4.6.3 as the ring sum of some or all of the four fundamental edge-cuts in Figure 4.6.5.

4.6.9 For Application 4.6.1, reverse the directions assigned to i_1, i_2, and i_3, and show that the solution for the actual currents does not change.

In Exercises 4.6.10 through 4.6.12, verify that Corollary 4.6.13 and Proposition 4.6.16 hold for the given edge subset of the following graph.

4.6.10^S $\{a, b, c, d\}$.

4.6.11 $\{a, b, e, f\}$.

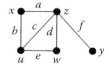

4.6.12 $\{a, b, e, d\}$.

4.6.13 Prove Proposition 4.6.1.

4.6.14 Prove Theorem 4.6.6.

4.6.15 Prove Proposition 4.6.9.

4.6.16 Prove Corollary 4.6.13.

4.6.17^S Test whether the cycle and edge-cut spaces for the graph in Figure 4.6.3 are orthogonal complements of the edge space $W_E(G)$.

4.6.18 Express the graph of Figure 4.6.3 as the ring sum of a subgraph in the cycle space with a subgraph in the edge-cut space of G.

4.7   MATROIDS AND THE GREEDY ALGORITHM

DEFINITION: A *hereditary subset system* (or *independence system*) $S = (E, \mathcal{I})$ is a finite set E together with a collection $\mathcal{I}$ of subsets of E closed under *inclusion* (i.e., if $A \in \mathcal{I}$ and $A' \subseteq A$, then $A' \in \mathcal{I}$). The subsets in the collection $\mathcal{I}$ are called **independent**, and any subset that is not in the collection is called **dependent**.

Our first example of a hereditary subset system suggests how the synonym "independence system" may have arisen.

Example 4.7.1: Let E be a finite set of vectors in a vector space V, and $\mathcal{B}$ the collection of all linearly independent subsets of E. For example, when $E = \{(1, 0, 0), (1, 1, 0), (1, 0, 1), (0, 0, 1)\} \subset \mathbb{R}^3$, one of the elements of $\mathcal{B}$ is the subset $\{(1, 0, 0), (1, 1, 0), (0, 0, 1)\}$. Then $S = (E, \mathcal{B})$ is a hereditary subset system since any subset of a linearly independent set of vectors is a linear independent set (possibly empty).

Example 4.7.2: Let G be a graph and $\mathcal{I}$ the set of subsets of E_G whose induced subgraphs are acyclic subgraphs of G. Then $S = (E_G, \mathcal{I})$ is an hereditary subset system.

Remark: For hereditary subset systems that consist of edge subsets of a graph G, no distinction will be made between the edge subset and the subgraph induced by that edge subset. In this context, if H is a subgraph, then $e \in H$ will mean $e \in E_H$.

Many combinatorial optimization problems are instances of the following maximization problem for hereditary subset systems.

DEFINITION: The **maximum-weight problem** associated with a hereditary subset system $S = (E, \mathcal{I})$ is the following: Let w be a *weight function* that assigns a nonnegative real number to each $e \in E$. Find an independent subset $I \in \mathcal{I}$ such that $\sum_{e \in I} w(e)$ is maximum.

Remark: The minimization version of the above problem is trivial, since the empty set is a minimum-weight independent subset. A truer analog is the **minimum-weight problem** of finding a minimum-weight maximal independent subset.

One approach to trying to solve the maximum-weight problem associated with a hereditary subset system $(E, \mathcal{I})$ is the following simple algorithm known as the **greedy algorithm**. To simplify the notation, let $I + e$ denote $I \cup \{e\}$.

Remark: Replacing the word "largest" with "smallest" in the while-loop results in a greedy algorithm for the minimum-weight problem.

Algorithm 4.7.1:   Greedy

   Input: a hereditary subset system $(E, \mathcal{I})$ and a nonnegative weight function w.
   Output: an independent set $I \in \mathcal{I}$
          Initialize $I := \emptyset$.
          Initialize $A := E$
          While $A \neq \emptyset$
                Choose $e \in A$ of largest weight.
                $A := A - e$
                If $I + e \in \mathcal{I}$ then $I := I + e$.
          Return I.

   The next two examples illustrate the greedy algorithm on two different graph optimization problems. The algorithm produces an optimal solution for the first problem but fails to do so for the second one. An examination of the underlying algebraic structure will reveal why.

Example 4.7.3: Consider the problem of finding a maximum-weight spanning tree for the weighted graph G shown in Figure 4.7.1.

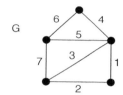

Figure 4.7.1   A weighted graph.

   Since a spanning tree is an acyclic subgraph, the problem may be viewed as a maximum-weight problem associated with the hereditary subset system of acyclic subgraphs. Thus, the problem is to find an element H of $\mathcal{I}$ (i.e., an edge subset whose induced subgraph is acyclic) whose total edge-weight is maximum.

   The greedy algorithm begins by choosing the edge of weight 7 and continues by choosing the edge of largest weight that does not create a cycle. It is easy to see that the resulting spanning tree, shown in Figure 4.7.2, has the maximum weight.

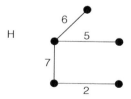

Figure 4.7.2   The greedy algorithm produces a maximum spanning tree.

Remark: This version of the greedy algorithm is known as **Kruskal's algorithm**, named for J. B. Kruskal [Kr56]. Since a spanning tree is an edge-maximal acyclic subgraph, the *minimum*-spanning-tree problem is an instance of the minimum-weight problem for a hereditary subset system, and as will be seen later, Kruskal's algorithm succeeds for both spanning-tree problems.

DEFINITION: A **matching** in a graph is a set of edges that have no endpoints in common.

The *maximum-matching problem* is to find a matching in a weighted graph, whose total edge-weight is a maximum. The hereditary subset system that corresponds to this maximum-weight problem is $(E_G, \mathcal{I})$, where $\mathcal{I}$ is the set of matchings in the graph G.

Example 4.7.4: Consider the maximum-matching problem for the graph G shown in Figure 4.7.3.

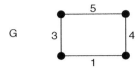

Figure 4.7.3

The greedy algorithm chooses the edge of weight 5 first and then must choose the edge of weight 1, resulting in a non-optimal solution. Thus, the greedy algorithm does *not* solve this version of the maximum-weight problem.

For Which Problems Will the Greedy Algorithm Perform Optimally?

DEFINITION: A hereditary subset system $M = (E, \mathcal{I})$ is called a **matroid** if it satisfies the following condition, which we refer to as the **augmentation property**.

Augmentation property: If $I, J \in \mathcal{I}$ and $|I| < |J|$, then there is an element $e \in J - I$ such that $I + e \in \mathcal{I}$.

The next proposition shows that the hereditary subset system associated with the maximum-spanning-tree problem is a matroid.

Proposition 4.7.1: *Let G be a graph. Then the hereditary subset system $M(G) = (E_G, \mathcal{I})$ where $\mathcal{I}$ is the collection of edge subsets whose induced subgraphs are acyclic, is a matroid.*

Proof: Suppose that $I, J \in \mathcal{I}$, with $|I| < |J|$. Let H be the subgraph induced on the edge subset $I \cup J$, and let F be a full spanning forest of H that contains I (F exists, by Proposition 4.5.2 applied to the components of H). Then F has at least as many edges as J, since J is acyclic. So there exists an edge $e \in F - I$. By the construction of H, $e \in J$. Furthermore, $I + e$ is a subgraph of F and, hence, is acyclic. This shows that the hereditary subset system satisfies the augmentation property, and the proof is complete. $\diamond$

Remark: One of the several alternative ways of defining a matroid is to specify its minimal dependent subsets, called *cycles*, and to require them to satisfy an additional property analogous to the augmentation property (see Exercises). The cycles of the matroid $M(G)$ are precisely the cycles in the graph G (see Exercises), which is why this matroid is called the cycle **matroid**.

Example 4.7.5: The hereditary subset system associated with the maximum matching problem is *not* a matroid. To see that it does not satisfy the augmentation property, consider the graph in Figure 4.7.3. The subsets $I = \{a\}$ and $J = \{b, c\}$ violate the augmentation property.

The final result of this chapter characterizes those problems for which the greedy algorithm produces an optimal solution for the maximum-weight problem.

Theorem 4.7.2: *Let $S = (E, \mathcal{I})$ be a hereditary subset system. Then the greedy algorithm solves every instance of the maximum-weight problem associated with S if and only if S is a matroid.*

Proof: *Necessity* $(\Rightarrow)$ Suppose that the augmentation property is not satisfied, and consider $I, J \in \mathcal{I}$, with $|I| < |J|$ such that there is no $e \in J - I$ for which $I + e \in \mathcal{I}$. Since S is a hereditary subset system, it can be assumed without loss of generality that $|J| = |I| + 1$. Let $|I| = p$, and consider the following weight function on E.

$$
w(e) = \begin{cases} p + 2, & \text{if } e \in I; \\ p + 1, & \text{if } e \in J - I; \\ 0, & \text{otherwise.} \end{cases}
$$

After the greedy algorithm chooses all the elements of I, it cannot increase the total weight, since the only edges left that have nonzero weight belong to J. Thus, the greedy algorithm ends with an independent subset whose weight is $p(p + 2)$. But the weight of subset J is at least $(p + 1)^2$, which shows that the greedy algorithm did not find a maximum-weight independent subset.

Sufficiency $(\Leftarrow)$ Let $I = \{e_1, e_2, \dots, e_n\}$, be the independent subset obtained when the greedy algorithm is applied, and let $J = \{f_1, f_2, \dots, f_l\}$ be any independent subset in $\mathcal{I}$. Also assume that the e_i's and f_i's are subscripted in descending order by weight. The augmentation property implies that $l \le n$, since otherwise there would be an f_i that should have been added to I by the greedy algorithm but was not. Now let $I_k = \{e_1, e_2, \dots, e_k\}$ and $J_k = \{f_1, f_2, \dots, f_k\}$, $k = 1, \dots, l$. To complete the proof, we first show, by way of induction,

$$
(*) \qquad \sum_{i=1}^{k} w(e_i) \doteq w(I_k) \ge w(J_k), \quad k = 1, \dots, l
$$

Since the algorithm starts by choosing a largest element, we have that $w(I_1) \ge w(J_1)$, which establishes the base. Assume $w(I_k) \ge w(J_k)$ for some $1 \le k < l$. By the augmentation property, there is an $x \in J_{k+1} - I_k$ such that $I_k + x \in \mathcal{I}$. Thus, $w(e_{k+1}) \ge w(x)$, since otherwise the algorithm would have chosen x instead of e_{k+1}. Also, $w(x) \ge w(f_{k+1})$, since the f_i's are subscripted in descending order. Therefore,

$$
w(I_{k+1}) = w(I_k) + w(e_{k+1}) \ge w(I_k) + w(f_{k+1}) \ge w(J_k) + w(f_{k+1}) = w(J_{k+1}),
$$

which completes the proof of $(*)$.

Finally

$$
w(I) = w(I_n) \ge w(I_l) \ge w(J_l) = w(J)
$$

which shows that I is a maximum-weight independent subset and completes the proof. $\diamondsuit$

Remark: An immediate consequence of Proposition 4.7.1 and Theorem 4.7.2 is that Kruskal's algorithm always finds a maximum-weight spanning forest (spanning *tree*, if the original graph is connected). Using an argument similar to the one given for the sufficiency part of Theorem 4.7.2, it is not hard to show that the version of Kruskal's algorithm that chooses edges of *smallest* possible weight finds the minimum-weight spanning tree (see Exercises). The algorithmic analysis and comparisons with Prim's algorithm may be found in one of the standard texts in data structures and algorithms (e.g., [AhUlHo83], [Se88]).

Using matroid theory, several results in combinatorial optimization can be unified, often resulting in simpler proofs. For those wanting to learn more about matroids, there are several excellent texts (e.g., [Ox92], [Wi96], [ThSw92], [La76], [We76]).

EXERCISES for Section 4.7

In Exercises 4.7.1 through 4.7.4, use Kruskal's algorithm to find a minimum-weight spanning tree of the specified graph.

4.7.1^S The graph of Exercise 4.3.1.   **4.7.2** The graph of Exercise 4.3.2.
4.7.3 The graph of Exercise 4.3.3.   **4.7.4** The graph of Exercise 4.3.4.

In Exercises 4.7.5 through 4.7.9, determine whether the given family of subsets $(E, \mathcal{I})$ is a hereditary subset system, and justify your answer. If the family is a hereditary system, then determine whether it is a matroid.

4.7.5^S $E = V_G$ for an arbitrary graph G, and $\mathcal{I}$ is the collection of all subsets of mutually non-adjacent vertices in graph G.

4.7.6 $E = V_G$ for an arbitrary graph G, and $\mathcal{I}$ is the collection of all subsets of mutually adjacent vertices in graph G.

4.7.7 $E = E_G$ for an arbitrary graph G, and $\mathcal{I}$ is the collection of all subsets of mutually non-adjacent edges in graph G.

4.7.8 $E = E_G$ for an arbitrary graph G, and $\mathcal{I}$ is the collection of all subsets of mutually adjacent edges in graph G.

4.7.9^S E is any set of n elements, and $\mathcal{I}$ is the collection of all subsets of k elements, where $1 \le k \le n$.

4.7.10 Show that when the greedy algorithm (Algorithm 4.7.1) has chosen k elements, this subset of k elements is of maximum weight among all independent subsets of k or fewer elements.

4.7.11 Let $M = (E, \mathcal{I})$ be a matroid. Show that all maximal independent subsets in M have the same number of elements.

4.7.12 Let G be a graph, and let $\mathcal{I}$ be the collection of subsets of E_G that do *not* contain an edge-cut.

a. Show that $H = (E_G, \mathcal{I})$ is a hereditary subset system.

b. Show that H is a matroid, by showing that it satisfies the augmentation property.

4.7.13^S Let $M = (E, \mathcal{I})$ be a matroid. Suppose that C_1 and C_2 are two different minimal *dependent* subsets in M and that both contain an element $e \in E$. Show that there exists a minimal dependent subset C_3 such that in $C_3 \subset C_1 \cup C_2$ and $e \notin C_3$.

4.7.14 Let E be a non-empty finite set, and let C be a collection of non-empty subsets of E satisfying the following two properties:

(1) if $C \in C$ and A is a proper subset of C, then $A \notin C$;

(2) if C_1 and C_2 are two different members of C and $e \in C_1 \cap C_2$, for some $e \in E$, then there exists a subset $C_3 \in C$ such that $C_3 \subset C_1 \cup C_2$ and $e \notin C_3$.

Prove that the subsets of E that are *not* members of C are the independent subsets of a matroid.

4.7.15 Show that the cycles of a graph G are the minimal dependent subsets of the cycle matroid $M(G)$.

4.7.16^S Let $M = (E, \mathcal{I})$ be a matroid. Prove that the "minimization" version of the greedy algorithm solves every instance of the minimum-weight problem associated with matroid M. This will confirm our claim that Kruskal's algorithm always finds a minimum-weight spanning tree for any connected graph (with nonnegative edge-weights).

4.7.17 Suppose that a set E of jobs are to be processed by a single machine. All jobs require the same processing time, and each job has a deadline. A subset of jobs is said to be **manageable** if there is a processing sequence for the jobs in the subset, such that each job is completed on time. Let $\mathcal{I}$ be the collection of all manageable subsets of jobs. Show that $(E, \mathcal{I})$ is a matroid.

4.7.18 [*Computer Project*] a. Rewrite Algorithm 4.7.1 as Kruskal's algorithm for finding a minimum spanning tree. b. Implement Kruskal's algorithm, and run the program on each of the instances of the problem given in Exercise 4.3.1 through 4.3.4 from §4.3.

4.8   SUPPLEMENTARY EXERCISES

4.8.1 Draw the isomorphism types of 6-vertex trees that cannot occur as spanning trees of $K_{3,3}$, and prove that they cannot.

4.8.2 In how many different spanning trees of the complete graph K_n does a given edge e lie?

4.8.3 Prove that the number of spanning trees in the wheel W_5 equals 45. (Hint: Partition into cases.)

4.8.4 Draw an example of a connected, simple graph G with a spanning tree T indicated by thickened edges, such that no matter what root or local ordering is selected, T could not possibly be either the BFS-tree or the DFS-tree. Explain briefly why not.

4.8.5 What is the maximum number of edges in a simple graph with a DFS-tree isomorphic to $K_{1,5}$. Prove your answer.

4.8.6 Show that if no two edges of a connected graph G have the same weight, then the minimum spanning tree is unique.

4.8.7 In the graph of Figure 4.8.1 (on next page),

   a. use Prim's algorithm to find a minimum spanning tree;

   b. use Dijkstra's algorithm to find a shortest s-t path.

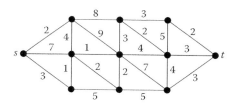

Figure 4.8.1

Glossary

augmentation property: see *matroid*.

back-arc relative to a rooted tree produced by an instance of Tree-Growing in a digraph: a non-tree arc directed from a vertex to one of its ancestors.

Betti number of a graph: synonym for *cycle rank*.

bfs-tree: shortened name for *breadth-first tree*.

breadth-first search: the version of Tree-Growing that selects, at each iteration, a frontier edge incident on the vertex having the smallest discovery number possible.

breadth-first tree: the output tree produced by a breadth-first search of a graph.

bridge: synonym for *cut-edge*.

bridge component BC of a connected G whose set of cut-edges bridges is B: a component of the subgraph $G - B$ (§2.4).

component of a vertex v in a graph G, denoted $C_G(v)$: the subgraph of G induced on the set of vertices that are reachable from v.

contraction of a subgraph H of a graph G to a vertex: the replacement of H by a single vertex k. Each edge that joined a vertex $v \in V_G - V_H$ to a vertex in H is replaced by an edge with endpoints v and k.

cross-arc relative to a rooted tree produced by an instance of Tree-Growing in a digraph: a non-tree arc directed from a vertex to a vertex to which it is not *related*.

—, left-to-right: directed from smaller discovery number to larger.

—, right-to-left: directed from larger discovery number to smaller.

cross-edge relative to a tree produced by an instance of Tree-Growing: a non-tree edge whose endpoints are not *related*.

cut-edge of a connected graph G: an edge e such that the edge-deletion subgraph $G - e$ has two components.

cycle-rank of a graph G: the number of edges in the complement of a full spanning forest of G; the quantity $|E_G| - |V_G|$ plus the number of components of G.

cycle space of a graph G: the subspace $W_C(G)$ consisting of the null graph $\emptyset$, all cycles in G, and all unions of edge-disjoint cycles of G.

dag: acronym for *directed acyclic graph*, i.e., it has no directed cycles.

depth-first tree: the output tree produced by a depth-first search of a graph.

depth-first search: the version of Tree-Growing that selects, at each iteration, a frontier edge incident on the vertex having the largest discovery number possible.

dfnumber of a vertex w for a depth-first search: the discovery number of vertex w; denoted *dfnumber(w)*.

dfs-path: a path in the depth-first tree produced by executing a depth-first search and stopping before you backtrack for the first time.

dfs-tree: shortened name for *depth-first tree.*

Dijkstra tree: the tree produced by Dijkstra's Shortest Path Algorithm 4.3.2.

discovery number of a vertex for an instance of Tree-Growing: the position of that vertex in the discovery order, starting with 0 for the start vertex.

discovery order for an instance of Tree-Growing: the order in which the vertices of G are added (discovered) as the tree is grown.

distance between two vertices s and t in a weighted graph: the length of a shortest s-t path.

edge-cut in a graph G: a set S of edges such that the edge-deletion subgraph $G - S$ has more components than G.

edge-cut rank of a graph G: the quantity $|V_G|$ minus the number of components of G; the number of edges in a spanning forest of G.

edge-cut space of a graph G: the subspace $W_S(G)$ consisting of the null graph $\emptyset$, all minimal edge-cuts of G, and all unions of edge-disjoint minimal edge-cuts of G.

edge space $W_E(G)$ of a graph G: the vector space over $GF(2)$ consisting of the collection of edge subsets of E_G under ring sum.

edge-weight of an edge in a weighted graph: the number assigned to that edge.

Eulerian graph: a graph that has an Eulerian tour.

Eulerian tour in a graph: a closed trail that contains every edge of that graph.

finish order of a depth-first search: the order in which the vertices are finished.

finished vertex during a depth-first search: a discovered vertex whose neighbors have all been discovered.

forward-arc relative to a rooted tree produced by an instance of Tree-Growing in a digraph: a non-tree arc directed from a vertex to one of its descendants.

frontier arc relative to a rooted tree T in a digraph: an arc whose tail is in T and whose head is not in T.

frontier edge for a given tree T: an edge with one endpoint in T and one endpoint not in T.

full spanning forest of a graph G: a spanning forest consisting of a collection of trees, such that each tree is a spanning tree of a different component of G.

fundamental cycle of a graph G associated with a full spanning forest F and an edge e not in F: the unique cycle that is created when the edge e is added to the forest F.

fundamental edge-cut of a graph G associated with a full spanning forest F and an edge e in F: the unique partition-cut $\langle V_1, V_2 \rangle$ of G, where V_1 and V_2 are the vertex-sets of the two components of the subgraph $F - e$.

fundamental system of cycles of a graph G associated with a full spanning forest F: the collection of fundamental cycles of the graph G that result from each addition of an edge to the spanning forest F.

fundamental system of edge-cuts of a graph G associated with a full spanning forest F: the collection of fundamental edge-cuts of the graph G that result from each removal of an edge from the spanning forest F.

greedy algorithm: an algorithm based on shortsighted greed; see Algorithm 4.7.1.

hereditary subset system $S = (E, \mathcal{I})$: a finite set E together with a collection $\mathcal{I}$ of subsets of E closed under set inclusion (i.e., if $A \in \mathcal{I}$ and $A' \subseteq A$, then $A' \in \mathcal{I}$).

independent subset of a subset system $S = (E, \mathcal{I})$: a subset in the collection $\mathcal{I}$.

matching in a graph: a set of edges that have no endpoints in common.

matroid: a hereditary subset system $M = (E, \mathcal{I})$ that satisfies the *augmentation property*: if $I, J \in \mathcal{I}$ and $|I| < |J|$, then there is an element $e \in J - I$ such that $I + e \in \mathcal{I}$.

—, cycle of a graph G: the collection of edge subsets whose induced subgraphs are acyclic (see Proposition 4.7.1).

maximum-weight problem associated with a hereditary subset system $S = (E, \mathcal{I})$: for a given weight function that assigns a nonnegative real number (weight) to each $e \in E$, the problem of finding an independent subset in the subset system S that has the largest total weight.

nextArc: the digraph analog of the function *nextEdge*.

—, dfs- : the digraph analog of *dfs-nextEdge*.

nextEdge(G,S) for a set S of frontier edges for a tree T in a graph G: a function that chooses and returns as its value the frontier edge in S that is to be added to tree T; used in various versions of Tree-Growing Algorithm 4.1.1.

—, bfs- : selects and returns as its value the frontier edge whose tree-endpoint has the smallest discovery number. If there is more than one such edge, then *bfs-nextEdge*(G, S) selects the one determined by the default priority.

—, dfs- : selects and returns as its value the frontier edge whose tree-endpoint has the largest discovery number. If there is more than one such edge, then *dfs-nextEdge*(G, S) selects the one determined by the default priority.

—, Dijkstra- : selects and returns as its value the frontier edge whose non-tree endpoint is closest to the start vertex s. If there is more than one such edge, then *Dijkstra-nextEdge*(G, S) selects the one determined by the default priority.

—, Prim- : selects and returns as its value the frontier edge with smallest edge-weight. If there is more than one such edge, then *Prim-nextEdge*(G, S) selects the one determined by the default priority.

non-tree edge: see *tree edge, non-*.

P-value $P(e)$ of a frontier edge e with tree endpoint x for a partial Dijkstra tree: the quantity given by $P(e) = dist[x] + w(e)$.

partition-cut $\langle X_1, X_2 \rangle$ of a graph G: the set of all edges of G having one endpoint in X_1 and the other endpoint in X_2, where X_1 and X_2 form a partition of V_G.

Prim tree: the tree produced by Prim's Minimum Spanning Tree Algorithm 4.3.1.

related vertices in a rooted tree: two vertices such that one is a (proper) descendant of the other.

relative complement of a subgraph H in a graph G, denoted $G - H$: the edge-deletion subgraph $G - E(H)$.

ring sum of two sets A and B: the set $(A - B) \cup (B - A)$, denoted $A \oplus B$.

sink in a dag: a vertex with outdegree 0.

skip-edge relative to a tree produced by an instance of Tree-Growing: a non-tree edge whose endpoints are *related*.

spanning forest of a graph G: an acyclic spanning subgraph of G.

—, full: a spanning forest consisting of a collection of trees, such that each tree is a spanning tree of a different component of G.

spanning tree of a (connected) graph: a spanning subgraph that is a tree.

source in a dag: a vertex with indegree 0.

Steiner-tree problem for a subset U of vertices in a connected edge-weighted graph G: the problem of finding a minimum-weight tree subgraph of G that contains all the vertices of U.

topological sort (or **topological order**) of an n-vertex dag G: a bijection ts from the set $\{1, 2, \dots, n\}$ to the vertex-set V_G such that every arc e is directed from a smaller number to a larger one, i.e., $ts(tail(e)) < ts(head(e))$.

tree edge relative to a given tree T in a graph G: an edge in tree T.

—, non- : an edge of G that is not in tree T.

tree endpoint of a frontier edge relative to a given tree T: the endpoint of that edge that is in tree T.

—, non- : the endpoint of that edge that is not in tree T.

updateFrontier(G, S) after a frontier edge is added to the current tree: the procedure that removes from S those edges that are no longer frontier edges and adds to S those that have become frontier edges; used in various versions of Tree-Growing Algorithm 4.1.1.

vertex space $W_V(G)$ of a graph G: the vector space over $GF(2)$ consisting of the collection of vertex subsets of V_G under ring sum.

weighted graph: a graph in which each edge is assigned a number, called its *edge-weight*.

Chapter 5

CONNECTIVITY

INTRODUCTION

Some connected graphs are "more connected" than others. For instance, some connected graphs can be disconnected by the removal of a single vertex or a single edge, whereas others remain connected unless more vertices or more edges are removed. Two numerical parameters, *vertex-connectivity* and *edge-connectivity*, are useful in measuring a graph's connectedness.

Determining the number of edges or vertices that must be removed to disconnect a given connected graph applies directly to analyzing the vulnerability of existing or proposed telecommunications networks, road systems, and other networks. Intuitively, a network's vulnerability should be closely related also to the number of alternative paths between each pair of nodes. There is a rich body of mathematical results concerning this relationship, many of which are variations of a classical result of Menger, and some of these extend well beyond graph theory.

5.1   VERTEX- AND EDGE-CONNECTIVITY

TERMINOLOGY: Let S be a subset of vertices or edges in a connected graph G. The removal of S is said to **disconnect** G if the deletion subgraph $G - S$ is not connected.

REVIEW FROM §2.4:

- A **vertex-cut** in a graph G is a vertex-set U such that $G - U$ has more components than G.

- A **cut-vertex** (or **cutpoint**) is a vertex-cut consisting of a single vertex.

- An **edge-cut** in a graph G is a set of edges D such that $G - D$ has more components than G.

- A **cut-edge** (or **bridge**) is an edge-cut consisting of a single edge.

- An edge is a cut-edge if and only if it is not a cycle-edge.

DEFINITION: The **vertex-connectivity** of a connected graph G, denoted $\kappa_v(G)$, is the minimum number of vertices, whose removal can either disconnect G or reduce it to a 1-vertex graph.

Thus, if G has at least one pair of non-adjacent vertices, then $\kappa_v(G)$ is the size of a smallest vertex-cut.

DEFINITION: A graph G is k-**connected** if G is connected and $\kappa_v(G) \geq k$. If G has non-adjacent vertices, then G is k-connected if every vertex-cut has at least k vertices.

DEFINITION: The **edge-connectivity** of a connected graph G, denoted $\kappa_e(G)$, is the minimum number of edges whose removal can disconnect G.

Thus, if G is a connected graph, the edge-connectivity $\kappa_e(G)$ is the size of a smallest edge-cut.

DEFINITION: A graph G is **k-edge-connected** if G is connected and every edge-cut has at least k edges (i.e., $\kappa_e(G) \geq k$].

Example 5.1.1: In the graph G shown in Figure 5.1.1, the vertex set $\{x, y\}$ is one of three different 2-element vertex-cuts, and it is easy to see that there is no cut-vertex. Thus, $\kappa_v(G) = 2$. The edge set $\{a, b, c\}$ is the unique 3-element edge-cut of graph G, and there is no edge-cut with fewer than three edges. Therefore, $\kappa_e(G) = 3$.

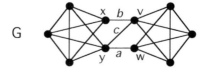

G

Figure 5.1.1   A graph G with $\kappa_v(G) = 2$ and $\kappa_e(G) = 3$.

Application 5.1.1:   *Network Survivability*   The connectivity measures κ_v and κ_e are used in a quantified model of *network survivability*, which is the capacity of a network to retain connections among its nodes after some edges or nodes are removed.

Remark: Since neither the vertex-connectivity nor the edge-connectivity of a graph is affected by the existence or absence of self-loops, we will assume that all graphs under consideration throughout this chapter are loopless, unless otherwise specified.

Proposition 5.1.1: *Let G be a graph. Then the edge-connectivity $\kappa_e(G)$ is less than or equal to the minimum degree $\delta_{\min}(G)$.*

Proof: Let v be a vertex of graph G, with degree $k = \delta_{\min}(G)$. Then the deletion of the k edges that are incident on vertex v separates v from the other vertices of G. ◇

REVIEW FROM §4.5: A **partition-cut** $\langle X_1, X_2 \rangle$ is an edge-cut each of whose edges has one endpoint in each of the vertex bipartition sets X_1 and X_2.

The following proposition characterizes the edge-connectivity of a graph in terms of the size of its partition-cuts.

Proposition 5.1.2: *A graph G is k-edge-connected if and only if every partition-cut contains at least k edges.*

Proof: ($\Rightarrow$) Suppose that graph G is k-edge-connected. Then every partition-cut of G has at least k edges, since a partition-cut is an edge-cut.

($\Leftarrow$) Suppose that every partition-cut contains at least k edges. By Proposition 4.5.4, every minimal edge-cut is a partition-cut. Thus, every edge-cut contains at least k edges. ◇

Relationship between Vertex- and Edge-Connectivity

The next few results concern the relationship between vertex-connectivity and edge-connectivity. They lead to a characterization of 2-connected graphs, first proved by Hassler Whitney in 1932.

Proposition 5.1.3: *Let e be any edge of a k-connected graph G, for $k \geq 3$. Then the edge-deletion subgraph $G - e$ is $(k-1)$-connected.*

Proof: Let $W = \{w_1, w_2, \ldots, w_{k-2}\}$ be any set of $k-2$ vertices in $G - e$, and let x and y be any two different vertices in $(G - e) - W$. It suffices to show the existence of an x-y walk in $(G - e) - W$.

First, suppose that at least one of the endpoints of edge e is contained in set W. Since the vertex-deletion subgraph $G - W$ is connected (in fact, 2-connected), there is an x-y path in $G - W$. This path cannot contain edge e and, hence, it is an x-y path in the subgraph $(G - e) - W$. Next, suppose that neither endpoint of edge e is in set W. Then there are two cases to consider.

Case 1: Vertices x and y are the endpoints of edge e. Graph G has at least $k+1$ vertices (since G is k-connected). So there exists some vertex $z \in G - \{w_1, w_2, \ldots, w_{k-2}, x, y\}$. Since graph G is k-connected, there exists an x-z path P_1 in the vertex-deletion subgraph $G - \{w_1, w_2, \ldots, w_{k-2}, y\}$ and a z-y path P_2 in the subgraph $G - \{w_1, w_2, \ldots, w_{k-2}, x\}$ (shown on the left in Figure 5.1.2). Neither of these paths contains edge e, and, therefore, their concatenation is an x-y walk in the subgraph $G - \{w_1, w_2, \ldots, w_{k-2}\}$.

Case 2: At least one of the vertices x and y, say x, is not an endpoint of edge e. Let u be an endpoint of edge e that is different from vertex x. Since graph G is k-connected, the subgraph $G - \{w_1, w_2, \ldots, w_{k-2}, u\}$ is connected. Hence, there is an x-y path P in $G - \{w_1, w_2, \ldots, w_{k-2}, u\}$ (shown on the right in Figure 5.1.2). It follows that P is an x-y path in $G - \{w_1, w_2, \ldots, w_{k-2}\}$ that does not contain vertex u and, hence, excludes edge e (even if P contains the other endpoint of e, which it could). Therefore, P is an x-y path in $(G - e) - \{w_1, w_2, \ldots, w_{k-2}\}$. ◇

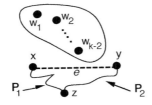 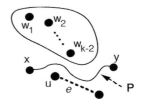

Figure 5.1.2   The existence of an x-y **walk in** $(G - e) - \{w_1, w_2, \ldots, w_{k-2}\}$.

Corollary 5.1.4: *Let G be a k-connected graph, and let D be any set of m edges of G, for $m \leq k - 1$. Then the edge-deletion subgraph $G - D$ is $(k - m)$-connected.*

Proof: The result follows by the iterative application of Proposition 5.1.3. $\Diamond$

Corollary 5.1.5: *Let G be a connected graph. Then $\kappa_e(G) \geq \kappa_v(G)$.*

Proof: Let $k = \kappa_v(G)$, and let S be any set of $k - 1$ edges in graph G. Since G is k-connected, the graph $G - S$ is 1-connected, by Corollary 5.1.4. Thus, edge subset S is not an edge-cut of graph G, which implies that $\kappa_e(G) \geq k$. $\Diamond$

Corollary 5.1.6: *Let G be a connected graph. Then $\kappa_v(G) \leq \kappa_e(G) \leq \delta_{\min}(G)$.*

Proof: The assertion simply combines Proposition 5.1.1 and Corollary 5.1.5. $\Diamond$

Internally Disjoint Paths and Vertex-Connectivity: Whitney's Theorem

A communications network is said to be *fault-tolerant* if it has at least two alternative paths between each pair of vertices. This notion actually characterizes 2-connected graphs, as the next theorem demonstrates. The theorem was proved by Hassler Whitney in 1932 and is a prelude to his more general result for k-connected graphs, which appears in §5.3.

TERMINOLOGY: A vertex of a path P is an **internal vertex** of P if it is neither the initial nor the final vertex of that path.

DEFINITION: Let u and v be two vertices in a graph G. A collection of u-v paths in G is said to be **internally disjoint** if no two paths in the collection have an internal vertex in common.

Theorem 5.1.7: [**Whitney's 2-Connected Characterization**] *Let G be a connected graph with three or more vertices. Then G is 2-connected if and only if for each pair of vertices in G, there are two internally disjoint paths between them.*

Proof: ($\Leftarrow$) Arguing by contrapositive, suppose that graph G is not 2-connected. Then let v be a cut-vertex of G. Since $G - v$ is not connected, there must be two vertices x and y such that there is no x-y path in $G - v$. It follows that v is an internal vertex of every x-y path in G.

($\Rightarrow$) Suppose that graph G is 2-connected, and let x and y be any two vertices in G. We use induction on the distance $d(x, y)$ to prove that there are at least two vertex-disjoint x-y paths in G. If there is an edge e joining vertices x and y, (i.e., $d(x, y) = 1$), then the edge-deletion subgraph $G - e$ is connected, by Corollary 5.1.4. Thus, there is an x-y path P in $G - e$. It follows that path P and edge e are two internally disjoint x-y paths in G.

Next, assume for some $k \geq 2$ that the assertion holds for every pair of vertices whose distance apart is less than k. Suppose $d(x, y) = k$, and consider an x-y path of length k. Let w be the vertex that immediately precedes vertex y on this path, and let e be the edge between vertices w and y. Since $d(x, w) < k$, the induction hypothesis implies that there are two internally disjoint x-w paths in G, say P and Q. Also, since G is 2-connected, there exists an x-y path R in G that avoids vertex w. This is illustrated in Figure 5.1.3 for the two possibilities for path Q: either it contains vertex y (as shown on the right), or it does not (as on the left).

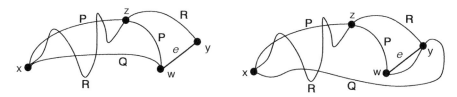

Figure 5.1.3

Let z be the last vertex on path R that precedes vertex y and is also on one of the paths P or Q (z might be vertex x). Assume, without loss of generality, that z is on path P. Then G has two internally disjoint x-y paths. One of these paths is the concatenation of the subpath of P from x to z with the subpath of R from z to y. If vertex y is not on path Q, then a second x-y path, internally disjoint from the first one, is the concatenation of path Q with the edge joining vertex w to vertex y. If y is on path Q, then the subpath of Q from x to y can be used as the second path.   ◇

Corollary 5.1.8: *Let G be a graph with at least three vertices. Then G is 2-connected if and only if any two vertices of G lie on a common cycle.*

Proof: This follows from Theorem 5.1.7, since two vertices x and y lie on a common cycle if and only if there are two internally disjoint x-y paths.   ◇

Remark: Theorem 5.1.7 is a prelude to Whitney's more general result for k-connected graphs, which appears in §5.3. Corollary 5.1.8 is used in Chapter 7 in the proof of Kuratowski's characterization of graph planarity.

The following theorem extends the list of characterizations of 2-connected graphs. Its proof uses reasoning similar to that used in the proof of the last two results (see Exercises).

Theorem 5.1.9: [**Characterization of 2-Connected Graphs**] *Let G be a connected graph with at least three vertices. Then the following statements are equivalent.*

1. *Graph G is 2-connected.*
2. *For any two vertices of G, there is a cycle containing both.*
3. *For any vertex and any edge of G, there is a cycle containing both.*
4. *For any two edges of G, there is a cycle containing both.*
5. *For any two vertices and one edge of G, there is a path from one of the vertices to the other, that contains the edge.*
6. *For any sequence of three distinct vertices of G, there is a path from the first to the third that contains the second.*
7. *For any three distinct vertices of G, there is a path containing any two of them which does not contain the third.*

EXERCISES for Section 5.1

5.1.1[S] Find the other two 2-element vertex-cuts in the graph of Example 5.1.1.

In Exercises 5.1.2 through 5.1.5, either draw a graph meeting the specifications or explain why no such graph exists.

5.1.2 A 6-vertex graph G such that $\kappa_v(G) = 2$ and $\kappa_e(G) = 2$.

5.1.3 A connected graph with 11 vertices and 10 edges and no cut-vertices.

5.1.4 A 3-connected graph with exactly one bridge.

5.1.5[S] A 2-connected 8-vertex graph with exactly two bridges.

In Exercises 5.1.6 through 5.1.11, determine the vertex- and edge-connectivity of the given graph.

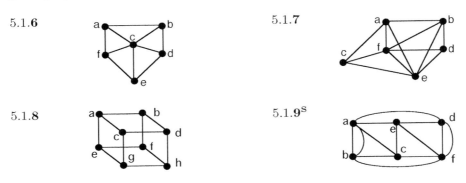

5.1.6

5.1.7

5.1.8

5.1.9[S]

5.1.10 Complete bipartite graph $K_{4,7}$.   **5.1.11** Hypercube graph Q_4.

5.1.12 Determine the vertex- and edge-connectivity of the complete bipartite graph $K_{m,n}$.

5.1.13[S] Determine the vertex-connectivity and edge-connectivity of the Petersen graph (§1.2), and justify your answer. (Hint: Use the graph's symmetry to reduce the number of cases to consider.)

In Exercises 5.1.14 through 5.1.17, give an example of a graph G satisfying the given conditions.

5.1.14 $\kappa_v(G) = \kappa_e(G) = \delta_{\min}(G)$ **5.1.15** $\kappa_v(G) = \kappa_e(G) < \delta_{\min}(G)$

5.1.16 $\kappa_v(G) < \kappa_e(G) = \delta_{\min}(G)$ **5.1.17**[S] $\kappa_v(G) < \kappa_e(G) < \delta_{\min}(G)$

5.1.18 Let $v_1, v_2, \ldots, v_k$ be k distinct vertices of a k-connected graph G, and let G^W be the graph formed from G by joining a new vertex w to each of the v_i's. Show that $\kappa_v(G^w) = k$.

5.1.19 Let G be a k-connected graph, and let v be a vertex not in G. Prove that the suspension $H = G + v$ (§2.4) is $(k+1)$-connected.

5.1.20 Prove that there exists no 3-connected simple graph with exactly seven edges.

5.1.21$^\text{S}$ Let a, b, and c be positive integers with $a \leq b \leq c$. Show that there exists a graph G with $\kappa_v(G) = a$, $\kappa_e(G) = b$, $\delta_{\min}(G) = c$.

DEFINITION: A **unicyclic** graph is a connected graph with exactly one cycle.

5.1.22 Show that the edge-connectivity of a unicyclic graph is no greater than 2.

5.1.23$^\text{S}$ Characterize those unicyclic graphs whose vertex-connectivity equals 2.

5.1.24 Prove the characterization of 2-connected graphs given by Theorem 5.1.9.

5.1.25 Prove that if G is a connected graph, then $\kappa_v(G) = 1 + \min_{v \in V}\{\kappa_v(G - v)\}$.

5.1.26$^\text{S}$ Find a lower bound on the number of vertices in a k-connected graph with diameter d (§1.4) and a graph that achieves that lower bound (thereby showing that the lower bound is *sharp*).

5.2   CONSTRUCTING RELIABLE NETWORKS

In this section we examine methods for constructing graphs with a prescribed vertex-connectivity. In light of earlier remarks, these graphs amount to blueprints for reliable networks.

Whitney's Synthesis of 2-Connected Graphs and 2-Edge-Connected Graphs

DEFINITION: A **path addition** to a graph G is the addition to G of a path between two existing vertices of G, such that the edges and internal vertices of the path are not in G. A **cycle addition** is the addition to G of a cycle that has exactly one vertex in common with G.

DEFINITION: A **Whitney-Robbins synthesis** of a graph G from a graph H is a sequence of graphs, $G_0, G_1, \ldots, G_l$, where $G_0 = H$, $G_l = G$, and G_i is the result of a path or cycle addition to G_{i-1}, for $i = 1, \ldots, l$. If each G_i is the result of a path addition *only*, then the sequence is called a **Whitney synthesis**.

Example 5.2.1: Figure 5.2.1 shows a 4-step Whitney synthesis of the cube graph Q_3, starting from the cycle graph C_4.

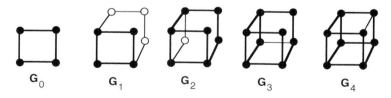

Figure 5.2.1   A Whitney synthesis of the cube graph Q_3.

Lemma 5.2.1: *Let H be a 2-connected graph. Then the graph G that results from a path addition to H is 2-connected.*

Proof: The property that every two vertices lie on a common cycle is preserved under path addition. Thus, by Corollary 5.1.8, graph G is 2-connected.   ◇

Theorem 5.2.2: [**Whitney Synthesis Theorem**] *A graph G is 2-connected if and only if G is a cycle or a Whitney synthesis from a cycle.*

Proof: ($\Leftarrow$) Suppose that $C = G_0, G_1, \ldots, G_l = G$ is a Whitney synthesis from a cycle C. Since a cycle is 2-connected, iterative application of Lemma 5.2.1 implies that graph G_i is 2-connected for $i = 1, \ldots, l$. In particular, $G = G_l$ is 2-connected.

($\Rightarrow$) Suppose that G is a 2-connected graph, and let C be any cycle in G. Consider the collection $\mathcal{H}$ of all subgraphs of G that are Whitney syntheses from cycle C. Since the collection $\mathcal{H}$ is nonempty ($C \in \mathcal{H}$), there exists a subgraph $H^* \in \mathcal{H}$ with the maximum number of edges.

Suppose that $H^* \neq G$. Then, the connectedness of G implies that there exists an edge $e = vw \in E_G - E_{H^*}$ whose endpoint v lies in H^*. Since G is 2-connected, every edge is a cycle-edge, from which it follows that there exists a cycle containing edge e. Moreover, since endpoint v is not a cut-vertex, there must be at least one such cycle, say C_e, that meets subgraph H^* at a vertex other than v. Let z be the first vertex on C_e at which the cycle returns to H^* (see Figure 5.2.2). Then the portion of C_e from v to z that includes edge e is a path addition to H^*. Thus, H^* is extendible by a path addition, contradicting the maximality of H^*. Therefore, $H^* = G$. $\diamond$

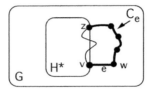

Figure 5.2.2   A path addition to H^*.

Using a similar strategy, we now establish Robbins' analogous characterization of 2-edge-connected graphs.

Lemma 5.2.3: *Let H be a 2-edge-connected (i.e., bridgeless) graph. Then the graph that results from a path or cycle addition to H is 2-edge-connected.*

Proof: The property that every edge is a cycle-edge is preserved under both path and cycle addition. $\diamond$

Theorem 5.2.4: [**Whitney-Robbins Synthesis Theorem**] *A graph G is 2-edge-connected if and only if G is a cycle or a Whitney-Robbins synthesis from a cycle.*

Proof: ($\Leftarrow$) Suppose that $C = G_0, G_1, \ldots, G_l = G$ is a Whitney-Robbins synthesis from a cycle C. Since a cycle is 2-edge-connected, iterative application of Lemma 5.2.3 implies that G is 2-edge-connected.

($\Rightarrow$) Suppose that G is a 2-edge-connected graph, and let C be any cycle in G. Among all subgraphs of G that are Whitney-Robbins syntheses from cycle C, let H^* be one with the maximum number of edges.

Suppose that $H^* \neq G$. As in the proof of Theorem 5.2.2, there exists an edge $e = vw \in E_G - E_{H^*}$ whose endpoint v lies in H^*. Moreover, edge e must be part of some cycle C_e (because G is 2-edge-connected). Again, let z be the first vertex at which the cycle returns to subgraph H^*. Because v can be a cut-vertex, there are now two possibilities, as shown in Figure 5.2.3.

Thus, H^* is extendible by a path addition or a cycle addition, contradicting the maximality of H^*. Therefore, $H^* = G$. $\diamond$

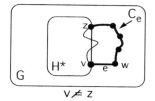 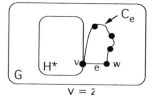

Figure 5.2.3   A path or cycle addition to H^*.

TERMINOLOGY: Path additions and cycle additions are often called, respectively, **open-ear** and **closed-ear additions**, since the new paths in drawings like Figure 5.2.3 are imagined to look like human ears. Moreover, the Whitney and Whitney-Robbins syntheses are called **ear decompositions**.

Tutte's Synthesis of 3-Connected Graphs

The next theorem is a constructive characterization of 3-connected graphs, analogous to Whitney's synthesis of 2-connected graphs. The result is due to W.T. Tutte, and a proof may be found in [Tu61]. Tutte's synthesis starts with a *wheel graph*.

REVIEW FROM §2.4: The **n-spoke wheel** (or **n-wheel**) W_n is the join $K_1 + C_n$ of a single vertex and an n-cycle. (The n-cycle forms the rim of the wheel, and the additional vertex is its hub.)

Example 5.2.2: The 5-spoke wheel W_5 is shown in Figure 5.2.4. It has six vertices.

Figure 5.2.4   The 5-spoke wheel W_5.

Theorem 5.2.5: [**Tutte Synthesis Theorem**] *A graph is 3-connected if and only if it is a wheel or can be obtained from a wheel by a sequence of operations of the following two types.*

1. *Adding an edge between two non-adjacent vertices.*
2. *Replacing a vertex v with degree at least 4 by two new vertices v^1 and v^2, joined by a new edge; each vertex that was adjacent to v in G is joined by an edge to exactly one of v^1 and v^2 so that $\deg(v^1) \geq 3$ and $\deg(v^2) \geq 3$.*   ◇ *([Tu61])*

Example 5.2.3: Figure 5.2.5 gives an illustration of Tutte's synthesis. The cube graph Q_3 is synthesized from the 4-spoke wheel W_4 in four steps, where all but the second step are operations of type 2.

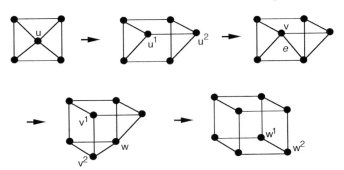

Figure 5.2.5   A 4-step Tutte synthesis of the cube graph Q_3.

Application 5.2.1:   *Construction of a Class of Reliable Networks*   In a communication network, the breakdown (deletion) of a vertex or edge from the graph representing that network is called a *fault*. Thus, the greater the vertex-connectivity and edge-connectivity, the more reliable the network. Naturally, the fewer the links used to achieve a given connectivity, the less costly the network is. This suggests the following optimization problem.

Given positive integers n and k, with $k < n$, find a k-connected n-vertex graph having the smallest possible number of edges.

   Let $h_k(n)$ denote the minimum number of edges that a k-connected graph on n vertices must have. Observe that $h_1(n) = n - 1$, since a tree is a connected graph with the minimum number of edges. The following proposition establishes a lower bound for $h_k(n)$.

DEFINITION: For any real number x, the **floor** of x, denoted $\lfloor x \rfloor$, is the greatest integer less than or equal to x, and $\lceil x \rceil$, the **ceiling** of x, is the smallest integer greater than or equal to x.

Proposition 5.2.6: *Let G be a k-connected graph on n vertices. Then the number of edges in G is at least $\left\lceil \frac{kn}{2} \right\rceil$. That is, $h_k(n) \geq \left\lceil \frac{kn}{2} \right\rceil$.*

Proof: Euler's Degree-Sum Theorem (§1.1) implies that $2\,|E_G| \geq n\delta_{\min}(G)$. The result now follows by Corollary 5.1.6.   ◇

Harary's Construction of an Optimal k-Connected Graph

   Frank Harary gave a procedure for constructing a k-connected graph $H_{k,n}$ on n vertices that has exactly $\left\lceil \frac{kn}{2} \right\rceil$ edges for $k \geq 2$ ([Ha62]). The construction of the **Harary graph** $H_{k,n}$ begins with an n-cycle graph, whose vertices are consecutively numbered $0, 1, 2, \ldots, n-1$ clockwise around its perimeter (see Figure 5.2.6 below).

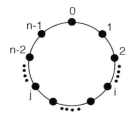

Figure 5.2.6   Harary construction starts with an n-cycle graph.

Adjacency between two vertices i and j in the graph $H_{k,n}$ is determined by the distance between i and j along the perimeter of the n-cycle. This distance is the length of the shorter of the two i-j paths on the perimeter and, hence, is the smaller of the two values $|j - i|$ and $n - |j - i|$.

DEFINITION: Let i and j be any two integers from the set $\{0, 1, \ldots, n-1\}$. The **mod** n **distance** between i and j, denoted $|j - i|_n$, is the smaller of the two values $|j - i|$ and $n - |j - i|$.

The construction of $H_{k,n}$ depends on the parity of k and n and falls into three cases. The construction for the first case (k even) is needed for the last two cases. Immediately following this description is an algorithm that encapsulates the construction for all three cases.

Case 1: k even

Let $k = 2r$. Vertices i and j are joined by an edge if $|j - i|_n \leq r$. Figure 5.2.7 shows the graphs $H_{4,8}$ and $H_{4,9}$.

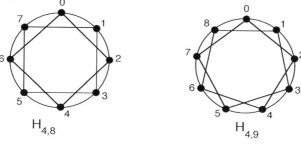

Figure 5.2.7   Harary graphs $H_{4,8}$ and $H_{4,9}$.

It is not hard to show that, $H_{2r,n}$ has exactly rn edges (see Exercises). Thus, $H_{2r,n}$ has the desired number of edges, since $rn = \frac{kn}{2} = \left\lceil \frac{kn}{2} \right\rceil$ (kn is even).

Case 2: k odd and n even

Let $k = 2r + 1$. Start with graph $H_{2r,n}$, and add the $\frac{n}{2}$ *diameters* of the original n-cycle. That is, an edge is drawn between vertices i and $i + \frac{n}{2}$, for $i = 0, \ldots, \frac{n}{2} - 1$. Thus, the total number of edges in $H_{2r+1,n}$ equals $rn + \frac{n}{2} = \frac{(2r+1)n}{2} = \frac{kn}{2} = \left\lceil \frac{kn}{2} \right\rceil$.

Graph $H_{5,8}$, shown in Figure 5.2.8 below, is obtained from $H_{4,8}$ by adding the four diameters.

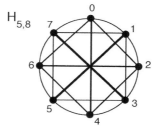

Figure 5.2.8   Harary graph $H_{5,8}$ *is* $H_{4,8}$ plus four diameters.

Case 3: k and n both odd

Let $k = 2r + 1$. Start with graph H_{2rn}, and add $\frac{n+1}{2}$ *quasi-diameters* as follows. First, draw an edge from vertex 0 to vertex $\frac{n-1}{2}$ and from vertex 0 to vertex $\frac{n+1}{2}$. Then draw an edge from vertex i to vertex $\left(i + \frac{n+1}{2}\right)$, for $i = 1, \ldots, \frac{n-3}{2}$. The total number of edges in $H_{2r+1,n}$ equals $rn + \frac{n+1}{2} = \frac{(2r+1)n+1}{2} = \left\lceil \frac{kn}{2} \right\rceil$ (since kn is odd).

Graph $H_{5,9}$, shown in Figure 5.2.9 below, is obtained from $H_{4,9}$ by adding five quasi-diameters.

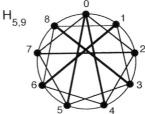

Figure 5.2.9   Harary graph $H_{5,9}$ is $H_{4,9}$ plus five quasi-diameters.

The following algorithm implements the Harary construction.

Algorithm 5.2.1:   Constructing an Optimal k-Connected n-Vertex Graph

Input: positive integers k and n, with $k < n$.
Output: Harary graph $H_{k,n}$ with the standard 0-based labeling.
   Initialize graph H to be n isolated vertices, with labels $0, \ldots, n$.
   Let $r = \left\lfloor \frac{k}{2} \right\rfloor$.
   For $i = 0$ to $n - 2$
      For $j = i + 1$ to $n - 1$
         If $j - i \le r$ OR $n + i - j \le r$
            Create an edge between vertices i and j.
   [*This completes the construction of $H_{2r,n}$.*]
   If k is even
      Return graph H.
   Else
      If n is even
         For $i = 0$ to $\frac{n}{2} - 1$
            Create an edge between vertex i and vertex $\left(i + \frac{n}{2}\right)$
      Else
         Create an edge from vertex 0 to vertex $\frac{n-1}{2}$.
         Create an edge from vertex 0 to vertex $\frac{n+1}{2}$
         For $i = 1$ to $\frac{n-3}{2}$
            Create an edge between vertex i and vertex $\left(i + \frac{n+1}{2}\right)$.
   Return graph H.

Theorem 5.2.7: *The Harary graph $H_{k,n}$ is k-connected.*

Proof: *Case 1: $k = 2r$*

Suppose that $2r - 1$ vertices $i_1, i_2, \ldots, i_{2r-1}$ are deleted from graph $H_{2r,n}$, and let x and y be any two of the remaining vertices. It suffices to show that there is an x-y path in the vertex-deletion subgraph $H_{2r,n} - \{i_1, i_2, \ldots, i_{2r-1}\}$.

The vertices x and y divide the perimeter of the original n-cycle into two sectors. One of these sectors contains the clockwise sequence of vertices from x to y, and the other sector contains the clockwise sequence from y to x. From one of these sequences, no more than

$r-1$ of the vertices $i_1, i_2, \ldots, i_{2r-1}$ were deleted. Assume, without loss of generality, that this is the sector that extends clockwise from x to y, and let S be the subsequence of vertices that remain between x and y in that sector after the deletions. The gap created between two consecutive vertices in subsequence S is the largest possible when the $r-1$ deleted vertices are consecutively numbered, say, from a to $a + r - 2$, using addition modulo n (see Figure 5.2.10 below). But even in this extreme case, the resulting gap is no bigger than r. That is, $|j - i|_n \leq r$, for any two consecutive vertices i and j in subsequence S.

Thus, by the construction, every pair of consecutive vertices in S is joined by an edge. Therefore, subsequence S is the vertex sequence of an x-y path in $H_{2r,n} - \{i_1, i_2, \ldots, i_{2r-1}\}$.   ◇ (Case 1)

Case 2: $k = 2r + 1$ and n even

Suppose that D is a set of $2r$ vertices that are deleted from graph $H_{2r+1,n}$, and let x and y be any two of the remaining vertices. It suffices to show that there is an x-y path in the vertex-deletion subgraph $H_{2r+1,n} - D$.

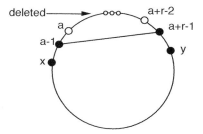

Figure 5.2.10

If one of the sectors contains fewer than r of the deleted vertices, then, as in Case 1, the subsequence of vertices left in that sector forms a path between vertices x and y. Therefore, assume both of the sectors contain exactly r of the deleted vertices. Further assume that both sets of deleted vertices are consecutively numbered, thereby creating the largest possible gap between consecutive vertices in each of the remaining subsequences.

Using addition modulo n, let $A = \{a, a+1, \ldots, a+r-1\}$ be the set of vertices that were deleted from one sector, and let $B = \{b, b+1, \ldots, b+r-1\}$ be the other set of deleted vertices. Each of the two subsets of remaining vertices is also consecutively numbered. One of these subsets starts with vertex $a + r$ and ends with vertex $b - 1$, and the other starts with $b + r$ and ends with $a - 1$ (see Figure 5.2.11).

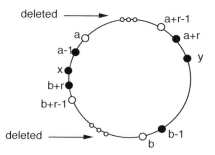

Figure 5.2.11   The four subsets of consecutively numbered vertices.

Let $l = |(b+r) - (a-1)|_n$ and let $w = (b+r) + \lfloor \frac{l}{2} \rfloor$. Then w is "halfway" between vertices $b+r$ and $a-1$, moving clockwise from $b+r$ to $a-1$. Now let $z = w + \frac{n}{2}$. Then z is halfway between $b+r$ and $a-1$, moving counterclockwise from $b+r$ to $a-1$. But both subsets of deleted vertices are of equal size r, which implies that z is in the other subset of remaining vertices (see Figure 5.2.12). Moreover, since $|z - w|_n = \frac{n}{2}$, there is an edge joining w and z, by the definition of $H_{2r+1,n}$ for Case 2.

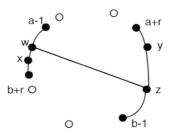

Figure 5.2.12   A diameter connecting the two remaining sectors.

Finally, there is a path from w to x, since both vertices are on a sector of the n-cycle left intact by the vertex deletions. By the same argument, there is a path from z to y. These two paths, together with the edge between w and z, show that there is an x-y path in the vertex-deletion subgraph $H_{2r+1,n} - D$.   ◇ (Case 2)

Case 3: $k = 2r + 1$ and n odd

The argument for Case 3 is similar to the one used for Case 2.   ◇ (*Exercises*)

The following result states that the Harary graphs are also optimal with respect to achieving a given edge-connectivity with a minimum number of edges.

Corollary 5.2.8: *The Harary graph $H(k,n)$ is a k-edge-connected, n-vertex graph with the fewest possible edges.*   ◇ (*Exercises*)

EXERCISES for Section 5.2

5.2.1[S] Find a Whitney synthesis of the 7-vertex wheel graph W_6.

5.2.2 Find a Whitney synthesis of the complete graph K_5.

5.2.3 Find a 3-step Whitney synthesis of the cube graph Q_3, starting with a 4-cycle.

5.2.4 For the 2-edge-connected graph shown, find a 3-step Whitney-Robbins synthesis from a cycle.

5.2.5[S] Find a 1-step Tutte synthesis of the complete bipartite graph $K_{3,3}$.

5.2.6 Use Tutte Synthesis Theorem 5.2.5 to show that the octahedral graph $\mathcal{O}_3$ is 3-connected.

5.2.7[S] Use Tutte Synthesis Theorem 5.2.5 to show that the Harary graph $H_{3,n}$ is 3-connected for all $n \geq 4$.

5.2.8 Show that the number of edges in the Harary graph $H_{2r,n}$ is rn.

In Exercises 5.2.9 through 5.2.12, use Algorithm 5.2.1 to construct the specified Harary graph.

5.2.9 $H_{4,7}$ **5.2.10** $H_{5,7}$ **5.2.11**^{S} $H_{6,8}$ **5.2.12** $H_{6,9}$

5.2.13 Show that Algorithm 5.2.1 correctly handles the three cases of the Harary construction.

5.2.14^S Prove Corollary 5.2.8.

5.2.15 Prove Case 3 of Theorem 5.2.7.

5.2.16 [*Computer Project*] Write a computer program whose inputs are two positive integers k and n, with $k < n$, and whose output is the adjacency matrix of the Harary graph $H_{k,n}$ with the vertices labeled $0, 1, \ldots, n-1$.

5.3   MAX-MIN DUALITY AND MENGER'S THEOREMS

Borrowing from operations research terminology, we consider certain *primal-dual pairs* of optimization problems that are intimately related. Usually, one of these problems involves the maximization of some objective function, while the other is a minimization problem. A feasible solution to one of the problems provides a bound for the optimal value of the other problem (this is sometimes referred to as *weak duality*), and the optimal value of one problem is equal to the optimal value of the other (*strong duality*). Menger's Theorems and their many variations epitomize this primal-dual relationship. The following terminology is used throughout this section.

DEFINITION: Let u and v be distinct vertices in a connected graph G. A vertex subset (or edge subset) S is u-v **separating** (or **separates** u and v), if the vertices u and v lie in different components of the deletion subgraph $G - S$.

Thus, a u-v separating vertex set is a vertex-cut, and a u-v separating edge set is an edge-cut. When the context is clear, the term u-v **separating set** will refer either to a u-v separating vertex set or to a u-v separating edge set.

Example 5.3.1: For the graph in Figure 5.3.1, the vertex-cut $\{x, w, z\}$ is a u-v separating set of vertices of minimum size, and the edge-cut $\{a, b, c, d, e\}$ is a u-v separating set of edges of minimum size. Notice that a minimize-size u-v separating set of edges (vertices) need not be a minimum-size edge-cut (vertex-cut). For instance the set $\{a, b, c, d, e\}$ is not a minimum-size edge-cut, because the set of edges incident on the 3-valent vertex y is an edge-cut of size 3.

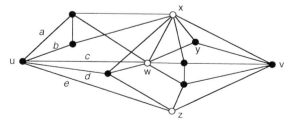

Figure 5.3.1   Vertex- and edge-cuts that are u-v separating sets.

A Primal-Dual Pair of Optimization Problems

The discussion in §5.1 suggests two different interpretations of a graph's connectivity. One interpretation is the number of vertices or edges it takes to disconnect the graph, and the other is the number of alternative paths joining any two given vertices of the graph.

Corresponding to these two perspectives are the following two optimization problems for two non-adjacent vertices u and v of a connected graph G.

Maximization Problem: Determine the maximum number of internally disjoint u-v paths in graph G.

Minimization Problem: Determine the minimum number of vertices of graph G needed to separate the vertices u and v.

Proposition 5.3.1: *(Weak Duality) Let u and v be any two non-adjacent vertices of a connected graph G. Let P_{uv} be a collection of internally disjoint u-v paths in G, and let S_{uv} be a u-v separating set of vertices in G. Then $|P_{uv}| \leq |S_{uv}|$.*

Proof: Since S_{uv} is a u-v separating set, each u-v path in P_{uv} must include at least one vertex of S_{uv}. Since the paths in P_{uv} are internally disjoint, no two of them can include the same vertex. Thus, the number of internally disjoint u-v paths in G is at most $|S_{uv}|$ (by the pigeonhole principle). $\diamond$

Corollary 5.3.2: *Let u and v be any two non-adjacent vertices of a connected graph G. Then the maximum number of internally disjoint u-v paths in G is less than or equal to the minimum size of a u-v separating set of vertices of G.*

The main result of this section is Menger's Theorem, which states that these two quantities are in fact equal. But even the weak duality result by itself provides *certificates of optimality*, as the following corollary shows. It follows directly from Proposition 5.3.1.

Corollary 5.3.3: **[Certificate of Optimality]** *Let u and v be any two non-adjacent vertices of a connected graph G. Suppose that P_{uv} is a collection of internally disjoint u-v paths in G, and that S_{uv} is a u-v separating set of vertices in G, such that $|P_{uv}| = |S_{uv}|$. Then P_{uv} is a maximum-size collection of internally disjoint u-v paths, and S_{uv} is a minimum-size u-v separating set.* $\diamond$ *(Exercises)*

Example 5.3.2: Consider the graph G shown in Figure 5.3.2. The vertex sequences $\langle u, x, y, t, v \rangle$, $\langle u, z, v \rangle$, and $\langle u, r, s, v \rangle$ represent a collection P of three internally disjoint u-v paths in G, and the set $S = \{y, \text{s}, z\}$ is a u-v separating set of size 3. Therefore, by Corollary 5.3.3, P is a maximum-size collection of internally disjoint u-v paths, and S is a minimum-size u-v separating set.

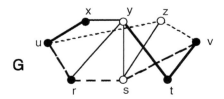

Figure 5.3.2

Menger's Theorem

The next theorem, proved by K. Menger in 1927, establishes a *strong duality* between the two optimization problems introduced earlier. The theorem and its variations are closely related (and in many cases, equivalent) to several other *max-min* duality results for graphs and directed networks and to several results outside graph theory as well. Some of these are stated in this section.

The proof, which involves several steps, appears at the end of this section so that we may first present a number of consequences and related results.

Theorem 5.3.4: [*Menger*] *Let u and v be distinct, non-adjacent vertices in a connected graph G. Then the maximum number of internally disjoint u-v paths in G equals the minimum number of vertices needed to separate u and v.*

Variations and Consequences of Menger's Theorem

The vertex-connectivity of a graph can be expressed in terms of the *local connectivity* between a given pair of vertices, and this relationship is used in the proof of Whitney's Theorem given below.

DEFINITION: Let s and t be non-adjacent vertices of a connected graph G. Then the **local vertex-connectivity** between s and t, denoted $\kappa_v(s,t)$, is the size of a smallest s-t separating vertex set in G.

Lemma 5.3.5: *Let G be a connected graph containing at least one pair of nonadjacent vertices. Then the vertex-connectivity $\kappa_v(G)$ is the minimum of the local vertex-connectivity $\kappa_v(s,t)$, taken over all pairs of non-adjacent vertices s and t.*

Proof: Since each s-t separating vertex set of the graph G is a vertex-cut, it follows that $\kappa_v(G) \leq \kappa_v(s,t)$ for all pairs of non-adjacent vertices s and t. Thus, $\kappa_v(G)$ is less than or equal to the minimum of $\kappa_v(s,t)$ over all non-adjacent s and t. On the other hand, if S is a vertex-cut of size $\kappa_v(G)$, then there are at least two vertices, say s and t, that lie in different components of the vertex-deletion subgraph $G - S$. But then $\kappa_v(s,t) \leq \kappa_v(G)$, which implies that the minimum of $\kappa_v(s,t)$ over all non-adjacent s and t is less than or equal to $\kappa_v(G)$. $\diamond$

The following variation of Menger's Theorem was published by Whitney in 1932. It generalizes the characterization of 2-connected graphs given by Theorem 5.1.7.

Theorem 5.3.6: [*Whitney's k-Connected Characterization*] *A nontrivial graph G is k-connected if and only if for each pair u, v of vertices, there are at least k internally disjoint u-v paths in G.*

Proof: If every two vertices in G are adjacent, then the special case of the vertex-connectivity definition applies, and the theorem assertion is immediately true. So assume that G has at least two non-adjacent vertices.

If G is k-connected, then there are at least k vertices in any vertex-cut of G. Thus, there are at least k vertices in any u-v separating set. Theorem 5.3.4 implies that the maximum number of internally disjoint u-v paths is at least k. Hence, there are at least k internally disjoint u-v paths.

Conversely, if for each pair of vertices u and v, there are at least k internally disjoint, u-v paths, then Proposition 5.3.1 implies that $\kappa_v(u,v) \geq k$, for each pair u, v of non-adjacent vertices. Therefore, $\kappa_v(G) \geq k$, by Lemma 5.3.5. $\diamond$

Corollary 5.3.7: *Let G be a k-connected graph and let $u, v_1, v_2, \ldots, v_k$ be any $k+1$ distinct vertices of G. Then there are paths P_i from u to v_i, for $i = 1, \ldots, k$, such that the collection $\{P_i\}$ is internally disjoint.*

Proof: Construct a new graph G^w from graph G by adding a new vertex w to G together with an edge joining w to v_i, for $i = 1, \ldots, k$, as in Figure 5.3.3.

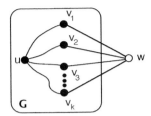

Figure 5.3.3   The graph G^w is constructed from G.

Since graph G is k-connected, it follows that graph G^w is also k-connected (by Exercise 5.1.18). By Theorem 5.3.6, there are k internally disjoint u-w paths in G^w. The u-v_i portions of these paths are k internally disjoint paths in G. $\Diamond$

The following theorem, proved by Dirac in 1960, generalizes one half of the characterization of 2-connected graphs given in Corollary 5.1.8.

Theorem 5.3.8: **[Dirac Cycle Theorem]** *Let G be a k-connected graph with at least $k+1$ vertices, for $k \geq 3$, and let U be any set of k vertices in G. Then there is a cycle in G containing all the vertices in U.*

Proof: Let C be a cycle in G that contains the maximum possible number of vertices of set U, and suppose that $\{v_1, \ldots, v_m\}$ is the subset of vertices of U that lie on C. By Corollary 5.1.8, $m \geq 2$. If there were a vertex $u \in U$ not on cycle C, then by Corollary 5.3.7, there would exist a set of internally disjoint paths P_i from u to v_i, one for each $i = 1, \ldots, m$. But then cycle C could be extended to include vertex u, by replacing the cycle edge between v_1 and v_2 by the paths P_1 and P_2 (see Figure 5.3.4), and this extended cycle would contradict the maximality of cycle C. $\Diamond$

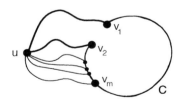

Figure 5.3.4   Extending cycle C to include vertex u.

Analogues of Menger's Theorem

Some of the results given at the beginning of this section have the following edge analogues. Each can be proved by mimicking the proof of the corresponding vertex version. The proofs are left as exercises.

Proposition 5.3.9: [*Edge Form of Certificate of Optimality*] *Let u and v be any two vertices of a connected graph G. Suppose P_{uv} is a collection of edge-disjoint u-v paths in G, and S_{uv} is a u-v separating set of edges in G, such that $|P_{uv}| = |S_{uv}|$. Then P_{uv} is the largest possible collection of edge-disjoint u-v paths, and S_{uv} is the smallest possible u-v separating set. In other words, each is an optimal solution to its respective problem.* ◇ *(Exercises)*

Theorem 5.3.10: [*Edge Form of Menger's Theorem*] *Let u and v be any two distinct vertices in a graph G. Then the minimum number of edges of G needed to separate u and v equals the maximum size of a set of edge-disjoint u-v paths in G.* ◇ *(Exercises)*

Theorem 5.3.11: [*Whitney's k-Edge-Connected Characterization*] *A nontrivial graph G is k-edge-connected if and only if for every two distinct vertices u and v of G, there are at least k edge-disjoint u-v paths in G.* ◇ *(Exercises)*

Example 5.3.3: For the graph shown in Figure 5.3.5, it is easy to find four edge-disjoint u-v paths and a u-v separating edge set of size 4. Thus, the maximum number of edge-disjoint u-v paths and the minimum size of a u-v separating set are both 4.

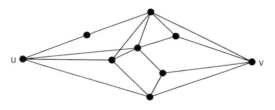

Figure 5.3.5

Proof of Menger's Theorem

The proof of Menger's Theorem given here is illustrative of a traditional-style proof in graph theory. A different proof, based on the theory of network flows, is given in Chapter 10, where various analogues of Menger's Theorem are also proved.

DEFINITION: Let W be a set of vertices in a graph G and x another vertex not in W. A **strict x-W path** is a path joining vertex x to a vertex in W and containing no other vertex of W. A **strict W-x path** is the reverse of a strict x-W path (i.e., its sequence of vertices and edges is in reverse order).

Example 5.3.4: Corresponding to the u-v separating set $W = \{y, s, z\}$ in the graph shown in Figure 5.3.6, the vertex sequences $\langle u, x, y \rangle$, $\langle u, r, y \rangle$, $\langle u, r, s \rangle$, and $\langle u, z \rangle$ represent the four strict u-W paths, and the three strict W-v paths are given by $\langle z, v \rangle$, $\langle y, t, v \rangle$, and $\langle s, v \rangle$.

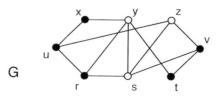

Figure 5.3.6

Proof: The proof uses induction on the number of edges. The smallest graph that satisfies the premises of the theorem is the path graph from u to v of length 2, and the theorem is trivially true for this graph. Assume that the theorem is true for all connected graphs having fewer than m edges, for some $m \geq 3$.

Now suppose that G is a connected graph with m edges, and let k be the minimum number of vertices needed to separate the vertices u and v. By Corollary 5.3.2, it suffices to show that there exist k internally disjoint u-v paths in G. Since this is clearly true if $k = 1$ (since G is connected), we may assume that $k \geq 2$.

Assertion 5.3.4a: If G contains a u-v path of length 2, then G contains k internally disjoint u-v paths.

Proof of 5.3.4a: Suppose that $P = \langle u, e_1, x, e_2, v \rangle$ is a path in G of length 2. Let W be a smallest u-v separating set for the vertex-deletion subgraph $G - x$. Since $W \cup \{x\}$ is a u-v separating set for G, the minimality of k implies that $|W| \geq k - 1$.

By the induction hypothesis, there are at least $k - 1$ internally disjoint $u - v$ paths in $G - x$. Path P is internally disjoint from any of these, and, hence, there are k internally disjoint u-v paths in G. $\diamond$ (Assertion 5.3.4a)

If there is a u-v separating set that contains a vertex adjacent to *both* vertices u and v, then Assertion 5.3.4a guarantees the existence of k internally disjoint u-v paths in G. The argument for *distance* $(u, v) \geq 3$ is broken into two cases, according to the kinds of u-v separating sets that exist in G.

In Case 1, there exists a u-v separating set W, as depicted on the left in Figure 5.3.7, where neither u nor v is adjacent to every vertex of W. In Case 2, no such separating set exists. Thus, in every u-v separating set for Case 2, either every vertex is adjacent to u or every vertex is adjacent to v, as shown on the right in the figure.

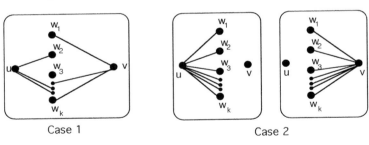

Case 1   Case 2

Figure 5.3.7   The two cases remaining in the proof of Menger's Theorem.

Case 1: There exists a u-v separating set $W = \{w_1, w_2, \ldots, w_k\}$ of vertices in G of minimum size k, such that neither u nor v is adjacent to every vertex in W.

Let G_u be the subgraph induced on the union of the edge-sets of all strict u-W paths in G, and let G_v be the subgraph induced on the union of edge-sets of all strict W-v paths (see Figure 5.3.8).

Assertion 5.3.4b: Both of the subgraphs G_u and G_v have more than k edges.

Proof of 5.3.4b: For each $w_i \in W$, there is a u-v path P_{w_i} in G on which w_i is the only vertex of W (otherwise, $W - \{w_i\}$ would still be a u-v separating set, contradicting the minimality of W). The u-w_i subpath of P_{w_i} is a strict u-W path that ends at w_i. Thus, the final edge of this strict u-W path is different for each w_i. Hence, G_u has at least k edges.

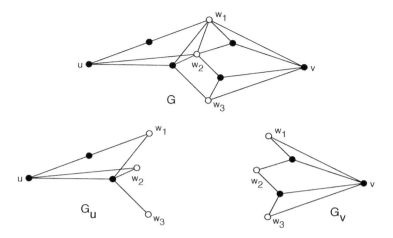

Figure 5.3.8   An example illustrating the subgraphs G_u and G_v.

The only way G_u could have exactly k edges would be if each of these strict u-W paths consisted of a single edge joining u and w_i, $i = 1, \ldots, k$. But this is ruled out by the condition for Case 1. Therefore, G_u has more than k edges. A similar argument shows that G_v also has more than k edges.   (Assertion 5.3.4b)

Assertion 5.3.4c: The subgraphs G_u and G_v have no edges in common.

Proof of 5.3.4c: By way of contradiction, suppose that the subgraphs G_u and G_v have an edge e in common. By the definitions of G_u and G_v, edge e is an edge of both a strict u-W path and a strict W-v path. Hence, at least one of its endpoints, say x, is not a vertex in the u-v separating set W (see Figure 5.3.9). But this implies the existence of a u-v path in $G - W$, which contradicts the definition of W. $\diamond$ (Assertion 5.3.4c)

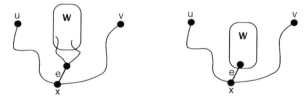

Figure 5.3.9   At least one of the endpoints of edge e lies outside W.

We now define two auxiliary graphs G_u^* and G_v^* : G_u^x is obtained from G by replacing the subgraph G_v with a new vertex v^* and drawing an edge from each vertex in W to v^* and G_v^* is obtained by replacing G_u with a new vertex u^* and drawing an edge from u^* to each vertex in W (see Figure 5.3.10).

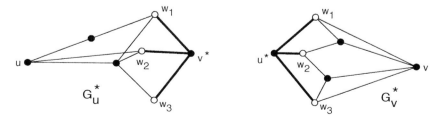

Figure 5.3.10   Illustration for the construction of graphs G_u^* and G_v^*.

Assertion 5.3.4d: Both of the auxiliary graphs G_u^* and G_v^* have fewer edges than G.

Proof of 5.3.4d: The following chain of inequalities shows that graph G_u^* has fewer edges than G.

$$|E_G| \geq |E_{G_u \cup G_u}| \quad \text{(since } G_u \cup G_v \text{ is a subgraph of } G)$$
$$= |E_{G_u}| + |E_{G_v}| \quad \text{(by Assertion 5.3.4c)}$$
$$> |E_{G_u}| + k \quad \text{(by Assertion 5.3.4b)}$$
$$= |E_{G_u^*}| \quad \text{(by the construction of } G_u^*)$$

A similar argument shows that G_v^* also has fewer edges than G. $\diamond$ (Assertion 5.3.4d)

By the construction of graphs G_u^* and G_v^*, every u-v^* separating set in graph G_u^* and every u^*-v separating set in graph G_v^* is a u-v separating set in graph G. Hence, the set W is a smallest u-v^* separating set in G_u^* and a smallest u^*-v separating set in G_v^*. Since G_u^* and G_v^* have fewer edges than G, the induction hypothesis implies the existence of two collections, P_u^* and P_v^*, of k internally disjoint u-v^* paths in G_u^* and k internally disjoint u^*-v paths in G_v^*, respectively (see Figure 5.3.11 below). For each w_i, one of the paths in P_u^* consists of a u-w_i path P_i' in G plus the new edge from w_i to v^*, and one of the paths in P_v^* consists of the new edge from u^* to w_i followed by a w_i-v path P_i'' in G.

Let P_i be the concatenation of paths P_i' and P_i'', for $i = 1, \ldots, k$. Then the set $\{P_i\}$ is a collection of k internally disjoint u-v paths in G. $\diamond$ (Case 1)

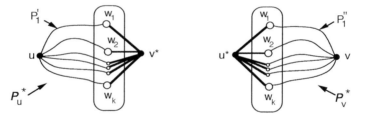

Figure 5.3.11   Each of the graphs G_u^* and G_v^* has k internally disjoint paths.

Case 2: Suppose that for each u-v separating set of size k, one of the vertices u or v is adjacent to all the vertices in that separating set.

Let $P = \langle u, e_1, x_1, e_2, x_2, \ldots, v \rangle$ be a shortest u-v path in G. By Assertion 5.3.4a, we can assume that P has length at least 3 and that vertex x_1 is not adjacent to vertex v. By Proposition 5.1.3, the edge-deletion subgraph $G - e_2$ is connected. Let S be a smallest u-v separating set in subgraph $G - e_2$ (see Figure 5.3.12). Then S is a u-v separating set in the vertex-deletion subgraph $G - x_1$ (since $G - x_1$ is a subgraph of $G - e_2$). Thus, $S \cup \{x_1\}$ is a u-v separating set in G, which implies that $|S| \geq k - 1$, by the minimality of k. On the other hand, the minimality of $|S|$ in $G - e_2$ implies that $|S| \leq k$, since every u-v separating set in G is also a u-v separating set in $G - e_2$.

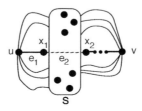

Figure 5.3.12   Completing Case 2 of Menger's Theorem.

If $|S| = k$, then, by the induction hypothesis, there are k internally disjoint u-v paths in $G - e_2$ and, hence, in G. If $|S| = k - 1$, then $x_i \notin S$, $i = 1, 2$ (otherwise $S - \{x_1\}$ would be a u-v separating set in G, contradicting the minimality of k). Thus, the sets $S \cup \{x_1\}$ and $S \cup \{x_2\}$ are both of size k and both u-v separating sets of G. The condition for Case 2 and the fact that vertex x_1 is not adjacent to v imply that every vertex in S is adjacent to vertex u. Hence, no vertex in S is adjacent to v (last there be a u-v path of length 2). But then condition for Case 2 applied to $S \cup \{x_2\}$ implies that vertex x_2 is adjacent to vertex u, which contradicts the minimality of path P and completes the proof. $\diamond$

Remark: Digraph versions of Menger's Theorem are also proved in Chapter 10. The assertions are the same, except that all paths are directed paths. The edge form and arc form of Menger's Theorem were proved by Ford and Fulkerson in 1955. Their approach, which is presented in Chapter 10, uses an algorithm to maximize the flow through a *capacitated network*, and a minimum cut for the network is a byproduct of the algorithm.

EXERCISES for Section 5.3

In Exercises 5.3.1 through 5.3.4, find the maximum number of internally disjoint u-v paths for the given graph, and use Certificate of Optimality (Corollary 5.3.3) to justify your answer.

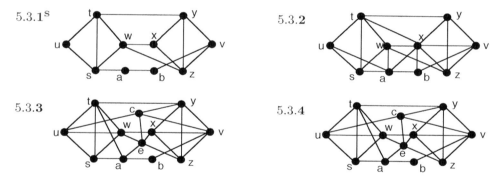

5.3.1^S

5.3.2

5.3.3

5.3.4

In Exercises 5.3.5 through 5.3.8, find the maximum number of edge-disjoint u-v paths for the specified graph, and use Edge Form of Certificate of Optimality (Proposition 5.3.9) to justify your answer.

5.3.5^S The graph of Exercise 5.3.1.   **5.3.6** The graph of Exercise 5.3.2.

5.3.7 The graph of Exercise 5.3.3.   **5.3.8** The graph of Exercise 5.3.4.

In Exercises 5.3.9 and 5.3.10, find the maximum number of arc-disjoint directed u-v paths for the given graph, and use the digraph version of Proposition 5.3.9 to justify your answer. That is, find k arc-disjoint u-v paths and a set of k arcs that separate vertices u and v, for some integer k.

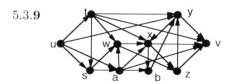

5.3.9

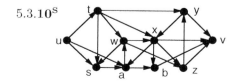

5.3.10^S

5.3.11 Prove Certificate of Optimality (Corollary 5.3.3).

5.3.12 Prove Edge Form of Certificate of Optimality (Proposition 5.3.9).

5.3.13 Prove Edge Form of Menger's Theorem 5.3.10.

5.3.14 Prove Whitney's k-Edge-Connected Characterization Theorem 5.3.11.

5.3.15^{S} Prove that Tutte Synthesis Theorem 5.2.5 always produces a 3-connected graph. (Hint: Develop an induction that uses Whitney's k-Connected Characterization Theorem 5.3.6.)

5.3.16 Let G be a simple k-connected graph with $k \geq 2$. Let S be a set of two edges and W a set of $k-2$ vertices. Prove that there exists a cycle in G containing the elements of S and W.

5.3.17 Let G be a simple k-connected graph. Let $W = \{w_1, w_2, \ldots, w_k\}$ be a set of k vertices and $v \in V_G - W$. Prove that there exists a v-w_i path P_i, $i = 1, \ldots, k$, such that the collection $\{P_i\}$ of k paths is internally disjoint.

5.4   BLOCK DECOMPOSITIONS

DEFINITION: A **block** of a loopless graph is a maximal connected subgraph H such that no vertex of H is a cut-vertex of H.

Thus, if a block has at least three vertices, then it is a maximal 2-connected subgraph. The only other types of blocks (in a loopless graph) are isolated vertices or dipoles (2-vertex graphs with a single edge or a multi-edge).

Remark: The blocks of a graph G are the blocks of the components of G and can therefore be identified one component of G at a time. Also, self-loops (or their absence) have no effect on the connectivity of a graph. For these reasons we assume throughout this section (except for the final subsection) that all graphs under consideration are loopless and connected.

Example 5.4.1: The graph in Figure 5.4.1 has four blocks; they are the subgraphs induced on the vertex subsets $\{t, u, w, v\}$, $\{w, x\}$, $\{x, y, z\}$, and $\{y, s\}$.

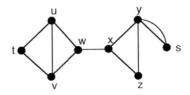

Figure 5.4.1   A graph with four blocks.

Remark: By definition, a block H of a graph G has no cut-vertices (of H), but H may contain vertices that are cut-vertices of G. For instance, in the above figure, the vertices w, x, and y are cut vertices of G.

The complete graphs K_n have no cut-vertices. The next result concerns the other extreme.

Proposition 5.4.1: *Every nontrivial connected graph G contains two or more vertices that are not cut-vertices.*

Proof: Choose two 1-valent vertices of a spanning tree of graph G.   ◇

Proposition 5.4.2: *Two different blocks of a graph can have at most one vertex in common.*

Proof: Let B_1 and B_2 be two different blocks of a graph G, and suppose that x and y are vertices in $B_1 \cap B_2$. Since the vertex-deletion subgraph $B_1 - x$ is a connected subgraph of B_1, there is a path in $B_1 - x$ between any given vertex w_1 in $B_1 - x$ and vertex y. Similarly, there is a path in $B_2 - x$ from vertex y to any given vertex w_2 in $B_2 - x$ (see Figure 5.4.2). The concatenation of these two paths is a w_1-w_2 walk in the vertex-deletion subgraph $(B_1 \cup B_2) - x$, which shows that x is not a cut-vertex of the subgraph $B_1 \cup B_2$. The same argument shows that no other vertex in $B_1 \cap B_2$ is a cut-vertex of $B_1 \cup B_2$. Moreover, none of the vertices that are in exactly one of the B_i's is a cut-vertex of $B_1 \cup B_2$, since such a vertex would be a cut-vertex of that block B_i. Thus, the subgraph $B_1 \cup B_2$ has no cut-vertices, which contradicts the maximality of blocks B_1 and B_2.   ◇

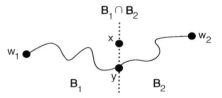

Figure 5.4.2   Two blocks cannot have more than one vertex in common.

The following assertions are immediate consequences of Proposition 5.4.2.

Corollary 5.4.3: *The edge-sets of the blocks of a graph G partition E_G.*
                                                                ◇ *(Exercises)*

Corollary 5.4.4: *Let x be a vertex in a graph G. Then x is a cut-vertex of G if and only if x is in two different blocks of G.*
                                                                ◇ *(Exercises)*

Corollary 5.4.5: *Let B_1 and B_2 be distinct blocks of a connected graph G. Let y_1 and y_2 be vertices in B_1 and B_2, respectively, such that neither is a cut-vertex of G. Then vertex y_1 is not adjacent to vertex y_2.*
                                                                ◇ *(Exercises)*

DEFINITION: The **block graph** of a graph G, denoted $BL(G)$, is the graph whose vertices correspond to the blocks of G, such that two vertices of $BL(G)$ are joined by a single edge whenever the corresponding blocks have a vertex in common.

Example 5.4.2: Figure 5.4.3 shows a graph G and its block graph $BL(G)$.

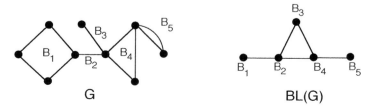

Figure 5.4.3   A graph and its block graph.

DEFINITION: A **leaf block** of a graph G is a block that contains exactly one cut-vertex of G.

The following result is used in §8.1 to prove *Brooks's Theorem* concerning the *chromatic number* of graph.

Proposition 5.4.6: *Let G be a connected graph with at least one cut-vertex. Then G has at least two leaf blocks.*   ◇ *(Exercises)*

Finding the Blocks of a Graph

In §4.4, depth-first search was used to find the cut-vertices of a connected graph (Algorithm 4.4.3). The following algorithm, which uses Algorithm 4.4.3 as a subroutine, finds the blocks of connected graph. Recall from §4.4 that $low(w)$ is the smallest *dfnumber* among all vertices in the depth-first tree that are joined to some descendant of vertex w by a non-tree edge.

Algorithm 5.4.1:   **Block-Finding**

  Input: a connected graph G.

  Output: the vertex-sets $B_1, B_2, \ldots, B_l$ of the blocks of G.

    Apply Algorithm 4.4.3 to find the set K of cut-vertices of graph G.

    Initialize the block counter $i := 0$.

    For each cut-vertex v in set K (in order of decreasing *dfnumber*)

      For each child w of v in depth-first search tree T

        If $low(w) \geq dfnumber(v)$

          Let T^w be the subtree of T rooted at w.

          $i := i + 1$

          $B_i := V_{T^w} \cup \{v\}$

          $T := T - V_{T^w}$

    Return sets $B_1, B_2, \ldots, B_i$.

COMPUTATIONAL NOTE: With some relatively minor modifications of Algorithm 5.4.1, the cut-vertices and blocks of a graph can be found in one pass of a depth-first search (see e.g., [AhHoUl83], [Ba83], [ThSw92]). A similar one-pass algorithm for finding the strongly connected components of a digraph is given in §9.5.

Block Decomposition of Graphs with Self-Loops

In a graph with self-loops, each self-loop and its endpoint are regarded as a distinct block, isomorphic to the bouquet B_1. The other blocks of such a graph are exactly the same as if the self-loops were not present. This extended concept of block decomposition preserves the property that the blocks partition the edge-set.

Example 5.4.3: The block decomposition of the graph G shown in Figure 5.4.4 contains five blocks, three of which are self-loops.

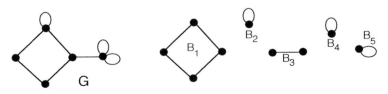

Figure 5.4.4   A graph G and its five blocks.

EXERCISES for Section 5.4

In Exercises 5.4.1 through 5.4.4, identify the blocks in the given graph and draw the block graph.

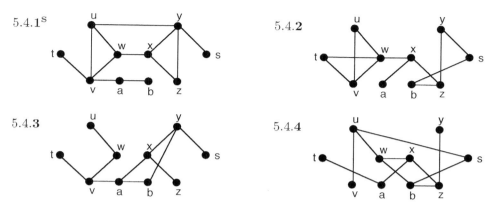

5.4.1^S

5.4.2

5.4.3

5.4.4

5.4.5^S Draw a graph whose block graph is the complete graph K_3.

5.4.6 Find two non-isomorphic connected graphs with six vertices, six edges, and three blocks.

5.4.7 Find a graph whose block graph is the n-cycle graph C_n.

5.4.8^S How many non-isomorphic simple connected graphs are there that have seven vertices, seven edges, and three blocks?

In Exercises 5.4.9 through 5.4.12, apply Algorithm 5.4.1 to the specified graph.

5.4.9 The graph of Exercise 5.4.1.   5.4.10 The graph of Exercise 5.4.2.

5.4.11 The graph of Exercise 5.4.3.   5.4.12 The graph of Exercise 5.4.4.

5.4.13 Prove or disprove: Every simple graph is the block graph of some graph.

5.4.14^S Prove or disprove: Two graphs are isomorphic if and only if their block graphs are isomorphic.

5.4.15 Prove Corollary 5.4.3.

5.4.16 Prove Corollary 5.4.4.

5.4.17 Prove Corollary 5.4.5.

DEFINITION: Let G be a simple connected graph with at least two blocks. The ***block-cutpoint graph*** $bc(G)$ of G is the bipartite graph with vertex bipartition $\langle V_b, V_c \rangle$, where the vertices in V_b bijectively correspond to the blocks of G and the vertices in V_c bijectively correspond to the cut-vertices of G, and where vertex V_b is adjacent to vertex v_c if cut-vertex c is a vertex of block b.

In Exercises 5.4.18 through 5.4.21, draw the block-cutpoint graph bc(G) of the specified graph G, and identify the vertices in bc(G) that correspond to leaf blocks of G.

5.4.**18**[S] The graph of Exercise 5.4.1.   5.4.**19** The graph of Exercise 5.4.2.

5.4.**20** The graph of Exercise 5.4.3.   5.4.**21** The graph of Exercise 5.4.4.

5.4.**22**[S] Let G be a simple connected graph with at least two blocks. Prove that the block-cutpoint graph $bc(\mathrm{G})$ is a tree.

5.4.**23** Prove Proposition 5.4.6. (**Hint**: See Exercise 5.4.22.)

5.4.**24** [*Computer Project*] Implement Algorithm 5.4.1 and run the program, using each of the graphs in Exercises 5.4.1 through 5.4.4 as input.

5.5   SUPPLEMENTARY EXERCISES

5.5.**1** Calculate the vertex-connectivity of $K_{4,7}$.

5.5.**2** Prove that the complete bipartite graph $K_{m,m}$ is m-connected.

5.5.**3** Prove that the vertex-connectivity of the hypercube graph Q_n is n.

5.5.**4** In the n-dimensional hypercube graph Q_n, find n edge-disjoint paths from $(0,0,\ldots,0)$ to $(1,1,\ldots,1)$.

5.5.**5** Prove that deleting two vertices from the graph below is insufficient to separate vertex s from vertex t.

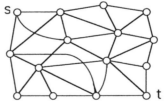

5.5.**6** How many vertices must be removed from the graph below to separate vertex s from vertex t?

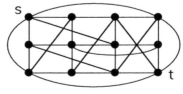

Glossary

block of a loopless graph G: a maximal connected subgraph H of G such that no vertex of H is a cut-vertex of H; in a graph with self-loops, each self-loop and its endpoint are regarded as a distinct block, isomorphic to the bouquet B_1.

—, **leaf:** a block that contains exactly one cut-vertex.

block-cutpoint graph $bc(G)$ of a simple connected graph G with at least two blocks: the bipartite graph with vertex bipartition $\langle V_b, V_c \rangle$, where the vertices in V_b bijectively correspond to the blocks of G and the vertices in V_c bijectively correspond to the cut-vertices of G, and where vertex v_b is adjacent to vertex v_c if cut-vertex c is a vertex of block b.

block graph $BL(G)$ of a graph G: the graph whose vertices correspond to the blocks of G, such that two vertices of $BL(G)$ are joined by a single edge whenever the corresponding blocks have a vertex in common.

bridge: synonym for *cut-edge*.

ceiling $\lceil x \rceil$ of a real number x: the smallest integer greater than or equal to x.

closed-ear addition: synonym for *cycle addition*.

k-connected graph G: a connected graph with $\kappa_v(G) \geq k$.

cut-edge of a graph: an edge whose removal increases the number of components.

cutpoint: a synonym for *cut-vertex*.

cut-vertex of a graph: a vertex whose removal increases the number of components.

cycle addition to a graph G: the addition of a cycle that has exactly one vertex in common with G.

disconnects a connected graph G: said of a subset of vertices or edges whose deletion from G results in a non-connected graph.

ear decomposition: a *Whitney* or *Whitney-Robbins synthesis*.

—, **closed-:** a *Whitney-Robbins synthesis*.

—, **open-:** a *Whitney synthesis*.

k-edge-connected graph G: a connected graph such that every edge-cut has at least k edges (i.e., $\kappa_e(G) \geq k$.

edge-connectivity of a connected graph G: the minimum number of edges whose removal can disconnect G; denoted $\kappa_e(G)$.

edge-cut in a graph G: a subset D of edges such that $G - D$ has more components than G.

floor $\lfloor x \rfloor$ of a real number x: the greatest integer less than or equal to x.

Harary graph $H_{k,n}$: a k-connected, n-vertex simple graph with the fewest edges possible, whose construction is due to Frank Harary (see §5.2).

internally disjoint paths: paths that have no internal vertex in common.

leaf block: see *block, leaf.*

local vertex-connectivity between non-adjacent vertices s and t: the size of a smallest s-t separating vertex set; denoted $\kappa_v(s,t)$.

mod n distance $|j - i|_n$ between integers $i, j \in \{0, 1, \ldots, n-1\}$: the smaller of the two values $|j - i|$ and $n - |j - i|$; used in Harary's construction of optimal k-connected graphs (§5.2).

open-ear addition: synonym for *path addition.*

path addition to a graph G: the addition to G of a path between two existing vertices of G, such that the edges and internal vertices of the path are new.

u-v separating set S of vertices or of edges: a set S such that the vertices u and v lie in different components of the deletion subgraph $G - S$.

strict x-W path: a path joining vertex x to a vertex in vertex set W and containing no other vertex of W. A strict W-x path is the reverse of a strict x-W path.

Tutte synthesis: a constructive characterization of 3-connected graphs that starts with a wheel graph; see Theorem 5.2.5 [Tutte Synthesis Theorem].

unicyclic graph: a connected graph with exactly one cycle.

vertex-connectivity of a connected graph G: the minimum number of vertices whose removal can disconnect G or reduce it to a 1-vertex graph; denoted $\kappa_v(G)$.

——, local between non-adjacent vertices s and t: the size of a smallest s-t separating vertex set; denoted $\kappa_v(s,t)$.

vertex-cut in a graph G: a subset U of vertices such that $G - U$ has more components than G.

(n-spoked) wheel graph W_n: the suspension of the n-vertex cycle graph from a new vertex x. That is, $W_n = C_n + x$.

Whitney synthesis: a Whitney-Robbins synthesis that uses path additions *only.*

Whitney-Robbins synthesis of a graph G from a graph H: a sequence of graphs, G_0, $G_1, \ldots, G_l$, where $G_0 = H$, $G_l = G$, and G_i is the result of a path or cycle addition to G_{i-1}, for $i = 1, \ldots, l$.

Chapter 6

OPTIMAL GRAPH TRAVERSALS

INTRODUCTION

Walks in a graph that use all the edges or all the vertices are often called *graph traversals*. A variety of practical problems can be posed in terms of graph traversals, and they tend to fall into one of two categories. The first half of the chapter deals with *Eulerian-type* problems, which require that each edge be traversed at least once. The second half is concerned with *Hamiltonian-type* problems, which involve traversals that must visit each vertex at least once.

Sometimes lurking beneath the surface of an application is some version of an Eulerian or Hamiltonian problem. Recognizing such a problem allows a host of strategies and algorithms to be brought to bear on its solution. A number of applications where this occurs are given. The classical optimization applications are known as the *Chinese Postman Problem* and the *Traveling Salesman Problem*.

On the surface, the two types of problems are similar, one involving an edge-based condition, the other an analogous, vertex-based condition. But the similarity ends there. Most Eulerian-type problems yield to polynomial-time algorithms, stemming from a simple characterization of Eulerian graphs in terms of vertex degrees. In sharp contrast, no simple characterization for Hamiltonian graphs is known, and most Hamiltonian-type problems are notoriously time-consuming to solve, classified as NP-hard. Practical problems where cost of solution is important call for strategies called *heuristics* that produce answers quickly, forfeiting the guarantee of optimality. Some of the most commonly used heuristics are presented in §6.4.

6.1 EULERIAN TRAILS AND TOURS

REVIEW FROM §1.5:

- An *Eulerian trail* in a graph is a trail that contains every edge of that graph.
- An *Eulerian tour* is a closed Eulerian trail.
- An *Eulerian graph* is a graph that has an Eulerian tour.

REVIEW FROM §4.5: [*Eulerian-Graph Characterization*] *The following statements are equivalent for a connected graph G.*

1. G is Eulerian.

2. The degree of every vertex in G is even.

3. E_G is the union of the edge-sets of a set of edge-disjoint cycles in G.

Königsberg Bridges Problem

In the town of Königsberg in what was once East Prussia, the two branches of the River Pregel converge and flow through to the Baltic Sea. Parts of the town were on an island and a headland that were joined to the outer river banks and to each other by seven bridges, as, shown in Figure 6.1.1. The townspeople wanted to know if it was possible to take a walk that crossed each of the bridges exactly once before returning to the starting point. The Prussian emperor, Frederick the Great, brought the problem to the attention of the famous Swiss mathematician Leonhard Euler, and Euler proved that no such walk was possible. His solution in 1736 is generally regarded as the origin of graph theory.

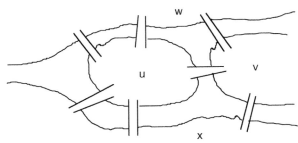

Figure 6.1.1   The seven bridges of Königsberg.

Euler's model for the problem was a graph with four vertices, one for each of the land masses u, v, w, and x, and seven edges, one for each of the bridges, as shown in Figure 6.1.2 below. In modern terminology, the problem was to determine whether that graph is Eulerian.

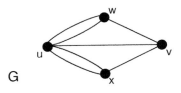

Figure 6.1.2   Graph representation of Königsberg.

Since the graph has vertices of odd degree, there is no Eulerian tour, by the Eulerian-Graph Characterization (§4.5). The argument given in the first part of the proof of that theorem is essentially the one that Euler used back in 1736.

The second and third parts of that proof show that a graph whose vertices all have even degree must be Eulerian. The essence of those arguments is embodied in the Eulerian Tour Algorithm given below, developed by Hierholzer in 1873, which constructs an Eulerian tour for any connected graph whose vertices are all of even degree.

Algorithm 6.1.1:   Eulerian Tour

    Input: a connected graph G whose vertices all have even degree.
    Output: an Eulerian tour T.
        Start at any vertex v, and construct a closed trail T in G.
        While there are edges of G not already in trail T
            Choose any vertex w in T that is incident with an unused edge.
            Starting at vertex w, construct a closed trail D of unused edges.
            Enlarge trail T by splicing trail D into T at vertex w.
        Return T.

COMPUTATIONAL NOTE: A modified *depth-first search* (§4.2), in which every unused edge remains in the stack, can be used to construct the closed trails.

Example 6.1.1: The key step in Algorithm 6.1.1 is enlarging a closed trail by combining it with a second closed trail — the *detour*. To illustrate, consider the closed trails $T = \langle t_1, t_2, t_3, t_4 \rangle$ and $D = \langle d_1, d_2, d_3 \rangle$ in the graph shown in Figure 6.1.3. The closed trail that results when detour D is spliced into trail T at vertex w is given by $T' = \langle t_1, t_2, d_1, d_2, d_3, t_3, t_4 \rangle$. At the next iteration, the trail $\langle e_1, e_2, e_3 \rangle$ is spliced into trail T', resulting in an Eulerian tour of the entire graph.

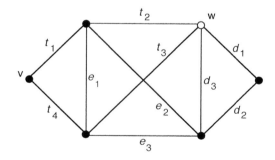

Figure 6.1.3   Graph representation of Königsberg.

Open Eulerian Trails

Certain applications require *open* Eulerian trails, which do not end at the starting vertex. The following characterization of graphs having an open Eulerian trail is obtained directly from the characterization of Eulerian graphs.

Theorem 6.1.1: *A connected graph G has an open Eulerian trail if and only if it has exactly two vertices of odd degree. Furthermore, the initial and final vertices of an open Eulerian trail must be the two vertices of odd degree.*

Proof: ($\Rightarrow$) Suppose that $\langle x, e_1, v_1, e_2, \ldots, e_m, y \rangle$ is an open Eulerian trail in G. Adding a new edge e joining vertices x and y creates a new graph $G^* = G + e$ with an Eulerian tour $\langle x, e_1, v_1, e_2, \ldots, e_m, y, e, x \rangle$. By the Eulerian-graph characterization, the degree in the Eulerian graph G^* of every vertex must be even, and thus, the degree in graph G of every vertex except x and y is even.

($\Leftarrow$) Suppose that x and y are the only two vertices of graph G with odd degree. If e is a new edge joining x and y, then all the vertices of the resulting graph G^* have even degree. It follows from the Eulerian-graph characterization that graph G^* has an Eulerian tour T. Hence, the trail $T - e$ obtained by deleting edge e from tour T is an open Eulerian trail of $G^* - e = G$. $\diamond$

Eulerian Trails in Digraphs

The use of the term *Eulerian* is identical for digraphs, except that all trails are directed. The characterizations of graphs that have Eulerian tours or open Eulerian trails apply to digraphs as well. Their proofs are essentially the same as the ones for undirected graphs and are left as exercises.

Theorem 6.1.2: [**Eulerian-Digraph Characterization**] *The following statements are equivalent for a connected digraph D.*

1. *Digraph D is Eulerian.*

2. *For every vertex v in D, $indegree(v) = outdegree(v)$.*

3. *The edge-set E_D is the union of the edge-sets of a set of edge-disjoint directed cycles in digraph D.* $\diamond$ *(Exercises)*

Theorem 6.1.3: *A connected digraph has an open Eulerian trail from vertex x to vertex y if and only if $indegree(x) + 1 = outdegree(x)$, $indegree(y) = outdegree(y) + 1$, and all vertices except x and y have equal indegree and outdegree.* $\diamond$ *(Exercises)*

Tarry's Maze-Searching Algorithm Revisited

In §4.4, we recognized that Tarry's maze-searching algorithm follows a depth-first search strategy. In the context of this section, the algorithm produces an Eulerian tour of the graph obtained by duplicating each edge of the input graph. Tarry's algorithm is just one of several maze-searching algorithms. For a more extensive study, see [F191].

DEFINITION: A **double tracing** is a closed walk that traverses every edge exactly twice. A double tracing is **bidirectional** if every edge is used once in each of its two directions.

Proposition 6.1.4: *Every connected graph has a bidirectional double tracing. In a tree, every double tracing is bidirectional* $\diamond$ *(Exercises)*

EXERCISES for Section 6.1

In Exercises 6.1.1 through 6.1.4, determine which graphs in the given graph family are Eulerian.

6.1.1[S] The complete graph K_n.

6.1.3 The n-vertex wheel W_n.

6.1.2 The complete bipartite graph $K_{m,n}$.

6.1.4 The hypercube graph Q_n.

6.1.5[S] Which platonic graphs (§1.2) are Eulerian?

In Exercises 6.1.6 through 6.1.9, apply Algorithm 6.1.1 to construct an Eulerian tour of the given graph. Begin the construction at vertex s.

6.1.6

6.1.7[S]

6.1.8

6.1.9

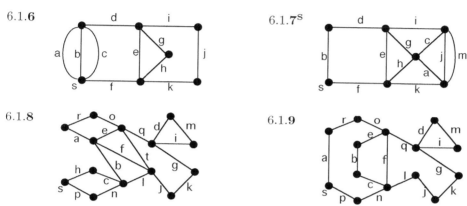

6.1.10[S] Use the strategy of the second half of the proof of Theorem 6.1.1 to design a variation of Algorithm 6.1.1 that constructs an open Eulerian trail in any graph that has exactly two vertices of odd degree.

In Exercises 6.1.11 through 6.1.14, use the modified version of Algorithm 6.1.1 from Exercise 6.1.10 to construct an open Eulerian trail in the given graph.

6.1.11

6.1.12

6.1.13[S]

6.1.14

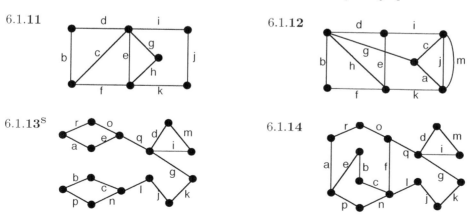

6.1.15[S] Design a variation of Algorithm 6.1.1 that constructs an Eulerian tour of any Eulerian digraph.

In Exercises 6.1.16 through 6.1.19, use an appropriate modification of Algorithm 6.1.1 to construct an Eulerian tour or open Eulerian trail in the given digraph.

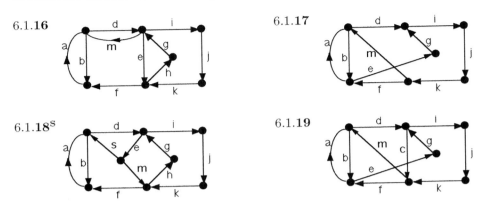

6.1.16   6.1.17   6.1.18^S 6.1.19

6.1.20 Referring to the initial step of Algorithm 6.1.1, why is it always possible to construct a closed trail, regardless of the choice of starting vertex?

6.1.21^S Referring to the while-loop in Algorithm 6.1.1, why is it always possible to find a vertex on trail T that is incident on an unused edge?

6.1.22 Prove Theorem 6.1.2. (Hint: See proof of Eulerian-graph characterization in §4.5.)

6.1.23 Prove Theorem 6.1.3. (Hint: Use Theorem 6.1.2.)

6.1.24 Prove Proposition 6.1.4.

6.1.25^S Prove that if a simple graph G is Eulerian, then its line graph $L(G)$ (§1.2) is Eulerian.

6.1.26 Prove or disprove: The line graph of *any* Eulerian graph is Eulerian.

6.1.27^S Prove or disprove: If the line graph of a graph G is Eulerian, then G is Eulerian.

6.1.28 Prove that if a connected graph G has $2k$ vertices of odd degree, then there is a set of k edge-disjoint trails that use all the edges of G.

6.1.29 [*Computer Project*] Implement Algorithm 6.1.1 and run the program on each of the graphs in Exercises 6.1.6 through 6.1.9.

6.1.30 [*Computer Project*] Implement a modified version of Algorithm 6.1.1 so that it finds an open Eulerian trail in a graph that has exactly two vertices of odd degree. Run the program on the graphs in Exercises 6.1.11 through 6.1.14.

6.2   DEBRUIJN SEQUENCES AND POSTMAN PROBLEMS

DEFINITION: A bitstring of length 2^n is called a $(2, n)$-***deBruijn sequence*** if each of the 2^n possible bitstrings of length n occurs *exactly once* as a substring, where wraparound is allowed.

Example 6.2.1: *The following four bitstrings are $(2, n)$-deBruijn sequences for the cases* $n = 1, 2, 3, 4$, *respectively.*

$$01 \qquad 0110 \qquad 01110100 \qquad 0000100110101111$$

Application 6.2.1: *Identifying the Position of a Rotating Drum* Suppose that a rotating drum has 16 different sectors. We can assign a 0 or 1 to each of the 16 sectors by putting conducting material in some of the sectors and nonconducting material in others. Sensors can then be placed in a fixed position to "read" a 4-bit string corresponding to the four sectors that are positioned there. For example, in Figure 6.2.1, the sensors read 1010.

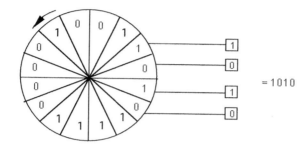

Figure 6.2.1   A rotating drum with 16 sectors and 4 sensors.

Observe that the bitstring of length 16 formed by starting at any sector of the drum and moving clockwise (or counterclockwise) around the drum is a $(2, 4)$-deBruijn sequence. This means that the position of the drum can be automatically determined by the 4-bit substring that the sensors pick up. This is more space-efficient than writing a 4-bit identifying code in each sector.

In general, $(2, n)$-deBruijn sequences can be constructed for any positive integer n. In fact, the construction outlined below can be generalized to produce a (p, n)-deBruijn sequence for any positive integers p and n, where a (p, n)-sequence consists of strings of characters from a set of p characters instead of just a binary set. The Dutch mathematician N. DeBruijn used a digraph to construct such sequences, based on an Eulerian tour.

DEFINITION: The $(2, n)$-**deBruijn digraph** $D_{2,n}$ consists of 2^{n-1} vertices, labeled by the bitstrings of length $n - 1$, and 2^n arcs, labeled by the bitstrings of length n. The arc from vertex $b_1 b_2 \cdots b_{n-1}$ to vertex $b_2 \cdots b_{n-1} b_n$ is labeled $b_1 b_2 \cdots b_n$.

Example 6.2.2: Figure 6.2.2 shows the $(2, 4)$-deBruijn digraph.

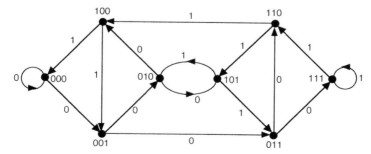

Figure 6.2.2   The $(2, 4)$-deBruijn digraph $D_{2,4}$.

To simplify the subsequent discussion, each arc in the figure is labeled with only the *leftmost bit* of its full label. Thus, the actual arc label is the string consisting of the single bit that appears on that arc, followed by the label of the vertex at the head of the arc. For instance, the actual label for the arc directed from 000 to 001 is 0001.

Proposition 6.2.1: *The $(2, n)$-deBruijn digraph $D_{2,n}$ is Eulerian.*

Proof: The deBruijn graph $D_{2,n}$ is strongly connected, since if $a_1 a_2 \cdots a_{n-1}$ and $b_1 b_2 \cdots b_{n-1}$ are any two vertices of $D_{2,n}$, then the vertex sequence:

$$a_1 a_2 \cdots a_{n-1}; \quad a_2 \cdots a_{n-1} b_1; \quad a_3 \cdots a_{n-1} b_1 b_2; \quad \ldots; \quad b_1 b_2 \cdots b_{n-1}$$

defines a directed trail from $a_1 a_2 \cdots a_{n-1}$ to $b_1 b_2 \cdots b_{n-1}$. Moreover, for every vertex $b_1 b_2 \cdots b_{n-1}$ of $D_{2,n}$, the only outgoing arcs from $b_1 b_2 \cdots b_{n-1}$ are $b_1 b_2 \cdots b_{n-1} 0$ and $b_1 b_2 \cdots b_{n-1} 1$, and the only incoming arcs to $b_1 b_2 \cdots b_{n-1}$ are $0 b_1 b_2 \cdots b_{n-1}$ and $1 b_1 b_2 \cdots b_{n-1}$. Thus, $indegree(b_1 b_2 \cdots b_{n-1}) = outdegree(b_1 b_2 \cdots b_{n-1}) = 2$, which implies, by Theorem 6.1.2, that $D_{2,n}$ is Eulerian. $\diamondsuit$

Algorithm 6.2.1:   **Constructing a $(2, n)$-deBruijn Sequence**
  Input: a positive integer n.
  Output: a $(2, n)$-deBruijn sequence S.
      Construct the $(2, n)$-deBruijn digraph $D_{2,n}$.
      Choose a vertex v.
      Construct a directed Eulerian tour T of $D_{2,n}$ starting at v.
      Initialize sequence S as the empty sequence $\langle \ \rangle$.
      For each arc e in tour T (taken in order of the tour sequence)
          Append the single-bit label of arc e to the right of sequence S.

Example 6.2.3: The construction of a $(2, 4)$-deBruijn sequence can begin with the $(2, 4)$-deBruijn digraph shown in Figure 6.2.2. It is easy to verify that the following sequence of vertices and arcs is an Eulerian tour.

$$000 \xrightarrow{0} 000 \xrightarrow{0} 001 \xrightarrow{0} 010 \xrightarrow{0} 100 \xrightarrow{1} 001 \xrightarrow{0} 011 \xrightarrow{0} 110 \xrightarrow{1} 101$$
$$\xrightarrow{1} 010 \xrightarrow{0} 101 \xrightarrow{1} 011 \xrightarrow{0} 111 \xrightarrow{1} 111 \xrightarrow{1} 110 \xrightarrow{1} 100 \xrightarrow{1} 000$$

The $(2, 4)$-deBruijn sequence 0000100110101111 is the sequence of the arc labels.

Remark: Traversing the tour starting from a vertex different from 000 would result in a $(2, 4)$-deBruijn sequence that is merely a *cyclic shift* of the sequence above. A different $(2, 4)$-deBruijn sequence (i.e., one that is not a cyclic shift) can be obtained by constructing a different Eulerian tour in $D_{2,4}$ (see Exercises).

The following proposition establishes the correctness of the construction of Algorithm 6.2.1.

Proposition 6.2.2: *The sequence of single-bit arc labels of any directed trail of length k, $1 \le k \le n$, in the $(2, n)$-deBruijn digraph, is precisely the string of the leftmost k bits of the full label of the initial arc of that trail.*

Proof: The definition of $D_{2,n}$ implies the assertion for $k = 1$. Assume for some k, $1 \le k \le n - 1$, that the assertion is true for any directed trail of length k, and let $\langle v_0, v_1, \ldots, v_{k+1} \rangle$ be any directed trail of length $k + 1$. By the induction hypothesis, the single-bit arc labels associated with the subtrail $\langle v_1, v_2, \ldots, v_{k+1} \rangle$ match the first k bits of the full label of the arc from v_1 to v_2. By the definition of the deBruijn digraph, these k bits are the first k bits of the label on vertex v_1, which means that they are also bits 2 through $k+1$ of the full label of the arc from v_0 to v_1. But the single-bit label on that arc is the leftmost bit of its full label, which completes the induction step.   ◇

Example 6.2.4: To illustrate Proposition 6.2.2 for the digraph $D_{2,4}$ in Figure 6.2.2, consider the directed subtrail given by

$$100 \overset{1}{\to} 001 \overset{0}{\to} 011 \overset{0}{\to} 110 \overset{1}{\to} 101$$

The bitstring 1001 formed from the single-bit arc labels is the full label for the arc from 100 to 001.

By the construction of $D_{2,4}$, each arc is labeled with a different bitstring of length 4. Thus, since the Eulerian tour uses each arc exactly once, Proposition 6.2.2 implies that the resulting sequence contains all possible bitstrings of length 4 and is, therefore, a $(2, 4)$-deBruijn sequence.

Guan's Postman Problem

The Chinese mathematician Meigu Guan [Gu62] introduced the problem of finding a shortest closed walk to traverse every edge of a graph at least once. Guan envisioned a letter carrier who wants to deliver the mail through a network of streets and return to the post office as quickly as possible. Jack Edmonds [Ed65a] dubbed this problem the **Chinese Postman Problem**.

DEFINITION: A **postman tour** in a graph G is a closed walk that uses each edge of G at least once.

DEFINITION: In a weighted graph, an **optimal postman tour** is a postman tour whose total edge-weight is a minimum.

The objective of the Chinese Postman Problem is to find an optimal postman tour in a given weighted graph, where the edge-weights represent some measurable quantity that depends on the application (e.g., distance, time, cost, etc.).

Of course, if each vertex of the graph has even degree, then any Eulerian tour is an optimal postman tour. Otherwise, some edges must be retraced (or *deadheaded*). Thus, the goal is to find a postman tour whose deadheaded edges have minimum total edge-weight. It is easy to see that every postman tour of a graph G corresponds to an Eulerian tour of a graph G^* formed from G by adding to G as many additional copies of each edge as the number of times it was deadheaded during the postman tour.

Edmonds and Johnson [EdJo73] solved the Chinese Postman Problem by using a polynomial-time algorithm, which appears on the next page as Algorithm 6.2.2.

REVIEW FROM §4.7: A **matching** in a graph G is a subset M of E_G such that no two edges in M have an endpoint in common.

DEFINITION: A **perfect matching** in a graph is a matching in which every vertex is an endpoint of one of the edges.

Matchings appear in a variety of applications, like machine scheduling, job assignment, interview pairings, etc. They are discussed in Chapter 10.

Algorithm 6.2.2:   Optimal Postman Tour

Input: a connected weighted graph G.

Output: an optimal postman tour W.

Find the set S of odd-degree vertices of G.

For each pair of odd-degree vertices u and v in S

  Find a shortest path P in G between vertices u and v.

  Let d_{uv} be the length of path P.

Form a complete graph K on the vertices of S.

For each edge e of the complete graph K

  Assign to edge e the weight d_{uv}, where u and v are the
      endpoints of e.

Find a perfect matching M in K whose total edge-weight is minimum.

For each edge e in the perfect matching M

  Let P be the corresponding shortest path in G between the endpoints
      of edge e.

  For each edge f on path P

    Add to graph G a duplicate copy of edge f, including its
        edge-weight.

Let G^* be the Eulerian graph formed by adding to graph G the edge
    duplications from the previous step.

Construct an Eulerian tour W in G^*.

The Eulerian tour W corresponds to an optimal postman tour of the
    original graph G.

The idea behind the Edmonds and Johnson algorithm is to retrace the edges on certain shortest paths between odd-degree vertices, where the edge-weights are regarded as distances. If the edges of a path between two odd-degree vertices are duplicated, then the degrees of those two vertices become even and the parity of the degree of each internal vertex on the path remains the same.

Example 6.2.5: The next few figures illustrate the application of 6.2.2 to the weighted graph G in Figure 6.2.3.

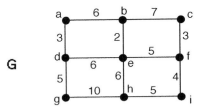

Figure 6.2.3   Weighted graph for the Chinese Postman Problem.

Figure 6.2.4 shows the weighted complete graph on the odd-degree vertices in G. For instance, the shortest path between odd-degree vertices b and d has length 8. The minimum-weight perfect matching is shown in bold.

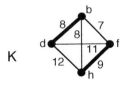

Figure 6.2.4   Minimum-weight perfect matching of complete graph K.

Each edge in the perfect matching obtained in Algorithm 6.2.2 represents a path in G. The edges on this path are the ones that are duplicated to obtain the Eulerian graph G^*, shown in Figure 6.2.5.

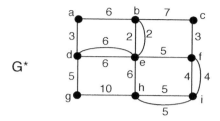

Figure 6.2.5   Duplicating edges to obtain an Eulerian graph G^*.

Finally, an Eulerian tour for graph G^* is an optimal postman tour for G. One such tour is given by the vertex sequence

$$\langle a, b, c, f, e, b, e, d, e, h, i, f, i, h, g, d, a \rangle$$

All but one of the steps of Algorithm 6.2.2 involve algorithms that have already been discussed. The exception is the step that finds an optimal perfect matching of the complete graph K. This problem can be transformed into the problem of finding an optimal perfect matching of an associated complete bipartite graph. Methods developed in Chapter 10 lead to a polynomial-time algorithm for this type of problem.

The optimality of the postman tour follows from having chosen a minimum-weight perfect matching of edges that correspond to shortest paths in G. (See Exercises.)

Remark: Edmonds and Johnson's polynomial-time solution applies equally well to the digraph version of the Chinese Postman Problem, but for *mixed graphs* (having both undirected and directed edges), the problem becomes NP-hard ([Pa76]). In their same paper, Edmonds and Johnson gave an approximate algorithm for the mixed-graph problem, and G. Frederickson [Fr79] showed that the solution obtained is never worse than twice the optimum. He also proposed some modifications that improved the worst-case performance.

Other Eulerian-Type Applications

Variations of the Chinese Postman Problem arise in numerous applications. These include garbage collection and street sweeping; snow-plowing; line-painting down the center of each street; police-car patrolling; census taking; and computer-driven plotting of a network.

Application 6.2.2:   *Street Sweeping*   Suppose that the digraph in Figure 6.2.6 represents a network of one-way streets, with the bold arcs representing streets to be swept. Each edge-weight gives the time required to traverse that street *without* sweeping (i.e., deadheading the street.). The time required to sweep a street is estimated as twice the deadheading time. What route minimizes the total time to sweep all the required streets, starting and ending at vertex z?

This is a digraph variation of the Chinese Postman Problem studied by Tucker and Bodin [TuBo83]. As in the postman problem for undirected graphs, if the subgraph to be swept (called the *sweep subgraph*) is Eulerian, then an Eulerian tour will be an optimal sweeping tour. Otherwise, a subset of arcs must be added to the sweep subgraph so that the resulting digraph is Eulerian; furthermore, the sum of the deadheading times of the additional arcs should be minimum. Adapting Edmonds and Johnson's algorithm, Tucker and Bodin modeled the problem of finding the minimum-weight set of deadheaded arcs as a special case of the *minimum-cost network flow problem* (network flows are discussed in Chapter 10). For the example here, it turns out that the trail given by $\langle z, y, x, y, v, u, v, u, x, y, v, w, z \rangle$ is an optimal route for the sweeper. The total sweeping time is 72, and the total deadheading time is 20.

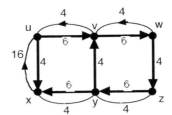

Figure 6.2.6   Edges in bold represent streets to be swept.

They also studied the more complex problem of having to route a fleet of sweepers for a large-scale network (they considered the streets of New York City). Their approach was to find an optimal postman tour and then to partition the edges of the solution into subtours, one for each sweeper.

Application 6.2.3:   *Mechanical Plotters*   Suppose that a mechanical plotter is to plot several thousand copies of the grid shown in Figure 6.2.7, and suppose that it takes twice as long to plot a horizontal edge as it takes to plot a vertical edge. The problem of routing the plotter so that the total time is minimized can be modeled as a postman problem [ReTa81]. (See Exercises).

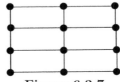

Figure 6.2.7

Application 6.2.4:   *Sequencing Two-Person Conferences*   Suppose certain pairs in a department of six people, $\{A, B, C, D, E, F\}$, must meet privately in a single available conference room. The matrix below indicates with a 1 each pair that must meet. Is it possible to sequence the two-person conferences so that one of the participants in each conference (except the last one) also participates in the next conference, but no one participates in three consecutive conferences?

	A	B	C	D	E	F
A	−	1	0	1	1	0
B	1	−	1	1	0	1
C	0	1	−	0	1	0
D	1	1	0	−	1	1
E	1	0	1	1	−	1
F	0	1	0	1	1	−

To solve this problem, first create a graph G with a vertex for each department member and an edge for each two-person conference (thus, the matrix is the adjacency matrix of graph G). Figure 6.2.8 shows this graph.

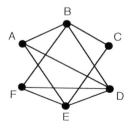

Figure 6.2.8   Graph model for two-person conferences.

Then an open Eulerian trail in G will correspond to a sequencing of conferences that meets the conditions. Since the graph has exactly two odd-degree vertices, Theorem 6.1.1 implies that the sequencing is possible, and a modified version of Algorithm 6.1.1 can be used to construct it.

Application 6.2.5:   *Determining an RNA Chain from its Fragments*   In   an   RNA (ribonucleic acid) chain, such as CGAGUGUACGAA, each link is one of four possible *nucleotides*: adenine (A), cytosine (C), guanine (G), or uracil (U). In trying to identify the RNA chain in an unknown sample, present technology does not permit direct identification of long chains. Our description of the method of fragmentation and subfragmentation of a long RNA chain into identifiable subchains is adapted from the work of George Hutchinson [Hu69].

Fragmentation: Fortunately, there are available two types of enzymes that can be used to break an RNA chain into identifiable subchains. A **G-enzyme** breaks an RNA chain after each G-link. A **UC-enzyme** breaks an RNA chain after each U-link and after each C-link. The resulting fragments of the original RNA chain are called **G-fragments** and **UC-fragments**, according to which enzyme created them.

An **abnormal G-fragment** is a G-fragment that does not end with a G nucleotide. Abnormal fragments can occur only at the end of a chain, because the only place the G-enzyme splits an RNA-chain is after a G-link. Similarly, an **abnormal UC-fragment** is a UC-fragment that does not end with a U or a C. In the description here, an abnormal fragment is indicated by underscoring its final link.

Illustration: If a sample of the chain CGAGUGUACGAA is subjected to the G-enzyme, then the result is the following **G-set** (i.e., of G-fragments):

$$CG, \ AG, \ UG, \ UACG, \ A\underline{A}$$

If another sample of the same chain is subjected to the UC-enzyme, then the result is the following **UC-set**:

$$C, \ GAGU, \ GU, \ AC, \ GA\underline{A}$$

Figure 6.2.9 illustrates the two enzymatic fragmentations of the given RNA chain.

Problem: Given the unordered set of G-fragments and the unordered set of UC-fragments from the same RNA-chain, reconstruct the RNA-chain, if unique; otherwise, specify the complete set of possible RNA-chains from which the G-set and the UC-set could have arisen.

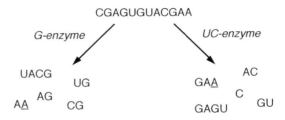

Figure 6.2.9   RNA-chain fragmentation under G-enzyme and UC-enzyme.

Small-Scale Approach: The abnormal G-fragment A<u>A</u> must go last in any plausible reconstruction of the RNA-chain from the G-fragments. Since there are four other fragments, there is a **G-list** of 24 = 4! possible chains (all cases of a permutation of the four normal G-fragments followed by A<u>A</u>) from which they might have arisen, starting alphabetically with AGCGUACGUGA<u>A</u>.

Similarly, the abnormal UC-fragment GA<u>A</u> must be last, and thus, there is a **UC-list** of 24 = 4! possible chains from which these fragments could have arisen, starting alphabetically with ACCGAGUGUGA<u>A</u>.

Any chain that appears in both lists is a candidate for the original RNA chain. One could sort both the G-list of 24 possible reconstructions and the UC-list of 24 possible reconstructions into alphabetical order, and then scan the two lists to find all chains that appear in both.

However, if a long RNA chain had 71 G-fragments and 82 UC-fragments, then there would be $70! \approx 10^{100}$ possible chains in the G-list, and $81! \approx 10^{118}$ in the UC-list. There is no feasible way to scan lists of such length.

Hutchinson's strategy is to construct an Eulerian digraph whose arcs are labeled by fragments, such that each Eulerian tour corresponds to an RNA chain that could be the unknown sample. Additional terminology is needed before describing his construction.

Subfragmentation: A **subfragment** is a nonempty subchain that results either when a UC-enzyme is applied to a G-fragment or when a G-enzyme is applied to a UC-fragment. A fragment is **irreducible** if it consists of a single subfragment. A subfragment within a G-fragment or within a UC-fragment is **internal** if it has subfragments to its left and right. Figure 6.2.10 below illustrates the two enzymatic subfragmentations of the given RNA chain. The G-fragments appear on the left, the UC-fragments are on the right, and dots are used as separators between subfragments.

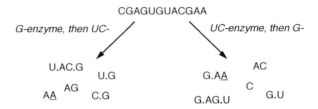

Figure 6.2.10   Subfragmentation under G-enzyme and UC-enzyme.

It is clear from the figure that the irreducible G-fragments are AG and A<u>A</u>, and that the irreducible UC-fragments are C and AC.

Also, the internal subfragment AC in the G-set is an irreducible UC-fragment, and the internal subfragment AG in the UC-set is an irreducible G-fragment. Moreover, the only irreducible fragments in either set that do not appear as internal subfragments in the other set are the first and last subfragments of the entire chain. Both of these properties hold in general for any chain that has at least two G-fragments and at least two UC-fragments (see Exercises). It implies for the example that the chain must begin with C and end with A<u>A</u>.

Constructing the Eulerian digraph: The vertices of the digraph correspond to a subset of the subfragments, and each vertex is labeled with a different subfragment. Each arc corresponds to a fragment and is labeled by that fragment. The correspondence is prescribed as follows:

1. Identify the first subfragment of the entire RNA chain, and draw a vertex labeled with that subfragment. (In the example, this is subfragment C.)

2. Identify the longest abnormal fragment, and draw a vertex labeled with the first subfragment of that abnormal fragment. (In the example, this is subfragment G.)

3. Identify all normal fragments that contain at least two subfragments.

4. Draw a vertex for each subfragment that appears at the beginning or at the end of at least one of the normal fragments identified in step 3.

5. For each fragment F identified in step 3, draw an arc from the vertex labeled with the first subfragment of F to the vertex labeled with the last subfragment of F; label the arc with F.

6. Draw an arc from the vertex drawn in step 2 to the vertex drawn in step 1, and label this arc with the longest abnormal fragment. Append an asterisk (*) to that label to indicate that it appears at the end of the entire chain and that it is to be the last arc in any corresponding Eulerian tour.

Illustration: The digraph for the example is shown in Figure 6.2.11.

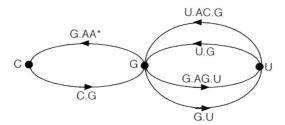

Figure 6.2.11 Digraph for determining the possible RNA chains.

Generating an RNA chain from a given Eulerian tour:

1. Starting at the vertex labeled with the subfragment known to be the first one of the entire chain, write down the entire fragment that labels the first arc of the Eulerian tour.

2. For each subsequent arc, including the last one, write down the fragment labeling that arc, excluding that fragment's first subfragment.

Illustration: There are four different Eulerian tours in the digraph of Figure 6.2.11, and their corresponding RNA chains are:

$$CGAGUGUACGAA$$
$$CGAGUACGUGAA$$
$$CGUACGAGUGAA$$
$$CGUGAGUACGAA$$

It is easy to verify that each of these chains will give rise to the same set of G-fragments and UC-fragments.

A number of details have been omitted. For example, why will the construction always produce an Eulerian digraph, and why are the Eulerian tours in one-to-one correspondence with the candidate RNA chains? These and other details may be found in [Hu69].

Application 6.2.6: *Information Encoding* In the last application, an RNA chain is treated like a string, where the individual nucleotides are its *letters*. More generally and under certain assumptions, the information carried by molecules depends only on the number of letters of each type and the frequency of each letter pair. J. Hutchinson and H. Wilf [HuWi75] considered a purely combinatorial problem that arises naturally from this perspective.

NOTATION: For a given string, let f_i be the number of occurrences of i, and let f_{ij} be the number of times j follows i.

Illustration: For the string 12321212, the nonzero values are:

$$f_1 = 3; \ f_2 = 4; \ f_3 = 1; \ f_{12} = 3; \ f_{21} = 2; \ f_{23} = f_{32} = 1$$

Problem: For a given set of f_i's and f_{ij}'s, $i, j = 1 \dots n$, how many different strings satisfy these prescribed frequency requirements?

It is also assumed that $f_{ii} = 0$ for all i, since a solution to the original problem can easily be deduced from this reduced problem (see [HuWi75]).

Strategy: Draw the digraph G whose adjacency matrix is (f_{ij}), and determine the number of Eulerian trails.

Observation: A string $i_1, i_2 \dots i_l$ will meet the requirements if and only if it corresponds to an Eulerian trail in G. The Eulerian trail is closed or open, depending on whether $i_1 = i_l$.

Illustration: The digraph shown in Figure 6.2.12 has an adjacency matrix whose entries match the f_{ij}'s in the previous illustration. It is easy to check that the digraph contains an open Eulerian trail from vertex 1 to vertex 2 and that the string 12321212 corresponds to the vertex sequence of one of the Eulerian trails. Moreover, there are two other Eulerian trails whose corresponding vertex sequences represent the only other strings satisfying the requirements.

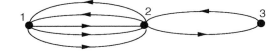

Figure 6.2.12   Digraph obtained from the f_{ij}'s.

The solution for the general case $i_1 \neq i_l$ is outlined below. The analysis for the case $i_1 = i_l$ is similar and left for the reader.

By Theorem 6.1.3, there are no solution-strings unless the corresponding digraph is connected (up to isolated vertices), and the f_i's and f_{ij}'s satisfy the following conditions.

$$\sum_{k=1}^{n} f_{ik} = \begin{cases} \sum_{k=1}^{n} f_{ki} + 1, & \text{if } i = i_1 \\ \sum_{k=1}^{n} f_{ki} - 1, & \text{if } i = i_l \\ \sum_{k=1}^{n} f_{ki}, & \text{otherwise} \end{cases}$$

$$f_i = \begin{cases} \sum_{k=1}^{n} f_{ki} + 1, & \text{if } i = i_1 \\ \sum_{k=1}^{n} f_{ki}, & \text{otherwise} \end{cases}$$

Each open Eulerian trail corresponds to a unique solution-string w, namely, the vertex sequence of the trail. Conversely, a solution-string w corresponds to the vertex sequence of many Eulerian trails. In particular, if T is an Eulerian trail corresponding to w, and i, j is fixed, then each permutation of the f_{ij} arcs from i to j results in a different Eulerian trail corresponding to w. Thus, w corresponds to precisely $\prod_{i \neq j} (f_{ij}!)$ different Eulerian trails.

It remains to count the open Eulerian trails of G. Let G_x be the digraph formed by adjoining a new vertex x to G, with arcs from x to i_1 and i_l to x.

Illustration: Figure 6.2.13 shows the digraph G_x for the example.

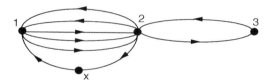

Figure 6.2.13   The Eulerian digraph G_x.

There is a one-to-one correspondence between Eulerian trails of G and Eulerian tours of G_x. The number of Eulerian tours in a directed graph has been determined by de Bruijn and Ehrenfest [DeEh51] and by Smith and Tutte [SmTu41], and for G_x, is given by

$$\prod_{i=1}^{n} (f_i - 1)! \, \det([a_{ij}])$$

where

$$a_{ij} = \begin{cases} f_i, & \text{if } i = j \\ -f_{ij}, & \text{otherwise} \end{cases}$$

Thus, the total number of solution-strings is

$$N = \left[\prod_{i=1}^{n} (f_i - 1)! \right] \left[\prod_{i,j=1}^{n} (f_{ij}!) \right]^{-1} \det([a_{ij}])$$

For the example string, this yields

$$2!3!0! \, [3!2!]^{-1} \det \begin{pmatrix} 3 & -3 & 0 \\ -2 & 4 & -1 \\ 0 & -1 & 1 \end{pmatrix} = 3$$

The three solution-strings are: 12321212;   12121232;   12123212.

Eulerian Digraphs and Software Testing[†]

In §1.6, we observed that a directed postman tour can be used in designing test input sequences to a computer program, that involve invoking all the transitions (actions) of a finite-state model of the program. Here, we give a small example to illustrate that software-testing method, after which, we consider the related problem of testing combinations of actions.

Example 6.2.6: Suppose that the digraph shown is the behavioral model of some computer program to be tested. The states, $\{S_1, S_2, S_3, S_4\}$, and actions, $\{a_1, a_2, \ldots, a_7\}$, of the program are represented by the vertices and arcs, respectively. A shortest directed postman tour is given by the arc sequence $\langle a_1, a_2, a_3, a_2, a_6, a_5, a_7, a_4, a_5, a_7 \rangle$. This was obtained by constructing a directed Eulerian tour of the *Eulerized* digraph that results from duplicating the arcs a_2, a_5, and a_7 of the original digraph.

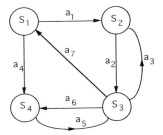

Figure 6.2.14   Finite-state model.

Now suppose we want to test all pairs of consecutive transitions that are possible. In the digraph model, each pair of consecutive transitions corresponds to a pair of arcs, say e_1 and e_2, such that $head(e_1) = tail(e_2)$. For instance, in the digraph in Figure 6.2.14, we seek a shortest closed directed walk that includes $a_3 a_2$, $a_2 a_3$, $a_2 a_6$, etc.

TERMINOLOGY: For a given finite-state machine (digraph), a sequence of arcs that includes all pairs of adjacent arcs is called a **length-2 switch-cover** for the digraph.

Our solution uses the following digraph version of a line graph (§1.2).

DEFINITION: The **line graph** of a digraph G is the digraph $L(G)$ whose vertex-set corresponds to the arc-set of G; an arc in $L(G)$ is directed from vertex e_1 to vertex e_2 if, in G, $head(e_1) = tail(e_2)$. (See Figure 6.2.15.)

A length-2 switch-cover is constructed as follows:

1. Create the line graph $L(G)$ of the original digraph G.
2. Construct an optimal directed postman tour of the line graph $L(G)$.
3. The vertex sequence of this tour corresponds to the desired arc sequence in the original digraph.

Example 6.2.7: Figure 6.2.15 shows the digraph from Example 6.2.6 on the left and its line graph on the right. An optimal postman tour of the line graph is given by the vertex sequence

$$\langle a_1, a_2, a_3, a_2, a_6, a_5, a_3, a_2, a_7, a_4, a_5, a_6, a_5, a_7 \rangle$$

Notice that this sequence of arcs in the original digraph is a length-2 switch-cover.

---

†The material in this subsection is adapted from H. Robinson, [Ro99].

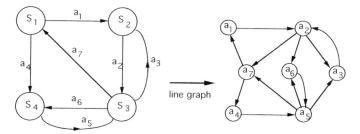

Figure 6.2.15   Finite-state model and its line graph.

EXERCISES for Section 6.2

6.2.1[S] Construct a $(2,4)$-deBruijn sequence that is different from the one obtained in Application 6.2.1 by finding a different Eulerian tour of $D_{2,4}$.

6.2.2 Draw the $(2,3)$-deBruijn digraph and use it to construct two different $(2,3)$-deBruijn sequences.

6.2.3 Which vertices of the deBruijn digraph have self-loops? Justify your answer.

6.2.4 Find an appropriate extension of the line-graph definition (§1.2) to digraphs, and show that the line graph of the deBruijn digraph $D_{2,3}$ is the deBruijn digraph $D_{2,4}$.

6.2.5[S] Prove that $L(D_{2,n}) = D_{2,n+1}$.

6.2.6 Give an alternative proof of Proposition 6.2.1 using line graphs.

6.2.7 A (p,n)-deBruijn sequence is a string or length p^n, containing as substrings (allowing wraparound) all possible sequences of length n of characters drawn from a p-element set. Generalize the procedure outlined in this section, and use it to construct the $(3,2)$-deBruijn digraph $D_{3,2}$ and a $(3,2)$-deBruijn sequence.

6.2.8[S] Construct the deBruijn digraph $D_{3,3}$ and use it to construct a $(3,3)$-deBruijn sequence.

6.2.9 Show that the line graph $L(D_{3,2})$ of the deBruijn digraph $D_{3,2}$ is the deBruijn digraph $D_{3,3}$.

6.2.10 Prove that $L(D_{p,n}) = D_{p,n+1}$.

6.2.11 Which vertices of the deBruijn digraph $D_{2,n}$ have self-loops? Justify your answer.

In Exercises 6.2.12 through 6.2.15, apply Algorithm 6.2.2 to find a minimum-weight postman tour for the given weighted graph.

6.2.12   **6.2.13**

6.2.14^S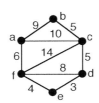

6.2.15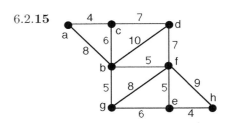

Exercises 6.2.16 through 6.2.18 refer to Algorithm 6.2.2.

6.2.16^S Prove that a perfect matching of K is guaranteed to exist.

6.2.17 Prove that G^* is Eulerian.

6.2.18 Show how the optimality of the postman tour follows from having chosen a minimum-weight perfect matching of edges that correspond to shortest paths in G.

6.2.19 Use a modified version of Algorithm 6.2.2 to solve the example problem in Application 6.2.3.

6.2.20 Solve the problem posed in Application 6.2.4.

In Exercises 6.2.21 through 6.2.24, use the method of Application 6.2.5 to find an RNA chain whose G- and UC-fragments are as given.

6.2.21 G-fragments: CCG, G, UCCG, AAAG;

UC-fragments: GGAAAG, GU, C, C, C, C.

6.2.22^S G-fragments: CUG, CAAG, G, UC;

UC-fragments: C, C, U, AAGC, GGU.

6.2.23 G-fragments: G, UCG, G, G, UU;

UC-fragments: GGGU, U, GU, C.

6.2.24 G-fragments: G, G, CC, CUG, G;

UC-fragments: GGGU, U, C, GC.

6.2.25 Referring to Application 6.2.5, construct a chain that contains no internal subfragments and is not irreducible.

6.2.26 Referring to Application 6.2.5, show that for any chain,

   a. every interior subfragment also appears as an irreducible fragment; and
   b. the only irreducible fragments that do not appear as internal subfragments are the first and last subfragments of the entire chain.

In Exercises 6.2.27 through 6.2.30, use the observation in Application 6.2.6 to determine whether there is a sequence of the numbers 1, 2, and 3 (or of 1, 2, 3, 4) that satisfies the given frequency and adjacency matrix (f_{ij}). If there is such a sequence, use an appropriate version of Algorithm 6.1.1 to find one.

6.2.27^S $f_1 = 3$; $f_2 = 3$; $f_3 = 3$.

$$\begin{pmatrix} 0 & 3 & 1 \\ 3 & 0 & 1 \\ 2 & 1 & 0 \end{pmatrix}$$

6.2.28 $f_1 = 4$; $f_2 = 4$; $f_3 = 4$.

$$\begin{pmatrix} 0 & 1 & 3 \\ 3 & 0 & 1 \\ 2 & 1 & 0 \end{pmatrix}$$

6.2.29 $f_1 = 5$; $f_2 = 4$; $f_3 = 3$.

$$\begin{pmatrix} 0 & 0 & 2 \\ 2 & 0 & 1 \\ 1 & 2 & 0 \end{pmatrix}$$

6.2.30 $f_1 = 3$; $f_2 = 3$; $f_3 = 4$; $f_4 = 4$.

$$\begin{pmatrix} 0 & 0 & 3 & 0 \\ 1 & 0 & 1 & 1 \\ 1 & 0 & 0 & 3 \\ 1 & 2 & 0 & 0 \end{pmatrix}$$

6.2.31 Apply the method illustrated in Example 6.2.7 to find a closed directed walk for the digraph shown, such that every pair of adjacent arcs is included in the arc sequence of the walk.

6.3   HAMILTONIAN PATHS AND CYCLES

DEFINITION: A **Hamiltonian path (cycle)** of a graph is a path (cycle) that contains all the vertices. For digraphs, the Hamiltonian path or cycle is directed.

DEFINITION: A **Hamiltonian graph** is a graph that has a Hamiltonian cycle.

Remark: Adding or deleting self-loops or extra adjacencies between two vertices does not change a graph from Hamiltonian to non-Hamiltonian.

Among the many contributions of the Irish mathematician Sir William Rowan Hamilton (1805–1865) was some of the earliest study of Hamiltonian graphs. A byproduct of this work was his invention of a puzzle known as the *Icosian Game*. One of several aspects of the Icosian Game was to find a Hamiltonian cycle on the dodecahedral graph, depicted in Figure 6.3.1.

Since a Hamiltonian cycle is analogous to an Eulerian tour, one might hope for a characterization of Hamiltonian graphs as convenient as the even-degree characterization of Eulerian graphs. No such characterization is known, nor is there a quick way of determining whether a given graph is Hamiltonian. In fact, the problem is NP-complete, making it extremely unlikely that there exists a polynomial algorithm that can answer the question "Is G Hamiltonian?" for an arbitrary graph G.

The absence of a polynomial-time algorithm that works for all graphs does not mean

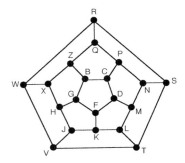

Figure 6.3.1   Dodecahedral graph for the Icosian Game.

the situation is hopeless. There are sufficient conditions for a graph to be Hamiltonian that apply to large classes of graphs. Also, there are some basic rules that help show certain graphs are not Hamiltonian.

Showing That a Graph Is Not Hamiltonian

The following rules are based on the simple observation that any Hamiltonian cycle must contain exactly two edges incident on each vertex. The strategy for applying these rules is to begin a construction of a Hamiltonian cycle and show at some point during the construction that it is impossible to proceed any further. The path that this kind of argument takes depends on the particular graph. There are three rules:

1. If a vertex v has degree 2, then both of its incident edges must be part of any Hamiltonian cycle.
2. During the construction of a Hamiltonian cycle, no cycle can be formed until all the vertices have been visited.
3. If during the construction of a Hamiltonian cycle two of the edges incident on a vertex v are shown to be required, then all other incident edges can be deleted.

Example 6.3.1: The graph in Figure 6.3.2 is not Hamiltonian.

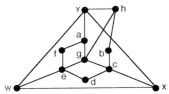

Figure 6.3.2   A non-Hamiltonian graph.

Proof: Rule 1 applied to vertices b, d, and f implies that edges bh, be, dc, de, fe, and fa must be on every Hamiltonian cycle. Rule 3 applied to vertices c, and e eliminates edges cg, ex, eg, and ew. In the resulting graph, Rule 1 implies that edges wv, vx, and xw must be on every Hamiltonian cycle. But these form a 3-cycle, which violates Rule 2. This implies the existence of two disjoint subcycles (the inner hexagon and the outer triangle), which violates Rule 2.   ◇

Example 6.3.2: The graph in Figure 6.3.3 is not Hamiltonian.

Proof: Rule 1 applied to vertices v, w, and x implies that all six vertical edges must be part of any Hamiltonian cycle. Rules 1 and 3 applied to vertex b imply that exactly one

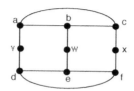

Figure 6.3.3   A non-Hamiltonian graph.

of the edges ab and be is part of the Hamiltonian cycle. If ab is on the cycle and be is not, then Rule 3 applied to vertex a implies that ae is not on the cycle. But then ex is the only edge on the cycle that is incident on e. Thus, by Rule 1, no Hamiltonian cycle exists in this case. By symmetry, a similar contradiction results if be is on the cycle and ab is not.
⬦

Sufficient Conditions for a Graph To Be Hamiltonian

The next two results lend precision to the notion that the more edges a simple graph has, the more likely it is Hamiltonian. The chronologically earlier of the two results, first proved by G. A. Dirac in 1952 ([Di52]), can now be deduced from the more recent result proved by O. Ore ([Or60]).

Theorem 6.3.1: [Ore, 1960] *Let G be a simple n-vertex graph, where $n \geq 3$, such that $\deg(x) + \deg(y) \geq n$ for each pair of non-adjacent vertices x and y. Then G is Hamiltonian.*

Proof: By way of contradiction, assume that the theorem is false, and let G be a maximal counterexample. That is, G is non-Hamiltonian and satisfies the conditions of the theorem, and the addition of any edge joining two non-adjacent vertices of G results in a Hamiltonian graph.

Let x and y be two non-adjacent vertices of G (G is not complete, since $n \geq 3$). To reach a contradiction, it suffices to show that $\deg(x) + \deg(y) \leq n - 1$.

Since the graph $G + xy$ contains a Hamiltonian cycle, G contains a Hamiltonian path whose endpoints are x and y. Let $\langle x = v_1, v_2, \ldots v_n = y \rangle$ be such a path (see Figure 6.3.4).

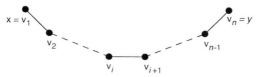

Figure 6.3.4   A Hamiltonian path in G.

For each $i = 2, \ldots, n-1$, at least one of the pairs v_1, v_{i+1} and v_i, v_n is non-adjacent, since otherwise, $\langle v_1, v_2, \ldots, v_i, v_n, v_{n-1}, \ldots, v_{i+1}, v_1 \rangle$ would be a Hamiltonian cycle in G (see Figure 6.3.5). This means that if $(a_{i,j})$ is the adjacency matrix for G, then $a_{1,i+1} + a_{i,n} \leq 1$, for $i = 2, \ldots, n - 2$.

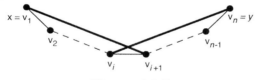

Figure 6.3.5

Thus,

$$\deg(x) + \deg(y) = \sum_{i=2}^{n-1} a_{1,i} + \sum_{i=2}^{n-1} a_{i,n}$$

$$= a_{1,2} + \sum_{i=3}^{n-1} a_{1,i} + \sum_{i=2}^{n-2} a_{i,n} + a_{n-1,n}$$

$$= 1 + \sum_{i=2}^{n-2} a_{1,i+1} + \sum_{i=2}^{n-2} a_{i,n} + 1$$

$$= 2 + \sum_{i=2}^{n-2} (a_{1,i+1} + a_{i,n})$$

$$\leq 2 + n - 3 = n - 1$$

which establishes the desired contradiction.   ◇

Corollary 6.3.2: [**Dirac, 1952**] *Let G be a simple n-vertex graph, where $n \geq 3$, such that $\deg(v) \geq \frac{n}{2}$ for each vertex v. Then G is Hamiltonian.*

The following two theorems are digraph versions of the results of Ore and Dirac. The proof of the first is considerably more difficult than its undirected-graph counterpart and may be found in [Wo72].

Theorem 6.3.3: *Let D be a simple n-vertex digraph. Suppose that for every pair of vertices v and w for which there is no arc from v to w, $outdeg(v) + indeg(w) \geq n$. Then D is Hamiltonian.*   ◇ ([Wo72])

Corollary 6.3.4: *Let D be a simple n-vertex digraph such that for every vertex v, $outdeg(v) \geq \frac{n}{2}$ and $indeg(v) \geq \frac{n}{2}$. Then D is Hamiltonian.*

EXERCISES for Section 6.3

In Exercises 6.3.1 through 6.3.4, determine which graphs in the given graph family are Hamiltonian.

6.3.1 The complete graph K_n.   6.3.2[S] The complete bipartite graph $K_{m,n}$.

6.3.3 The n-wheel W_n.   6.3.4 Trees on n vertices.

6.3.5 Which platonic graphs are Hamiltonian?

In Exercises 6.3.6 through 6.3.10, draw the specified graph or prove that it does not exist.

6.3.6[S] An 8-vertex simple graph with more than 8 edges that is both Eulerian and Hamiltonian.

6.3.7 An 8-vertex simple graph with more than 8 edges that is Eulerian but not Hamiltonian.

6.3.8 An 8-vertex simple graph with more than 8 edges that is Hamiltonian but not Eulerian.

6.3.9 An 8-vertex simple Hamiltonian graph that does not satisfy the conditions of Ore's theorem.

6.3.10 A 6-vertex connected simple graph with 10 edges that is not Hamiltonian.

6.3.11^S Prove that the **Grötzsch graph**, shown below, is Hamiltonian.

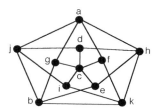

6.3.12 Prove that a bipartite graph that is Hamiltonian must have an even number of vertices.

In Exercises 6.3.13 through 6.3.18, either construct a Hamiltonian cycle in the given graph, or prove that the graph is not Hamiltonian.

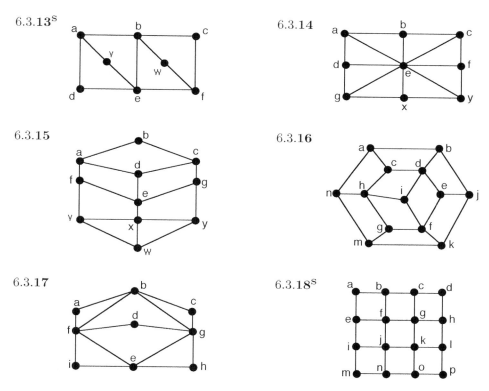

6.3.13^S

6.3.14

6.3.15

6.3.16

6.3.17

6.3.18^S

6.3.19 Show that the Petersen graph is not Hamiltonian.

6.3.20 One version of the Icosian Game was to find a Hamiltonian cycle that started with a given five letters (see Figure 6.3.1). Find all Hamiltonian cycles that begin with the letters BCPNM.

6.3.21[S] Characterize the graphs for which the following properties hold.
   a. A depth-first search produces a Hamiltonian path, regardless of the starting vertex.
   b. A breadth-first search produces a Hamiltonian path, regardless of the starting vertex.

DEFINITION: A **knight's tour** of a chessboard is a sequence of knight moves that visits each square exactly once and returns to its starting square with one more move.

6.3.22 The problem of determining whether a knight's tour exists actually predates the work of Hamilton. Pose the knight's tour problem as one of determining whether a certain graph is Hamiltonian.

6.3.23 Show that if the requirement in Dirac's result (Corollary 6.3.2) is relaxed to "$\deg(v) \geq \frac{n-1}{2}$" (from "$\deg(v) \geq \frac{n}{2}$"), then the assertion of the theorem is false.

6.3.24[S] Show that the sufficient condition in Ore's Theorem 6.3.1 is not a necessary condition, by giving an example of a Hamiltonian graph that does not satisfy the conditions of the theorem.

6.3.25 Consider the complete graph K_n with vertices labeled $1, 2, \ldots, n$.
   a. Find the number of different Hamiltonian cycles. Two cycles that differ only in where they start and end should be counted as the same cycle.
   b. Find the number of Hamiltonian paths from vertex 1 to vertex 2.
   c. Find the number of open Hamiltonian paths.

6.3.26 Prove or disprove each of the following statements.
   a. There exists a 6-vertex Eulerian graph that is not Hamiltonian.
   b. There exists a 6-vertex Hamiltonian graph that is not Eulerian.

6.3.27 Show that for any odd prime n, the edges of K_n can be partitioned into $\frac{n-1}{2}$ edge-disjoint Hamiltonian cycles. (Hint: Arrange the vertices $1, 2, \ldots, n$ around a circle.)

6.3.28 Suppose that 19 world leaders are to dine together at a circular table during a conference. It is desired that each leader sit next to a pair of different leaders for each dinner. How many consecutive dinners can be scheduled? (Hint: See the preceding exercise.)

6.3.29 A mouse eats its way through a $3 \times 3 \times 3$ cube of cheese by tunneling through all of the 27 $1 \times 1 \times 1$ subcubes. If the mouse starts at one corner and always moves on to an uneaten subcube, can it finish at the center of the cube?

6.4   GRAY CODES AND TRAVELING SALESMAN PROBLEMS

Initially, the word "code" suggested secrecy. But with the advances in digital techniques, now applied routinely to represent analog signals in digital form (e.g., music on compact disks), the word "code" has acquired much broader meaning.

DEFINITION: A **Gray code of order** n is an ordering of the 2^n bitstrings of length n such that consecutive bitstrings (and the first and last bitstring) differ in precisely one bit position.

Example 6.4.1: The sequence $\langle 000, 100, 110, 010, 011, 111, 101, 001 \rangle$ is a Gray code of order 3.

REVIEW FROM §1.2: The **n-dimensional hypercube** Q_n is the n-regular graph whose vertex-set is the set bitstrings of length n, such that there is an edge between two vertices if and only if they differ in exactly one bit.

Thus, a Gray code of order n corresponds to a Hamiltonian cycle in the hypercube graph Q_n. The hypercube Q_3 is shown in Figure 6.4.1, and the edges drawn in bold form a Hamiltonian cycle.

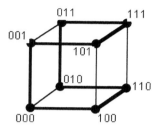

Figure 6.4.1   A Hamiltonian cycle in the hypercube Q_3.

The next result shows that Q_n is Hamiltonian for all n, thereby proving that Gray codes of all orders exist. The proof relies on the following inductive construction of the hypercube. The $(n+1)$-dimensional hypercube Q_{n+1} can be obtained from two copies of the n-dimensional hypercube Q_n, as follows: adjoin a 0 to the right of each n-bit sequence of one of the copies of Q_n, adjoin a 1 to the right of each sequence of the other copy, and join by an edge the two vertices labeled by corresponding sequences.

Remark: This construction is a special case of the **Cartesian product** of two graphs, which is defined in §2.7.

A closer look at the Hamiltonian cycle shown in Figure 6.4.1 previews the inductive step in the proof. Figure 6.4.2 suggests how a Hamiltonian cycle in Q_3 can be constructed from two oppositely oriented Hamiltonian cycles in two copies of Q_2. The vertices of Q_3 are obtained by adding a 0 or 1 as a third bit to the corresponding vertex of the bottom and top Q_2 graphs, respectively.

Starting at 000, traverse all but the last edge of the Hamiltonian cycle in the bottom Q_2; traverse the vertical edge to get to the corresponding vertex in the top Q_2 (011); then follow the top Hamiltonian cycle in the opposite direction until reaching 001; finally, complete the Hamiltonian cycle for Q_3 by returning to 000 via the vertical edge.

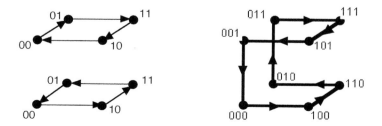

Figure 6.4.2   Gray code of order 3 from a Gray code of order 2.

Theorem 6.4.1: *The n-dimensional hypercube Q_n is Hamiltonian for all $n \geq 2$.*

Proof: Q_2 is a 4-cycle and, hence, Hamiltonian. Assume for some $n \geq 2$ that there exists a Hamiltonian cycle $\langle b_1, b_2, \ldots, b_{2n}, b_1 \rangle$ in Q_n. Then

$$\langle b_1 0, b_2 0, \ldots, b_{2n} 0, b_{2^n} 1, b_{2^n-1} 1, \ldots b_1 1, b_1 0 \rangle$$

is a Hamiltonian cycle in Q_{n+1}.   ◇

Application 6.4.1:   *Transmitting Photographs from a Spacecraft*   A spacecraft transmits a picture back to earth using long sequences of numbers, where each number is a darkness value for one of the dots in the picture. The advantage of using a Gray code to encode the picture is that if an error from "cosmic noise" causes one binary digit in a sequence to be misread by the receiver, then the mistaken sequence will often be interpreted as a darkness value that is almost the same as the true darkness number. For example, if the Gray code given in Example 6.4.1 were used to encode the darkness values 1 through 8, respectively, and if 011 ($= 5$) were transmitted, then the sequence that results from an error in the first or last bit (reading right to left) would be interpreted as a darkness value of 4 or 6, respectively.

Traveling Salesman Problem

One version of the **Traveling Salesman Problem** (**TSP**) is to minimize the total airfare for a traveling salesman who wants to make a tour of n cities, visiting each city exactly once before returning home. Figure 6.4.3 shows a weighted graph model for the problem; the vertices represent the cities and the edge-weights give the airfare between each pair of cities. A solution requires finding a minimum-weight Hamiltonian cycle. The graph can be assumed to be complete by assigning arbitrarily large weight to edges that do not actually exist.

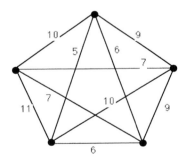

Figure 6.4.3   Weighted graph for a TSP.

The TSP has a long history that has stimulated considerable research in *combinatorial optimization*. The earliest work related to the TSP dates back to 1759, when Euler published a solution to the Knight's Tour Problem (see Exercises). Other early efforts were made by A.T. Vandermonde (1771), T.P. Kirkman (1856), and of course W.R. Hamilton (1856).

The problem of finding a minimum-weight Hamiltonian cycle appears to have been first posed as a TSP by H. Whitney in 1934. M. Flood of Rand Corporation recognized its importance in the context of the then young field of *operations research*, and in 1954, three of his colleagues at Rand, G. B. Dantzig, D.R. Fulkerson, and S.M. Johnson

([DaFuJo54]), achieved the first major breakthrough by finding a provably optimal tour of 49 cities (Washington, D.C. and the capitals of the 48 contiguous states). Their landmark paper used a combination of linear programming and graph theory, and it was probably the earliest application of what are now two of the standard tools in integer programming, *branch-and-bound* and *cutting planes*.

The next dramatic success occurred in 1980 with the publication by Crowder and Padberg of a provably optimal solution to a 318-city problem ([CrPa80]). To enumerate the problem's approximately 10^{655} tours at the rate of 1 billion tours per second, it would take a computer 10^{639} years. The Crowder-Padberg solution took about 6 minutes by computer, using a combination of branch-and-bound and *facet-defining inequalities*.

Heuristics and Approximate Algorithms for the TSP

DEFINITION: A *heuristic* is a guideline that helps in choosing from among several possible alternatives for a decision step.

Heuristics are what human experts apply when it is difficult or impossible to evaluate every possibility. In chess, the stronger the player, the more effective that player's heuristics in eliminating all but a few of the possible moves, without evaluating each legal move. Of course, this has the risk of sometimes missing what may be the best move.

DEFINITION: A *heuristic algorithm* is an algorithm whose steps are guided by heuristics. In effect, the heuristic algorithm is forfeiting the guarantee of finding the best solution, so that it can terminate quickly.

Since the TSP is *NP*-hard, there is a trade-off between heuristic algorithms that run quickly and those that guarantee finding an optimal solution. Time constraints in many applications usually force practitioners to opt for the former. An excellent survey and detailed analysis of heuristics are found in [LaLeKaSh85]. A few of the more commonly used ones are given here.

The simplest TSP heuristic is **nearest neighbor**. Its philosophy is one of shortsighted greed: from wherever you are, pick the cheapest way to go somewhere else. Thus, the nearest-neighbor algorithm (appearing on the next page) is simply a depth-first traversal where ties are broken by choosing the edge of smallest weight.

Algorithm 6.4.1:   Nearest Neighbor
Input: a weighted complete graph.
Output: a sequence of labeled vertices that forms a Hamiltonian cycle.
    Start at any vertex v.
    Initialize $l(v) = 0$.
    Initialize $i = 0$.
    While there are unlabeled vertices
        $i := i + 1$
        Traverse the cheapest edge that joins v to an unlabeled vertex, say w.
        Set $l(w) = i$.
        $v := w$

As is typical of greedy algorithms, the nearest-neighbor heuristic is very fast, and it is easy to implement. The algorithm sometimes performs quite well; for instance, for the weighted graph in Figure 6.4.3, it produces the optimal solution if it starts at the top vertex. However, in general, it can produce arbitrarily bad (costly) Hamiltonian cycles, as it does for the graph in Figure 6.4.4.

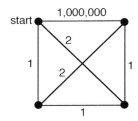

Figure 6.4.4   The nearest-neighbor heuristic can be arbitrarily bad.

The following theorem shows that performance guarantees for approximate algorithms for the general TSP are extremely unlikely.

Theorem 6.4.2: *If there exists a polynomial-time approximate algorithm whose solution to every instance of the general TSP is never worse than some constant r times the optimum, then $P = NP$.*   ◇ *([SaGo76])*

However, for TSPs satisfying the *triangle inequality*, there are such algorithms. But the algorithm based on the nearest-neighbor heuristic is not one of them, as Theorem 6.4.3 demonstrates.

DEFINITION: Let G be a weighted simple graph with vertices labeled $1, 2, \ldots, n$, such that edge ij has weight c_{ij}. Then G is said to satisfy the **triangle inequality** if $c_{ij} \leq c_{ik} + c_{kj}$ for all i, j, and k.

Theorem 6.4.3: *For every $r > 1$, there exists an instance of the TSP, obeying the triangle inequality, for which the solution obtained by the nearest-neighbor algorithm is at least r times the optimal value.*   ◇ *([RoStLe77])*

Two Heuristic Algorithms That Have Performance Guarantees

The next two heuristic algorithms have low-order polynomial complexity, and both have performance guarantees for graphs satisfying the triangle inequality. Both algorithms find a minimum spanning tree, create an Eulerian tour of an associated graph, and then extract a Hamiltonian cycle from the Eulerian tour by taking shortcuts.

Algorithm 6.4.2:   Double the Tree

Input: a weighted complete graph G.

Output: a sequence of vertices and edges that forms a Hamiltonian cycle.

    Find a minimum spanning tree T^* of G.

    Create an Eulerian graph H by using two copies of each edge of T^*.

    Construct an Eulerian tour W of H.

    Construct a Hamiltonian cycle in G from W as follows:

        Follow the sequence of edges and vertices of W until the next edge in the sequence is joined to an already visited vertex. At that point, skip to the next unvisited vertex by taking a shortcut, using an edge that is not part of W. Resume the traversal of W, taking shortcuts whenever necessary, until all the vertices have been visited. Complete the cycle by returning to the starting vertex via the edge joining it to the last vertex.

Example 6.4.2: Suppose that the tree shown in Figure 6.4.5(i) is a minimum spanning tree for a 7-vertex weighted graph. The Eulerian tour of the doubled edges of the tree is shown in Figure 6.4.5(ii), and the vertex labels indicate the order of their first visits. Figure 6.4.5(iii) shows the Hamiltonian cycle that results when the Eulerian tour is modified by shortcuts via non-tree edges. The triangle inequality implies that these shortcuts are indeed shortcuts.

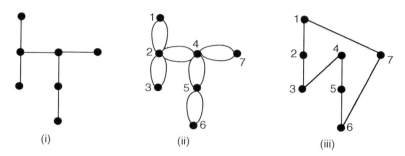

(i)   (ii)   (iii)

Figure 6.4.5   Illustration of Algorithm 6.4.2.

Theorem 6.4.4: *For all instances of the TSP that obey the triangle inequality, the solution produced by Algorithm 6.4.2 is never worse than twice the optimal value.*

Proof: Let C be the Hamiltonian cycle produced by Algorithm 6.4.2, and let C^* and T^* be a minimum-weight Hamiltonian cycle and a minimum-weight spanning tree, respectively. The total edge-weight of the Eulerian tour is $2 \times wt(T^*)$, and since each shortcut is an edge that joins the initial and terminal points of a path of length at least 2, the triangle inequality implies that $wt(C) \le 2 \times wt(T^*)$. But C^* minus one of its edges is a spanning tree, which implies $2 \times wt(T^*) \le 2 \times wt(C^*)$. ◇

One of the key steps of Algorithm 6.4.2 is the creation of an Eulerian graph by duplicating each edge of the minimum spanning tree. N. Christofides took this idea one step further by recognizing that an Eulerian graph can be created without having to duplicate all of the tree's edges. His idea is based on the strategy Edmonds and Johnson used in their solution of the Chinese Postman Problem (§6.2). That is, find a minimum-weight matching of the odd-degree vertices of T^*, and add those edges to T^* to obtain an Eulerian graph. Christofides's algorithm, outlined below, achieves the best known performance guarantee of any approximate algorithm for the TSP.

Algorithm 6.4.3:   **Tree and Matching**
Input: a weighted complete graph G.
Output: a sequence of vertices and edges that forms a Hamiltonian cycle.
    Find a minimum spanning tree T^* of G.
    Let O be the subgraph of G induced on the odd-degree vertices in T^*.
    Find a minimum-weight perfect matching M^* in O.
    Create an Eulerian graph H by adding the edges in M^* to T^*.
    Construct an Eulerian tour W of H, as in Algorithm 6.4.2.
    Construct a Hamiltonian cycle in G from W, as in Algorithm 6.4.2.

The details of Christofides's algorithm, including a proof of the following performance bound, can be found in [LaLeKaSh85].

Theorem 6.4.5: *For all instances of the TSP that obey the triangle inequality, the solution produced by Christofides's algorithm is never worse than $\frac{3}{2}$ times the optimal.*

Remark: Performance guarantees must consider the worst-case behavior of a heuristic, and they may not reflect how well the heuristic actually performs in practice. Thus, performance guarantees should not be the only criterion in evaluating a heuristic. Run-time, ease of implementation, and empirical analysis are at least as important for the practitioner.

TSPs in Disguise

Recognizing when a given problem can be transformed to the TSP does not make the problem's difficulty vanish, but it does make available a variety of methods developed over the last century. Here are three commonly encountered variations of the TSP that can be transformed to a standard TSP. Let G be a weighted digraph on vertices $1, 2 \ldots, n$, where c_{ij} is the weight of arc ij.

Variation 1. Find a minimum-weight Hamiltonian path.

Transformation 1: Form a new graph G^* by adding to G an artificial vertex 0 and arcs to and from each of the other vertices, with $c_{0j} = c_{j0} = 0$ for $j = 1, 2 \ldots, n$ (see Figure 6.4.6). Then an optimal Hamiltonian path is obtained by deleting vertex 0 from an optimal Hamiltonian cycle of G^*.

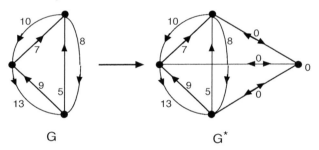

Figure 6.4.6 Transformation 1.

Variation 2. Find a minimum-weight Hamiltonian path in digraph G, from a specified vertex s to a specified vertex t.

Transformation 2: Solve the standard TSP for a digraph G^*, formed from digraph G by replacing vertices s and t by a single vertex u and defining the edge-weights for G^* by

$$c_{ij}^* = \begin{cases} c_{sj}, & \text{if } i = u, \\ c_{it}, & \text{if } j = u, \\ c_{ij}, & \text{otherwise} \end{cases}$$

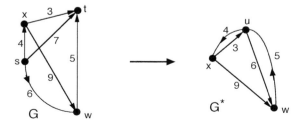

Figure 6.4.7 Transformation 2.

Variation 3. Find a minimum-weight closed walk in digraph G that visits each vertex *at least* once.

Transformation 3: Replace each c_{ij} by the length of a shortest path from i to j (in the original graph).

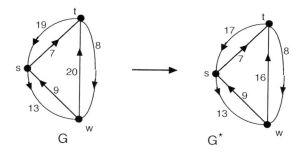

Figure 6.4.8   Transformation 3.

Application 6.4.2:   *Job Sequencing on a Single Machine*   Suppose there are n jobs that must be processed on a single machine. The time required to process job j *immediately after* job i is c_{ij}. How should the jobs be sequenced so that the total time is minimized?

Draw an n-vertex digraph whose vertices correspond to the jobs; the arc directed from vertex i to vertex j is assigned the weight c_{ij}. Then an optimal sequencing of the jobs corresponds to a minimum-weight Hamiltonian path.

Application 6.4.3:   *Circuit Design*   A computer or other digital system consists of a number of modules and several pins located on each module. The physical position of each module has already been determined, and a given subset of pins has to be connected by wires. Because of the small size of the pins, at most two wires are to be attached to any pin. Furthermore, to minimize noise (stray signals from external sources) and to improve the ease of wiring, the total wire length should be minimized. An optimal wiring design will correspond to minimum-weight Hamiltonian path.

The next application is a special instance of *vehicle routing*. The general vehicle routing problem is to determine which vehicles should serve which customers and in what order. Typically, constraints include capacities of the vehicles as well as time windows for the customers.

Application 6.4.4:   *School Bus Routing*   Each weekend, a private school transports n children to m bus stops across the state. The parents then meet their children at the bus stops. The school owns k buses, and the jth bus has seating capacity $c_j, j = 1, \ldots, k$. How should the buses be routed so that the total cost is minimized?

The ith bus stop is associated with a subset S_i of children, $i = 1, \ldots, m$. Let the vertices $v_1, \ldots, v_m$ represent the m bus stops, and let vertex v_0 represent the school location. An assignment $\{v_{j_1}, v_{j_2} \ldots, v_{j_l}\}$ of l bus stops to the jth bus is *feasible* if $\sum_{k=1}^{l} |S_{j_k}| \le c_j$. The cost of a feasible assignment is the cost of a minimum-weight Hamiltonian cycle on the vertices $\{v_0, v_{j_1}, v_{j_2} \ldots, v_{j_l}\}$ where edge-weight represents distance. Thus, the problem is to partition the m bus stops into k subsets, such that the sum of all the buses' optimal tours is minimized.

Notice that calculating the cost of a given partition requires solving k TSP subproblem. But a number of details have been omitted that actually make the problem even more complicated. For example, the distances that parents must travel to get to their children's bus stops are not being considered. Also, the operating costs of the chartered buses are likely to be more than those of the school-owned buses. Furthermore, there are limitations on how late each bus can make its last stop. If these are handicapped students, then there may be an additional requirement of one adult chaperone per bus. A final complication is that the m bus stops need not be predetermined. That is, the school may be able to choose which and how many bus stops from among hundreds of potential ones.

EXERCISES for Section 6.4

6.4.1^S Use the inductive step in the proof of Theorem 6.4.1 to obtain a Gray code of order 4 from the Gray code of order 3 in Example 6.4.1.

In Exercises 6.4.2 through 6.4.5, apply each of Algorithms 6.4.1 through 6.4.3 to the given graph, starting at vertex a and resolving ties alphabetically. Indicate the vertex sequence and total edge-weight for each of the three outputs.

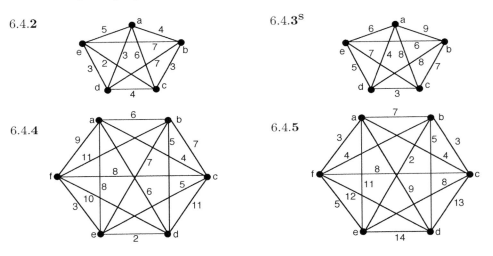

6.4.2

6.4.3^S

6.4.4

6.4.5

In Exercises 6.4.6 through 6.4.9, apply a modified version of each of Algorithms 6.4.1 through 6.4.3 to the specified graph to try to find a minimum-weight open Hamiltonian path that starts at vertex a. Base your modification on Transformation 1. Indicate the vertex sequence and total edge-weight for each of the three outputs.

6.4.6 The graph of Exercise 6.4.2.   **6.4.7^S** The graph of Exercise 6.4.3.

6.4.8 The graph of Exercise 6.4.4.   **6.4.9** The graph of Exercise 6.4.5.

6.4.10 Prove or disprove: In the n-dimensional hypercube Q_n (§1.2), the initial and terminal vertices of every open Hamiltonian path are adjacent.

In Exercises 6.4.11 through 6.4.14, apply a modified version of each of Algorithms 6.4.1 through 6.4.3 to the specified graph to try to find a minimum-weight Hamiltonian path from vertex a to vertex d. Base your modification on Transformation 2. Indicate the vertex sequence and total edge-weight for each of the three outputs.

6.4.11 The graph of Exercise 6.4.2.   **6.4.12^S** The graph of Exercise 6.4.3.

6.4.13 The graph of Exercise 6.4.4.   **6.4.14** The graph of Exercise 6.4.5.

In Exercises 6.4.15 and 6.4.16, apply a modified version of each of Algorithms 6.4.1 through 6.4.3 to the specified graph to try to find a minimum-weight closed walk that starts at vertex a and visits each vertex at least once. Base your modification on Transformation 3. Indicate the vertex sequence and total edge-weight for each of the three outputs.

6.4.15 The graph of Exercise 6.4.2.   **6.4.16^S** The graph of Exercise 6.4.3.

In Exercises 6.4.17 and 6.4.18, do each of the following for the given digraph.

a. Draw the digraph obtained by applying each of Transformations 1, 2, and 3 to the given digraph.

b. Give the vertex sequence and total edge-weight of the Hamiltonian path (in the original digraph) from vertex d to vertex a that is obtained by applying each of Algorithms 6.4.1 and 6.4.2 to the digraph from part (a), and use exhaustive analysis to determine if either of these is a minimum-weight one.

c. Give the vertex sequence and total edge-weight of the closed walk (in the original digraph) that uses each vertex at least once that is obtained by applying each of Algorithms 6.4.1 and 6.4.2 to the digraph from part (a), and use exhaustive analysis to determine if either of these is a minimum-weight one.

6.4.17   **6.4.18**

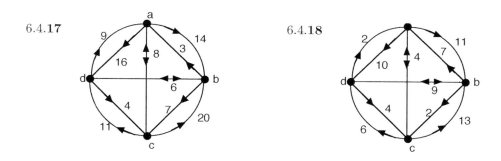

6.4.19 Suppose each of n jobs requires processing by m machines in the same order. Each machine can work on at most one job at a time, and once it begins work on a job, it must work on it to completion, without interruption. The amount of processing time that job i requires on machine h is p_{hi}. Also, once the processing of a job is complete on machine h, its processing must begin immediately on machine $h+1$ (this is called a *flow shop with no wait in process*). Show that the problem of sequencing the jobs so that the last job is completed as early as possible can be formulated as a standard $(n+1)$-vertex TSP. (Hint: Create a dummy job requiring zero units of processing, and let c_{ij} represent the amount of idle time required on machine 1 if job j immediately follows job i.)

6.4.20 [*Computer Project*] Implement Algorithm 6.4.1 and run the program on each of the graphs in Exercises 6.4.2 through 6.4.5.

6.5   SUPPLEMENTARY EXERCISES

DEFINITION: A *deadhead path* in a postman tour is one that is retraced.

6.5.1 Draw a minimum-length postman tour in the graph of Figure 6.5.1(a), and show that the set of deadhead paths is of minimum cost.

6.5.2 Draw a minimum-length postman tour in the graph of Figure 6.5.1(b), and show that the set of deadhead paths is of minimum cost.

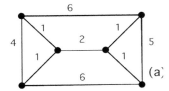 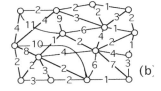

Figure 6.5.1

6.5.3 Calculate the number of Hamiltonian tours in the complete graph K_5.

6.5.4 Give a pair of graphs one of which is Eulerian, the other non-Eulerian, with the same number of vertices, such that all the vertex-deleted subgraphs of both graphs are non-Eulerian. This would establish the non-reconstructibility of the Eulerian property.

6.5.5 Two Eulerian graphs A and B are amalgamated across a subgraph C. Prove that the resulting graph G is Eulerian if and only if the subgraph C is Eulerian.

6.5.6 Decide whether the join $2K_1 + K_{1,4}$ is Hamiltonian.

6.5.7 Decide whether the join $3K_1 + K_{1,4}$ is Hamiltonian.

6.5.8 Decide whether $K_{3,4}$ is Hamiltonian.

6.5.9 Give an example of two connected graphs, each with at least two vertices, whose join is non-Hamiltonian.

6.5.10 Draw a Hamiltonian graph H such that two disjoint copies of H can be amalgamated across K_3, so that the result is non-Hamiltonian. Explain why the result is non-Hamiltonian.

6.5.11 Show that the graph of Figure 6.5.2(a) has a Hamiltonian path but no Hamiltonian cycle.

6.5.12 Show that the graph of 6.5.2(b) has a Hamiltonian path but no Hamiltonian cycle.

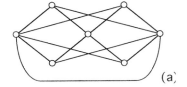 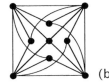

Figure 6.5.2

6.5.13 Decide whether the graph in Figure 6.5.3(a) is Hamiltonian.

6.5.14 Decide whether the graph in 6.5.3(b) is Hamiltonian.

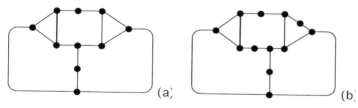

Figure 6.5.3

6.5.15 Prove or disprove: In the n-dimensional hypercube, Q_n, the initial and terminal vertex of any open Hamiltonian trail must be adjacent.

6.5.16 Suppose that two non-adjacent vertices are chosen in a Hamiltonian graph H, and two disjoint copies of H are amalgamated across this vertex pair. Give examples to show that the resulting graph can be Hamiltonian or non-Hamiltonian.

6.5.17 Give a pair of graphs one of which is Hamiltonian, the other non-Hamiltonian, with the same number of vertices, such that all the vertex-deleted subgraphs of both graphs are non-Hamiltonian. This would establish the non-reconstructibility of the Hamiltonian property.

Glossary

deadhead path in a postman tour: a path that is retraced.

deBruijn digraph $D_{2,n}$: the digraph consisting of 2^{n-1} vertices, each labeled by a different bitstring of length $n-1$, and 2^n arcs such that the arc from vertex $b_1 b_2 \cdots b_{n-1}$ to vertex $b_2 \cdots b_{n-1} b_n$ is labeled by the bitstring $b_1 b_2 \cdots b_n$ referred to as the $(2, n)$-deBruijn digraph.

deBruijn sequence of length 2^n: a bitstring of length 2^n in which each of the 2^n possible bitstrings of length n occurs *exactly once* as a substring, where wraparound is allowed; referred to as a $(2, n)$-deBruijn sequence.

double tracing in a graph: a closed walk that traverses every edge exactly twice.

—, bidirectional: a double tracing such that every edge is traversed in both directions.

Eulerian graph: a graph that has an Eulerian tour.

Eulerian tour: a closed Eulerian trail.

Eulerian trail in a graph: a trail that contains every edge of that graph.

Gray code of order n: an ordering of the 2^n bitstrings of length n such that consecutive bitstrings differ in precisely one bit position.

Hamiltonian graph: a graph that has a Hamiltonian cycle.

Hamiltonian path (cycle) in a graph or digraph: a path (cycle) that contains all the vertices; for digraphs, the Hamiltonian path or cycle is directed.

heuristic: a guideline that helps in choosing from among several possible alternatives for a decision step.

heuristic algorithm: an algorithm whose steps are guided by heuristics.

knight's tour of a chessboard: a sequence of knight moves that visits each square exactly once and returns to its starting square with one more move.

line graph of a digraph G: the digraph $L(G)$ whose vertex-set corresponds to the arc-set of G; an arc in $L(G)$ is directed from vertex e_1 to vertex e_2 if, in G, $head(e_1) = tail(e_2)$.

line graph $L(G)$ **of a graph** G: the graph that has a vertex for each edge of G, such that two of these vertices are adjacent if and only if the corresponding edges in G have a vertex in common.

matching in a graph G: a subset M of E_G such that no two edges in M have an endpoint in common.

—, perfect: a matching in which every vertex of the graph is an endpoint of one of the edges.

postman tour in a graph G: a closed walk that uses each edge of G at least once.

—, optimal: a postman tour whose total edge-weight is a minimum.

switch cover, length 2 for a digraph: a sequence of arcs that includes all pairs of adjacent arcs.

triangle inequality: in a weighted simple graph whose vertices are labeled $1, 2, \ldots, n$: a condition on the edge-weights c_{ij}, given by $c_{ij} \leq c_{ik} + c_{kj}$ for all i, j, and k.

Chapter 7

PLANARITY AND KURATOWSKI'S THEOREM

INTRODUCTION

The central theme of this chapter is the topological problem of deciding whether a given graph can be drawn in the plane or sphere with no edge-crossings. Some planarity tests for graphs are in the form of algebraic formulas (see §7.5) based on the numbers of vertices and edges. These are the easiest tests to apply, yet they are one-way tests, and there are difficult cases in which they are inconclusive.

A celebrated result of the Polish mathematician Kasimir Kuratowski transforms the planarity decision problem into the combinatorial problem of calculating whether the given graph contains a subgraph *homeomorphic* (defined in §7.2) to the complete graph K_5 or to the complete bipartite graph $K_{3,3}$. Directly searching for K_5 and $K_{3,3}$ would be quite inefficient, but this chapter includes a simple, practical algorithm (see §7.6) to test for planarity.

The relationship of planarity to topological graph theory is something like the relationship of plane geometry to what geometers call geometry, where mathematicians long ago began to develop concepts and methods to progress far beyond the plane. Whereas planarity consists mostly of relatively accessible ideas that were well understood several decades ago, topological graph theorists use newer methods to progress to all the other surfaces.

7.1   PLANAR DRAWINGS AND SOME BASIC SURFACES

Our approach to drawings of graphs on surfaces begins with our intuitive notions of what we mean by a drawing and a surface. This chapter provides some precise examples of a surface.

Planar Drawings

Consistent with our temporary informality, we introduce the main topic of this chapter with the following definition.

DEFINITION: A **planar drawing** of a graph is a drawing of the graph in the plane without edge-crossings.

DEFINITION: A graph is said to be **planar** if there exists a planar drawing of it.

Example 7.1.1: Two drawings of the complete graph K_4 are shown in Figure 7.1.1. The planar drawing on the right shows that K_4 is a planar graph.

Figure 7.1.1   A nonplanar drawing and a planar drawing of K_4.

Example 7.1.2: An instance of the problem of determining whether a given graph is planar occurs in the form of a well-known puzzle, called the **utilities problem**, in which three houses are on one side of a street and three utilities (electricity, gas, and water) are on the other. The objective of the puzzle is to join each of the three houses to each of the three utilities without having any crossings of the utility lines. Later in this section, we show that this is impossible, by proving that $K_{3,3}$ is nonplanar.

Remark: A graph G and a given drawing of G are categorically different objects. That is, a graph is combinatorial and a drawing is topological. In particular, the *vertices* and *edges* in a drawing of a graph are actually *images* of the vertices and edges in that graph. Yet, to avoid excessive formal phrasing, these distinctions are relaxed when it is discernable from context what is intended.

TERMINOLOGY: Intuitively, we see that in a planar drawing of a graph, there is exactly one *exterior* (or *infinite*) region whose area is infinite.

Example 7.1.3: The exterior region, R_e, and the three *finite* regions of a planar drawing of a graph are shown in Figure 7.1.2.

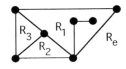

Figure 7.1.2   The four regions of a planar drawing of a graph.

Remark: If we consider a planar drawing of a graph on a piece of paper, then the intuitive notion of *region* corresponds to the pieces of paper that result from cutting the paper along the length of every edge in the drawing. If one adds new edges to a graph without crossing an existing edge, then a region may be subdivided into more regions.

TERMINOLOGY NOTE: We restrict the use of the word "regions" to the case of crossing-free drawings, since many assertions that are true in that case may be untrue when there are edge-crossings.

Three Basic Surfaces

All of the surfaces under consideration in this chapter are subsets of Euclidean 3-space. Although we assume that the reader is familiar with the following surfaces, it is helpful to think about their mathematical models.

DEFINITION: A *plane* in Euclidean 3-space $\mathbb{R}^3$ is a set of points (x, y, z) such that there are numbers a, b, c, and d with $ax + by + cz = d$.

DEFINITION: A *sphere* is a set of points in $\mathbb{R}^3$ equidistant from a fixed point, the *center*.

DEFINITION: The *standard torus* is the surface of revolution obtained by revolving a circle of radius 1 centered at $(2, 0)$ in the xy-plane around the y-axis in 3-space, as depicted in Figure 7.1.3. The solid inside is called the *standard donut*.

Figure 7.1.3   Creating a torus.

DEFINITION: The circle of intersection, as in Figure 7.1.4(a), of the standard torus with the half-plane $\{(x, y, z) \mid x = 0,\ z \geq 0\}$ is called the *standard meridian*. We observe that the standard meridian bounds a disk inside the standard donut.

DEFINITION: The circle of tangent intersection of the standard torus with the plane $y = 1$ is called the *standard longitude*. Figure 7.1.4(b) illustrates the standard longitude. We observe that the standard longitude bounds a disk in the plane $y = 1$ that lies outside the standard donut.

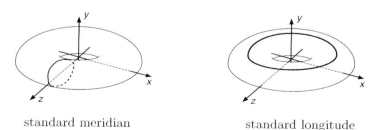

standard meridian   standard longitude

Figure 7.1.4

TERMINOLOGY: Any closed curve that circles the torus once in the meridian direction (the "short" direction) without circling in the longitude direction (the "long" direction) is called a **meridian**. Any closed curve that circles the torus once in the longitude direction without circling in the meridian direction is called a **longitude**.

Remark: Surfaces generally fall into two infinite sequences. The three surfaces described above are all relatively uncomplicated and lie in the sequence of orientable surfaces. Surfaces such as the *Möbius band* and the *Klein bottle* are examples of unorientable surfaces.

Riemann Stereographic Projection

Riemann observed that deleting a single point from a sphere yields a surface that is equivalent to the plane for many purposes, including that of drawing graphs.

DEFINITION: The **Riemann stereographic projection** is the function ρ that maps each point w of the unit-diameter sphere (tangent at the origin $(0,0,0)$ to the xz-plane in Euclidean 3-space) to the point $\rho(w)$ where the ray from the north pole $(0,1,0)$ through point w intersects the xz-plane.

Under the Riemann projection, the "southern hemisphere" of the sphere is mapped continuously onto the unit disk. The "northern hemisphere" (minus the north pole) is mapped continuously onto the rest of the plane. The points nearest to the north pole are mapped to the points farthest from the origin. Figure 7.1.5 illustrates the construction.

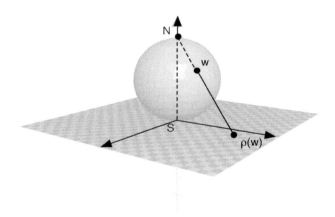

Figure 7.1.5   The Riemann stereographic projection.

Remark: Whereas a planar drawing of a graph has one exterior region (containing the "point at infinity"), a crossing-free graph drawn on the sphere has one region that contains the north pole. Given a graph drawn on the sphere, the Riemann stereographic projection enables us to move the drawing to the plane so that the point at infinity is deleted from whatever region we choose. That is, we simply rotate the sphere so that a point in the designated region is at the north pole. Accordingly, the choice of exterior region is usually irrelevant to understanding anything about the drawings of a particular graph in the plane. Moreover, a graph drawing on a flat piece of paper may be conceptualized as a drawing on a sphere whose radius is so large that its curvature is imperceptible.

Proposition 7.1.1: *A graph is planar if and only if it can be drawn without edge-crossings on the sphere.*

Proof: This is an immediate consequence of the Riemann stereographic projection.   ◇

Representing a Torus as a Rectangle

Akin to the way that the Riemann stereographic projection enables us to represent drawings on a sphere by drawings on a flat piece of paper, there is a way to represent toroidal drawings on a flat piece of paper.

The rectangle in Figure 7.1.6 represents a torus. The top edge and the bottom edge of the rectangle are both marked with the letter "a." This indicates that these two edges are to be identified with each other, that is, pasted together. This pasting of edges creates a cylinder.

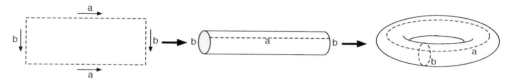

Figure 7.1.6   Folding a rectangle into a torus.

Once the top and bottom edges are pasted together, the left and right edges become closed curves, both marked with the letter "b." When the left end of the cylinder is pasted to the right end, the resulting surface is a torus.

Remark: Pasting the left edge to the right edge converts line "a" into a closed curve. Indeed, on the resulting torus, what was once edge "a" has become a longitude, and what was once edge "b" has become a meridian.

Jordan Separation Property

The objective of the definitions in this subsection is to lend precision to the intuitive notion of using a closed curve to separate a surface. Most of these definitions can be generalized.

DEFINITION: By a **Euclidean set**, we mean a subset of any Euclidean space $\mathbb{R}^n$.

DEFINITION: An **open path** from s to t in a Euclidean set X is the image of a continuous bijection f from the unit interval $[0,1]$ to a subset of X such that $f(0) = s$ and $f(1) = t$. (One may visualize a path as the trace of a particle traveling through space for a fixed length of time.)

DEFINITION: A **closed path** or **closed curve** in a Euclidean set is the image of a continuous function 1 from the unit interval $[0,1]$ to a subset of that space such that $f(0) = f(1)$, but which is otherwise a bijection. (For instance, this would include a "knotted circle" in space.)

Example 7.1.4: Figure 7.1.7 shows an open path and a closed curve in the plane.

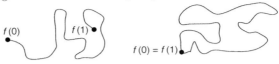

Figure 7.1.7   Open path and closed curve (= closed path).

Example 7.1.5: In a crossing-free drawing of a graph on a surface, a cycle of the graph is a closed curve.

DEFINITION: A Euclidean set X is **connected**[†] if for every pair of points $s, t \in X$, there exists a path within X from s to t.

DEFINITION: The Euclidean set X **separates** the connected Euclidean set Y if there exist a pair of points s and t in $Y - X$, such that every path in Y from s to t intersects the set X.

Example 7.1.6: Figure 7.1.8 shows a meandering closed curve in the plane. It separates the white part from the shaded part.

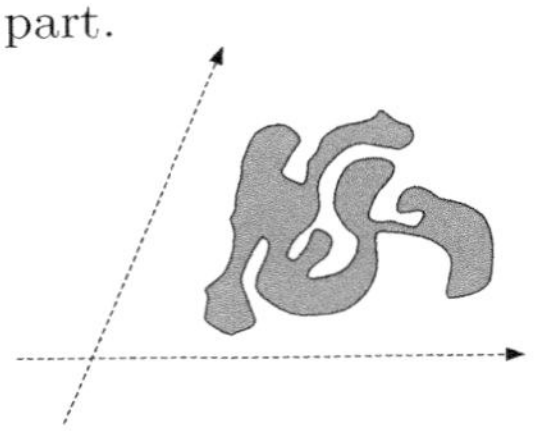

Figure 7.1.8   A closed curve separating the plane.

What makes the plane and the sphere the simplest surfaces for drawing graphs is the *Jordan separation property*. By invoking this property, we can prove that K_5 and $K_{3,3}$ cannot be drawn without edge-crossings in the plane or the sphere.

DEFINITION: A Euclidean set X has the **Jordan separation property** if every closed curve in X separates X.

Theorem 7.1.2: [**Jordan Curve Theorem**] *Every closed curve in the sphere (plane) has the Jordan separation property, that is, it separates the sphere (plane) into two regions, one of which contains the north pole (contains "infinity").*[‡] ◇ *(proof omitted)*

Corollary 7.1.3: *A path from one point on the boundary of a disk through the interior to another point on the boundary separates the disk.* ◇ *(proof omitted)*

We again emphasize that the Jordan separation property distinguishes the plane and sphere from other surfaces. For instance, although it is possible to draw a closed curve on a torus that separates the surface into two parts (e.g., just draw a little circle around a point), a meridian does not, nor does a longitude, as one sees clearly in Figure 7.1.4.

The Jordan Curve Theorem is quite difficult to prove in full generality; in fact, Jordan himself did it incorrectly in 1887. The first correct proof was by Veblen in 1905. A proof for the greatly simplified case in which the closed curve consists entirely of straight-line segments, so that it is a closed polygon, is given by Courant and Robbins in *What Is Mathematics?*.

[†] A topologist would say *path-connected*.

[‡] By the Schönfliess theorem, both regions on a sphere and the interior region in a plane are topologically equivalent to open disks. For a proof, see [Ne54].

Applying the Jordan Curve Theorem to the Nonplanarity of K_5 and $K_{3,3}$

It is possible to prove that a particular graph is nonplanar directly from the Jordan Curve Theorem. In what follows, we continue to rely temporarily on an intuitive notion of the *regions* of a crossing-free drawing of a graph on a sphere or plane. A more formal definition of *region* is given at the beginning of §7.3.

Theorem 7.1.4: *Every drawing of the complete graph K_5 in the sphere (or plane) contains at least one edge-crossing.*

Proof: Label the vertices $0, \ldots, 4$. By the Jordan Curve Theorem, any drawing of the cycle $(1, 2, 3, 4, 1)$ separates the sphere into two regions. Consider the region with vertex 0 in its interior as the "inside" of the cycle. By the Jordan Curve Theorem, the edges joining vertex 0 to each of the vertices 1, 2, 3, and 4 must also lie entirely inside the cycle, as illustrated in Figure 7.1.9. Moreover, each of the 3-cycles $(0, 1, 2, 0)$, $(0, 2, 3, 0)$, $(0, 3, 4, 0)$, and $(0, 4, 1, 0)$ also separates the sphere, and, hence, edge 24 must lie to the exterior of the cycle $(1, 2, 3, 4, 1)$, as shown.

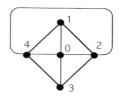

Figure 7.1.9   Drawing most of K_5 in the sphere.

It follows that the cycle formed by edges 24, 40, and 02 separates vertices 1 and 3, again by the Jordan Curve Theorem. Thus, it is impossible to draw edge 13 without crossing an edge of that cycle.   ◇

Theorem 7.1.5: *Every drawing of the complete bipartite graph $K_{3,3}$ in the sphere (or plane) contains at least one edge-crossing.*

Proof: Label the vertices of one partite set 0, 2, 4, and of the other 1, 3, 5. By the Jordan Curve Theorem, cycle $(2, 3, 4, 5, 2)$ separates the sphere into two regions, and, as in the previous proof, we regard the region containing vertex 0 as the "inside" of the cycle. By the Jordan Curve Theorem, the edges joining vertex 0 to each of the vertices 3 and 5 lie entirely inside that cycle, and each of the cycles $(0, 3, 2, 5, 0)$ and $(0, 3, 4, 5, 0)$ separates the sphere, as illustrated in Figure 7.1.10.

Figure 7.1.10   Drawing most of $K_{3,3}$ in the sphere.

Thus, there are three regions: the exterior of cycle $(2, 3, 4, 5, 2)$, and the inside of each of the other two cycles. It follows that no matter which region contains vertex 1, there must be some even-numbered vertex that is not in that region, and, hence, the edge from vertex 1 to that even-numbered vertex would have to cross some cycle edge.   ◇

Corollary 7.1.6: *If either K_5 or $K_{3,3}$ is a subgraph of a graph G, then every drawing of G in the sphere (or plane) contains at least one edge-crossing.*

DEFINITION: The complete graph K_5 and the complete bipartite graph $K_{3,3}$ are called the **Kuratowski graphs**.

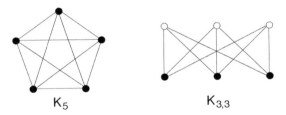

Figure 7.1.11   The Kuratowski graphs.

Example 7.1.7: *Figure 7.1.12 illustrates that K_5 can be drawn without edge-crossings on the torus, even though this is impossible on the sphere. It can be shown that for every graph there is a surface on which a crossing-free drawing is possible.*

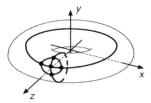

Figure 7.1.12   A crossing-free drawing of K_5 on the torus.

EXERCISES for Section 7.1

Each of the Exercises 7.1.1 through 7.1.4 gives a point w on the sphere that serves as the domain of the Riemann stereographic projection. Calculate the coordinates of the point $\rho(w)$ in the xz-plane to which it is projected. (Hint: Write the locus of the line through the north pole and w.)

7.1.1^S $\left(\frac{1}{2}, \frac{1}{2}, 0\right)$.

7.1.2 $\left(\frac{1}{4}, \frac{3}{4}, \frac{\sqrt{2}}{4}\right)$.

7.1.3 $\left(\frac{1}{5}, \frac{9}{10}, \frac{\sqrt{5}}{10}\right)$.

7.1.4 $\left(\frac{1}{3}, \frac{3}{4}, \frac{\sqrt{11}}{12}\right)$.

Each of the Exercises 7.1.5 through 7.1.8 gives a point w in the xz-plane. Calculate the co-ordinates of the point $p^{-1}(w)$ in the unit sphere under the inverse Riemann stereographic projection. (Hint: Use analytic geometry, as in the previous four exercises.)

7.1.5^S $\left(\frac{\sqrt{3}}{3}, 0, 0\right)$.

7.1.6 $(3, 0, 4)$.

7.1.7 $(1, 0, 1)$.

7.1.8 $\left(\frac{1}{2}, 0, 0\right)$.

7.1.9 Explain why every complete graph K_n, $n \geq 5$, and every complete bipartite graph $K_{m,n}$, $m, n \geq 3$, is not planar.

7.1.10 Determine if the hypercube Q_3 is planar.

7.1.11 Determine if the complete tripartite graph $K_{2,2,2}$ is planar.

7.1.12 Determine if the complete tripartite graph $K_{2,2,3}$ is planar.

7.1.13 Determine if the complete 4-partite graph $K_{2,2,2,2}$ is planar.

7.2   SUBDIVISION AND HOMEOMORPHISM

The graph in Figure 7.2.1 looks a lot like the Kuratowski graph $K_{3,3}$, but one of the "edges" has an intermediate vertex. Clearly, if this graph could be drawn on a surface without any edge-crossings, then so could $K_{3,3}$. Intuitively, it would be an easy matter of erasing the intermediate vertex and splicing the "loose ends" together.

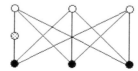

Figure 7.2.1   Subdividing an edge.

Placing one or more intermediate vertices on an edge is formally known as *subdivision*. The relationship between the two graphs is known as *homeomorphism*. The concept of homeomorphism originated as an equivalence relation of topological spaces. It is defined in topology to be a one-to-one, onto, continuous function with a continuous inverse. This section introduces the graph-theoretic analogue of this topological notion.

Graph Subdivision

DEFINITION: Let e be an edge in a graph G. **Subdividing the edge** e means that a new vertex w is added to V_G, and that edge e is replaced by two edges. If e is a proper edge with endpoints u and v, then it is replaced by e' with endpoints u and w and e'' with endpoints w and v. If e is a self-loop incident on vertex u, then e' and e'' form a multi-edge between w and u.

Figure 7.2.2   Subdividing an edge.

DEFINITION: Let w be a 2-valent vertex in a graph G, such that two proper edges e' and e'' are incident on w. **Smoothing away** (or **smoothing out**) vertex w means replacing w and edges e' and e'' by a new edge e. If e' and e'' have only the endpoint w in common, then e is a proper edge joining the other endpoints of e' and e''. If e' and e'' have endpoints w and u in common (i.e., they form a multi-edge), then e is a self-loop incident on u.

Figure 7.2.3   Smoothing away a vertex.

Example 7.2.1: The $(n+1)$-cycle graph can be obtained from the n-cycle graph by subdividing any edge, as illustrated in Figure 7.2.4. Inversely, smoothing away any vertex on the $(n+1)$-cycle, for $n \geq 1$, yields the n-cycle. It is not permitted to smooth away the only vertex of the 1-cycle C_1 ($=$ the bouquet B_1).

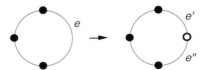

Figure 7.2.4   Subdividing an edge of the 3-cycle yields the 4-cycle.

DEFINITION: **Subdividing a graph** G means performing a sequence of edge-subdivision operations. The resulting graph is called a **subdivision** of the graph G.

Example 7.2.2: Performing any k subdivisions on the n-cycle graph C_n yields the $(n+k)$-cycle graph C_{n+k}, and C_n can be obtained by any k smoothing operations on C_{n+k}. Thus, C_{n+k} is a subdivision of C_k, for all $n \geq k \geq 1$.

Proposition 7.2.1: *A subdivision of a graph can be drawn without edge-crossings on a surface if and only if the graph itself can be drawn without edge-crossings on that surface.*

Proof: When the operations of subdivision and smoothing are performed on a copy of the graph already drawn on the surface, they neither introduce nor remove edge-crossings.
$\diamond$

Barycentric Subdivision

The operation of subdivision can be used to convert a general graph into a simple graph. This justifies a brief digression on our path toward a proof of Kuratowski's theorem.

DEFINITION: The **(first) barycentric subdivision** of a graph is the subdivision in which one new vertex is inserted in the interior of each edge.

Example 7.2.3: The concept of barycentric subdivision and the next three propositions are illustrated by Figure 7.2.5.

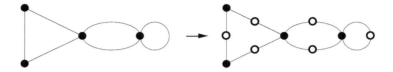

Figure 7.2.5   A graph and its barycentric subdivision.

Proposition 7.2.2: *The barycentric subdivision of any graph is a bipartite graph.*

Proof: Let G' denote the barycentric division of graph G. One endpoint of each edge in G' is an "old" vertex (i.e., from V_G), and the other endpoint is "new" (i.e., from subdividing).
$\diamond$

Proposition 7.2.3: *Barycentric subdivision of any graph yields a loopless graph.*

Proof: By Proposition 7.2.2, a barycentric subdivision of a graph is bipartite, and a bipartite graph has no self-loops.   ◇

Proposition 7.2.4: *Barycentric subdivision of any loopless graph yields a simple graph.*

Proof: Clearly, barycentric subdivision of a loopless graph cannot create loops. If two edges of a barycentric subdivision graph have the same "new" vertex w as an endpoint, then the other endpoints of these two edges must be the distinct "old" vertices that were endpoints of the old edge subdivided by w.   ◇

DEFINITION: For $n \geq 2$, the nth **barycentric subdivision** of a graph is the first barycentric subdivision of the $(n-1)^{st}$ barycentric subdivision.

Proposition 7.2.5: *The second barycentric subdivision of any graph is a simple graph.*

Proof: By Proposition 7.2.3, the first barycentric subdivision is loopless, and thus, by Proposition 7.2.4, the second barycentric subdivision is simple. See Figure 7.2.6.   ◇

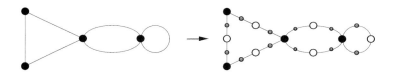

Figure 7.2.6   A graph and its second barycentric subdivision.

Graph Homeomorphism

DEFINITION: The graphs G and H are **homeomorphic graphs** if there is an isomorphism from a subdivision of G to a subdivision of H.

Example 7.2.4: Graphs G and H in Figure 7.2.7 cannot be isomorphic, since graph G is bipartite and graph H contains a 3-cycle. However, they are homeomorphic, because if edge d in graph G and edge e in graph H are both subdivided, then the resulting graphs are isomorphic.

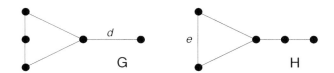

Figure 7.2.7   Two non-isomorphic graphs that are homeomorphic.

Remark: Notice in Example 7.2.4 that no subdivision of graph G is isomorphic to graph H, and that no subdivision of graph H is isomorphic to graph G.

Proposition 7.2.6: *Let G and H be homeomorphic graphs. Then G can be drawn without edge-crossings on a surface S if and only if H can be drawn on S without edge-crossings.*

Proof: This follows from an iterated application of Proposition 7.2.1.

Proposition 7.2.7: *Every graph is homeomorphic to a bipartite graph.*

Proof: By Proposition 7.2.2, the barycentric subdivision of a graph is bipartite. Of course, a graph is homeomorphic to a subdivision of itself. $\diamond$

Subgraph Homeomorphism Problem

Deciding whether a graph G contains a subgraph that is homeomorphic to a target graph H is a common problem in graph theory.

Example 7.2.5: Figure 7.2.8 shows that the complete bipartite graph $K_{3,3}$ contains a subgraph that is homeomorphic to the complete graph K_4. Four of the edges of K_4 are represented by the cycle $0 - 1 - 2 - 3 - 0$, and the other two are represented by the two paths of length 2 indicated by broken lines.

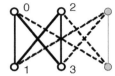

Figure 7.2.8   A homeomorphic copy of K_4 in $K_{3,3}$.

Notice in Figure 7.2.8 that each side of the bipartition of $K_{3,3}$ contains two of the images of vertices of K_4. A homeomorphic copy of K_4 in $K_{3,3}$ cannot have three vertex images on one side of the bipartition. To see this, suppose (without loss of generality) that vertex images 0, 1, and 2 are on one side, and that vertex 3 is on the other side, as illustrated in Figure 7.2.9 below. A homeomorphic copy of K_4 would require three internally disjoint paths joining the three possible pairs of vertices 0, 1, and 2. But this is impossible, since there are only two remaining vertices on the other side that can be used as internal vertices of these paths.

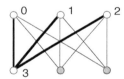

Figure 7.2.9   An impossible way to place K_4 into $K_{3,3}$.

Example 7.2.6: Figure 7.2.10 shows that the hypercube graph Q_4 contains a subgraph that is homeomorphic to the complete graph K_5. The five labeled white vertices represent vertices of K_5. The ten internally disjoint paths joining the ten different pairs of these vertices are given bold edges. Gray vertices are internal vertices along such paths.

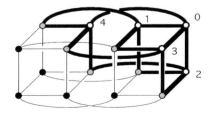

Figure 7.2.10   A homeomorphic copy of K_5 in Q_4.

EXERCISES for Section 7.2

For Exercises 7.2.1 through 7.2. 4, for the given pair of graphs, either prove that they are homeomorphic or prove that they are not.

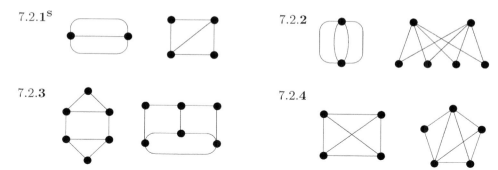

7.2.1[S]

7.2.2

7.2.3

7.2.4

7.2.5[S] What is the minimum number of subdivision vertices required to make the bouquet B_n simple? (Hint: The second barycentric subdivision is not the minimum such subdivision.)

7.2.6 Show that two graphs are homeomorphic if and only if they are isomorphic to subdivisions of the same graph.

For Exercises 7.2.7 through 7.2.10, prove that a homeomorphic copy of the first of the given graphs is contained in the second graph.

7.2.7[S] K_4 in $K_{3,4}$.   **7.2.8** $K_{3,3}$ in $K_{4,5}$.

7.2.9 W_4 in $K_{4,5}$.   **7.2.10** K_5 in $K_{4,5}$.

For Exercises 7.2.11 through 7.2.14, prove that the second given graph does NOT contain a homeomorphic copy of the first graph.

7.2.11[S] K_4 in $K_{2,3}$.   **7.2.12** $K_{3,3}$ in $K_{4,4} - 4K_2$.

7.2.13 K_5 in $K_{4,4}$.   **7.2.14** Q_3 in $K_{3,5}$.

7.3   EXTENDING PLANAR DRAWINGS

We continue to regard a *drawing* of a graph on a surface as an intuitive notion. A more precise definition appears in the next chapter. The following definitions become mathematically precise as soon as the notion of a *drawing* is precise.

DEFINITION: An **imbedding of a graph** G on a surface S is a drawing without any edge-crossings. We denote the imbedding $\iota : G \to S$.

NOTATION: When the surface of the imbedding is a plane or sphere, we use S_0 to denote the surface.

Thus, a planar drawing of a graph G is an imbedding, $\iota : G \to S_0$, of G on the plane S_0.

DEFINITION: A **region** of a graph imbedding $\iota : G \to S$ is a component of the Euclidean set that results from deleting the image of G from the surface S.

DEFINITION: The **boundary of a region** of a graph imbedding $\iota : G \to S$ is the subgraph of G that comprises all vertices that abut that region and all edges whose interiors abut the region.

Remark: When a 2-connected graph is imbedded on the sphere, the boundary of every region is a cycle on the perimeter of the region. However, it is a dangerous misconception to imagine that this is the general case.

DEFINITION: A **face** of a graph imbedding $\iota : G \to S$ is the union of a region and its boundary.

Remark: A drawing does *not* have faces unless it is an imbedding.

Planar Extensions of a Planar Subgraph

A standard way to construct a planar drawing of a graph is to draw a subgraph in the plane, and then to extend the drawing by adding the remaining parts of the graph. This section gives the terminology and some basic results about the planar extensions of a subgraph. These results are helpful in proving Kuratowski's theorem and in specifying a planarity algorithm in the final two sections of this chapter.

Proposition 7.3.1: *A planar graph G remains planar if a multiple edge or self-loop is added to it.*

Proof: Draw graph G in the plane. Alongside any existing edge e of the drawing, another edge can be drawn between the endpoints of e, sufficiently close to e that it does not intersect any other edge. Moreover, at any vertex v, inside any region with v on its boundary, it is possible to draw a self-loop with endpoint v, sufficiently small that it does not intersect another edge. $\diamond$

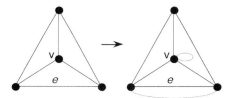

Figure 7.3.1 **Adding a self-loop or a multi-edge preserves planarity.**

Corollary 7.3.2: *A nonplanar graph remains nonplanar if a self-loop is deleted or if one edge of a multi-edge is deleted.*

Proposition 7.3.3: *A planar graph G remains planar if any edge is subdivided or if a new edge e is attached to a vertex $v \in V_G$ (with the other endpoint of e added as a new vertex).*

Proof: Clearly, placing a dot in the middle an edge in a planar drawing of G to indicate subdivision does not create edge-crossings. Moreover, it is easy enough to insert edge e into any region that is incident on vertex v. (See Figure 7.3.2.) $\diamond$

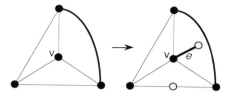

Figure 7.3.2   Subdividing an edge or adding a spike preserves planarity.

The following proposition applies to imbeddings in all surfaces, not just to imbeddings in the sphere.

Proposition 7.3.4: *Let $\iota : G \to S$ be a graph imbedding on a sphere or on any other surface. Let d be an edge of G with endpoints u and v. Then the imbedding of the graph $G - d$ obtained by deleting edge d from the imbedding $\iota : G \to S$ has a face whose boundary contains both of the vertices u and v.*

Proof: In the imbedding $\iota : G \to S$, let f be a face whose boundary contains edge d. When edge d is deleted, face f is merged with whatever face lies on the other side of edge d, as illustrated in Figure 7.3.3. On surface S, the boundary of the merged face is the union of the boundaries of the faces containing edge d, minus the interior of edge d. (This is true even when the same face f lies on both sides of edge d.) Vertices u and v both lie on the boundary of that merged face, exactly as illustrated.   ◇

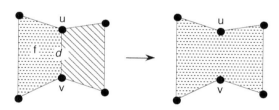

Figure 7.3.3   Merging two regions by deleting an edge.

Amalgamating Planar Graphs

We recall from §2.7 that amalgamating two graphs means pasting a subgraph in one graph to an isomorphic subgraph in the other.

Proposition 7.3.5: *Let f be a face of a planar drawing of a connected graph G. Then there is a planar drawing of G in which the boundary walk of face f bounds a disk in the plane that contains the entire graph G. That is, in the new drawing, face f is the "outer" face.*

Proof: Copy the planar drawing of G onto the sphere so that the north pole lies in the interior of face f. Then apply the Riemann stereographic projection.   ◇

Proposition 7.3.6: *Let f be a face of a planar drawing of a graph H, and let $u_1, \ldots, u_n$ be a subsequence of vertices in the boundary walk of f. Let f' be a face of a planar drawing of a graph J, and let $w_1, \ldots, w_n$ be a subsequence of vertices in the boundary walk of f'. Then the amalgamated graph $(H \cup J)/\{u_1 = w_1 \cdots, u_n = w_n\}$ is planar.*

Proof: Planar drawings of graphs H and J with $n = 3$ are illustrated in Figure 7.3.4.

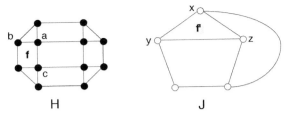

Figure 7.3.4   Two planar graph drawings.

First redraw the plane imbedding of H so that the unit disk lies wholly inside face f. Next redraw graph J so that the boundary walk of face f' surrounds the rest of graph J, which is possible according to Proposition 7.3.5. Figure 7.3.5 shows these redrawing of graphs H and J.

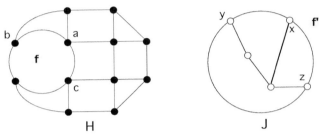

Figure 7.3.5   Redrawings of the two planar graph imbeddings.

We may assume that the cyclic orderings of the vertex sequences $\{u_1, \ldots, u_n\}$ and $\{w_1, \ldots, w_n\}$ are consistent with each other, since the drawing of graph J can be reflected, if necessary, to obtain cyclic consistency. Now shrink the drawing of graph J so that it fits inside the unit disk in face f, as shown on the left in Figure 7.3.6 below. Then stretch the small copy of J outward, as illustrated on the right side of that figure, thereby obtaining a crossing-free drawing of the amalgamation $(H \cup J)/\{a = x, b = y, c = z\}$. $\diamond$

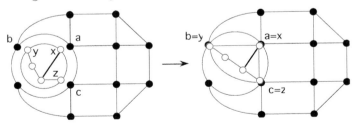

Figure 7.3.6   Position the two graphs and then amalgamate by stretching.

Corollary 7.3.7: *Let H and J be planar graphs. Let U be a set of one, two, or three vertices in the boundary of a face f of the drawing of H, and let W be a set of the same number of vertices in the boundary of a face f' of the drawing of J. Then the amalgamated graph $(H \cup J)/\{U = W\}$ is planar.*

Proof: Whenever there are at most three vertices in the vertex subsequences to be amalgamated, there are only two possible cyclic orderings. Since reflection of either of the drawings is possible, the vertices of sets U and W in the boundaries of faces f and f', respectively, can be aligned to correspond to any bijection $U \to W$. $\diamond$

Remark: The requirement that vertices of amalgamation be selected in their respective graphs from the boundary of a single face cannot be relaxed. Amalgamating two planar graphs across two arbitrarily selected vertices may yield a nonplanar graph. For instance, Figure 7.3.7 shows how the nonplanar graph K_5 can be derived as the amalgamation of the planar graphs G and K_2. This does not contradict Corollary 7.3.7, because the two vertices of amalgamation in G do not lie on the same face of any planar drawing of G.

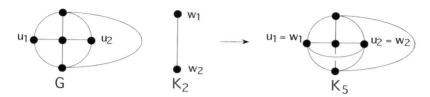

Figure 7.3.7   A 2-vertex amalgamation of two planar graphs into K_5.

Remark: Amalgamating two planar graphs across sets of four or more vertices per face may yield a nonplanar graph, as shown in Figure 7.3.8. The resulting graph shown is $K_{3,3}$ with two doubled edges. The bipartition of $K_{3,3}$ is shown in black and white.

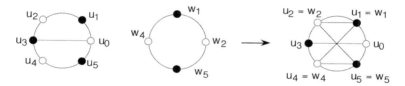

Figure 7.3.8   A 4-vertex amalgamation of two planar graphs into $K_{3,3}$.

Appendages to a Subgraph

In an intuitive sense, subgraph H of a graph G *separates* two edges if it is impossible to get from one edge to the other without going through H. The following definition make this precise.

DEFINITION: Let H be a subgraph of a connected graph G. Two edges e_1 and e_2 of $E_G - E_H$ are **unseparated by subgraph** H if there exists a walk in G that contains both e_1 and e_2, but whose internal vertices are not in H.

Remark: The relation *unseparated by subgraph H* is an equivalence relation on $E_G - E_H$; that is, it is reflexive, symmetric, and transitive.

DEFINITION: Let H be a subgraph of a graph G. Then an **appendage to subgraph** H is the induced subgraph on an equivalence class of edges of $E_G - E_H$ under the relation *unseparated by H.*

DEFINITION: Let H be a subgraph of a graph. An appendage to H is called a **chord** if it contains only one edge. Thus, a chord joins two vertices of H, but does not lie in the subgraph H itself.

DEFINITION: Let H be a subgraph of a graph, and let B be an appendage to H. Then a **contact point** of B is a vertex of both B and H.

Example 7.3.1: In the graph G of Figure 7.3.9, subgraph H is the bold cycle. Appendage B_1 is the broken-edge subgraph with contact points a_1, a_2, and a_3. Appendage B_2 is the exterior subgraph with contact points a_3 and a_4. Appendage B_3 is the interior subgraph with contact points a_1, a_5, and a_6. Appendage B_4 is an exterior chord. Notice that within each of the three non-chord appendages, it is possible to get from any edge to any other edge by a walk in which none of the internal vertices is a contact point.

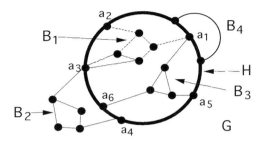

Figure 7.3.9   A subgraph H and its appendages B_1, B_2, B_3, and B_4.

Also notice that each non-chord appendage contains a single component of the deletion subgraph $G - V_H$. In addition to a component of $G - V_H$, a non-chord appendage also contains every edge extending from that component to a contact point, and the contact point as well.

Remark: Every subgraph of a graph (not just cycles) has appendages. Even a subgraph comprising a set of vertices and no edges would have appendages.

Overlapping Appendages

The construction of a planar drawing of a connected graph G commonly begins with the selection of a "large" cycle to be the subgraph whose appendages constitute the rest of G. In this case, some special terminology describes the relationship between two appendages, in terms of their contact points.

DEFINITION: Let C be a cycle in a graph. The appendages B_1 and B_2 of C **overlap** if either of these conditions holds:

  i. Two contact points of B_1 alternate with two contact points of B_2 on cycle C.
  ii. B_1 and B_2 have three contact points in common.

Example 7.3.2: Both possibilities (i) and (ii) for overlapping appendages are illustrated in Figure 7.3.10.

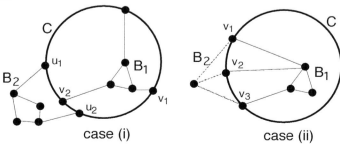

Figure 7.3.10   The two ways that appendages can overlap.

Proposition 7.3.8: *Let C be a cycle in a planar drawing of a graph, and let B_1 and B_2 be overlapping appendages of C. Then one appendage lies inside cycle C and the other outside.*

Proof: Overlapping appendages on the same side of cycle C would cross, by Corollary 7.1.3 to the Jordan Curve Theorem. $\diamond$

TERMINOLOGY: Let C be a cycle of a connected graph, and suppose that C has been drawn in the plane. Relative to that drawing, an appendage of C is said to be *inner* or *outer*, according to whether that appendage is drawn inside or outside of C.

Example 7.3.3: In both parts of Figure 7.3.10, appendage B_1 is an inner appendage and appendage B_2 is an outer appendage.

EXERCISES for Section 7.3

For Exercises 7.3.1 through 7.3.4, the given graphs are to be amalgamated so that each vertex labeled u_i is matched to the vertex v_i with the same subscript. Draw a planar imbedding of the resulting graph.

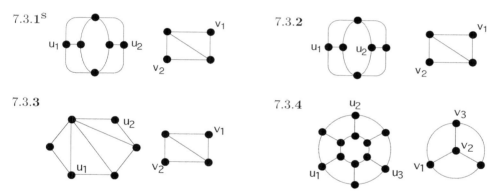

For Exercises 7.3.5 through 7.3.8, the given graphs are to be amalgamated so that each vertex labeled u_i is matched to the vertex v_i with the same subscript. Either draw a planar imbedding of the resulting graph or prove that it is nonplanar.

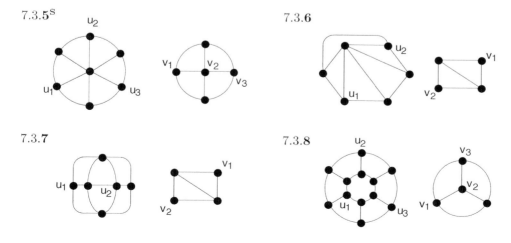

For Exercises 7.3.9 through 7.3.12, relative to the subgraph shown in bold,
   a) specify each appendage by listing its vertex-set and its edge-set;
   b) list the contact-point sets of each appendage;
   c) determine which appendages are overlapping.

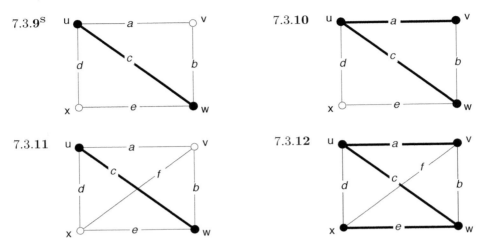

7.3.13 Given two graph imbeddings $\iota : G \to S_0$ and $\iota' : H \to S_0$ in the sphere, suppose that two vertices u_1 and u_2 are selected from G that do not lie on the same face boundary of the imbedding of G, and two vertices v_1 and v_2 are selected from H that do not lie on the same face boundary of the imbedding of H. Decide whether the amalgamation of G and H across these two pairs of vertices must be nonplanar. Explain your answer.

7.4   KURATOWSKI'S THEOREM

One of the landmarks of graph theory is Kuratowski's characterization of planarity in terms of two forbidden subgraphs, K_5 and $K_{3,3}$.

TERMINOLOGY: Any graph homeomorphic either to K_5 or to $K_{3,3}$ is called a **Kuratowski subgraph**.

Theorem 7.4.1: [**Kuratowski, 1930**] *A graph is planar if and only if it contains no subgraph homeomorphic to K_5 or to $K_{3,3}$.*

Proof: Theorems 7.1.4 and 7.1.5 have established that K_5 and $K_{3,3}$ are both nonplanar. Proposition 7.2.6 implies that a planar graph cannot contain a homeomorphic copy of either. Thus, containing no Kuratowski subgraph is necessary for planarity. The rest of this section is devoted to proving sufficiency.

If the absence of Kuratowski subgraphs were not sufficient, then there would exist nonplanar graphs with no Kuratowski subgraphs. If there were any such counterexamples, then some counterexample graph would have the minimum number of edges among all counterexamples. The strategy is to derive some properties that this minimum counterexample would have to have, which ultimately establish that it could not exist. The main steps within this strategy are proofs of three statements:

Step 1: The minimum counterexample would be simple and 3-connected.

Step 2: The minimum counterexample would contain a cycle with three mutually overlapping appendages.

Step 3: Any configuration comprising a cycle and three mutually overlapping appendages must contain a Kuratowski subgraph.

Step 1: A Minimum Counterexample Would Be Simple and 3-Connected

Assertion 1.1: *Let G be a nonplanar connected graph with no Kuratowski subgraph and with the minimum number of edges for any such graph. Then G is a simple graph.*

Proof of Assertion 1.1: Suppose that G is not simple. By Corollary 7.3.2, deleting a self-loop or one edge of a multi-edge would result in a smaller nonplanar graph, still with no Kuratowski subgraph, contradicting the minimality of G. $\diamond$ (Assertion 1.1)

Assertion 1.2: *Let G be a nonplanar connected graph with no Kuratowski subgraph and with the minimum number of edges for any such graph. Then graph G has no cut-vertex.*

Proof of Assertion 1.2: If graph G had a cut-vertex v, then every appendage of v would be planar, by minimality of G. By iterative application of Corollary 7.3.7, graph G itself would be planar, a contradiction. $\diamond$ (Assertion 1.2)

Assertion 1.3: *Let G be a nonplanar connected graph containing no Kuratowski subgraph, with the minimum number of edges for any such graph. Let $\{u, v\}$ be a vertex-cut in G, and let L be a non-chord appendage of $\{u, v\}$. Then there is a planar drawing of L with a face whose boundary contains both vertices u and v.*

Proof of Assertion 1.3: Since $\{u, v\}$ is a vertex-cut of graph G, the graph $G - \{u, v\}$ has at least two components. Thus, the vertex set $\{u, v\}$ must have at least one more non-chord appendage besides L. Since, by Assertion 1.2, $\{u, v\}$ is a minimal cut, it follows that both u and v must be contact points of this other non-chord appendage. Since this other non-chord appendage is connected and has no edge joining u and v, it must contain a u-v path P of length at least 2, as illustrated in Figure 7.4.1.

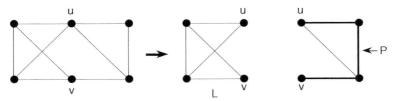

Figure 7.4.1 A u-v path P in an appendage other than L.

The subgraph $H = (V_L \cup V_P, E_L \cup E_P)$ (obtained by adding path P to subgraph L) has no Kuratowski subgraph, because it is contained in graph G, which by premise contains no Kuratowski subgraphs. Let d be the edge obtained from path P by smoothing away all the internal vertices. Then the graph $L + d$ contains no Kuratowski subgraphs (because $L + d$ is homeomorphic to subgraph H). Moreover, the graph $L + d$ has fewer edges than the minimal counterexample G (because P has at least one internal vertex). Thus, the graph $L + d$ is planar.

In every planar drawing of the graph $L + d$, vertices u and v both lie on the boundary of each face containing edge d. Discarding edge d from any such planar drawing of $L + d$ yields a planar drawing of appendage L such that vertices u and v lie on the same face (by Proposition 7.3.4). $\diamond$ (Assertion 1.3)

Assertion 1.4: *Let G be a nonplanar connected graph containing no Kuratowski sub-graph, with the minimum number of edges for any such graph. Then graph G has no vertex-cut with exactly two vertices.*

Proof of Assertion 1.4: Suppose that graph G has a minimal vertex-cut $\{u, v\}$. Clearly, a chord appendage of $\{u, v\}$ would have a planar drawing with only one face, whose boundary contains both the vertices u and v. By Assertion 1.3, every non-chord appendage of $\{u, v\}$ would also have a planar drawing with vertices u and v on the same face boundary. Figure 7.4.2 illustrates the possible decomposition of graph G into appendages.

By Proposition 7.3.6 , when graph G is reassembled by iteratively amalgamating these planar appendages at vertices u and v, the result is a planar graph. $\diamond$ (Assertion 1.4)

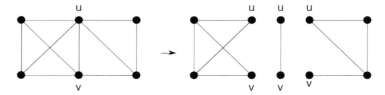

Figure 7.4.2   Decomposition of a graph at vertices u and v.

Completion of Step 1: *Let G be a nonplanar connected graph containing no Kura-towski subgraph and with the minimum number of edges for any such graph. Then graph G is simple and 3-connected.*

Proof: This summarizes the four preceding assertions. The connectedness premise serves only to avoid isolated vertices. $\diamond$ (Step 1)

Step 2: Finding a Cycle with Three Mutually Overlapping Appendages

Let G be a nonplanar graph containing no Kuratowski subgraph and with the minimum number of edges for any such graph. Let e be any edge of graph G, say with endpoints u and v, and consider a planar drawing of $G - e$. Since graph G is 3-connected (by Step 1), it follows (see §5.1) that $G - e$ is 2-connected. This implies (see §5.1) that there is a cycle in $G - e$ through vertices u and v. Among all such cycles, choose cycle C, as illustrated in Figure 7.4.3, so that the number of edges "inside" C is as large as possible. The next few assertions establish that cycle C must have two overlapping appendages in $G - e$ that both overlap edge e.

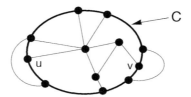

Figure 7.4.3   Cycle C has the maximum number of edges inside it.

Assertion 2.1: *Cycle C has at least one outer appendage.*

Proof of Assertion 2.1: Otherwise, edge e could be drawn in the outer region, thereby completing a planar drawing of G. $\diamond$ (Assertion 2.1)

Assertion 2.2: *Let B be an appendage of C that has only two contact points, neither of which is u or v. Then appendage B is a chord.*

Proof of Assertion 2.2: If B were a non-chord appendage, then those two contact points of B would separate vertices u and v in G from the other vertices of B, which would contradict the 3-connectivity of G.   ◇ (Assertion 2.2)

Assertion 2.3: *Let d be an outer appendage of cycle C. Then d is a chord, and its endpoints alternate on C with u and v, so that edges e and d are overlapping chords of cycle C.*

Proof of Assertion 2.3: Suppose that two of the contact points, say a_1 and a_2, of appendage d do not alternate with vertices u and v on cycle C. Then appendage d would contain a path P between vertices a_1 and a_2 with no contact point in the interior of P. Under such a circumstance, cycle C could be "enlarged" by replacing its arc between a_1 and a_2 by path P. The enlarged cycle would still pass through vertices u and v and would have more edges inside it than cycle C, as shown in Figure 7.4.4, thereby contradicting the choice of cycle C. Moreover, if outer appendage d had three or more contact points, then there would be at least one pair of them that does not alternate with u and v. Thus, appendage d has only two contact points, and they alternate with u and v. By Assertion 2.2, such an appendage is a chord.   ◇ (Assertion 2.3)

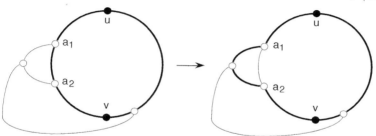

Figure 7.4.4   Using a hypothetical outer appendage to extend cycle C.

Assertion 2.4: *There is an inner appendage of cycle C that overlaps edge e and also overlaps some outer chord.*

Proof of Assertion 2.4: One or more inner appendages must overlap chord e, because otherwise edge e could be drawn inside cycle C without crossing any inner appendages, thereby completing a planar drawing of graph G. Moreover, at least one inner appendage that overlaps e also overlaps some outer appendage d. Otherwise, every such inner appendage could be redrawn outside cycle C, which would permit edge e to be drawn inside, thereby completing the drawing of G. By Assertion 2.3, outer appendage d is necessarily a chord.   ◇ (Assertion 2.4)

Completion of Step 2: *Let G be a nonplanar graph containing no Kuratowski subgraph and with the minimum number of edges for any such graph. Then graph G contains a cycle that has three mutually overlapping appendages, two of which are chords.*

Proof: The cycle C selected at the start of Step 2 meets the requirements of this concluding assertion. In particular, one of the appendages of cycle C is the chord e designated at the start of Step 2. Assertion 2.4 guarantees the existence of a second appendage B that not only overlaps chord e, but also overlaps some outer chord d. By Assertion 2.3, chord d also overlaps chord e. Thus, the three appendages e, B, and d are mutually overlapping.   ◇ (Step 2)

Step 3: Analyzing the Cycle-and-Appendages Configuration

The concluding step in the proof of Kuratowski's theorem is to show that the cycle-and-appendages configuration whose existence is guaranteed by Step 2 must contain a Kuratowski subgraph.

Step 3: *Let C be a cycle in a connected graph G. Let edge e be an inner chord, edge d an outer chord, and B an inner appendage, such that e, d, and B are mutually overlapping. Then graph G has a Kuratowski subgraph.*

Proof: Let u and v be the contact points of inner chord e, and let x and y be the contact points of outer chord d. These pairs of contact points alternate on cycle C, as shown in Figure 7.4.5. Observe that the union of cycle C, chord e, and chord d forms the complete graph K_4. There are two cases to consider, according to the location of the contact points of appendage B.

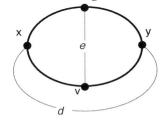

Figure 7.4.5   Forming K_4 with cycle C and chords e and d.

Case 1. Suppose that appendage B has at least one contact point s that differs from u, v, x, and y.

By symmetry of the C-e-d configuration, it suffices to assume that contact point s lies between vertices u and x on cycle C. In order to overlap chord e, appendage B must have a contact point t on the other side of u and v from vertex s. In order to overlap chord d, appendage B must have a contact point t' on the other side of x and y from vertex s.

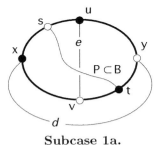 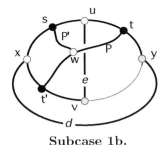

Subcase 1a.   Subcase 1b.

Figure 7.4.6

Subcase 1a. Contact point t lies in the interior of the arc between v and y on cycle C, in which case it may be considered that $t' = t$, as in Figure 7.4.6. In this subcase, let P be a path in appendage B between contact points s and t. Then the union of cycle C, path P, and chords e and d forms a homeomorphic copy of $K_{3,3}$. $\diamond$ (Subcase 1a)

Subcase 1b. Contact point t lies on arc uy, with $t \neq u$ (so that appendage B overlaps chord e), and contact point t' lies on arc xv with $t' \neq x$ (so that B overlaps chord d), as in Figure 7.4.6. In this subcase, let P be a path in appendage B between contact points t' and t. Let w be an internal vertex on path P such that there is a w-s path P' in B with no internal vertices in P. (Such a path P' exists, because appendage B is connected.) Then this configuration contains a homeomorph of $K_{3,3}$ in which each of the vertices $w, x,$ and u is joined to each of the vertices $s, t,$ and t'. As is apparent in Figure 7.4.6, it does not matter if $t = y$ or if $t' = v$.   ◇ (Subcase 1b)

Case 2. Appendage B has no contact points other than $u, v, x,$ and y.

Vertices x and y must be contact points of B so that it overlaps chord e. Vertices u and v must be contact points of B, so that it overlaps chord d. Thus, all four vertices $u, v, x,$ and y must be contact points of appendage B. Appendage B has a u-v path P and an x-y path Q whose internal vertices are not contact points of B.

Subcase 2a.   Subcase 2b.

Figure 7.4.7

Subcase 2a. Paths P and Q intersect in a single vertex w, as shown in Figure 7.4.7. Then the union of the C-e-d configuration with paths P and Q yields a configuration homeomorphic to K_5, as illustrated, in which the five mutually linked vertices are $u, v, y, x,$ and w.   ◇ (Subcase 2a)

Subcase 2b. Paths P and Q intersect in more than one vertex.

Let s be the intersection nearest on path P to contact point u, and t the intersection nearest on P to contact point v, as in Figure 7.4.7. Also, let P' be the us-subpath of P and P'' the tv-subpath of P. Then the union of chords e and d with the paths $Q, P',$ and P'' and with arcs vx and uy on cycle C form a subgraph homeomorphic to $K_{3,3}$ with bipartition $(\{u, x, t\}, (\{v, y, s\})$.   ◇ (Subcase 2b) ◇ (Theorem 7.4.1)

The following Corollary sometimes expedites planarity testing.

Corollary 7.4.2: Let G be a nonplanar graph formed by the amalgamation of subgraphs H and J at vertices u and v, such that J is planar. Then the graph obtained from graph H by joining vertices u and v is nonplanar.

Proof: By Theorem 7.4.1, graph G must contain a Kuratowski subgraph K. All the vertices of the underlying Kuratowski graph would have to lie on the same side of the cut $\{u, v\}$, because K is 3-connected. At most, a subdivided edge of the Kuratowski subgraph crosses through u and back through v. The conclusion follows.   ◇

Finding $K_{3,3}$ or K_5 in Small Nonplanar Graphs

A simple way to find a homeomorphic copy of $K_{3,3}$ in a small graph is to identify two cycles C and C' (necessarily of length at least 4) that meet on a path P, such that there is a pair of paths between $C - P$ and $C' - P$ that "cross," thereby forming a subdivided Möbius ladder ML_3, as illustrated in Figure 7.4.8(a).

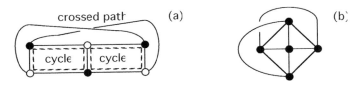

Figure 7.4.8   Finding a $K_{3,3}$ (a) or a K_5 (b) by visual inspection.

To find a homeomorphic copy of K_5, look for a 4-wheel with a pair of disjoint paths joining two pair of vertices that alternate on the rim, as illustrated in Figure 7.4.8(b).

Example 7.4.1: The bold edges form paths joining the three white vertices with the three black vertices of Figure 7.4.9, thereby forming a subdivided $K_{3,3}$.

Figure 7.4.9   Finding a $K_{3,3}$, example.

Example 7.4.2: The bold edges form paths joining the three white vertices with the three black vertices of Figure 7.4.10.

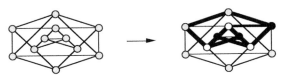

Figure 7.4.10   Finding a $K_{3,3}$, example.

Example 7.4.3: The bold edges form paths joining the five black vertices of Figure 7.4.11, thereby forming a subdivided K_5.

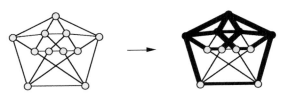

Figure 7.4.11   Finding a K_5, example.

EXERCISES for Section 7.4

In all of the following exercises, find a Kuratowski subgraph in the given graph.

7.4.1

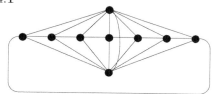

7.4.2^S

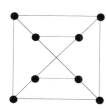

7.4.3

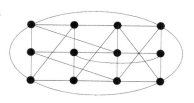

7.4.4

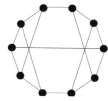

7.4.5

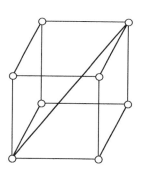

7.4.6

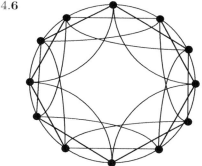

7.4.7^{S} $K_9 - K_{4,5}$.

7.4.9 $C_4 + K_2$.

7.4.11 $K_2 \times K_4$.

7.4.13 $Q_3 + K_1$.

7.4.8 $(K_4 - K_2) \times P_3$.

7.4.10 $(K_5 - K_3) \times P_2$.

7.4.12 $C_5 \times C_5$.

7.4.14 $K_7 - C_7$.

7.5 ALGEBRAIC TESTS FOR PLANARITY

The most basic type of graph-imbedding problem asks whether a particular graph can be imbedded in a particular surface. In this chapter, the surface of particular concern is the sphere (or plane). Transforming such problems into algebraic problems is the inspired approach of algebraic topology. This section derives the fundamental relations used in algebraic analysis of graph-imbedding problems.

The algebraic methods now introduced apply to all surfaces. Although our present version of the Euler polyhedral equation is only for the sphere, it can be generalized to all surfaces.

About Faces

DEFINITION: The **face-set of a graph imbedding** $\iota : G \to S$ is the set of all faces of the imbedding, formally denoted F_ι. Informally, when there is no ambiguity regarding the imbedding, the face-set is often denoted F_G or F.

Example 7.5.1: The graph imbedding in Figure 7.5.1 has four faces, labeled f_1, f_2, f_3, and f_4. Observe that the boundary of face f_4 is not a cycle.

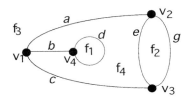

Figure 7.5.1   A graph imbedding in the sphere.

DEFINITION: The **boundary walk of a face** f of a graph imbedding $\iota : G \to S$ is a closed walk in graph G that corresponds to a complete traversal of the perimeter of the polygonal region within the face.

Remark: Vertices and edges can reoccur in a boundary walk. One visualizes one's hand tracing along the edges and vertices in the walk, from just slightly within the region. Thus, if both sides of an edge lie on a single region, the edge is retraced on the boundary walk.

DEFINITION: The **size of a face** or **face-size** means the number of edge-steps in the boundary walk (which may be more than the number of edges on the face boundary, due to retracings). A face of size n is said to be n-**sided**.

DEFINITION: A **monogon** is a 1-sided region.

DEFINITION: A **digon** is a 2-sided region.

Example 7.5.2: In Figure 7.5.1, face f_1 is 1-sided (a monogon) with boundary walk $\langle d \rangle$, face f_2 is 2-sided (a digon) with boundary walk $\langle e, g \rangle$, and face f_3 is 3-sided with boundary walk $\langle a, g, c \rangle$. Even though only five edges lie on the boundary of face f_4, face f_4 is 6-sided with boundary walk $\langle a, b, d, b, c, e \rangle$. Two sides of face f_4 are pasted together across edge b, and that edge is a cut-edge of the graph.

Proposition 7.5.1: *Let* $\iota : G \to S$ *be a planar graph imbedding. Let* e *be an edge of* G *that occurs twice on the boundary walk of some face* f*. Then* e *is a cut-edge of* G*.*

Proof: Choose any point x in the interior of edge e. Let C be a curve in face f from the instance of point x on one occurrence of edge e to the instance of x on the other occurrence of edge e, as illustrated in Figure 7.5.2.

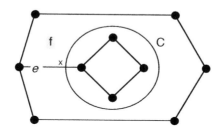

Figure 7.5.2   A curve C from one instance of point x to the other.

Since curve C starts and stops at the same point x, it is a closed curve, so it separates the plane (by the Jordan curve theorem). Thus, curve C separates whatever subgraph of G lies on one side of curve C from whatever subgraph lies on the other side. Since curve C intersects graph G only in edge e, it follows that deleting edge e would separate graph G. In other words, edge e is a cut-edge.   ◇

The number of faces of a drawing of G can vary from surface to surface, as we observe in the following example.

Example 7.5.3: When the graph of Figure 7.5.1 is imbedded in the torus, as shown in Figure 7.5.3, the number of faces drops from four to two. One face is the monogon $\langle d \rangle$ shaded with wavy lines. The other face is 11-sided, with boundary walk $\langle a, b, d, b, c, e, g, c, a, e, g \rangle$. The surface is not the plane, so it should not be surprising that even though edge a occurs twice on the boundary of the 11-sided face, edge a is not a cut-edge.

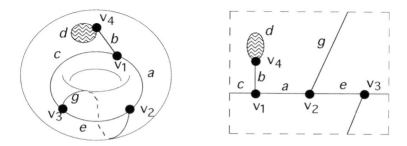

Figure 7.5.3   Two views of the same graph imbedding in the torus.

Face-Size Equation

Theorem 7.5.2: [**Face-Size Equation**] *Let $\iota : G \to S$ be an imbedding of graph G into surface S. Then*

$$2\,|E_G| = \sum_{f \in F} size(f)$$

Proof: Each edge either occurs once in each of two different face boundary walks or occurs twice in the same boundary walk. Thus, by definition of face-size, each edge contributes two sides to the sum. $\diamond$

Example 7.5.4: The graph in Figure 7.5.4 has six edges, so the value of the left side of the equation is $2\,|E| = 12$. The value of the right side of the equation is

$$\sum_{f \in F} size(f) = size(f_1) + size(f_3) + size(f_4)$$

$$= 1 + 2 + 3 + 6 = 12$$

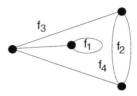

Figure 7.5.4   A graph imbedding in the sphere.

Edge-Face Inequality

DEFINITION: A ***proper walk*** in a graph is a walk in which no edge occurs immediately after itself.

Proposition 7.5.3: *Every proper closed walk W in a graph contains a subwalk (possibly W itself) that is a cycle.*

Proof: If $length(W) = 1$, then walk W itself is a cycle. If $length(W) = 2$, then walk W has the form

$$W = \langle v, d, v', e, v \rangle$$

where $d \neq e$, since walk W is proper. If $v' = v$, then the subwalk v, d, v of length 1 is a cycle. Otherwise, walk W itself is a 2-cycle.

Now suppose that $length(W) \geq 3$. If walk W contains no repeated vertex, then walk W itself is a cycle. Otherwise, let v be a vertex with two occurrences on walk W such that the subwalk between them contains no repeated occurrences of any vertex and no additional occurrences of v. Then that subwalk is a cycle. $\diamond$

Proposition 7.5.4: *Let f be a face of an imbedding $\iota : G \to S$ whose boundary walk W contains no 1-valent vertices and no self-loops. Then W is a proper closed walk.*

Proof: Suppose that the improper subwalk

$$\langle \ldots, u, e, v, e, w, \ldots \rangle$$

occurs in boundary walk W. This means that face f meets itself on edge e, and that the two occurrences of e are consecutive sides of f when f is regarded as a polygon. Clearly, the direction of the second traversal of edge e is opposite to that of the immediately preceding traversal, since e is not a self-loop. Moreover, since edge e has only one endpoint other than v, it follows that $u = w$. This situation is illustrated by Figure 7.5.5(i) below.

Consider a closed curve C in surface S that begins at some point x on edge e near vertex v, and runs down the end of e around vertex v and back out the other side of that end of e to that same point x on the other side of edge e. Since all of curve C except point x lies in the interior of face f, it follows that no edge other than e intersects curve C. Since curve C separates surface S (one side is a topological neighborhood of an end of edge e, and the other is the rest of surface S), it follows that no edge except e is incident on v, as shown in Figure 7.5.5(ii). Thus, vertex v is 1-valent, which contradicts the premise. ◇

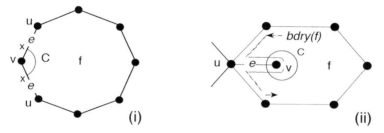

Figure 7.5.5   Surface configuration of an improper boundary walk.

We recall that the **girth** of a non-acyclic graph G is defined to be the length of a smallest cycle in G.

Proposition 7.5.5: *The number of sides of each face f of an imbedding $\iota : G \to S$ of a connected graph G that is not a tree is at least as large as $girth(G)$.*

Proof: (Case 1) If the boundary walk of face f is a proper walk, then its length is at least as large as the girth, by Proposition 7.5.3.

(Case 2) If the boundary walk of face f contains a self-loop, then

$$girth(G) = 1 \le size(f)$$

since 1 is the minimum possible size of any face.

(Case 3) By Proposition 7.5.4, the remaining case to consider is when the boundary walk contains one or more 1-valent vertices and no self-loop. In this case, consider the imbedding $\iota' : G' \to S'$ that results from iteratively contracting every edge with a 1-valent endpoint, until no such edges remain, with f' denoting the face that results from deleting all these edges that "spike" into face f. Then

$$
\begin{aligned}
size(f) &\ge size(f') \\
&\ge girth(G') &&\text{by case 1} \\
&= girth(G) &&\text{since contracting a cut-edge preserves the girth} \qquad ◇
\end{aligned}
$$

Theorem 7.5.6: [***Edge-Face Inequality***] *Let G be a connected graph that is not a tree, and let $\iota : G \to S$ be an imbedding. Then*

$$2\,|E| \geq girth(G) \cdot |F|$$

Proof: In the Face-Size Equation

$$2\,|E| = \sum_{f \in F} size(f)$$

each term in the sum on the right side is at least $girth(G)$, and there are exactly $|F|$ terms.   ◇

Euler Polyhedral Equation

We observe that a tetrahedron has 4 vertices, 6 edges, and 4 faces, and we calculate that $4 - 6 + 4 = 2$. We also observe that a cube has 8 vertices, 12 edges, and 6 faces, and that $8 - 12 + 6 = 2$.

Remark: Robin Wilson [Wi04] offers an historical account of the generalization of this observation. In 1750, Euler communicated the formula $|V| - |E| + |F| = 2$ by letter to Goldbach. The first proof, in 1794, was by Legedre, who used metrical properties. Lhuilier gave a topological proof in 1811, and he also derived the formula $|V| - |E| + |F| = 0$ for graphs on a torus. Poincare generalized the formula to other surfaces in a series of papers of 1895–1904.

Theorem 7.5.7: [***Euler Polyhedral Equation***] *Let $\iota : G \to S$ be an imbedding of a connected graph G into a sphere. Then*

$$|V| - |E| + |F| = 2$$

Proof: If the cycle rank $\beta(G)$ is zero, then graph G is a tree, because $\beta(G)$ equals the number of non-tree edges for any spanning tree in G. Of course, a tree contains no cycles. Thus, in a drawing of graph G on the sphere, the original single region is not subdivided. It follows that $|F| = 1$. Thus,

$$|V| - |E| + |F| = |V| - (|V| - 1) + |F| \qquad \text{by Prop. 3.1.3}$$
$$= |V| - (|V| - 1) + 1 = 2$$

This serves as the base case for an induction on cycle rank. Next suppose that the Euler Polyhedral Equation holds for $\beta(G) = n$, for some $n \geq 0$.

Then (consider an imbedding in which graph G has cycle rank $\beta(G) = n + 1$. Let T be a spanning tree in G, and let $e \in E_G - E_T$, so that $\beta(G - e) = n$. Clearly,

(1) $|V_G| = |V_{G-e}|$

and

(2) $|E_G| = |E_{G-e}| + 1$

Figure 7.5.6 illustrates the relationship between the face sets.

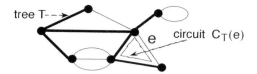

Figure 7.5.6   Erasing a non-tree edge in a sphere.

By Theorem 3.1.8, the subgraph $T + e$ contains a unique cycle, $C_T(e)$. By the Jordan curve theorem, the cycle $C_T(e)$ separates the sphere, which implies that two different faces meet at edge e. Thus, deleting edge e from the imbedding merges those two faces. This yields an imbedding of the connected graph $G - e$ in the sphere. Therefore,

(3) $$|F_G| = |F_{G-e}| + 1$$

Accordingly,

$$\begin{aligned}
|V_G| - |E_G| + |F_G| &= |V_{G-e}| - |E_G| + |F_G| && \text{by (1)}\\
&= |V_{G-e}| - (|E_{G-e}| + 1)\,|F_G| && \text{by (2)}\\
&= |V_{G-e}| - (|E_{G-e}| + 1) + (|F_{G-e}| + 1) && \text{by (3)}\\
&= |V_{G-e}| - |E_{G-e}| + |F_{G-e}| && \\
&= 2 && \text{(by the induction hypothesis)} \quad \diamond
\end{aligned}$$

Theorem 7.5.7 specifies the Euler equation only for an imbedding of a connected graph in the sphere, not for other kinds of drawings. Figure 7.5.7 illustrates two drawings for which the equation does not apply.

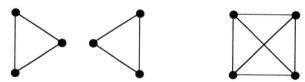

Figure 7.5.7   (a) Non-connected graph; (b) drawing with edge-crossings.

Example 7.5.5: The non-connected graph $2K_3$ in Figure 7.5.7(a) has 6 vertices and 6 edges. When it is imbedded in the plane, there are 3 faces, so the Euler equation does not hold. We observe that there is a face (the exterior face) with two boundary components, which cannot happen with a connected graph in the plane. We also observe that the equation does not hold for the drawing in Figure 7.5.7(b), because it has edge-crossings and, thus, is not an imbedding.

Poincaré Duality

The French mathematician Henri Poincaré [1900] invented a method by which, for every (*cellular*) imbedding $\iota : G \to S$ of any graph G in any (*closed*) surface S, one can construct a new imbedding $\iota^* : G^* \to S$ of a new graph C^* in that same surface S. The formal definition of **cellular imbedding** and **closed surface** is beyond the scope of this chapter. The sphere is closed, and an imbedding of a connected graph on the sphere is cellular.

DEFINITION: Whatever graph imbedding $\iota : G \to S$ is to be supplied as input to the duality process (whose definition follows below) is called the **primal graph imbedding**. Moreover, the graph G is called the **primal graph**, the vertices of G are called **primal vertices**, the edges of G are called **primal edges**, and the faces of the imbedding of G are called **primal faces**.

DEFINITION: Given a primal graph imbedding $\iota : G \to S$, the **(Poincaré) duality construction** of a new graph imbedding is a two-step process, one for the dual vertices and one for the dual edges:

dual vertices: Into the interior of each primal face f, insert a new vertex f^* (as in Figure 7.5.8), to be regarded as dual to face f. The set $\{f^* | f \in F_\iota\}$ is denoted V^*.

dual edges: Through each primal edge e, draw a new edge e^* (as in Figure 7.5.9), joining the dual vertex in the primal face on one side of that edge to the dual vertex in the primal face on the other side, to be regarded as dual to edge e. If the same primal face f contains both sides of primal edge e, then dual edge e^* is a self-loop through primal edge e from the dual vertex f^* to itself. The set $\{e^* | e \in E_G\}$ is denoted E denoted E^*.

Figure 7.5.8   Inserting dual vertices into a primal imbedding.

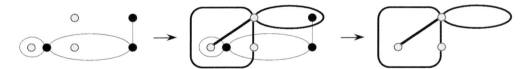

Figure 7.5.9   Inserting dual edges completes the dual imbedding.

DEFINITION: In Poincaré duality, the **graph** is the graph G^* with vertex-set V^* and edge-set E^*. (The vertices and edges of the primal graph are not part of the dual graph.)

DEFINITION: In Poincaré duality, the **dual imbedding** is the imbedding $\iota^* : G^* \to S$ of the dual graph G^* that is constructed in the dualization process.

DEFINITION: In the Poincaré duality construction, the **dual faces** are the faces of the dual imbedding. The dual face containing the vertex v is denoted v^*. (For each vertex $v \in V_G$, there is only one such face.) The set $\{v^* | v \in V_G\}$ of all dual faces is denoted F^*.

NOTATION: Occasionally, a dual graph (or its vertex-set, edge-set, or face-set) is subscripted to distinguish it from the dual graph of other imbeddings of the same graph.

Proposition 7.5.8: *(Numerics of Duality) Let $\iota : G \to S$ be a cellular graph imbedding. Then the following numerical relations hold.*

    (i) $|V^*| = |F|$

   (ii) $|E^*| = |E|$

  (iii) $|F^*| = |V|$

Proof: Parts (i) and (ii) are immediate consequences of the definition of the Poincaré duality construction. Part (iii) holds because the dual edges that cross the cyclic sequence of edges emanating from a primal vertex v of graph G form a closed walk that serves as boundary of the dual face v^* that surrounds primal vertex v, as illustrated in Figure 7.5.10.

$\diamondsuit$

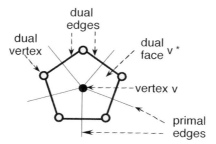

Figure 7.5.10   A dual face surrounding its primal vertex.

Example 7.5.6: The same graph may have many different cellular imbeddings in a given surface and many more over the full range of possible surfaces. Figure 7.5.11 shows a different imbedding of the primal graph from the imbedding of Figure 7.5.9. The dual graph of the previous imbedding has degree sequence $\langle 1, 2, 5 \rangle$, and the dual graph of Figure 7.5.11 has degree sequence $\langle 1, 3, 4 \rangle$, so the dual graphs are not isomorphic.

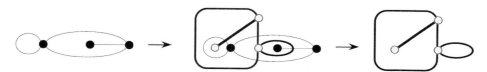

Figure 7.5.11   A different dual of the same primal graph.

Remark: Applying the Poincaré duality construction to the dual graph imbedding simply reconstructs the primal graph and the primal imbedding. This restoration of the original object through redualization is an essential feature of any "duality" construction.

Remark: The Edge-Face Inequality (Theorem 7.5.6) is dual to Euler's theorem on degree-sum (Theorem 1.1.2).

Algebraic Proofs of Nonplanarity

The Euler polyhedral equation $|V| - |E| + |F| = 2$ amounts to an algebraization of the Jordan property. Upon the Euler equation, we can construct inequalities that serve as quick tests for planarity, which we can apply to graphs far too large for ad hoc application of the Jordan property. These are one-way tests, and there are difficult cases in which they are inconclusive.

This algebraic approach to such non-imbeddability proofs illustrates one of the dominant themes of 20th-century mathematics — the transformation of various kinds of other problems, especially topological and geometric problems, into algebraic problems. The formula derived in the following theorem illustrates this theme, by providing a necessary algebraic condition for graph planarity.

Theorem 7.5.9: *Let G be any connected simple graph with at least three vertices and an imbedding $\iota : G \to S_0$ in the sphere or plane. Then*

$$|E_G| \leq 3\,|V_G| - 6$$

Proof: The Euler Polyhedral Equation is the starting point.

(1) $|V_G| - |E_G| + |F_\iota| = 2$ Euler Polyhedral Equation

(2) $girth(G) \cdot |F_\iota| \leq 2\,|E_G|$ Edge-Face Inequality

(3) $3\,|F_\iota| \leq 2\,|E_G|$ G simple bipartite $\Rightarrow 3 \leq girth(G)$

(4) $|F_\iota| \leq \dfrac{2\,|E_G|}{3}$

(5) $|V_G| - |E_G| + \dfrac{2\,|E_G|}{3} \geq 2$ substitute (4) into (1)

(6) $|V_G| - \dfrac{|E_G|}{3} \geq 2$

(7) $\dfrac{|E_G|}{3} \leq |V_G| - 2$

(8) $|E_G| \leq 3\,|V_G| - 6$ $\Diamond$

Theorem 7.5.10: *The complete graph K_5 is nonplanar.*

Proof: Since $3\,|V| - 6 = 9$ and $|E| = 10$, the inequality $|E| > 3\,|V| - 6$ is satisfied. According to Theorem 7.5.9, the graph K_5 must be nonplanar. $\Diamond$

Remark: The Jordan Curve Theorem was used to derive the Euler Polyhedral Equation, which was used, in turn, to derive the planarity criterion $|E_G| \leq 3\,|V_G| - 6$. It remains the ultimate reason why K_5 is not imbeddable in the sphere.

Whereas an ad hoc proof of the nonplanarity of K_5 establishes that fact alone, the inequality of Theorem 7.5.9, derived by the algebraic approach, can be used to prove the nonplanarity of infinitely many different graphs. The next proposition is another application of that inequality.

Proposition 7.5.11: *Let G be any 8-vertex simple graph with $|E_G| \geq 19$. Then graph G has no imbeddings in the sphere or plane.*

Proof: Together, the calculation $3\,|V_G| - 6 = 3 \cdot 8 - 6 = 18$ and the premise $|E_G| \geq 19$ imply that $|E_G| > 3\,|V_G| - 6$. By Theorem 7.5.9, graph G is nonplanar. $\Diamond$

Remark: According to Table A1 of [Ha69], there are 838 connected simple 8-vertex graphs with 19 or more edges. Proposition 7.5.11 establishes their nonplanarity collectively with one short proof. An ad hoc approach could not so quickly prove the nonplanarity of even one such graph, much less all of them at once.

A More Powerful Nonplanarity Condition for Bipartite Graphs

For a bipartite graph, the same algebraic methods used in Theorem 7.5.9 yield an inequality that can sometimes establish nonplanarity when the more general inequality is not strong enough.

Example 7.5.7: The graph $K_{3,3}$ has 6 vertices and 9 edges, Thus, $|E| \leq 3\,|V| - 6$, even though $K_{3,3}$ is not planar.

Theorem 7.5.12: *Let G be any connected bipartite simple graph with at least three vertices and an imbedding $\iota : G \to S_0$ in the sphere or plane. Then*

$$|E_G| \le 2\,|V_G| - 4$$

Proof: This proof is exactly like that of the previous theorem.

(1) $|V_G| - |E_G| + |F_\iota| = 2$ Euler Polyhedral Equation

(2) $girth(G) \cdot |F_\iota| \le 2\,|E_G|$ Edge-Face Inequality

(3) $4\,|F_\iota| \le 2\,|E_G|$ G simple bipartite $\Rightarrow 4 \le girth(G)$

(4) $|F_\iota| \le \dfrac{2\,|E_G|}{4}$

(5) $|V_G| - |E_G| + \dfrac{2\,|E_G|}{4} \ge 2$ substitute (4) into (1)

(6) $|V_G| - \dfrac{|E_G|}{2} \ge 2$

(7) $\dfrac{|E_G|}{2} \le |V_G| - 2$

(8) $|E_G| \le 2\,|V_G| - 4$ ◇

Theorem 7.5.13: *The complete bipartite graph $K_{3,3}$ is nonplanar.*

Proof: Since $2\,|V| - 4 = 8$ and $|E| = 9$, the inequality $|E| > 2\,|V| - 4$ is satisfied. According to Theorem 7.5.12, the graph $K_{3,3}$ must be nonplanar.   ◇

REVIEW FROM §7.1: The complete graph K_5 and the complete bipartite graph $K_{3,3}$ are called the **Kuratowski graphs**.

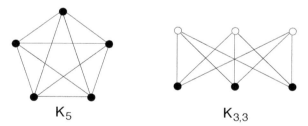

K_5 $K_{3,3}$

Figure 7.5.12   The Kuratowski graphs.

Nonplanar Subgraphs

Another way to prove a graph is nonplanar, besides applying Theorem 7.5.9 or Theorem 7.5.12, is to establish that it contains a nonplanar subgraph.

Theorem 7.5.14: *If a graph contains a nonplanar subgraph, then that graph is nonplanar.*

Proof: Every planar drawing of a graph remains free of edge-crossings after the deletion (e.g., by erasure) of any set of edges and vertices. This establishes the contrapositive, that every subgraph of a planar graph G is planar.   ◇

Example 7.5.8: The graph $G = K_7 - E(K_4)$ is obtained by deleting the six edges joining any four designated vertices in K_7. Since $|V_G| = 7$ and $|E_G| = 21 - 6 = 15$, it follows that

$$|E_G| \leq 3\,|V_G| - 6$$

Thus, the planarity formula does not rule out the planarity of $K_7 - E(K_4)$. However, $K_7 - E(K_4)$ contains the Kuratowski graph $K_{3,3}$ as indicated by Figure 7.5.13. Hence, $K_7 - E(K_4)$ is nonplanar, by Theorem 7.5.14.

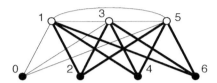

Figure 7.5.13 $K_7 - E(K_4)$ **contains** $K_{3,3}$**.**

Proposition 7.5.15: *Planarity is invariant under subdivision.*

Proof: Suppose that G is a planar graph and that graph H is homeomorphic to G. Then, in a planar drawing of G, smooth whatever vertices are needed and subdivide whatever edges are needed to transform G into H. These operations simultaneously transform the planar drawing of G into a planar drawing of H. $\diamond$

Proposition 7.5.16: *Suppose that a graph G contains a subgraph H that is homeomorphic to a nonplanar graph. Then G is nonplanar.*

Proof: This follows immediately from the preceding theorem and proposition. $\diamond$

As important as the nonplanarity inequalities of Theorem 7.5.9 and Theorem 7.5.12 may be, they provide only one-sided tests. That is, if a graph has too many edges according to either theorem, then it is nonplanar; however, not having too many edges does not imply that the graph is planar. The usual ways to prove planarity are to exhibit a planar drawing or to apply a planarity algorithm, as in §7.6.

Heuristic 7.5.17: *To construct a nonplanar, non-bipartite graph with a specified number of vertices and edges, start with a Kuratowski graph and add vertices, add edges, and subdivide edges.*

Example 7.5.9: To construct an 8-vertex nonplanar, non-bipartite graph with at most 13 edges, simply subdivide K_5 or $K_{3,3}$, as shown in Figure 7.5.14. Observe that the two graphs resulting from these subdivisions have too few edges to make a conclusive application of the numerical nonplanarity formulas.

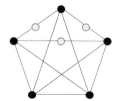 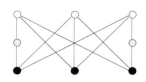

Figure 7.5.14   Two graphs with too few edges for the nonplanarity formula.

EXERCISES for Section 7.5

For Exercises 7.5.1 through 7.5.4, do the following for the given graph:
    a) redraw the given graph;
    b) write the face-size into each region;
    c) show that the sum of the face sizes equals twice the number of edges.

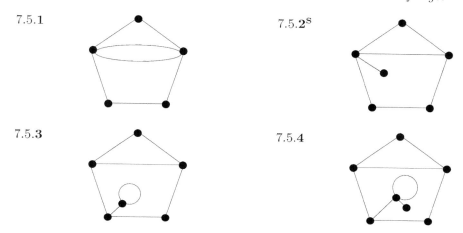

7.5.1

7.5.2^S

7.5.3

7.5.4

For Exercises 7.5.5 through 7.5.8, do the following for the given graph:
    a) calculate the girth of the graph;
    b) show that the product of the girth and the number of faces is less than or equal to twice the number of edges.

7.5.5 The graph of Exercise 7.5.1.   **7.5.6**^{S} The graph of Exercise 7.5.2.

7.5.7 The graph of Exercise 7.5.3.   **7.5.8** The graph of Exercise 7.5.4.

For Exercises 7.5.9 through 7.5.12, do the following for the given graph:
    a) calculate $|V|$, $|E|$, and $|F|$;
    b) show that the Euler polyhedral equation $|V| - |E| + |F| = 2$ holds.

7.5.9 The graph of Exercise 7.5.1.   **7.5.10**^{S} The graph of Exercise 7.5.2.

7.5.11 The graph of Exercise 7.5.3.   **7.5.12** The graph of Exercise 7.5.4.

For Exercises 7.5.13 through 7.5.16, draw the Poincaré dual of the given planar graph imbedding.

7.5.13 The graph of Exercise 7.5.1.   **7.5.14**^{S} The graph of Exercise 7.5.2.

7.5.15 The graph of Exercise 7.5.3.   **7.5.16** The graph of Exercise 7.5.4.

For Exercises 7.5.17 through 7.5.20, draw two additional planar imbeddings of the given imbedded graph, so that the degree sequences of all three duals (including the given imbedding) of the indicated graph are mutually distinct.

7.5.17 The graph of Exercise 7.5.1.   **7.5.18^{S}** The graph of Exercise 7.5.2.

7.5.19 The graph of Exercise 7.5.3.   **7.5.20** The graph of Exercise 7.5.4.

DEFINITION: A graph G is **self-dual** in the sphere if there is an imbedding of G such that the Poincaré dual graph G^* for that imbedding is isomorphic to G.

For Exercises 7.5.21 through 7.5.26, either draw an imbedding establishing self-duality of the given graph, or prove that no such imbedding is possible.

7.5.21 K_4.   **7.5.22^{S}** $K_4 - K_2$.

7.5.23 D_2.   **7.5.24** $K_5 - K_2$.

7.5.25 $B_1 + K_1$.   **7.5.26** $K_6 - 3K_2$.

For Exercises 7.5.27 through 7.5.32, prove that the given graph is nonplanar.

7.5.27 $K_8 - 4K_2$.   **7.5.28** $K_{3,6} - 3K_2$.

7.5.29^{S} $K_8 - 2C_4$.   **7.5.30** $K_8 - K_{3,3}$.

7.5.31 $K_2 + Q_3$.   **7.5.32** $K_{6,6} - C_{12}$.

For Exercises 7.5.33 through 7.5.40, either draw a plane graph that meets the given description or prove that no such graph exists.

7.5.33^{S} A simple graph with 6 vertices and 13 edges.

7.5.34 A non-simple graph with 6 vertices and 13 edges.

7.5.35 A bipartite simple graph with 7 vertices and 11 edges.

7.5.36 A bipartite non-simple graph with 7 vertices and 11 edges.

7.5.37 A simple planar graph with 144 vertices and 613 edges.

7.5.38 A simple planar graph with 1728 vertices and 5702 edges.

7.5.39 A bipartite simple graph with 36 vertices and 68 edges.

7.5.40 A planar simple graph with 36 vertices and 102 edges.

7.5.41 Draw a bipartite graph G with 15 vertices and 18 edges that is not planar, even though it satisfies the formula $|E_G| \leq 2\,|V_G| - 4$.

For Exercises 7.5.42 through 7.5.45, show that the given graph satisfies the planarity inequality $|E_G| \leq 3\,|V_G| - 6$, and that it is nonetheless nonplanar.

7.5.42^{S} $K_{3,3} + K_1$.   **7.5.43** $K_{3,3} \times K_2$.

7.5.44 $K_8 - (K_5 \cup K_3)$.   **7.5.45** $W_5 \times C_4$.

7.6   PLANARITY ALGORITHM

Several good planarity algorithms have been developed. The planarity algorithm described here (from [DeMaPe64]) has been chosen for simplicity of concept and of implementability. It starts by drawing a cycle, and adds to it until the drawing is completed or until further additions would force an edge-crossing in the drawing. This section first gives a prose description of this algorithm, then pseudocode for the algorithm.

Blocked and Forced Appendages of a Subgraph

The body of the main loop of the planarity algorithm attempts to extend a subgraph already drawn in the plane by choosing a plausible appendage of that subgraph to be used to augment the drawing.

DEFINITION: Let H be a subgraph of a graph. In a planar drawing of H, an appendage of H is **undrawable in region** R if the boundary of region R does not contain all the contact points of that appendage.

DEFINITION: Let H be a subgraph of a graph. In a planar drawing of H, an appendage of H is **blocked** if that appendage is undrawable in every region.

Example 7.6.1: In Figure 7.6.1, a subgraph isomorphic to $K_{2,3}$ is shown with black vertices and solid edges, and its only appendage is shown with a white vertex and broken edges. The appendage is blocked, because no region boundary contains all three contact points.

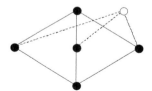

Figure 7.6.1   **A subgraph with a blocked appendage.**

Proposition 7.6.1: Let $\iota : H \to S_0$ be a planar drawing of a subgraph H of a graph G, such that H has a blocked appendage B. Then it is impossible to extend $\iota : H \to S_0$ to a planar drawing of G.

Proof: Let u be a contact point in B, and let e be an edge in B incident on u. In any extension of $\iota : H \to S_0$ to a planar drawing of G, e must be drawn in some region R whose boundary contains u. Since B is a blocked appendage, it contains a contact point v that does not lie on the boundary of R and an edge e' incident on v. It follows that e' must drawn in a different region than R. Also, there is a walk W in B containing edges e and e', such that no internal vertex of W is in H. Since e and e' lie in different regions, that walk must cross some edge on the boundary of R, by the Jordan Curve Theorem.

DEFINITION: Let H be a subgraph of a graph. In a drawing of H on any surface, an appendage of H is **forced into region** R if R is the only region whose boundary contains all the contact points of that appendage.

Example 7.6.2: In Figure 7.6.2, a subgraph isomorphic to $K_{2,3}$ is shown with black vertices and solid edges, and its only appendage is shown with a white vertex and broken edges. The appendage is forced into region R, because it is the only region whose boundary contains all three contact points.

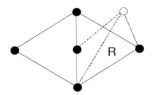

Figure 7.6.2   A subgraph with a forced appendage.

Algorithmic Preliminaries

By iterative application of Corollary 7.3.7, a graph is planar if and only if all its blocks are planar. Performing a prior decomposition into blocks (see §5.4) frees the main part of the algorithm from a few cluttersome details, and permits a focus on the most interesting case, in which the graph to be tested for planarity is 2-connected.

In testing the planarity of a 2-connected graph G, the basis for a Whitney synthesis (see §5.2) is to find a cycle and to draw it in the plane. (If there is no cycle, then the graph is a tree, and it is planar.) The cycle is designated as the basis subgraph G_0.

Selecting the Next Path Addition

Before each iteration of the body of the main loop, there is a two-part exit test. One exit condition is that the entire graph is already drawn in the plane, in which case it is decided, of course, that the graph is planar. The other exit condition is that some appendage of subgraph G_j is blocked. If so, then graph G is declared nonplanar and the algorithm terminates. If neither exit is taken, then there are two possible cases of appendage selection.

In the main loop, graph G_{j+1} is obtained from graph G_j by choosing an appendage of G_j and adding a path joining two contact points of that appendage to the planar drawing of G_j. The main technical concern is that of choosing an appropriate appendage at each iteration.

In the first case, it is determined that some appendage is forced. If so, then a path between two of its contact points is drawn into the one region whose boundary contains all contact points of that forced appendage, thereby extending the drawing of G_j to a drawing of G_{j+1}.

In the second case, there are no forced appendages. Under this circumstance, an arbitrary appendage is chosen, and a path between two of its contact points is drawn into any region whose boundary contains all contact points of that appendage, thereby extending the drawing of G_j to a drawing of G_{j+1}.

After extension to a drawing of G_{j+1}, the loop returns to the two-fold exit test and possibly continues with the next attempt at extending the drawing.

The Algorithm

Naively searching for a homeomorphic copy of K_5 or of $K_{3,3}$ in an n-vertex graph would require exponentially many steps. The planarity algorithm given here requires $O(n^2)$ execution steps.

Algorithm 7.6.1:   Planarity-Testing for a 2-Connected Graph
Input: a 2-connected graph G
Output: a planar drawing of G, or the decision FALSE.
    {Initialize} Find an arbitrary cycle G_0 in G, and draw it in the plane.
    While $G_j \neq G$ {this exit implies that G is planar}
        If any appendage is blocked
            return FALSE
        Else
            If some appendage is forced
                $B := $ that appendage
            Else
                $B := $ any appendage whatever
                $R := $ any region whose boundary contains all contact points of B
                Select any path between two contact points of B
                Draw that path into region R to obtain G_{j+1}
                Continue with next iteration of while-loop
    {End of while-loop body}
    Return (planar drawing of G)

Outline of Correctness of Planarity Algorithm

To establish the correctness of this algorithm, there is a two-step proof that when no appendage is forced, the particular choice of an appendage to extend the subgraph G_j and a region in which to draw it does not affect the eventual decision of the algorithm. The first step, of identifying the configuration under which there is an appendage A with no forced choice between two regions R and R', is illustrated in Figure 7.6.3.

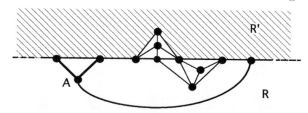

Figure 7.6.3   Appendage A has option of region R or region R'.

The second step is to demonstrate that regardless of which region is chosen, the eventual algorithmic decision on planarity is the same. For further details, see §9.8 of [BoMu76] or §6.3 of [We0l].

Hopcroft and Tarjan [HoTa74] have published a linear-time planarity algorithm. Topological arguments of Gross and Rosen [GrRo8l] enabled them to achieve a linear time reduction of planarity-testing for *2-complexes* [GrRO79] to a planarity-test for graphs.

EXERCISES for Section 7.6

For Exercises 7.6.1 through 7.6.12, apply Algorithm 7.6.1 to the designated graph.

7.6.1 The graph of Exercise 7.4.1.   **7.6.2**[S] The graph of Exercise 7.4.2.

7.6.3 The graph of Exercise 7.4.3.   **7.6.4** The graph of Exercise 7.4.4.

7.6.5 The graph of Exercise 7.4.5.   **7.6.6** The graph of Exercise 7.4.6.

7.6.7 $K_{3,3}$.   **7.6.8** K_5.

7.6.9 $K_3 + K_1$.   **7.6.10** $(K_4 - K_2) \times P_3$.

7.6.11 $K_9 - K_{4,5}$.   **7.6.12** $(K_5 - K_3) \times P_2$.

7.6.13 [Computer Project] Implement Algorithm 7.6.1 on a computer.

7.7 CROSSING NUMBERS AND THICKNESS

Two quantifications of the nonplanarity of a graph involve only the plane, without going to higher-order surfaces. One is to determine the minimum number of edge-crossings needed in a drawing of that graph in the plane. The other partitions the edge-set of a graph into "layers" of planar subgraphs and establishes the minimum number of layers needed.

Crossing Numbers

Although it is easy to count the edge-crossings in a graph drawing, it is usually very difficult to prove that the number of crossings in a particular drawing is the smallest possible. The context of most such calculations is that the drawing has been *normalized* so that no edge crosses another more than once and that at most two edges cross at any point.

DEFINITION: Let G be a simple graph. The **crossing number** $cr(G)$ is the minimum number of edge-crossings that can occur in a normal drawing of G in the plane.

The usual way to calculate the crossing number of a graph G has two parts. A lower bound $cr(G) \geq b$ is established by a theoretical proof. An upper bound $cr(G) \leq b$ is established by exhibiting a drawing of G in the plane with b crossings. If one is able to draw the given graph with only one edge-crossing, then calculating the crossing number reduces to a planarity test.

Proposition 7.7.1: $cr(K_5) = 1$.

Proof: By Theorem 7.1.4, every drawing of K_5 in the plane must have at least one crossing. Thus, $cr(K_5) \geq 1$. Since the drawing of K_5 in Figure 7.7.1 has only one crossing, it follows that $cr(K_5) \leq 1$.   ◇

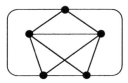

Figure 7.7.1 A drawing of K_5 in the plane with only one crossing.

Proposition 7.7.2: $cr(K_{3,3}) = 1$.

Proof: By Theorem 7.1.5, it follows that $cr(K_{3,3}) \geq 1$. The drawing of $K_{3,3}$ in Figure 7.7.2 with only one crossing implies that $cr(K_{3,3}) \geq 1$. $\diamond$

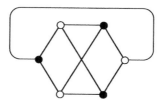

Figure 7.7.2   A drawing of $K_{3,3}$ in the plane with only one crossing.

Lower Bounds for Crossing Numbers

The main interest in crossing numbers of graphs is for cases in which the minimum number is more than 1.

Theorem 7.7.3: *Let G be a connected simple graph. Then*

$$cr(G) \geq |E_G| - 3|V_G| + 6$$

Proof: Let H be a planar spanning subgraph of G with the maximum number of edges. Theorem 7.5.9 implies that $|E_H| \leq 3|V_H| - 6$, from which it follows that $|E_H| \leq 3|V_G| - 6$, since $V_H = V_G$. Since H is maximum planar, we infer that no matter how each of the remaining $|E_G| - |E_H|$ edges is added to a planar drawing of H, it crosses at least one edge of H. Thus, $cr(G) \geq |E_G| - |E_H| \geq |E_G| - 3|V_G| + 6$. $\diamond$

Proposition 7.7.4: $cr(K_6) = 3$.

Proof: Since $|E(K_6)| = 15$ and $|cr(K_6)| = 6$, Theorem 7.7.3 implies that

$$cr(K_6) \geq 15 - 3 \cdot 6 + 6 = 3$$

The drawing of K_6 in Figure 7.7.3 with three crossings establishes that $cr(K_6) \leq 3$. $\diamond$

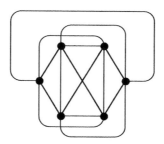

Figure 7.7.3   A drawing of K_6 in the plane with three crossings.

Theorem 7.7.5: *Let G be a connected simple bipartite graph. Then*

$$cr(G) \geq |E_G| - 2|V_G| + 4$$

Proof: Let H be a planar spanning subgraph of G with the maximum number of edges. Theorem 7.5.12 implies that $|E_H| \leq 2|V_H| - 4$, from which it follows that $|E_H| \leq 2|V_G| - 4$, since $V_H = V_G$. Since H is maximum planar, it follows that no matter how each of remaining $|E_G| - |E_H|$ edges is added to a planar drawing of H, it crosses at least one edge of H. Thus, $cr(G) \geq |E_G| - |E_H| \geq |E_G| - 2|V_G| + 4$.   ◇

Proposition 7.7.6: $cr(K_{3,4}) = 2$.

Proof: Since $|E(K_{3,4})| = 12$ and $|V(K_{3,4})| = 7$, Theorem 7.7.5 implies that

$$cr(K_{3,4}) \geq 12 - 2 \cdot 7 + 4 = 2$$

The drawing of $K_{3,4}$ in Figure 7.7.4 with two crossings implies that $cr(K_{3,4}) \leq 2$. ◇

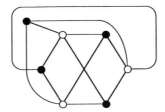

Figure 7.7.4   A drawing of $K_{3,4}$ in the plane with two crossings.

Example 7.7.1: The requirement that no more than two edges cross at the same intersection is necessary for these results to hold. Figure 7.7.5 shows that if a triple intersection is permitted, then $K_{3,4}$ can be drawn in the plane with only one crossing.

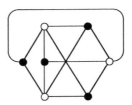

Figure 7.7.5   A drawing of $K_{3,4}$ in the plane with one triple-crossing.

Thickness

Application 7.7.1:   *Partitioning a Network into Planar Layers*   One technique involved in miniaturizing a nonplanar electronic network is to sandwich layers of insulation between planar layers of uninsulated wires, which connect nodes that pierce through all the layers. Minimizing the number of layers tends to reduce the size of the chip. Moreover, in applying the technique to mass production, minimizing the number of layers tends to reduce the cost of fabrication.

DEFINITION: The **thickness** of a simple graph G, denoted $\theta(G)$, is the smallest cardinality of a set of planar spanning subgraphs of G whose edge-sets partition E_G.

Example 7.7.2: The two planar graphs in Figure 7.7.6 below both span K_8. The union of their edge-sets is E_{K_8}. Thus, a chip implementing K_8 would need only two planar layers.

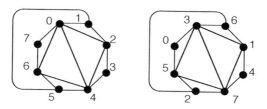

Figure 7.7.6   Two planar spanning subgraphs of K_8.

Theorem 7.7.7: *Let G be a connected simple graph. Then*

$$\theta(G) \geq \left\lceil \frac{|E_G|}{3\,|V_G| - 6} \right\rceil$$

Proof: By Theorem 7.5.9, at most $3\,|V_G| - 6$ edges lie in a planar subgraph of G.   ◇

Proposition 7.7.8: $\theta(K_8) = 2$.

Proof: By Theorem 7.7.7, $\theta(K_8) \geq 2$. Figure 7.7.6 shows two planar graphs whose union is K_8, which implies that $\theta(K_8) \leq 2$.   ◇

Theorem 7.7.9: *Let G be a connected simple bipartite graph. Then*

$$\theta(G) \geq \left\lceil \frac{E_G}{2\,|V_G| - 4} \right\rceil$$

Proof: By Theorem 7.5.12, at most $2\,|V_G| - 4$ edges lie in a planar subgraph of G.   ◇

Straight-Line Drawings of Graphs

In circuit design (either single-layer or multi-layer), it is convenient for wires to be straight-line segments, rather than having curves or bends.

DEFINITION: A ***straight-line drawing*** of a graph is a planar drawing in which every edge is represented by a straight-line segment.

Wagner [Wa36] and Fary [Fa48] proved the following result, which is known as Fary's theorem. Thomassen [Th80, Th81] developed a proof of Kuratowski's theorem that simultaneously yields Fary's Theorem.

Theorem 7.7.10: *Let G be a simple planar graph. Then G has a straight-line drawing without crossings.*   ◇

EXERCISES for Section 7.7

7.7.1[S] Derive this lower bound for the crossing number of the complete graph K_n.

$$cr(K_n) \geq \left\lceil \frac{(n-3)(n-4)}{2} \right\rceil$$

7.7.2 Derive this lower bound for the crossing number of the complete bipartite graph $K_{m,n}$.

$$cr(K_{m,n}) \geq (m-2)(n-2)$$

7.7.3 Draw the Petersen graph in the plane with 2 (normal) crossings.

7.7.4 Draw K_7 in the plane with 9 (normal) crossings.

Exercises 7.7.5 through 7.7.8 require calculating the crossing number of a given graph.

7.7.5[S] $cr(K_{4,4})$.   **7.7.6** $cr(K_{4,5})$.

7.7.7 $cr(Q_3 + K_1)$.   **7.7.8** $cr(C_4 + K_2)$.

7.7.9 Prove that $\theta(K_{m,m}) \geq \left\lceil \frac{mn}{2(m+n-2)} \right\rceil$.

7.7.10[S] Prove that $\theta(K_{6,6}) = 2$.

7.7.11 Prove that $\theta(K_9) = 3$.

Application 7.7.1, continued: A straightforward approach to multi-layer circuit design is to use "simultaneous straight-line drawings" of the spanning subgraphs that induce the edge-set partition. This means that the nodes in each layer are in a fixed location in the plane, so that each layer is given a straight-line drawing. Figure 7.7.7 shows a two-layer partition of the complete graph K_5 by simultaneous straight-line drawings.

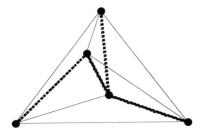

Figure 7.7.7   **Two-layer straight-line partition of K_5.**

Exercises 7.7.12 through 7.7.15 require partitioning the given graph by simultaneous straight-line drawings of spanning subgraphs into the prescribed number of layers.

7.7.12[S] Partition K_6 into two simultaneous straight-line layers.

7.7.13 Partition $K_{4,4}$ into two simultaneous straight-line layers.

7.7.14 Partition $K_8 - K_4$ into three simultaneous straight-line layers.

7.7.15 Partition K_8 into four simultaneous straight-line layers.

7.8   SUPPLEMENTARY EXERCISES

7.8.1 Consider a plane drawing of an n-vertex simple graph G with fewer than $3n - 6$ edges.
   a) Use the Euler Polyhedral Equation and the Edge-Face Inequality to prove that there exists a face f with more than 3 sides.
   b) Use the Jordan curve theorem to prove that there are two non-consecutive vertices on the boundary of face f that are not adjacent in graph G.

7.8.2 Draw a self-dual imbedding of a planar simple graph with degree sequence $\langle 4, 4, 4, 3, 3, 3, 3 \rangle$.

7.8.3 Draw a non-self-dual imbedding of a planar simple graph with degree sequence $\langle 4, 4, 4, 3, 3, 3, 3 \rangle$.

7.8.4 A connected graph G with p vertices and q edges is imbedded in the sphere. Its dual is drawn. The dual vertex in each primal face is joined to every vertex on the face boundary, and every crossing of a primal and dual edge is converted into a new vertex, so that the result is a graph $G^{\#}$ imbedded in the sphere.
   a) How many vertices and edges does $G^{\#}$ have?
   b) For $G = C_3$, draw the edge-complement of $G^{\#}$.

7.8.5 Draw a 4-regular simple 9-vertex planar graph.

7.8.6 Suppose that a simple graph has degree sequence $\langle 5, 5, 5, 5, 5, 5, 4, 4 \rangle$. Prove that it cannot be planar.

7.8.7 Prove that every simple planar graph with at least 4 vertices has at least 4 vertices of degree less than 6.

7.8.8 Can a 3-connected simple graph of girth 3 be drawn in the plane so that every face has at least four sides? Either give an example or prove it is impossible.

7.8.9 Prove that the Petersen graph has no 3-cycles or 4-cycles. Then use the Euler Polyhedral Equation and the Edge-Face Inequality to show that the Petersen graph is nonplanar.

7.8.10 Let H be a 4-regular simple graph, and let G be the graph obtained by joining H to K_1. Decide whether G can be planar. Either give an example or prove impossibility.

7.8.11 Draw each of the four isomorphism types of a simple 6-vertex graph that contains K_5 homeomorphically, but does not contain $K_{3,3}$.

7.8.12 a. Draw a 9-vertex tree T such that $K_2 + T$ is planar.
b. Prove that there is only one such tree.

7.8.13 Decide whether the following graphs contain a homeomorphic copy of $K_{3,3}$, of K_5, of both, or of neither.

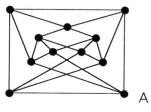
A

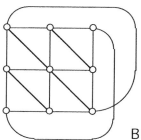
B

7.8.14 Decide which of the graphs below are planar.

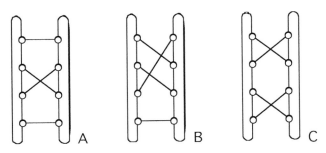

7.8.15 Decide which of the graphs below are planar.

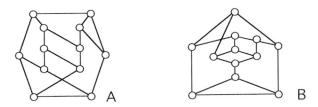

7.8.16 a. Show how to add one edge to graph A so that the resulting graph is nonplanar (and prove nonplanarity). b. Show how to delete one edge from graph B so that the resulting graph is planar.

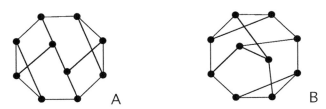

7.8.17 Decide which of the graphs below are planar.

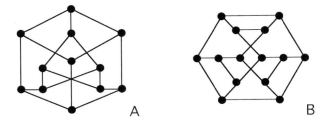

7.8.18 Use the planarity algorithm to show that the circulant graph $circ(7 : 1, 2)$ is nonplanar. Find a Kuratowski subgraph.

7.8.19 Use the planarity algorithm to show that the circulant graph $circ(9 : 1, 3)$ is nonplanar. Find a Kuratowski subgraph.

7.8.20 Redraw the following graph with only two crossings.

7.8.21 Calculate the crossing number of the following graph.

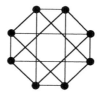

7.8.22 Calculate the crossing number of $K_6 - K_2$.

7.8.23 In a planar drawing of $K_6 - K_2$, in which (as usual) at most two edges cross at a point, what is the minimum number of edges that must cross at least one other edge. (Hint: Be careful; for $K_{3,3}$, the answer is two, not one, since two edges cross at the single crossing.)

7.8.24 Suppose that a triple edge-crossing were permitted. Prove that even then, it would still be impossible to draw $K_6 - K_2$ in the plane with only one crossing. (Hint: The graph obtained by inserting a new vertex at the triple-intersection point would have to be planar.)

7.8.25 Prove that $\theta(K_n) \geq \lfloor \frac{n+7}{6} \rfloor$.

7.8.26 Calculate $\theta(K_{12})$.

Glossary

appendage to a subgraph H: the induced subgraph on a set of edges such that each pair lies on a path with no internal vertices in H and is maximal with respect to this property.

—, **blocked:** for a subgraph H drawn on a surface, an appendage that is undrawable in every region.

—, **forced** into region R: an appendage B such that R is the only region whose boundary contains all the contact points of B.

—, **overlapping pair for a cycle C:** appendages B_1 and B_2 such that either of these conditions holds:

  (i) Two contact points of B_1 alternate with two contact points of B_2 on cycle C.
  (ii) B_1 and B_2 have three contact points in common.

—, **undrawable** for region R: for a subgraph H drawn on a surface, an appendage B such that the boundary of region R does not contain all the contact points of B.

barycentric subdivision, (first) of a graph G: the subdivision of G such that every edge of G is subdivided exactly once.

—, n^{th}: the first barycentric subdivision of the $(n-1)^{st}$ barycentric subdivision.

boundary of a region of an imbedding of a graph G: the subgraph of G that comprises all vertices that abut the region and all edges whose interiors abut the region.

boundary walk of a face of an imbedding of a graph G: a closed walk in graph G that corresponds to a complete traversal of the perimeter of the polygonal region within the face.

cellular imbedding $\iota : G \to S$: a graph imbedding such that every region is topologically equivalent to an open disk.

chord of a subgraph H: an appendage of H that consists of a single edge.

connected Euclidean set: see *Euclidean set*.

contact point of an appendage B to a subgraph H: a vertex in both B and H.

crossing number $cr(G)$ of a simple graph G: the minimum number of edge-crossings that can occur in a normal planar drawing of G.

curve, closed or open in a Euclidean set: see *path*.

digon: a 2-sided region.

dual edges: the edges of the dual graph.

dual faces: the faces of the dual imbedding.

dual graph and dual imbedding: the new graph and its imbedding derived, starting with a cellular imbedding of a graph, by inserting a new vertex into each existing face, and by then drawing through each existing edge a new edge that joins the new vertex in the region on one side to the new vertex in the region on the other side.

dual vertices: the vertices of the dual graph.

edges separated in a graph G by the subgraph H: two edges e_1 and e_2 of $E_G - E_H$ such that every walk in G from an endpoint of e_1 to an endpoint of e_2 contains a vertex of H.

edges unseparated in a graph G by the subgraph H: two edges e_1 and e_2 of $E_G - E_H$ that are not separated.

Euclidean set X a subset of any Euclidean space $\mathbb{R}^n$.

—, connected a Euclidean set X such that for every pair of points $s, t \in X$, there exists a path within X from s to t.

—, separation of X by a subset $W \subset X$: the condition that there exists a pair of points s and t in $X - W$, such that every path in X from s to t intersects the set W.

face of a graph imbedding $\iota : G \to S$:] the union of a region and its boundary. (A drawing does *not* have faces unless it is an imbedding.)

face-set of a graph imbedding $\iota : G \to S$: the set of all faces of the imbedding, denoted F_ι, F_G, or F, depending on the context.

face-size: the number of edge-steps in the closed walk around its boundary (in which some edges may count twice). A face of size n is said to be *n-sided*.

girth of a non-acyclic graph: the length of the smallest closed cycle (undefined for an acyclic graph).

homeomorphic graphs: a pair of graphs G and H such that there is an isomorphism from a *subdivision* of G to a subdivision of H.

imbedding of a graph G on a surface S: a drawing without any edge-crossings.

Jordan separation property: for a Euclidean set X, the property that every closed curve in X separates X.

Kuratowski graphs: the complete graph K_5 and the complete bipartite graph $K_{3,3}$.

Kuratowski subgraph of a graph: a subgraph homeomorphic either to K_5 or $K_{3,3}$.

longitude on the standard torus: a closed curve that bounds a disk in the space exterior to the solid donut (and goes around in the "long" direction).

meridian on the standard torus: a closed curve that bounds a disk inside the solid donut it surrounds (and goes around the donut in the "short" direction).

monogon: a 1-sided region.

path from s to t in a Euclidean set: the image of a continuous function f from the unit interval $[0,1]$ to a subset of that space such that $f(0) = s$ and $f(1) = t$, that is a bijection on the interior of $[0,1]$. (One may visualize a path as the trace of a particle traveling through space for a fixed length of time.)

—, **closed:** a path such that $f(0) = f(1)$, i.e., in which $s = t$. (For instance, this would include a "knotted circle" in space.)

—, **open:** a path such that $f(0) \neq f(1)$, i.e., in which $s \neq t$.

planar drawing of a graph: a drawing of the graph in the plane without edge-crossings.

planar graph: a graph such that there exists a planar drawing of it.

plane in Euclidean 3-space $\mathbb{R}^3$: a set of points (x, y, z) such that there are numbers a, b, c, and d with $ax + by + cz = d$.

Poincaré dual: see *dual graph and dual imbedding*.

primal graph and primal imbedding: whatever graph and imbedding are supplied as input to the Poincaré duality process.

primal vertices, primal edges, and primal faces: the vertices, edges, and faces of the primal graph and primal imbedding.

proper walk: a walk in which no edge occurs immediately after itself.

region of a graph imbedding $\iota : G \to S$: a component of the Euclidean set $S - \iota(G)$ which does *not* contain its boundary.

Riemann stereographic projection: a function ρ that maps each point w of the unit-diameter sphere (tangent at the origin $(0,0,0)$ to the xz-plane in Euclidean 3-space) to the point $\rho(x)$ where the ray from the north pole $(0,1,0)$ through the point w intersects the xz-plane.

self-dual graph: a graph G with an imbedding in the sphere (or sometimes another surface) such that the Poincaré dual graph G^* for that imbedding is isomorphic to G.

separated Euclidean set: see *Euclidean set*.

size of a face: see *face-size*.

smoothing out a 2-valent vertex v: replacing two different edges that meet at v by a new edge that joins their other endpoints.

sphere: a set of points in $\mathbb{R}^3$ equidistant from a fixed point.

standard donut: the surface of revolution obtained by revolving a disk of radius 1 centered at $(2,0)$ in the xy-plane around the y-axis in 3-space.

standard torus: the surface of revolution obtained by revolving a circle of radius 1 centered at $(2,0)$ in the xy-plane around the y-axis in 3-space. It is the surface of the standard donut.

straight-line drawing of a graph: a planar drawing in which every edge is represented by a straight-line segment.

subdividing an edge e: with endpoints u and v:] replacing that edge e by a path u, e', w, e'', v, where edges e' and e'' and vertex w are new to the graph.

subdividing a graph: performing a sequence of edge-subdivision operations.

thickness $\theta(G)$ of a simple graph G: the smallest cardinality of a set of planar spanning subgraphs of G whose edge-sets partition E_G.

Chapter 8

GRAPH COLORINGS

INTRODUCTION

The first known mention of coloring problems was in 1852, when Augustus De Morgan, Professor of Mathematics at University College, London, wrote Sir William Rowan Hamilton in Dublin about a problem posed to him by a former student, named Francis Guthrie. Guthrie noticed that it was possible to color the counties of England using four colors so that no two adjacent counties were assigned the same color. The question raised thereby was whether four colors would be sufficient for all possible decompositions of the plane into regions.

The Poincaré duality construction transforms this question into the problem of deciding whether it is possible to color the vertices of every planar graph with four colors so that no two adjacent vertices are assigned the same color. Wolfgang Haken and Kenneth Appel provided an affirmative solution in 1976.

Numerous practical applications involving graphs and some form of coloring are described in this chapter. Factorization is included here, because, like edge-coloring, it involves a partitioning of the edge-set of a graph.

8.1   VERTEX-COLORINGS

In the most common kind of graph coloring, colors are assigned to the vertices. From a standard mathematical perspective, the subset comprising all the vertices of a given color would be regarded as a cell of a partition of the vertex-set. Drawing the graph with colors on the vertices is simply an intuitive way to represent such a partition.

NOTATION: The maximum degree of a vertex in a graph G is denoted $\delta_{\max}(G)$, or simply by $\delta_{\max}$ when the graph of reference is evident from context.

The Minimization Problem for Vertex-Colorings

Most applications involving *vertex-colorings* are concerned with determining the minimum number of colors required under the condition that the endpoints of an edge cannot have the same color.

DEFINITION: A **vertex k-coloring** is an assignment $f : V_G \to C$ from its vertex-set onto a k-element set C whose elements are called *colors* (typically, $C = \{1, 2, \ldots k\}$). For any k, such an assignment is called a **vertex-coloring**.

TERMINOLOGY: Since vertex-colorings arise more frequently than edge-colorings or map-colorings, one often says *coloring*, instead of *vertex-coloring*, when the context is clear.

DEFINITION: A **color class** in a vertex-coloring of a graph G is a subset of V_G containing all the vertices of a given color.

DEFINITION: A **proper vertex-coloring** of a graph is a vertex-coloring such that the endpoints of each edge are assigned two different colors.

Remark: Quite commonly, it is implicit from context that the colorings under consideration are proper, in which case each color class is an independent set of vertices.

DEFINITION: A graph is said to be **vertex k-colorable** if it has a proper vertex k-coloring.

Example 8.1.1: The vertex-coloring shown in Figure 8.1.1 demonstrates that the graph is vertex 4-colorable.

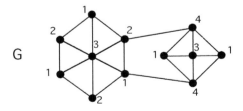

Figure 8.1.1   A proper vertex 4-coloring of a graph.

DEFINITION: The **(vertex) chromatic number** of a graph G, denoted $\chi(G)$, is the minimum number of different colors required for a proper vertex-coloring of G. A graph G is **(vertex) k-chromatic** if $\chi(G) = k$.

Thus, $\chi(G) = k$, if graph G is k-colorable but not $(k-1)$-colorable.

Example 8.1.1, continued: The 3-coloring in Figure 8.1.2 shows that the graph G in Figure 8.1.1 is 3-colorable, which means that $\chi(G) \leq 3$. However, graph G contains three mutually adjacent vertices and hence is not 2-colorable. Thus, G is 3-chromatic.

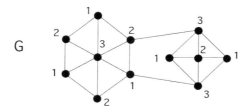

Figure 8.1.2   A proper 3-coloring of graph G.

Example 8.1.2: The 4-coloring of the graph G shown in Figure 8.1.3 establishes that $\chi(G) \leq 4$, and the K_4-subgraph (drawn in bold) shows that $\chi(G) \geq 4$. Hence, $\chi(G) = 4$.

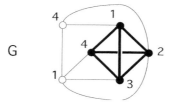

Figure 8.1.3   Graph G has chromatic number 4.

Remark: The study of vertex-colorings of graphs is customarily restricted to simple, connected graphs for the following reasons:

i. A graph with a self-loop is uncolorable, since the endpoint of a self-loop is adjacent to itself.

i. A multi-edge requires two different colors for its endpoints in the same way that a single edge does.

i. The chromatic number of a graph is simply the maximum of the chromatic numbers of its components.

Modeling Applications as Vertex-Coloring Problems

When an application is modeled as a vertex-coloring problem, the vertices in each color class typically represent individuals or items that do not compete or conflict with each other.

Application 8.1.1:   *Assignment of Radio Frequencies*   Suppose that the vertices of a graph G represent transmitters for radio stations. In this model, two stations are considered adjacent when their broadcast areas overlap, which would result in interference if they broadcast at the same frequency. Then two "adjacent" stations should be assigned different transmission frequencies. Regarding the frequencies as colors transforms the situation into a graph-coloring problem, in which each color class contains vertices representing stations with no overlap. In this model, $\chi(G)$ equals the minimum number of transmission frequencies required to avoid broadcast interference. (See Application 1.3.6.)

Application 8.1.2: *Separating Combustible Chemical Combinations* Suppose that the vertices of a graph represent different kinds of chemicals needed in some manufacturing process. For each pair of chemicals that might explode if combined, there is an edge between the corresponding vertices. The chromatic number of this graph is the number of different storage areas required so that no two chemicals that mix explosively are stored together.

Application 8.1.3: *University Course Scheduling* Suppose that the vertices of a simple graph G represent the courses at a university. In this model, two vertices are adjacent if and only if at least one student preregisters for both of the corresponding classes. Clearly, it would be undesirable for two such courses to be scheduled at the same time. Then the vertex-chromatic number $\chi(G)$ gives the minimum number of time periods in which to schedule the classes so that no student has a conflict between two courses.

Application 8.1.4: *Fast-Register Allocation for Computer Programming* In some computers, there are a limited number of special "registers" that permit faster execution of arithmetic operations than ordinary memory locations. The program variables that are used most often can be declared to have the "register" storage class. Unfortunately, if the programmer declares more "register" variables than the number of hardware registers available, then the program execution may waste more time swapping variables between ordinary memory and the fast registers than is saved by using the fast registers. One solution is for the programmer to control the register designation, so that variables that are simultaneously active are assigned to different registers. The graph model has one vertex for each variable, and two vertices are adjacent if the corresponding variables can be simultaneously active. Then the chromatic number equals the number of registers needed to avoid the overswapping phenomenon.

Sequential Vertex-Coloring Algorithm

There is a naive (brute-force) algorithm to decide, for some fixed k, whether a given n-vertex graph is k-colorable. Just check to see if any of the k^n possible k-colorings is proper and return YES if so. By iterating this decision procedure, starting with $k = 1$, until a YES is returned, one obtains an algorithm for calculating the exact value of the chromatic. However, its running time is exponential in the number of vertices.

Alternatively, Algorithm 8.1.1 is a *sequential* algorithm that quickly produces a proper coloring of any graph; yet the coloring it produces is unlikely to be a minimum one. Moreover, it seems unlikely that *any* polynomial-time algorithm can accomplish this, since the problem of calculating the chromatic number of a graph is known to be NP-hard [GaJo79]. In fact, deciding whether a graph has a 3-coloring is an NP-complete problem.

Algorithm 8.1.1:   **Sequential Vertex-Coloring**
Input: a graph G with vertex list $v_1, v_2, \ldots, v_p$.
Output: a proper vertex-coloring $f : V_G \to \{1, 2, \ldots\}$.
    For $i = 1, \ldots, p$
        Let $f(v_i) :=$ the smallest color number not used on any of the smaller-subscripted neighbors of v_i.
    Return vertex-coloring f.

Example 8.1.3: When the sequential vertex-coloring algorithm is applied to the graph of Figure 8.1.4, the result is a 4-coloring.

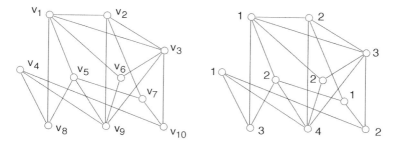

Figure 8.1.4   A graph and a sequential 4-coloring.

However, if the coloring assignment relaxes its obsessiveness with using the smallest possible color number at vertex v_4 and at vertex v_7, then a 3-coloring can be obtained for the same graph, as shown in Figure 8.1.5.

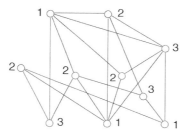

Figure 8.1.5   A nonsequential 3-coloring for the same graph.

Remark: Notice that if the vertices of the graph in Figure 8.1.5 are ordered so that v_1, v_2, v_3 are the vertices with color 1, vertices v_4, v_5, v_6, v_7 are the ones with color 2, and v_8, v_9, v_{10} are the vertices with color 3, then the sequential coloring algorithm would have produced that minimum coloring. More generally, for a given minimum coloring for a graph G (obtained by whatever means), if the vertices of G are linearly ordered so that all the vertices with color i precede all the vertices with color j whenever $i < j$, then, when G along with this vertex ordering is supplied as input to the sequential-coloring algorithm, the result is a minimum coloring. Thus, for any n-vertex graph, there is at least one vertex ordering (of the $n!$ orderings) for which Algorithm 8.1.1 will produce a minimum coloring.

Basic Principles for Calculating Chromatic Numbers

A few basic principles recur in many chromatic-number calculations. They combine to provide a direct approach involving two steps, as already seen for Examples 8.1.2 and 8.1.3.

- *Upper Bound*: Show $\chi(G) \leq k$, most often by exhibiting a proper k-coloring of G.

- *Lower Bound*: Show $\chi(G) \geq k$, most especially, by finding a subgraph that requires k colors.

The following result is an easy upper bound for $\chi(G)$; it is complemented by the easy lower bound of clique number $\omega(G)$. Brooks's Theorem, appearing later in this section, sharpens this upper bound for a large class of graphs.

Proposition 8.1.1: *Let G be a simple graph. Then $\chi(G) \leq \delta_{\max}(G) + 1$.*

Proof: The sequential coloring algorithm never uses more than $\delta_{\max}(G) + 1$ colors, no matter how the vertices are ordered, since a vertex cannot have more than $\delta_{\max}(G)$ neighbors. $\diamondsuit$

Proposition 8.1.2: *Let G be a graph that has k mutually adjacent vertices. Then $\chi(G) \geq k$.*

Proof: Using fewer than k colors on graph G would result in a pair from the mutually adjacent set of k vertices being assigned the same color. $\diamondsuit$

REVIEW FROM §2.3: A ***clique*** in a graph G is a maximal subset of V_G whose vertices are mutually adjacent. The ***clique number*** $\omega(G)$ of a graph G is the number of vertices in a largest clique in G.

Corollary 8.1.3: *Let G be a graph. Then $\chi(G) \geq \omega(G)$.*

REVIEW FROM §2.3: The ***independence number*** $\alpha(G)$ of a graph G is the number of vertices in an independent set in G of maximum cardinality.

Proposition 8.1.4: *Let G be any graph. Then*

$$\chi(G) \geq \left\lceil \frac{|V_G|}{\alpha(G)} \right\rceil$$

Proof: Since each color class contains at most $\alpha(G)$ vertices, the number of different color classes must be at least $\left\lceil \frac{|V_G|}{\alpha(G)} \right\rceil$. $\diamondsuit$

Example 8.1.4: Consider the graph $G = circ(7 : 1, 2)$, shown in Figure 8.1.6 (§1.2). If G had an independent set of size 3 (i.e., if $\alpha(G) \geq 3$), then its edge-complement $\overline{G}$ would contain a 3-cycle. However, $\overline{G} = circ(7 : 3)$ is a 7-cycle. Thus, $\alpha(G) = 2$, and, consequently, by Proposition 8.1.4,

$$\chi(G) \geq \left\lceil \frac{|V_G|}{\alpha(G)} \right\rceil = \left\lceil \frac{7}{2} \right\rceil = 4$$

There is a proper 4-coloring with color classes $\{0, 3\}$, $\{1, 4\}$, $\{2, 5\}$, and $\{6\}$.

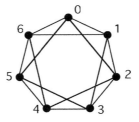

Figure 8.1.6 $circ(7 : 1, 2)$.

Proposition 8.1.5: *Let H be a subgraph of graph G. Then $\chi(G) \geq \chi(H)$.*

Proof: Whatever colors are used on the vertices of subgraph H in a minimum coloring of graph G can also be used in a coloring of H by itself. ◇

REVIEW FROM §2.7: The ***join*** $G + H$ of the graphs G and H is obtained from the graph union $G \cup H$ by adding an edge between each vertex of G and each vertex of H.

Proposition 8.1.6: *The join of graphs G and H has chromatic number*

$$\chi(G + H) = \chi(G) + \chi(H)$$

Proof: *Lower Bound.* In the join $G + H$, no color used on the subgraph G can be the same as a color used on the subgraph H, since every vertex of G is adjacent to every vertex of H. Since $\chi(G)$ colors are required for subgraph G and $\chi(H)$ colors are required for subgraph H, it follows that $\chi(G + H) \geq \chi(G) + \chi(H)$.

Upper Bound. Just use any $\chi(G)$ colors to properly color the subgraph G of $G + H$, and use $\chi(H)$ different colors to color the subgraph H. ◇

Chromatic Numbers for Common Graph Families

By using the basic principles given above, it is straightforward to establish the chromatic numbers of graphs in some of the most common graph families, which are summarized in Table 8.1.1.

Table 8.1.1: Chromatic numbers for common graph families

Graph G	$\chi(G)$
trivial graph	1
bipartite graph	2
nontrivial path graph P_n	2
nontrivial tree T	2
cube graph Q_n	2
even cycle graph C_{2n}	2
odd cycle graph C_{2n+1}	3
even wheel W_{2n}	3
odd wheel W_{2n+1}	4
complete graph K_n	n

Proposition 8.1.7: *A graph G has $\chi(G) = 1$ if and only if G has no edges.*

Proof: The endpoints of an edge must be colored differently. ◇

Proposition 8.1.8: *A bipartite graph G has $\chi(G) = 2$, unless G is edgeless.*

Proof: A 2-coloring is obtained by assigning one color to every vertex in one of the bipartition parts and another color to every vertex in the other part. If G is not edgeless, then $\chi(G) \geq 2$, by Proposition 8.1.2. ◇

Corollary 8.1.9: *Path Graphs:* $\chi(P_n) = 2$, *for* $n \geq 2$.

Proof: P_n is bipartite. Apply Proposition 8.1.8.   ◇

Remark: Just as a bipartite graph is a graph whose vertices can be partitioned into two color classes, we define a **multipartite graph** for any number of color classes.

DEFINITION: A k-**partite graph** is a loopless graph whose vertices can be partitioned into k independent sets, which are sometimes called the **partite sets** of the partition.

Corollary 8.1.10: *Trees:* $\chi(T) = 2$, *for any nontrivial tree* T.

Proof: Trees are bipartite.   ◇

Corollary 8.1.11: *Cube Graphs:* $\chi(Q_n) = 2$.

Proof: The cube graph Q_n is bipartite.   ◇

Corollary 8.1.12: *Even Cycles:* $\chi(C_{2n}) = 2$.

Proof: An even cycle is bipartite.   ◇

Proposition 8.1.13: *Odd Cycles:* $\chi(C_{2n+1}) = 3$.

Proof: Clearly, $\alpha(C_{2n+1}) = n$. Thus, by Proposition 8.1.4,

$$\chi(C_{2n+1}) \geq \left\lceil \frac{|V(C_{2n+1})|}{\alpha(V(C_{2n+1}))} \right\rceil = \left\lceil \frac{2n+1}{n} \right\rceil = 3$$

◇

REVIEW FROM §2.4: The wheel graph $W_n = K_1 + C_n$ is called an **odd wheel** if n is odd, and an **even wheel** if n is even.

Proposition 8.1.14: *Even Wheels:* $\chi(W_{2m}) = 3$.

Proof: Using the fact that $W_{2m} = C_{2m} + K_1$, Proposition 8.1.6 and Corollary 8.1.12 imply that $\chi(W_{2m}) = \chi(C_{2m}) + \chi(K_1) = 2 + 1 = 3$.   ◇

Proposition 8.1.15: *Odd Wheels:* $\chi(W_{2m+1}) = 4$, *for all* $m \geq 1$.

Proof: Using the fact that the wheel graph W_{2m+1} is the join $C_{2m+1} + K_1$, Propositions 8.1.6 and 8.1.13 imply that $\chi(W_{2m+1}) = \chi(C_{2m+1}) + \chi(K_1) = 3 + 1 = 4$, for all $m \geq 1$.
◇

Proposition 8.1.16: *Complete Graphs:* $\chi(K_n) = n$.

Proof: By Proposition 8.1.1, $\chi(K_n) \leq n$ and by Proposition 8.1.2, $\chi(K_n) \geq n$.   ◇

Chromatically Critical Subgraphs

DEFINITION: A connected graph G is **(chromatically) k-critical** if $\chi(G) = k$ and the edge-deletion subgraph $G - e$ is $(k-1)$-colorable, for every edge $e \in E_G$ (i.e., if G is an edge-minimal k-chromatic graph).

Example 8.1.5: An odd cycle is 3-critical, since deleting any edge yields a path, thereby reducing the chromatic number from 3 to 2.

Example 8.1.6: The odd wheel W_5 is 4-chromatic, by Proposition 8.1.15, and Figure 8.1.7 illustrates how removing either a spoke-edge or a rim-edge reduces the chromatic number to 3. Thus, W_5 is 4-critical. The argument that all odd wheels W_{2m+1} with $m \geq 1$ are 4-critical is essentially the same (see Exercises).

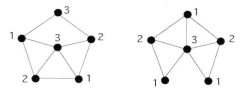

Figure 8.1.7   The graph $W_5 - e$ is 3-colorable, for any edge e.

Proposition 8.1.17: *Let G be a chromatically k-critical graph, and let v be any vertex of G. Then the vertex-deletion subgraph $G - v$ is $(k-1)$-colorable.* $\Diamond$ *(Exercises)*

Theorem 8.1.18: *Let G be a chromatically k-critical graph. Then no vertex of G has degree less than $k - 1$.*

Proof: By way of contradiction, suppose a vertex v had degree less than $k-1$. Consider a $(k-1)$-coloring of the vertex-deletion subgraph $G - v$, which exists by Proposition 8.1.17. The colors assigned to the neighbors of v would not include all the colors of the $(k-1)$-coloring, because vertex v has fewer than $k-1$ neighbors. Thus, if v were restored to the graph, it could be colored with any one of the $k-1$ colors that was not used on any of its neighbors. This would achieve a $(k-1)$-coloring of G, which is a contradiction.
$\Diamond$

Obstructions to k-Chromaticity

For a relatively small graph G, it is usually easier to calculate a plausible upper bound for $\chi(G)$ than a lower bound. Finding a configuration, called an *obstruction*, that forces the number of colors to exceed some value is the general approach to establishing a lower bound. The presence of either a subgraph or a graph property can serve as an obstruction.

DEFINITION: An **obstruction to k-chromaticity** (or k-**obstruction**) is a subgraph that forces every graph that contains it to have chromatic number greater than k.

Example 8.1.7: The complete graph K_{k+1} is an obstruction to k-chromaticity. (This is simply a restatement of Proposition 8.1.2.)

Proposition 8.1.19: *Every $(k+1)$-critical graph is an edge-minimal obstruction to k-chromaticity.*

Proof: If any edge is deleted from a $(k+1)$-critical graph, then, by definition, the resulting graph is not an obstruction to k-chromaticity. $\Diamond$

DEFINITION: A set $\{G_j\}$ of chromatically $(k+1)$-critical graphs is a **complete set of obstructions** if every $(k+1)$-chromatic graph contains at least one member of $\{G_j\}$ as a subgraph.

Example 8.1.8: The singleton set $\{K_2\}$ is a complete set of obstructions to 1-chromaticity.

REVIEW FROM §1.5: [Characterization of bipartite graphs] A graph is bipartite if and only if it contains no cycles of odd length.

Example 8.1.9: The bipartite-graph characterization above implies that the family $\{C_{2j+1} \mid j = 1, 2, \ldots\}$ of all odd cycles is a complete set of 2-obstructions.

Example 8.1.10: Although the odd-wheel graphs W_{2m+1} with $m \geq 1$ are 4-critical, they do not form a complete set of 3-obstructions, since there are 4-chromatic graphs that contain no such wheel. For example, the graph G in Figure 8.1.8 does not contain an odd wheel, but its independence number is 2, which implies, by Proposition 8.1.4, that $\chi(G) \geq 4$. The 4-coloring shown in Figure 8.1.8 shows that $\chi(G) = 4$.

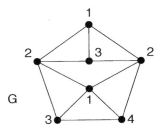

Figure 8.1.8   A 4-chromatic graph that contains no odd wheel.

Brooks's Theorem

REVIEW FROM §5.1:   The **vertex-connectivity** of a connected graph G, denoted $\kappa_v(G)$, is the minimum number of vertices, whose removal can either disconnect G or reduce it to a 1-vertex graph.

REVIEW FROM §5.4:

- A **block** of a loopless graph is a maximal connected subgraph H such that no vertex of H is a cut-vertex of H.

- A **leaf block** of a graph G is a block that contains exactly one cut-vertex of G.

Lemma 8.1.20: *Let G be a non-complete, k-regular 2-connected graph with $k \geq 3$. Then G has a vertex x with two non-adjacent neighbors y and z such that $G - \{y, z\}$ is a connected graph.*

Proof: Let w be any vertex in graph G. First suppose that subgraph $G - w$ is 2-connected. Since the graph G is regular and non-complete, there is a vertex z at distance 2 from vertex w. If x is a vertex adjacent to both w and z, then vertices x, w, and z satisfy the conditions of the assertion (with w playing the role of y).

If $G - w$ is not 2-connected, then it has at least one cut-vertex. By Proposition 5.4.6, there exist two leaf blocks B_1 and B_2 of $G - w$. Since graph G has no cut-vertices, it follows that vertex w is adjacent to some vertex y_1 in B_1 that is not a cut-vertex of $G - w$ and, likewise, adjacent to some vertex y_2 in B_2 that is not a cut-vertex of $G - w$. By Corollary 5.4.5, vertex y_1 is not adjacent to vertex y_2. Then w, y_1, and y_2 satisfy the conditions of the assertion (with w playing the role of x).   ◇

Theorem 8.1.21: [**Brooks, 1941**] *Let G be a non-complete, simple connected graph with maximum vertex degree $\delta_{\max}(G) \geq 3$. Then $\chi(G) \leq \delta_{\max}(G)$.*

Proof: In this proof of [Lo75], the sequential coloring algorithm enables us to make some simplifying reductions. Suppose that $|V_G| = n$.

Case 1. G is not regular.
First choose vertex v_n to be any vertex of degree less than $\delta_{\max}(G)$. Next grow a spanning tree (see §4.1) from v_n, assigning indices in decreasing order. In this ordering of V_G, every vertex except v_n has a higher-indexed neighbor along the tree path to v_n. Hence, each vertex has at most $\delta_{\max}(G) - 1$ lower-index neighbors. It follows that the sequential coloring algorithm uses at most $\delta_{\max}(G)$ colors.

Case 2. G is regular, and G has a cut-vertex x.
Let $C_1, \ldots, C_m$ be the components of the vertex-deletion subgraph $G - x$, and let G_i be the subgraph of G induced on the vertex-set $V_{C_i} \cup \{x\}$, $i = 1, \ldots, m$. Then the degree of vertex x in each subgraph G_i is clearly less than $\delta_{\max}(G)$. By Case 1, each subgraph G_i has a proper vertex coloring with $\delta_{\max}(G)$ colors. By permuting the names of the colors in each such subgraph so that vertex x is always assigned color 1, one can construct a proper coloring of G with $\delta_{\max}(G)$ colors.

Case 3. G is regular and 2-connected.
By Lemma 8.1.20, graph G has a vertex, which we call v_n, with two non-adjacent neigbhors, which we call v_1 and v_2, such that $G - \{v_1, v_2\}$ is connected. Grow a spanning tree from v_n in $G - \{v_1, v_2\}$, assigning indices in decreasing order. As in Case 1, each vertex except v_n has at most $\delta_{\max}(G) - 1$ lower-indexed neighbors. It follows that the sequential coloring algorithm uses at most $\delta_{\max}(G)$ colors on all vertices except v_n. Moreover, the sequential coloring algorithm assigns the same colors to v_1 and v_2 and at most $k - 2$ colors on the other $k - 2$ neighbors of v_n. Thus, one of the $\delta_{\max}(G)$ colors is available for v_n.   ◇

Heuristics for Vertex-Coloring

Many vertex-coloring heuristics for graphs are based on the intuition that a vertex of large degree will be more difficult to color later than one of smaller degree [Ma81, Ca86, Br79], and that if two vertices have equal degree, then the one having the denser *neighborhood subgraph* will be harder to color later. These three references discuss variations and combinations of these ideas.

The following algorithm is one such combination. Assume that the colors are named $1, 2, \ldots$. As the algorithm proceeds, the *colored degree* of a vertex v is the number of different colors that have thus far been assigned to vertices adjacent to v.

Algorithm 8.1.2:   **Vertex-Coloring: Largest Degree First**
Input: an n-vertex graph G.
Output: a vertex-coloring f of graph G.
    While there are uncolored vertices of G
        Among the uncolored vertices with maximum degree,
           choose vertex v with maximum colored degree.
           Assign smallest possible color k to vertex v : $f(v) := k$.
        Return graph G with vertex-coloring f.

Example 8.1.11: Descending degree order for the graph of Figure 8.1.9 (left) is

$$v_1, \ v_3, \ v_9, \ v_2, \ v_5, \ v_4, \ v_6, \ v_7, \ v_8, \ v_{10}$$

Applying the largest-degree-first algorithm yields a 3-coloring, as shown in 8.1.9 (right).

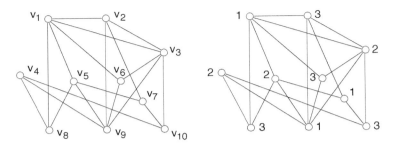

Figure 8.1.9   A largest-degree-first sequential coloring.

Coloring the Vertices of an Edge-Weighted Graph

Application 8.1.5:   *Timetabling with Unavoidable Conflicts* [KiYe92] Suppose that the evening courses at a certain school must be scheduled in k timeslots, $t_1, t_2, \ldots, t_k$. The school would like to schedule the courses so that, whenever at least three students are preregistered for both of a particular pair of courses, the two courses are assigned different timeslots. The graph model has a vertex-set corresponding to the set of courses, and has an edge between a pair of vertices if the corresponding pair of courses should be scheduled in different timeslots.

Typically, the chromatic number of such a graph is greater than k, which means that it is impossible to avoid *conflicts* (i.e., assigning the same timeslot to certain pairs of courses that should have been scheduled in different timeslots). A more realistic objective under such circumstances is to schedule the courses so that the total number of conflicts is minimized. Moreover, since some conflicts have greater impact than others, the number of students that preregistered for a given pair of courses is assigned as a weight to the corresponding edge. A further refinement of the model may adjust these edge-weights according to other factors not related to numbers of students. For instance, an edge corresponding to a pair of courses that are both required for graduation by some students or are to be taught by the same instructor should be assigned a large enough positive integer so as to preclude their being assigned to the same timeslot.

Suppose that $w(e)$ denotes the positive integer indicating the deleterious impact of scheduling the two courses corresponding to the endpoints of edge e in the same time-slot. For a given k-coloring f (not necessarily *proper*), let z_f be the total edge-weight of those edges whose endpoints are assigned the same color. Then the objective is to find a k-coloring f for which z_f is as small as possible.

[KiYe92] develops a number of heuristics based on coloring the "most difficult" vertices as early as possible, where "most difficult" means having the greatest impact on the remaining vertices to be colored.

[WeYe14] extends the graph model to 2-component edge-weights in order to include a secondary objective, creating compact schedules.

EXERCISES for Section 8.1

8.1.1 The chromatic number of an acquaintance network tells the minimum number of groups into which the persons in that network must be partitioned so that no two persons in a group have prior acquaintance. Calculate the chromatic number of the following acquaintance network.

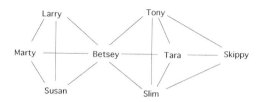

8.1.2 Calculate the number of different radio frequencies needed to avoid interference among the stations in the following configuration, from §1.3, in which two stations interfere if they are within 100 miles of each other.

	B	C	D	E	F	G
A	55	110	108	60	150	88
B		87	142	133	98	139
C			77	91	85	93
D				75	114	82
E					107	41
F						123

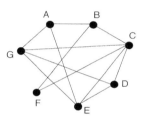

For Exercises 8.1.3 through 8.1.9, assign a minimum vertex-coloring to the given graph, and prove that it is a minimum coloring.

8.1.3ˢ   **8.1.4**

8.1.5

8.1.6   **8.1.7**ˢ

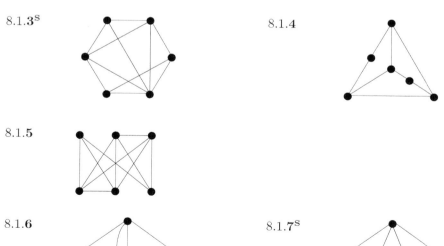

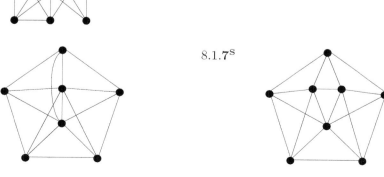

8.1.**8**   8.1.**9**

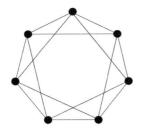   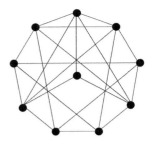

For Exercises 8.1.10 through 8.1.16, apply the largest-degree-first heuristic to the given graph.

8.1.**10**^S ~~8.1.10^S~~ The graph of Exercise 8.1.3.

8.1.**11** The graph of Exercise 8.1.4.

8.1.**12** The graph of Exercise 8.1.5.

8.1.**13** The graph of Exercise 8.1.6.

8.1.**14**^S ~~8.1.14^S~~ The graph of Exercise 8.1.7.

8.1.**15** The graph of Exercise 8.1.8.

8.1.**16** The graph of Exercise 8.1.9.

8.1.**17** Apply the sequential vertex-coloring algorithm to this bipartite graph.

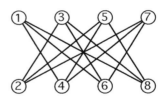

8.1.**18** Give orderings to the vertices of the graphs of Figure 8.1.7 so that the sequential vertex-coloring algorithm will yield 3-colorings.

8.1.**19**^S ~~8.1.19^S~~ Prove that the sequential vertex-coloring algorithm always colors a complete bipartite graph with two colors, regardless of the order of its vertices in the input list.

8.1.**20** Prove that adding an edge to a graph increases its chromatic number by at most one.

8.1.**21** Prove that deleting a vertex from a graph decreases the chromatic number by at most one.

8.1.**22**^S ~~8.1.22^S~~ Apply the sequential vertex-coloring algorithm to this graph, with vertices in order of ascending index.

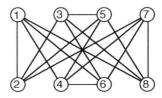

8.1.23 Apply the largest-degree-first heuristic to the graph of Exercise 8.1.22.

8.1.24 Prove that the chromatic number of an *interval graph* (§1.2) equals its clique number. (Hint: Apply the sequential vertex-coloring algorithm.)

8.1.25 Prove Proposition 8.1.17. (Hint: Use Proposition 8.1.5.)

8.1.26 a) Label the vertices of the following 3-chromatic graph so that the sequential vertex-coloring algorithm uses 4 colors.

b) Label the vertices of the following 3-chromatic graph so that the sequential vertex-coloring algorithm uses 5 colors.

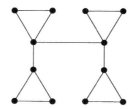

8.1.27 Give an inductive argument, based on the strategy used in Exercise 8.1.26, to show that the number of colors used by the sequential vertex-coloring algorithm can be arbitrarily larger than the chromatic number of a graph.

8.1.28^{S} Prove that the wheel W_4 is not chromatically critical.

8.1.29 Prove that all odd wheels are chromatically 4-critical.

8.1.30 Prove that for any graph G, G is chromatic k-critical if and only if $G + K_1$ is chromatic $(k+1)$-critical.

8.1.31 Prove that $K_7 - C_7$ is not 4-critical.

8.1.32 Prove that the join of two chromatically critical graphs is a chromatically critical graph.

For Exercises 8.1.33 through 8.1.37, calculate the independence number of the given graph.

8.1.33^{S} The graph of Exercise 8.1.3.

8.1.34 The graph of Exercise 8.1.4.

8.1.35 The graph of Exercise 8.1.5.

8.1.36 The graph of Exercise 8.1.6.

8.1.37^{S} The graph of Exercise 8.1.7.

8.1.38 Describe how to construct a connected graph G with independence number $\alpha(G) = a$ and chromatic number $\chi(G) = c$, for arbitrary values $a \geq 1$ and $c \geq 2$.

DEFINITION: The **domination number** of a graph G, denoted $\gamma(G)$ is the cardinality of a minimum set S of vertices such that every vertex of G is either in S or a neighbor of a vertex in S.

For Exercises 8.1.39 through 8.1.43, calculate the domination number of the given graph.

8.1.**39**^{S} The graph of Exercise 8.1.3.

8.1.**40** The graph of Exercise 8.1.4.

8.1.**41** The graph of Exercise 8.1.5.

8.1.**42** The graph of Exercise 8.1.6.

8.1.**43**^{S} The graph of Exercise 8.1.7.

8.1.**44** Construct a graph with chromatic number 5 and domination number 2.

8.1.**45** Construct a graph with domination number 5 and chromatic number 2.

8.1.**46** Describe how to construct, for arbitrary values $c \geq 2$ and $m \geq 1$, a connected graph G with chromatic number $\chi(G) = c$ and domination number $\gamma(G) = m$.

DEFINITION: A graph G is **perfect** if for every *induced subgraph* (§2.3) H in G, the chromatic number $\chi(H)$ equals the clique number $\omega(H)$.

8.1.**47**^{S} Prove that every bipartite graph is perfect.

8.1.**48** Prove that an odd cycle of length at least 5 is not perfect.

8.1.**49** [*Computer Project*] Implement Algorithms 8.1.1 and 8.1.2 and compare their results on the graphs of Exercises 8.1.3 through 8.1.9.

For Exercises 8.1.50 and 8.1.51, find a vertex 2-coloring of the given weighted graph such that the total weight of the edges whose endpoints are assigned the same color is minimized.

8.1.**50**   8.1.**51**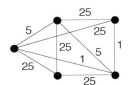

Computer Project Use the ideas discussed in the last subsection of this section to design and write a computer program to construct a vertex k-coloring of a weighted graph that tries to minimize the total weight of the edges whose endpoints receive the same color. Test your program on various edge-weighted graphs, including the ones in Exercises 8.1.48 and 8.1.49, and compare its vertex-colorings to the ones you obtained by hand.

8.1.**52** **Application 8.1.6:** *Examination Scheduling* Suppose that four final exams are to be scheduled in three timeslots. It would be ideal if no two of the exams were assigned the same timeslot, but that would require four timeslots. The table shown below assigns a number to each exam pair, indicating the penalty for scheduling those two exams in the same timeslot. Find an examination schedule that minimizes the total penalty.

$$
\begin{array}{c}
\quad\quad e_1 \quad e_2 \quad e_3 \quad e_4 \\
\begin{array}{c} e_1 \\ e_2 \\ e_3 \\ e_4 \end{array}
\left(
\begin{array}{cccc}
- & 4 & 16 & 4 \\
4 & - & 4 & 16 \\
1 & 16 & - & 4 \\
4 & 1 & 4 & -
\end{array}
\right)
\end{array}
$$

8.2 MAP-COLORINGS

Francis Guthrie, a South African mathematician, found that exactly four colors were needed to color a map of the English counties, so that no two counties that shared a border received the same color. In 1852, he proposed that four colors sufficed for *any* map.

For every closed surface S, there is a minimum number $chr(S)$ of colors sufficient so that every map on S can be colored *properly* with $chr(S)$ colors, which means that no color meets itself across an edge. The methods needed to establish a narrow range of possibilities for that minimum sufficient number are elementary enough to be presented in this book. Tightening the range to a single value was the substance of two of the outstanding mathematical problems solved in the 20th century, the Four-Color map problem for the plane (and sphere) and the Heawood map problem for all the other closed surfaces. In this section, the possibilities for $chr(S)$ are narrowed down to four and five.

Dualizing Map-Colorings into Vertex-Colorings

DEFINITION: A **map on a surface** is an imbedding of a graph on that surface.

TERMINOLOGY: Whereas the term *mapping* is a generic synonym for function, the term map refers to a function from a graph to a surface.

DEFINITION: A **map k-coloring** for an imbedding $\iota : G \to S$ of a graph on a surface is an assignment $f : F \to C$ from the face-set F onto the set $C = \{1, \ldots, k\}$, or onto another set of cardinality k, whose elements are called *colors*. For any k, such an assignment is called a **map-coloring**.

DEFINITION: A map-coloring is **proper** if for each edge $e \in E_G$, the regions that meet on edge e are colored differently.

DEFINITION: The **chromatic number of a map** $\iota : G \to S$ is the minimum number $chr(\iota)$ of colors needed for a proper coloring.

Example 8.2.1: Figure 8.2.1 below shows a proper 4-coloring of a planar map. Observe that in a proper coloring of this map, no two regions can have the same color, since every pair of regions meets at an edge. Thus, this map requires four colors.

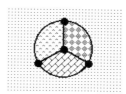

Figure 8.2.1 A 4-colored planar map.

TERMINOLOGY: A region is said to **meet itself on edge** e if edge e occurs twice in the region's boundary walk. It is said to **meet itself on vertex** v if vertex v occurs more than once in the region's boundary walk.

Remark: A map cannot be properly colored if a region meets itself.

Example 8.2.2: The map in Figure 8.2.2 has no proper coloring, since region f_2 meets itself on an edge (which cannot occur in a geographic map). Observe also that region f_1 meets region f_4 in two distinct edges.

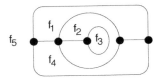

Figure 8.2.2   A map with no proper coloring.

Long ago, it was realized that a coloring problem for the regions of a map on any closed surface can be converted by Poincaré duality (see §7.5) into a vertex-coloring problem for the dual graph. The dual of the map in Figure 8.2.2 is the graph in Figure 8.2.3, with dual vertex v_j corresponding to primal face f_j, for $j = 1, \ldots, 5$. Thus, the self-adjacent region f_2 dualizes to the self-adjacent vertex v_2.

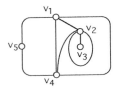

Figure 8.2.3   The dual graph for the map of Figure 9.2.2.

Proposition 8.2.1: *The chromatic number of a map equals the chromatic number of its dual graph.*   ◇

Geographic Maps

When the surface is the plane, sometimes a collection of contiguous regions are colored, and the other regions are ignored.

Application 8.2.1:   *Political Cartography*   In the cartography of political maps, various interesting configurations arise. For instance, France and Spain meet on two distinct borders, one from Andorra to the Atlantic Ocean, the other from Andorra to the Mediterranean Sea, as represented in Figure 8.2.4. Similar configurations occur where Switzerland meets Austria twice around Liechtenstein. Moreover, India and China have a triple adjacency around Nepal and Bhutan. Multiple adjacency of regions does not affect the rules for coloring a map.

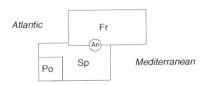

Figure 8.2.4   Double adjacency of France and Spain around Andorra.

Remark: Two faces that meet at a vertex but not along an edge may have the same color in a proper map coloring. Thus, a checkerboard configuration such as the Four Corners, USA, representation in Figure 8.2.5, may be properly colored with only two colors.

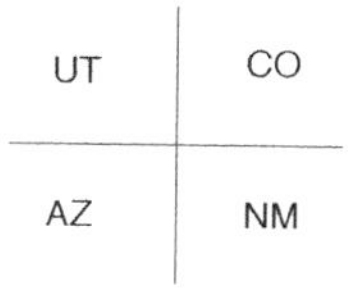

UT
CO
AZ
NM

Figure 8.2.5 A proper 2-coloring at Four Corners, USA.

Example 8.2.3: The chromatic number of the map of the countries of South America is equal to 4. In the dual graph, which appears in Figure 8.2.6, there is a 5-wheel with Bolivia as the hub and a 3-wheel with Paraguay as hub. Thus, by Proposition 8.1.15, South America requires at least four colors. We leave it as an exercise to give a 4-coloring of South America (see Exercises).

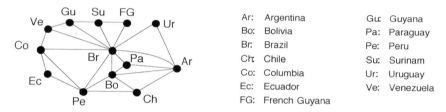

Gu
Su
FG
Ve
Ur
Co
Br
Pa
Ar
Ec
Bo
Pe
Ch
Ar: Argentina
Bo: Bolivia
Br: Brazil
Ch: Chile
Co: Columbia
Ec: Ecuador
FG: French Guyana
Gu: Guyana
Pa: Paraguay
Pe: Peru
Su: Surinam
Ur: Uruguay
Ve: Venezuela

Figure 8.2.6 The dual graph of the map of South America.

Example 8.2.4: The chromatic number of the map of the United States of America is four. The graph of the USA contains three odd wheels, as illustrated in Figure 8.2.7 below. West Virginia and Nevada are each encircled by five neighbors, and Kentucky is encircled by seven neighbors.

PA
OH
MD
WV
KY
VA
OR
CA
ID
NV
AZ
UT
IN
IL
OH
MO
KY
WV
TN
VA

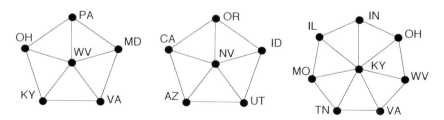

Figure 8.2.7 The three odd wheels in the map of the USA.

Remark: Utah meets five other states across an *edge* in the map, but these five do not quite encircle Utah, since Arizona and Colorado do not meet at an *edge*, even though they do meet at Four Corners. Thus, the Utah configuration is the join $P_5 + K_1$, and not the wheel W_5.

Five-Color Theorem for Planar Graphs and Maps

An early investigation of the Four Color Problem by A. B. Kempe [1879] introduced a concept that enabled Heawood [1890] to prove without much difficulty that five colors are sufficient. Heawood's proof is the main concern of this section.

DEFINITION: The $\{i,j\}$-*subgraph* of a graph G with a vertex-coloring that has i and j in its color set is the subgraph of G induced on the subset of all vertices that are colored either i or j.

DEFINITION: A *Kempe i-j chain* for a vertex-coloring of a graph is a component of the $\{i,j\}$-subgraph.

Example 8.2.5: Figure 8.2.8 illustrates two Kempe 1-3 chains in a graph coloring. The edges in the Kempe chains are dashed.

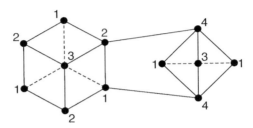

Figure 8.2.8   A graph coloring with two Kempe 1-3 chains.

The next theorem gives an upper bound on the average degree $\delta_{avg}(G)$ of a graph G imbedded in the sphere.

Theorem 8.2.2: *For any connected simple planar graph G, with at least three vertices, $\delta_{avg}(G) < 6$.*

Proof: By Theorem 7.5.9,

$$\frac{2\,|E_G|}{|V_G|} \le 6 - \frac{12}{|V_G|}$$

Therefore,

$$\delta_{avg}(G) = \frac{\sum_{v \in V_G} deg(v)}{|V_G|} \qquad \text{definition of average}$$

$$= \frac{2\,|E_G|}{|V_G|} \qquad\qquad \text{by Theorem 1.1.2}$$

$$\le 6 - \frac{12}{|V_G|} \qquad\qquad \text{by Theorem 7.5.9}$$

$$< 6 \qquad\qquad\qquad\qquad\qquad\qquad \diamondsuit$$

Theorem 8.2.3: [**Heawood, 1890**] *The chromatic number of a planar simple graph is at most 5.*

Proof: Starting with an arbitrary planar graph G, edges and vertices can be removed until a chromatically critical subgraph is obtained having the same chromatic number as G. Therefore, we may assume, without loss of generality that G is chromatically critical. We may also assume G is connected by the remark in Section §8.1. It suffices to prove that G is 5-colorable.

By Theorem 8.2.2, there is a vertex $w \in V_G$ of degree at most 5 and therefore, by Theorem 8.1.18, $\chi(G) \le 6$. Since G is chromatically critical, the vertex-deletion subgraph $G - w$ is 5-colorable, by Proposition 8.1.17.

Next, consider any 5-coloring of subgraph $G - w$. If not all five colors were used on the neighbors of vertex w, then the 5-coloring of $G - w$ could be extended to graph G by assigning to w a color not used on the neighbors of w. Thus, we can assume that all five colors are assigned to the neighbors of vertex w. Moreover, there is no loss of generality in assuming that these colors are consecutive in counterclockwise order, as shown on the left in Figure 8.2.9. Consider the $\{2,4\}$-subgraph shown on the left in Figure 8.2.9 with dashed edges, and let K be the Kempe 2-4 chain that contains the *2-neighbor* of vertex w (i.e., the neighbor that was assigned color 2).

Case 1. Suppose that Kempe chain K does not also contain the 4-neighbor of vertex w. Then colors 2 and 4 can be swapped in Kempe chain K, as shown on the right in Figure 8.2.9. The result is a 5-coloring of $G - w$ that does not use color 2 on any neighbor of w. This 5-coloring extends to a 5-coloring of graph G when color 2 is assigned to vertex w.

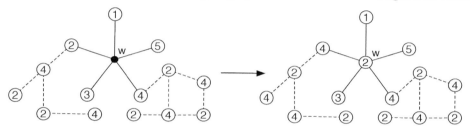

Figure 8.2.9   Swapping colors in a Kempe 2-4 chain.

Case 2. Suppose that Kempe chain K contains both the 4-neighbor and the 2-neighbor of vertex w. Then there is a path in Kempe chain K from the 2-neighbor to the 4-neighbor, as illustrated with a bold broken path on the left in Figure 8.2.10. Appending the edges between vertex w and both these neighbors extends that path to a cycle, as depicted on the right in Figure 8.2.10. By the Jordan Curve Theorem (§7.1), this cycle separates the plane.

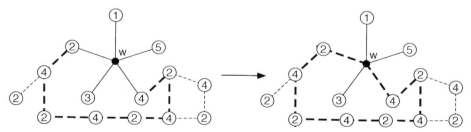

Figure 8.2.10   Extending a path in a Kempe 2-4 chain to a cycle.

Since the 3-neighbor and 5-neighbor of w are on different sides of the separation, it follows that the Kempe 3-5 chain L containing the 5-neighbor cannot also contain the 3-neighbor. Thus, it is possible to swap colors in Kempe chain L and assign color 5 to vertex w, thereby completing a 5-coloring of G.   ◇

Theorem 8.2.4: [***Appel and Haken, 1976***] *Every planar graph is 4-colorable.*   ◇

Remark: The proof by Appel and Haken [ApHa76] of the Four Color Theorem is highly specialized, intricate, and long. Following an approach initiated by Heesch, Appel and Haken first reduced the seemingly infinite problem of considering every planar graph to checking a finite, unavoidable set of (over 1900) reducible configurations. Over 1200 hours of computer time were used. Eventually, a more concise proof was derived by Robertson, Sanders, Seymour, and Thomas [RoSaSeTh97].

EXERCISES for Section 8.2

8.2.1 Draw a minimum proper coloring of the following map, excluding the exterior region, and prove it is a minimum coloring.

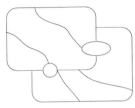

8.2.2^S Draw a minimum proper coloring of the map of Exercise 8.2.1, including the exterior region, and prove it is a minimum coloring.

8.2.3 Draw a minimum proper coloring of the following map, excluding the exterior region, and prove it is a minimum coloring.

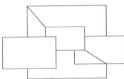

8.2.4 Draw a minimum proper coloring of the map of Exercise 8.2.3, including the exterior region, and prove it is a minimum coloring.

8.2.5^S Is it possible for a map in the plane to be 4-chromatic when the exterior region is included, but 3-chromatic when it is excluded? Explain your answer.

8.2.6 Draw the dual graph of the map of Exercise 8.2.3, and give it a minimum vertex-coloring.

8.2.7 Draw a minimum proper coloring of the following map, excluding the exterior region, and prove it is a minimum coloring.

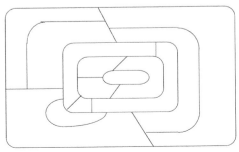

8.2.8^{S} Find the Kempe 1-4 chains in the following graph.

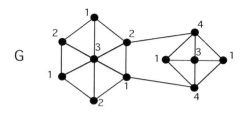

8.2.9 Find the Kempe 1-3 chains in the following graph.

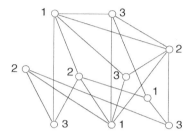

REVIEW FROM §7.1: The *Riemann stereographic projection* provides a correspondence between the plane and the sphere.

8.2.10 Convert the Four Corners map to a sphere map, by redrawing it so that the "infinite lines" all meet at the restored *infinity* point.

UT	CO
AZ	NM

8.2.11 Give a 4-coloring of the map (see Figure 8.2.6) of the countries of South America, or of the dual graph.

8.2.12 Consult a map of Europe, if necessary, and draw a map representing the following countries: France, Belgium, Netherlands, Luxembourg, Germany, and Switzerland. What is the chromatic number of this map?

8.2.13 Calculate the chromatic number of the map of the countries of North America.

8.2.14^{S} Calculate the chromatic number of the map of the countries of Africa.

8.2.15 Calculate the chromatic number of the map of the countries of Asia.

8.2.16[S] Calculate the chromatic number of the following map on the torus.

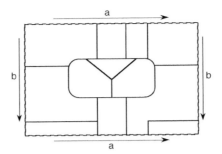

8.2.17 Calculate the chromatic number of the following map on the torus.

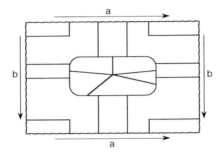

8.3   EDGE-COLORINGS

For certain problems, the most natural graph model for a problem might involve edge-colorings instead of vertex-colorings. Analogous to vertex-colorings, an edge-coloring partitions the edge-set of a graph into color classes. Although the *line-graph* transformation converts an edge-coloring problem into a vertex-coloring problem, the theory of edge-colorings has some special aspects.

The Minimization Problem for Edge-Colorings

REVIEW FROM §1.1: Two different edges are **adjacent** if they have at least one endpoint in common.

DEFINITION: An **edge k-coloring** of a graph G is an assignment $f : E_G \to C$ from its edge-set onto a k-element set C whose elements are called *colors* (typically, $C = \{1, 2, \ldots k\}$). For any k, such an assignment is called an **edge coloring**.

DEFINITION: An **edge color class** in an edge-coloring of a graph G is a subset of E_G containing all the edges of a given color.

DEFINITION: A **proper edge-coloring** of a graph is an edge-coloring such that adjacent edges are assigned different colors.

Remark: Whereas multi-edges have no bearing on the proper vertex-colorings of a graph, they have an obvious effect on the proper edge-colorings and cannot be ignored. Graphs with self-loops are excluded from the present discussion.

DEFINITION: A graph is said to be **edge k-colorable** if it has a proper edge k-coloring.

Example 8.3.1: A proper edge 5-coloring of a graph is shown in Figure 8.3.1.

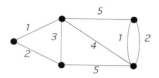

Figure 8.3.1   A graph with a proper edge 5-coloring.

DEFINITION: The **edge-chromatic number** of a graph G, denoted $\chi'(G)$, is the minimum number of different colors required for a proper edge-coloring of G. A graph G is edge k-**chromatic** if $\chi'(G) = k$.

Thus, $\chi'(G) = k$ if graph G is edge k-colorable but not edge $(k-1)$-colorable.

Example 9.3.1, continued: The proper edge-coloring in Figure 8.3.2 improves on the one in Figure 9.3.1, since it uses only four colors. Moreover, the graph is not edge 3-colorable, since it contains four mutually adjacent edges. Thus, $\chi'(G) = 4$.

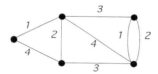

Figure 8.3.2   A proper edge 4-coloring of the graph from Figure 8.3.1.

Modeling Applications as Edge-Coloring Problems

Application 8.3.1:   *Circuit Boards*   Some electronic devices $x_1, x_2, \ldots, x_n$ are on a board. The connecting wires emerging from each device must be colored differently, so that they can be distinguished. The least number of colors required is the edge-chromatic number of the associated network.

Application 8.3.2:   *Scheduling Class Times*   A high school has teachers $t_1, \ldots, t_m$ to teach courses $s_1, \ldots, s_n$. In particular, teacher t_j must teach $s_{j,k}$ sections of course s_k. PROBLEM: Calculate the minimum number of time periods required to schedule all the courses so that no two sections of the same course are taught at the same time. SOLUTION: Form a bipartite graph on the two sets $\{t_1, \ldots, t_m\}$ and $\{s_1, \ldots, s_n\}$ so that there are $s_{j,k}$ edges joining t_j and s_k, for all j and k, as shown in Figure 8.3.3.

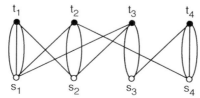

Figure 8.3.3   Representing a scheduling problem by a bipartite graph.

A matching of teachers to courses can be realized in a time period. If each edge-color represents a timeslot in the schedule, then an edge-coloring of the bipartite graph represents a feasible timetable for sections of courses. A minimum edge-coloring, as shown in Figure 8.3.4, uses the smallest number of time periods.

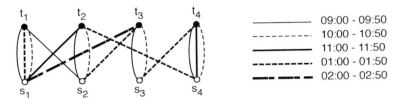

————————	09:00 - 09:50
- - - - - - - - -	10:00 - 10:50
————————	11:00 - 11:50
– - – - – - – -	01:00 - 01:50
— — — — —	02:00 - 02:50

Figure 8.3.4   A minimum proper edge-coloring for the graph of Figure 8.3.3.

Sequential Edge-Coloring Algorithm

There is a sequential edge-coloring algorithm analogous to the sequential vertex-coloring algorithm of §8.1.

DEFINITION: A **neighbor of an edge** e is another edge that shares one or both of its endpoints with e.

Algorithm 8.3.1:   **Sequential Edge-Coloring**
Input: a graph with edge list e_1, e_2, $\ldots$, e_p.
Output: a proper edge-coloring f, with positive integers as colors
    For $i = 1, \ldots, p$
        Let $f(e_i) :=$ the smallest color number not used on any of the
        smaller-subscripted neighbors of e_i.
    Return edge-coloring f.

Basic Principles for Calculating Edge-Chromatic Numbers

Edge-chromatic-number calculations are largely based on a few simple principles, mostly analogous to those used in the two-step vertex-chromatic-number calculations.

- *Upper Bound*: Show $\chi'(G) \leq k$ by exhibiting a proper edge k-coloring of G.

- *Lower Bound*: Show $\chi'(G) \geq k$ by using properties of graph G.

The next three results help in establishing a lower bound for the edge-chromatic number. They are immediate consequences of the definitions.

Proposition 8.3.1: *Let G be a graph that has k mutually adjacent edges. Then $\chi'(G) \geq k$.* $\diamond$

Corollary 8.3.2: *For any graph G, $\chi'(G) \geq \delta_{\max}(G)$.* $\diamond$

Proposition 8.3.3: *Let H be a subgraph of graph G. Then $\chi'(G) \geq \chi'(H)$.* $\diamond$

Example 8.3.2: *The edge 5-coloring of the graph G in Figure 8.3.5 establishes that $\chi'(G) \leq 5$, and the existence of a 5-valent vertex shows that $\chi'(G) \geq 5$, by Corollary 8.3.2. Hence, $\chi'(G) = 5$.*

G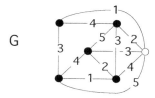

Figure 8.3.5   Graph G is edge 5-colorable.

Matchings

A *matching* is the edge analogue of an independent vertex set. It can be used to obtain another lower bound for the edge-chromatic number.

DEFINITION: A **matching** (or **independent set of edges**) of a graph G is a subset of edges of G that are mutually non-adjacent.

DEFINITION: A **maximum matching** in a graph is a matching with the maximum number of edges.

NOTATION: The cardinality of a maximum matching in a graph G is denoted $\alpha'(G)$, analogous to the independence number $\alpha(G)$ for the vertices.

Remark: It follows immediately from the definition that each color class of a proper edge-coloring of a graph G is a matching of G. The following proposition provides a lower bound on the edge-chromatic number that is based on the size of a maximum matching.

Proposition 8.3.4: *For any graph G, $\chi'(G) \geq \left\lceil \frac{|E_G|}{\alpha'(G)} \right\rceil$.*   ◇ *(Exercises)*

The following algorithm constructs a proper edge-coloring by iteratively finding maximum matchings.

Algorithm 8.3.2:   Edge-Coloring by Maximum Matching
Input: a graph G.
Output: a proper edge k-coloring f.
    Initialize color number $k := 0$.
    While $E_G \neq \emptyset$
        $k := k + 1$
        Find a maximum matching M of graph G.
        For each edge $e \in M$
            $f(e) := k$
        $G := G - M$ (edge-deletion subgraph)
    Return edge-coloring f.

COMPUTATIONAL NOTE: There are low-order polynomial-time algorithms that find maximum matchings, based largely on the work of Jack Edmonds. Bipartite matching is discussed in §10.4, and references to algorithms and applications of matchings in general graphs are given there.

Edge-Chromatic Numbers for Common Graph Families

It is now possible to derive the edge-chromatic numbers for the same graph families, summarized in Table 8.3.1, for which the vertex-chromatic numbers were derived in §8.1. The first of the following six results are analogous to their companion results for vertex-chromatic number and are left as exercises.

Proposition 8.3.5: *A graph G has $\chi'(G) = 1$ if and only if $\delta_{\max}(G) = 1$.* ◇ *(Exercises)*

Proposition 8.3.6: *Path Graphs: $\chi'(P_n) = 2$, for $n \geq 3$.*   ◇ *(Exercises)*

Proposition 8.3.7: *Even Cycle Graphs: $\chi'(C_{2n}) = 2$.*   ◇ *(Exercises)*

Proposition 8.3.8: *Odd Cycle Graphs: $\chi'(C_{2n+1}) = 3$.*   ◇ *(Exercises)*

Proposition 8.3.9: *Trees: $\chi'(T) = \delta_{\max}(T)$, for any tree T.*   ◇ *(Exercises)*

Proposition 8.3.10: *Hypercube Graphs: $\chi'(Q_n) = n$.*   ◇ *(Exercises)*

Proposition 8.3.11: *Wheel Graphs: $\chi'(W_n) = n$, for $n \geq 3$.*   ◇ *(Exercises)*

Example 8.3.3: Figure 8.3.6 illustrates an edge 5-coloring of the wheel W_5.

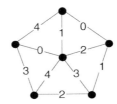

Figure 8.3.6   A proper edge 5-coloring of the wheel W_5.

Proposition 8.3.12: *Odd Complete Graphs: $\chi'(K_n) = n$ for all odd $n \geq 3$.*

Proof: *Upper bound:* Draw the complete graph K_n so that its vertices are the vertices of a regular n-gon, labeled $0, 1, 2, \ldots, n-1$ clockwise around the n-gon (illustrated in Figure 8.3.7 for $n = 7$). Observe that the edge joining vertices 0 and 1 along with all the other edges whose endpoints sum to 1 (mod n) (depicted as bold edges in Figure 8.3.7) form a matching and so they can all be assigned the color 1.

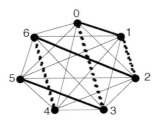

Figure 8.3.7   Two matchings in K_7.

Similarly, the edges whose endpoints sum to 3 (mod n) form a matching (the dashed edges in Figure 8.3.7) and can be assigned the color 3. In all, there are n sets $S_1, S_2, \ldots, S_n$, where S_k is the matching consisting of those edges whose endpoints sum to k (mod n).

Thus, if each of the edges in set S_k is assigned color k, then a proper edge n-coloring of K_n is obtained.

Lower bound: The size of a maximum matching in K_n with n odd is $\frac{n-1}{2}$. Since K_n contains $\binom{n}{2} = \frac{n(n-1)}{2}$ edges, Proposition 8.3.4 implies that $\chi'(K_n) \geq n$. $\diamond$

Corollary 8.3.13: *Even Complete Graphs:* $\chi'(K_n) = n - 1$ *for all even* n.

Proof: The even complete graph K_n is the join of the odd complete graph K_{n-1} with a single vertex x. The proof of Proposition 8.3.12 constructs a proper edge n-coloring of K_{n-1} in which each edge-color is missing at exactly one vertex. Thus, the edge-coloring of K_{n-1} can be extended to an edge-coloring of K_n by assigning the missing color at each vertex v in K_{n-1} to the edge joining vertex v to vertex x (see Figure 8.3.8 below). $\diamond$

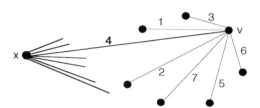

Figure 8.3.8   Extending the edge n-coloring from K_{n-1} to K_n.

Table 8.3.1 below summarizes the edge-chromatic numbers of the common graph families considered here, and also of the bipartite graphs, which we discuss later in this section.

Table 8.3.1: Edge-chromatic numbers for common graph families

Graph G	$\chi'(G)$
graph G with $\delta_{\max}(G) = 1$	1
path graph P_n, $n \geq 3$	2
even cycle graph C_{2n}	2
odd cycle graph C_{2n+1}	3
bipartite graph G	$\delta_{\max}(G)$
tree T	$\delta_{\max}(T)$
cube graph Q_n	n
wheel W_n, $n \geq 3$	n
even complete graph K_{2n}	$n - 1$
odd complete graph K_{2n+1}	n

Chromatic Incidence

The next few definitions pertain to *all* edge-colorings, not just to proper ones.

DEFINITION: For a given edge-coloring of a graph, color i is an **incident edge-color** on vertex v if some edge incident on v has been assigned color i. Otherwise, color i is an **absent edge-color** at vertex v.

DEFINITION: The **chromatic incidence at** v of a given edge-coloring f is the number of different edge-colors incident on vertex v. It is denoted $ecr_v(f)$.

DEFINITION: The **total chromatic incidence** for an edge-coloring f of a graph G, denoted $ecr(f)$, is the sum of the chromatic incidences of all the vertices. That is,

$$ecr(f) = \sum_{v \in V_G} ecr_v(f)$$

Example 8.3.4: For the edge-colorings shown in Figure 8.3.9 below, the three different edge-colors are represented by dashed, regular, and bold edges, instead of color numbers. This is to avoid confusion with the chromatic incidence numbers on the vertices. For the edge-coloring f on the left, the total chromatic incidence is 13, and for edge-coloring g on the right, the total chromatic incidence is 15.

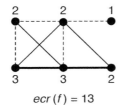 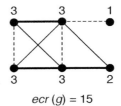

$$ecr(f) = 13 \qquad\qquad ecr(g) = 15$$

Figure 8.3.9   Total-chromatic-incidence calculations.

The following four assertions are immediate consequences of the definitions.

Proposition 8.3.14: *Let f be any edge-coloring of a graph G. Then for every $v \in V_G$,*

$$ecr_v(f) \le deg(v) \qquad\qquad \diamondsuit$$

Corollary 8.3.15: *Let f be any edge-coloring of a graph G. Then*

$$\sum_{v \in V_G} ecr_v(f) \le \sum_{v \in V_G} deg(v) \qquad\qquad \diamondsuit$$

Proposition 8.3.16: *An edge-coloring f of a graph G is proper if and only if for every vertex $v \in V_G$,*

$$ecr_v(f) = deg(v) \qquad\qquad \diamondsuit$$

Corollary 8.3.17: *An edge-coloring f of a graph G is proper if and only if*

$$\sum_{v \in V_G} ecr_v(f) \le \sum_{v \in V_G} deg(v) \qquad\qquad \diamondsuit$$

Edge-Coloring of Bipartite Graphs

Deriving a formula for the edge-chromatic number of a bipartite graph G is not quite as easy as for the vertex-chromatic number. Nonetheless, the eventual formula is uncomplicated: $\chi'(G) = \delta_{\max}(G)$. The characterization of bipartite graphs as the graphs without odd cycles is crucial to the derivation.

The following two lemmas establish facts about the chromatic degree that are used in the derivation. They involve edge-colorings that are *not* assumed to be proper. The first lemma makes use of the properties of an *Eulerian graph*.

REVIEW FROM §1.5: An **Eulerian tour** in a graph is a closed trail that contains every edge of that graph. An **Eulerian graph** is a graph that has an Eulerian tour.

Lemma 8.3.18: *Let G be a connected graph with at least two edges. If G is not an odd-cycle graph, then G has an edge 2-coloring such that both colors are incident on every vertex of degree at least 2.*

Proof: *Case* 1: G is an even cycle. The edge 2-coloring obtained by assigning two edge-colors that alternate around the cycle meets the requirement.

Case 2: G is Eulerian but not a cycle. By Theorem 4.5.11, every vertex in G has even degree and since G is not a cycle, it has a vertex v with degree at least 4. Consider an Eulerian tour that starts (and ends) at v. Assign color 1 to the edges that occur as odd terms in the edge sequence of the tour, and assign color 2 to the even-term edges. Then the two colors are incident at least once on each internal vertex of the tour, since each such vertex is an endpoint of both an odd-term edge and an even-term edge. Moreover, since the start vertex has degree at least 4, it also occurs on the tour as an internal vertex. Thus, both colors are incident on every vertex.

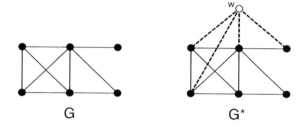

Figure 8.3.10   Constructing the auxiliary graph for Case 3.

Case 3: G is not Eulerian. Construct an auxiliary graph G^* by joining a new vertex w to every odd-degree vertex of G, thereby making each such vertex have even-degree in G^* (see Figure 8.3.10 above). By Corollary 1.1.3, every graph has an even number of odd-degree vertices, so vertex w has even degree. Thus, the auxiliary graph G^* is Eulerian, by Theorem 4.5.11. Now let f be an edge 2-coloring of graph G^*, as specified in Case 2. Then it is easy to verify that the edge-coloring $f|_G$ of f restricted to the edges of graph G is an edge-coloring such that both colors are incident on each vertex of G of degree at least 2.   ◇

DEFINITION: In a graph G with a (possibly improper) edge-coloring, a **Kempe i-j edge-chain** is a component of the subgraph of G induced on all the i-colored and j-colored edges.

Example 8.3.5: An edge 4-coloring is shown in Figure 8.3.11, and the two Kempe 1-2 edge-chains are shown as dashed edges.

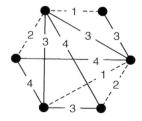

Figure 8.3.11   An edge 4-coloring with two Kempe 1-2 edge-chains.

Lemma 8.3.19: *Let f be an edge k-coloring of a graph G with the largest possible total chromatic incidence. Let v be a vertex on which some color i is incident at least twice and on which some color j is not incident at all. Then the Kempe i-j edge-chain K containing vertex v is an odd cycle.*

Proof: By Lemma 8.3.18, if the Kempe i-j edge-chain K incident on vertex v were not an odd cycle, then we could rearrange edge colors i and j within K so that the chromatic incidence of the coloring of K would be 2 at every vertex. The edge-coloring for G thereby obtained would have higher chromatic incidence at vertex v and at-least-equal chromatic incidence at every other vertex of G. This would contradict the premise that edge-coloring f has the maximum possible total chromatic incidence.   ◇

Theorem 8.3.20: *[König, 1916] Let G be a bipartite graph. Then $\chi'(G) = \delta_{\max}(G)$.*

Proof: Let $\Delta = \delta_{\max}(G)$, and, by way of contradiction, suppose that $\chi'(G) \neq \Delta$. Then, by Corollary 8.3.2, $\Delta < \chi'(G)$. Next, let f be an edge Δ-coloring of graph G for which the total chromatic incidence $ecr_G(f)$ is maximum. Since f is not a proper edge-coloring, there is a vertex v such that $ecr_v(f) < deg(v)$ (by Proposition 8.3.16). Thus, some color occurs on at least two edges incident on v. But there are $\Delta - 1$ other colors and at most $\Delta - 2$ other edges incident on v, which means that some other color is not incident on vertex v. It follows by Lemma 8.3.19, that graph G contains an odd cycle, which contradicts the fact that G is bipartite.   ◇

Vizing's Theorem

Complementing the lower bound $\chi'(G) \geq \delta_{\max}(G)$ for a simple graph, provided by Corollary 8.3.2, Vizing's theorem provides a sharp upper bound that narrows the range for $\chi'(G)$ to two possible values.

DEFINITION: Let G be a graph, and let f be a proper edge k-coloring of a subset S of the edges of E_G. Then f is a **blocked partial edge k-coloring** if for each uncolored edge e, every color has already been assigned to the edges that are adjacent to e. Thus, f cannot be extended to any edge outside subset S.

Example 8.3.6: In Figure 8.3.12, the edge 5-coloring of all but one of the edges of K_5 is blocked, since all five colors have been assigned to the neighbors of the uncolored edge.

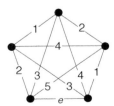

Figure 8.3.12   Attempted edge 5-coloring of K_5 that is blocked at edge e.

Lemma 8.3.21: *Let i and j be two of the colors used in a proper edge-coloring of a graph. Then every Kempe i-j edge-chain K is a path (open or closed).*

Proof: Every vertex of Kempe chain K has degree at most 2 (since the edge-coloring is proper), and, by definition, K is a connected subgraph.   ◇

Theorem 8.3.22:   [*Vizing, 1964, 1965*] [*Gupta, 1966*] *Let G be a simple graph.
Then there exists a proper edge-coloring of G that uses at most $\delta_{\max}(G) + 1$ colors.*

Proof: To construct a $(\delta_{\max}(G) + 1)$-edge-coloring, start by successively coloring edges,
using any method (e.g., Algorithm 8.3.1) until the coloring is blocked or complete. If
the set of uncolored edges is empty, then the construction is complete. Otherwise, there
is some edge e, with endpoints u and v, that remains uncolored. It will be shown that
by recoloring some edges, the blocked coloring can be transformed into one that can be
extended to edge e. The process can then be repeated until all edges have been colored.

Since the number of colors exceeds $\delta_{\max}(G)$, it follows that at each vertex, at least one of
the colors is absent. Let c_0 be a color absent at vertex u, and c_1 a color absent at vertex
v. Color c_1 cannot also be absent at vertex u, since if it were, edge e would not have
remained uncolored. (For the same reason, color c_0 must occur at vertex v.) So let e_1 be
the c_1-edge incident on vertex u, and let v_1 be its other endpoint. Next, let c_2 be a color
absent at v_1. If c_2 is also absent at vertex u, then the color of edge e_1 can be changed
from c_1 to c_2, thereby permitting the assignment of color c_1 to edge e, as illustrated
in Figure 8.3.13. A missing color c at a vertex is indicated by placing c alongside that
vertex. Several missing colors may be grouped in braces.

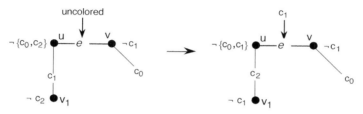

Figure 8.3.13   Extending an edge-coloring to edge e by recoloring edge e_1.

If color c_2 does occur at vertex u, then let e_2 be the c_2-edge incident on vertex u, let
v_2 be its other endpoint, and let c_3 be a color absent at vertex v_2. Continue iteratively
in this way, so that at the jth iteration, e_j is the c_j-edge incident on vertex u, v_j is its
other endpoint, and c_{j+1} is the color absent at vertex v_j. Let ℓ be the smallest j such
that vertex v_l has a missing color c_{l+1} and that c_{l+1} is also absent at vertex u or is one
of the colors in the list $c_1, \ldots, c_l$ (such an l exists, since the set of colors is finite).

Case 1: Color c_{l+1} is absent at both vertex v_l and vertex u.   *Color Shift.*
Then perform the following *color shift*: for $j = 1, \ldots, l$, change the color of edge e_j from
c_j to c_{j+1}. This releases color c_1 from edge e_1, so that it can be reassigned to edge e. The
color shift is illustrated in Figure 8.3.14. Notice that it maintains a proper edge-coloring,
because, by the construction, color c_{j+1} was absent at both endpoints of edge e_j before
the shift.

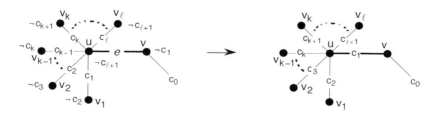

Figure 8.3.14   Case 1: *color shift* to free color c_1 for edge e.

Case 2: Color $c_{l+1} = c_k$, where $1 \le k \le l$.   *Swap and Shift.*
Let K be the Kempe c_0-c_k edge-chain incident on vertex v_l. By definition, K includes the c_0-edge incident on v_l, but there is no c_k-edge incident on vertex v_l (by definition of l). By Lemma 8.3.21, Kempe chain K is a path, and one end of this path is vertex v_l. There are three subcases to consider, according to where the other end of the path is. In each of the three subcases, the two colors are swapped so that a Case 1 color shift can then be performed.

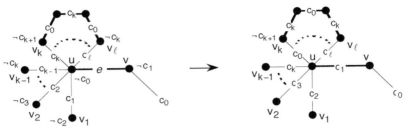

Figure 8.3.15   Case 2a: swap and shift.

Case 2a. Path K reaches vertex v_k.
Then swap colors c_0 and c_k along path K. As a result of the swap, color c_k no longer occurs at vertex u. This configuration permits a Case 1 color shift that releases color c_1 for edge e. The swap and shift are illustrated in Figure 8.3.15 above.

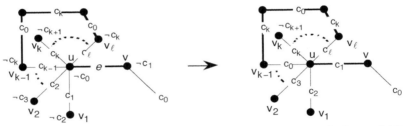

Figure 8.3.16   Case 2b: swap, recolor edge e_1, and then shift.

Case 2b. Path K reaches vertex v_{k-1}.
Then swap colors c_0 and c_k along path K. As a result of the swap, color c_0 no longer occurs at vertex v_{k-1}. Thus, edge e_{k-1} can be recolored c_0, as in Figure 8.3.16 above. A color shift can now be performed to release color c_1 for edge e.

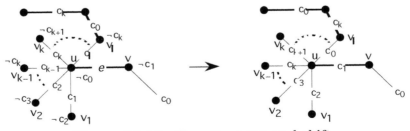

Figure 8.3.17   Case 2c: swap and shift.

Case 2c. Path K never reaches vertex v_{k-1} or vertex v_k.
Since color c_0 does not occur at vertex u, and since color c_k occurs at u only on the edge from v_k, it follows that path K does not reach vertex u. Then swap colors c_0 and c_k along path K, so that color c_0 no longer occurs at vertex v_l. Now perform a Case 1 color shift that releases color c_1 for edge e, as in Figure 8.3.17 above.   ◇

Corollary 8.3.23: *Let G be a simple graph. Then either $\chi'(G) = \delta_{\max}(G)$ or $\chi'(G) = \delta_{\max}(G) + 1$.*

Proof: This follows immediately from Vizing's theorem and Corollary 8.3.2.   ◇

DEFINITION: ***Class 1*** is the set of non-empty simple graphs G such that $\chi'(G) = \delta_{\max}(G)$. ***Class 2*** is the set of simple graphs G such that $\chi'(G) = \delta_{\max}(G) + 1$.

COMPUTATIONAL NOTE: Deciding whether a simple graph is in Class 1 is an *NP*-complete problem [Ho81].

DEFINITION: The ***multiplicity*** $\mu(G)$ of a graph G is the maximum number of edges joining two vertices of G.

Remark: A more general result of Vizing, beyond simple graphs, which applies to every loopless graph G, is that $\delta_{\max}(G) \le \chi'(G) \le \delta_{\max}(G) + \mu(G)$. The edge-chromatic number achieves the upper bound of Vizing's general formula when all three vertex pairs of a "fat triangle," as illustrated by Figure 8.3.18, are joined by the same multiplicity of edges.

Figure 8.3.18   A symmetric "fat triangle" requires $\delta_{\max} + \mu$ edge colors.

Line Graphs

A line graph can be used to convert an edge-coloring problem into a vertex-coloring problem.

REVIEW FROM §1.2: The ***line graph*** of a graph G is the graph $L(G)$ whose vertices correspond bijectively to the edges of G, and such that two of these vertices are adjacent if and only if their corresponding edges in G have a vertex in common.

Example 8.3.7: The line graph of the complete graph K_4 is the octahedron graph $\mathcal{O}_3$, as illustrated in Figure 8.3.19.

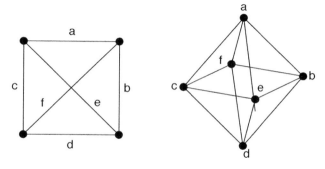

Figure 8.3.19   The complete graph K_4 and its line graph $\mathcal{O}_3$.

Proposition 8.3.24: *The edge-chromatic number of a graph G equals the vertex-chromatic number of its line graph L(G).*

Proof: This follows immediately from the definitions.   ◇

Remark: Beineke [Be68] proved that a simple graph G is a line graph of some simple graph if and only if G does not contain any of the graphs in Figure 8.3.20 as an induced subgraph. Since it is an *NP*-complete problem to decide this subgraph problem, much of the theory of edge-colorings has prospered separately from the theory of vertex-colorings.

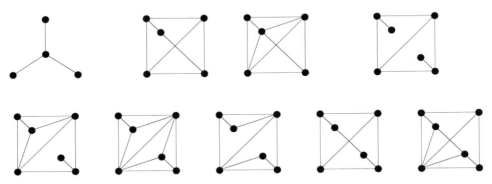

Figure 8.3.20   The nine forbidden induced subgraphs of line graphs.

EXERCISES for Section 8.3

For Exercises 8.3.1 through 8.3.12, assign a minimum edge-coloring and prove that it is a minimum coloring.

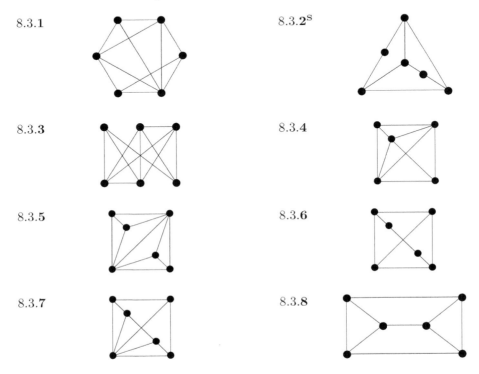

8.3.1

8.3.2^S

8.3.3

8.3.4

8.3.5

8.3.6

8.3.7

8.3.8

8.3.9

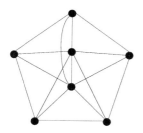

8.3.10

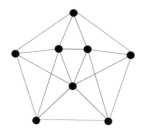

8.3.11^{S}

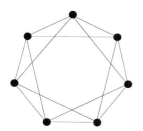

8.3.12

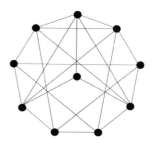

8.3.13 Prove Proposition 8.3.4. Let $\alpha'(G)$ be the size of a maximum matching in a graph G. Then $\chi'(G) \geq \left\lceil \frac{|E_G|}{\alpha'(G)} \right\rceil$.

8.3.14 Prove Proposition 8.3.5. A graph G has $\chi'(G) = 1$ if and only if $\delta_{\max}(G) = 1$.

8.3.15 Prove Proposition 8.3.6. $\chi'(P_n) = 2$, for $n \geq 3$.

8.3.16^{S} Prove Proposition 8.3.7. $\chi'(C_{2n}) = 2$.

8.3.17 Prove Proposition 8.3.8. $\chi'(C_{2n+1}) = 3$.

8.3.18 Prove Proposition 8.3.9. $\chi'(T) = \delta_{\max}(T)$, for any tree T.

8.3.19 Prove Proposition 8.3.10. $\chi'(Q_n) = n$.

8.3.20 Prove Proposition 8.3.11. $\chi'(W_n) = n$, for $n \geq 3$.

8.3.21^{S} Prove that adding an edge to a graph increases its edge-chromatic number by at most 1.

8.3.22 Show that K_3 and $K_{1,3}$ have isomorphic line graphs.

8.3.23 Show that the edge-complement of $L(K_5)$ is isomorphic to the Petersen graph.

8.3.24 Explain how iteratively subdividing graph G ultimately produces a graph whose edge-chromatic number equals $\delta_{max}(G)$.

8.3.25^{S} Give an example of a graph G and an edge $e \in E_G$ such that subdividing e causes the edge-chromatic number to increase by 1.

8.3.26 Let G be a graph such that $|E_G| \geq \delta_{\max}(G) \cdot \alpha'(G)$. Prove that this implies that G is in Class 2.

8.3.27 Let G be a regular graph with an odd number of vertices. Prove that G is in Class 2.

8.3.28[S] Let G be a 3-regular graph with a cut-edge. Prove that G is in Class 2.

8.3.29 Give an example of a graph for which Algorithm 8.3.1 does not produce a minimum edge-coloring.

8.3.30 [*Computer Project*] Implement Algorithms 8.3.1 and 8.3.2 and compare their results on the graphs in Exercises 8.3.1 through 8.3.12.

8.4   FACTORIZATION

We observe that an edge-coloring of a graph is a partitioning of the edge-set into cells of mutually nonadjacent edges, and we now finish this chapter by considering an additional topic concerned with partitioning an edge-set, called *factorization*. In a factorization, each cell of the partition of the edge-set induces a spanning subgraph. We are able to prove a classical result of W. Tutte. However, since some of the key theorems on factorization are most easily proved with the aid of flows, we defer their proofs until Chapter 10.

Factors

DEFINITION: A **factor** of a graph is a spanning subgraph.

DEFINITION: A **factorization** of a graph G is a set of factors whose edge-sets form a partition of the edge-set E_G.

Example 8.4.1: Figure 8.4.1 shows a factorization of the complete graph K_6 into three spanning paths.

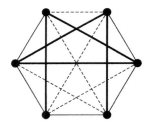

Figure 8.4.1   Factorization of K_6 into three paths.

DEFINITION: A k-**factor** of a graph G is a k-regular factor of G.

DEFINITION: A k-**factorization** of a graph G is a factorization into k-factors.

Example 8.4.2: Figure 8.4.2 shows two 2-factorizations of the complete graph K_7. The factors of factorization (a) are all spanning cycles. In factorization (b), two factors are spanning cycles, but the factor represented by bold dashes is the union of a 3-cycle and a 4-cycle.

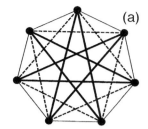 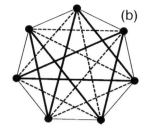

Figure 8.4.2   Two factorizations of K_7 into 2-factors.

TERMINOLOGY: A 1-factor of a graph G is also called a **perfect matching**. Two vertices are said to be **matched** with respect to a 1-factor if they are the endpoints of an edge in the 1-Factor.

Tutte's 1-Factor Theorem

M. Plummer [Pl04] characterizes Tutte's 1-Factor Theorem as the most influential theorem in the study of 1-Factors. The definitions and lemmas that precede Tutte's theorem help to simplify the proof.

DEFINITION: An **odd component of a graph** is a component with an odd number of vertices.

DEFINITION: **Tutte's condition** on a graph G is that for every subset $S \subset V_G$, the number of odd components of $G - S$ does not exceed $|S|$.

Lemma 8.4.1: *Tutte's condition is preserved under edge addition.*

Proof: Let G be a graph that satisfies Tutte's condition, and let e be an edge added to G between two nonadjacent vertices. To show that $G + e$ satisfies Tutte's condition, let $S \subset V_G$. If either endpoint of e lies in S, then the components of the graph $(G + e) - S$ are exactly the components of $G - S$. If both endpoints of e lie in the same component of $G - S$, then the number of odd components is unchanged. This reduces our consideration to the circumstance that e joins two components of $G - S$. If both endpoints of e are in even components of $G - S$, or if one endpoint of e is in an odd component and the other in an even component, then the number of odd components stays the same, that is, less than or equal to $|S|$ (while the number of even components decreases by 1). If both endpoints of e are in odd components, then the number of odd components decreases by 2 (while the number of even components increases by 1).   ◇

Lemma 8.4.2: *Let G be a connected graph with evenly many vertices and evenly many edges, with one vertex v of degree 3, one vertex u of degree 1, and all other vertices of degree 2. Then G has a 1-factor.*

Proof: By Theorem 6.1.1, G has an Eulerian trail from v to u. If we color the edges alternately red and blue, starting with red at vertex v, then the trail terminates with a blue edge at u (since there are evenly many edges). We observe that at every vertex except u, both colors are present, because by whatever color edge the trail enters a vertex, it leaves by an edge of the other color. (Thus, there are two red edges at v and one blue edge.) It follows that the blue edges form a 1-factor.   ◇

Example 8.4.3: A graph that satisfies the premises of Lemma 8.4.2 looks something like a *polygon kite*, as illustrated in Figure 8.4.3. In order for the number of vertices in the graph to be even, the number of edges in the tail and the number of edges in the polygon must have the same parity. The dark edges indicate the 1-factor promised by Lemma 8.4.2.

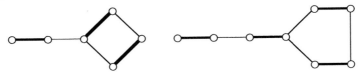

Figure 8.4.3   Two polygon kites with their 1-factors.

DEFINITION: Let M and N be spanning subgraphs of the same graph G. The **symmetric difference** $M \triangle N$ is the spanning subgraph of G whose edge-set is $(E_M \cup E_N) - (E_M \cap E_N)$.

Tutte's 1-Factor Theorem characterizes graphs having a 1-factor as those satisfying *Tutte's condition*. The following proof is due to Lovász [L075].

Theorem 8.4.3: [***Tutte's 1-Factor Theorem***] *A nontrivial graph G has a 1-factor if and only if for every subset $S \subset V_G$, the number of odd components of $G - S$ does not exceed $|S|$.*

Proof: ($\Rightarrow$) Suppose that G has a 1-factor. Then in each odd component of $G - S$, there is at least one vertex that is not matched to another vertex within that component, and hence, such a vertex is matched to a vertex of S. It follows that $|S|$ is at least as large as the number of components of $G - S$.

($\Leftarrow$) By way of contradiction, suppose that there exists a graph that satisfies Tutte's condition but has no 1-factor. By adding edges, one at a time, until it is impossible to do so without creating a 1-Factor, we obtain a graph H that still satisfies Tutte's condition (by Lemma 8.4.1) and is edge maximal with respect to having no 1-Factor. We now pursue the contradiction that H does contain a 1-factor.

Let W be the vertex subset given by

$$W = \{\, w \in V_H \,|\, w \text{ is adjacent to every other vertex of } H \,\}$$

Case 1. Suppose that every component of $H - W$ is a complete graph. Then we may match the vertices of $H - W$ in pairs, except for one leftover vertex in each odd component of $H - W$. Every vertex in W is adjacent to everyone of these leftover vertices (by the definition of set W). Moreover, Tutte's condition implies that $|W|$ is at least as large as the number of odd components of $H - W$. Hence, we may pair each of the leftover vertices from the odd components with a vertex of W. A 1-factor H will exist if the remaining (unpaired) vertices of W can be matched into pairs. These remaining vertices are mutually adjacent (again, by the definition of W) and hence, can be matched if there are evenly many of them. Since we have previously matched evenly many vertices of H, it suffices to show that $|V_H|$ is even. Tutte's condition, with $S = \emptyset$, implies that H has no odd components, which implies that H has evenly many vertices.

Case 2. Suppose that some component of $H - W$ is not a complete graph. In this case, there is a pair of non-adjacent vertices u and v in that component with a common neighbor $y \notin W$. Moreover, since $y \notin W$, it follows that some vertex z of H is not adjacent to y. By the definition of graph H, adding an edge creates a 1-factor. In particular, let M and N be 1-factors in the graphs $H + uv$ and $H + yz$, respectively. We shall show that the symmetric difference $M \Delta N$ has a 1-factor that contains neither uv nor yz, and is, thus, also a 1-factor of H.

Since every vertex of H has degree 1 in M and degree 1 in N, it follows that every vertex of H has degree 0 or 2 in $M \Delta N$, which implies that the components of $M \Delta N$ are cycles and isolated vertices. Moreover, since M and N are 1-factors, the M-edges and N-edges must alternate on each cycle component, implying that all the cycles are even.

Let C be the cycle of $M \Delta N$ that contains the edge uv. (Of course, the 1-factor M contains uv, because we have specified that H has no 1-factor.) If cycle C does not also contain the edge yz, then the union of the set of the N-edges in cycle C with the set of M-edges not in cycle C forms a 1-factor of H.

If cycle C does contain the edge yz, then, using the fact that we chose vertices u and v to have the common neighbor y, we consider the subgraph $J = (C + uy + vy - uv - yz)$ of the graph H. See Figure 8.4.4.

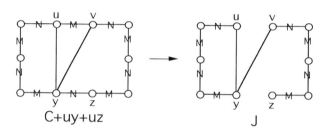

Figure 8.4.4   Constructing the subgraph J.

Since C has evenly many vertices and edges, so does J. Moreover, in subgraph J, vertex y has degree 3, vertex z has degree 1, and all other vertices have degree 2. It follows from Lemma 8.4.2 that the subgraph J has a 1-factor. By combining the 1-factor in subgraph J with the edges of M that are not part of cycle C, we obtain a 1-factor of the graph H.
$\diamondsuit$

Petersen's 1-Factor Theorem

NOTATION: The number of odd components of a graph G is denoted $oc(G)$.

Theorem 8.4.4: [**Petersen's 1-Factor Theorem**] *Every 2-edge-connected 3-regular graph G has a 1-factor.*

Proof: By Tutte's 1-Factor Theorem, it suffices to show that G satisfies Tutte's condition. Let S be an arbitrary subset of vertices of G, and let k be the number of edges between S and the odd components of $G - S$. We first observe that

$$k \leq 3\,|S|$$

because each vertex of S has degree 3, since G is 3-regular. For any odd component H (odd H) of $G - S$, let k_H be the number of edges joining H with S. Summing over all the odd components,

$$\sum_{H \text{ odd}} k_H = k$$

The sum of the vertex degrees in H is $3|V_H| - k_H$. By Euler's Degree-Sum Theorem (Theorem 1.1.2), applied to the graph H, this sum is even. Since $|V_H|$ is odd, so is $3|V_H|$, which implies that k_H is odd. Since G has no cut-edge, we have $k_H > 1$, from which it now follows that $k_H \geq 3$, and in turn, that

$$3oc(G - S) \leq \sum_{H \text{ odd}} k_H = k$$

Thus,

$$3oc(G - S) \leq 3|S|$$

which implies Tutte's condition $oc(G - S) \leq |S|$. $\Diamond$

Corollary 8.4.5: *Every 2-edge-connected 3-regular graph G has a 2-factor.*

Proof: By Petersen's 1-Factor Theorem, the graph G has a 1-factor. Since G is 3-regular, the edge-complement of that 1-factor is a 2-factor. $\Diamond$

Remark: F. Bäbler ([Bä38]) proved that every $(r-1)$-edge-connected r-regular graph has a 1-factor, and also that every 2-edge-connected $(2r+1)$-regular graph without self-loops has a 1-factor.

Remark: The following two results will be proved in Chapter 10. Their proofs use *Hall's Theorem for Bipartite Graphs*, which is also proved in Chapter 10 with the aid of network flows.

- **König's 1-Factorization Theorem** [Kö16] Every r-regular bipartite graph G with $r > 0$ is 1-factorable.

- **Petersen's 2-Factorization Theorem** [Pe189I]. Every regular graph G of even degree is 2-factorable.

EXERCISES for Section 8.4

8.4.1 Draw a 3-regular simple graph that has no 1-factor or 2-factor.

8.4.2 Draw a 5-regular simple graph that has no 1-factor or 2-factor.

8.4.3^{S} Prove that if two graphs G and H each have a k-factor, then their join has a k-factor.

8.4.4 Draw a connected simple graph that is decomposable into a 2-factor and a 1-factor, but is non-Hamiltonian.

8.4.5 Prove that if a graph has a k-factor, then its Cartesian product with any other graph has a k-factor.

8.4.6 Prove that the complete graph K_{2r} is 1-factorable. Hint: Use induction on r.

8.4.7 Prove that an r-regular bipartite graph G can be decomposed into k-factors if and only if k divides r.

8.4.8^{S} Describe a 2-factorable graph whose edge-complement contains no 2-factor.

8.4.9 Suppose that two graphs with a 1-factor are amalgamated across an edge. Does the resulting graph necessarily have a 1-factor?

8.4.10 Draw two graphs that have no 1-factor, but whose Cartesian product does have a 1-factor.

8.4.11$^\text{S}$ Draw two graphs that have no 2-factor, but whose Cartesian product does have a 2-factor.

8.5   SUPPLEMENTARY EXERCISES

Exercises 8.5.1 through 8.5.6 concern the set S of connected graphs with 8 vertices and 17 edges.

8.5.1 Prove that no graph in S is 2-chromatic.

8.5.2 Draw a 3-chromatic graph from S, and prove it is 3-chromatic.

8.5.3 Draw a 4-chromatic graph from S, and prove it is 4-chromatic.

8.5.4 Draw a 5-chromatic graph from S, and prove it is 5-chromatic.

8.5.5 Draw a 6-chromatic graph from S, and prove it is 6-chromatic.

8.5.6 Prove that no graph in S is 7-chromatic.

8.5.7 Prove that for any two graphs G and H, $\chi(G + H) = \chi(G) + \chi(H)$.

8.5.8 Draw a simple 6-vertex graph G such that $\chi(G) + \chi(\overline{G}) = 5$, where $\overline{G}$ is the edge-complement of G.

8.5.9 Draw a simple 6-vertex graph G such that $\chi(G) + \chi(\overline{G}) = 7$.

8.5.10 Construct a non-complete graph of chromatic number 6 that is chromatically critical.

8.5.11 Draw a minimum vertex-coloring and a minimum edge-coloring of the graph A in Figure 8.5.1 and prove their minimality.

8.5.12 Draw a minimum vertex-coloring and a minimum edge-coloring of the graph B in Figure 8.5.1 and prove their minimality.

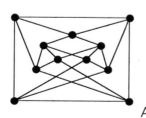
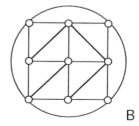

A   B

Figure 8.5.1

8.5.**13** Calculate the vertex chromatic number and edge chromatic number of the graph below. Is it vertex chromatically critical? Is it edge chromatically critical?

8.5.**14** Prove that the minimum chromatic number among all 4-regular 9-vertex graphs is three. (Hint: First prove that chromatic number two is impossible, and then draw a 3-chromatic 4-regular 9-vertex graph.)

8.5.**15** Calculate the maximum possible number of edges of a simple 3-colorable planar graph on 12 vertices. Be sure to prove that your number is achievable.

8.5.**16** Calculate the chromatic number of the circulant graph $circ(13 : 1, 5)$.

8.5.**17** a. Prove that the graph below has chromatic number 5.
b. Mark two edges whose removal would make the graph 3-chromatic.

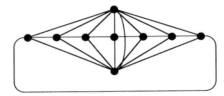

8.5.**18** Calculate the chromatic number of the graph $C_3 \times C_3 \times C_3 \times C_3 \times C_3$.

8.5.**19** Give a proper 4-coloring of the **Grötzsch graph** (a Mycielski graph), shown below. Why must at least 3 different colors be used on the outer cycle in a proper coloring? Why must at least 3 different colors be used on the five hollow vertices? (This implies that a fourth color is needed for the central vertex.)

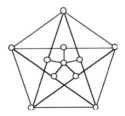

Figure 8.5.2

8.5.**20** Calculate the chromatic number of the product graph $K_4 \times K_4$.

8.5.**21** Among all graphs with 8 vertices and independence number 4, determine the largest possible number of edges and draw such a graph.

8.5.**22** Among all graphs with 8 vertices and clique number 4, draw a graph with the largest possible number of edges. Write the number.

8.5.**23** Calculate the independence number of $C_5 \times C_5$.

8.5.24 Calculate the clique number of the circulant graph $circ(9 : 1, 3, 4)$.

8.5.25 Calculate the independence number and chromatic number of the graph A in Figure 8.5.3.

8.5.26 Calculate the independence number and chromatic number of the graph B in Figure 8.5.3.

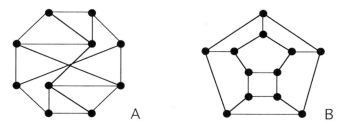

Figure 8.5.3

8.5.27 Prove that every Hamiltonian 3-regular graph is 3-edge-colorable.

8.5.28 Draw as many copies of the cube Q_3 as needed, each with a different 1-factor, to give a complete repetition-free list of all the possible 1-factors.

Glossary

adjacent edges: different edges with at least one endpoint in common.

block of a loopless graph: a maximal connected subgraph H such that no vertex of H is a cut-vertex of H.

k-chromatic graph: a graph whose vertex chromatic number is k.

chromatic incidence of an edge-coloring f at a vertex v: the number of different edge-colors present at v, denoted by $ecr_v(f)$.

chromatic number:

—, **of a graph** G: the minimum number of different colors required for a *proper vertex-coloring* of G, usually denoted by $\chi(G)$.

—, **of a map:** the minimum number of colors needed for a *proper map-coloring*.

—, **of a surface** S: the maximum of the chromatic numbers of the maps on S, or equivalently, of the graphs (without self-loops) that can be imbedded in S, denoted by $chr(S)$.

chromatically k-critical graph: a graph whose chromatic number would decrease if any edge were deleted.

class one: the class of graphs containing every non-empty graph whose edge-chromatic number equals its maximum degree.

class two: the class of graphs containing every graph whose edge-chromatic number is one more than its maximum degree.

clique in a graph: a maximal subset of mutually adjacent vertices.

clique number of a graph: the number of vertices in the largest clique.

color class in a vertex-coloring of a graph G: a subset of V_G containing all the vertices of some color.

k-colorable graph: a graph that has a proper vertex k-coloring.

coloring of a graph: usually refers to a vertex-coloring.

k-coloring of a graph: a vertex-coloring that uses exactly k different colors.

k-coloring of a map: a map-coloring that uses exactly k different colors.

colors of vertices or faces: a set, usually of integers $1, 2, \ldots$, to be assigned to the vertices of a graph or the regions of a map.

complete set of obstructions to k-chromaticity: a set $\{G_j\}$ of chromatically $(k + 1)$-critical graphs such that every $(k+1)$-chromatic graph contains at least one graph G_j as a subgraph.

domination number $\gamma(G)$ of a graph G: the cardinality of a minimum set S of vertices such that every vertex of G is either in S or a neighbor of a vertex in S.

edge k-chromatic graph G: a graph with $\chi'(G) = k$.

edge-chromatic number $\chi'(G)$ of a graph: the minimum number of different colors required for a proper edge-coloring of a graph G.

edge k-colorable graph: a graph that has a proper edge k-coloring.

edge-coloring of a graph: an assignment to its edges of "colors" from any set.

edge k-coloring of a graph: an edge coloring that uses exactly k different colors.

—, blocked partial: the circumstance in which a proper subset of edges has a proper edge k-coloring, and every uncolored edge is adjacent to at least one edge of each of the k colors.

edge-independence number $ind_E(G)$ of a graph G: the maximum cardinality of an independent set of edges.

Eulerian graph: a graph that has an Eulerian tour.

Eulerian tour in a graph: a closed trail that contains every edge of that graph.

even wheel: a wheel graph $W_n = K_1 + C_n$ such that n is even.

factor of a graph: a spanning subgraph.

k-factor of a graph: a spanning subgraph of that is regular of degree k.

factorization of a graph G: a set of factors whose edge-sets form a partition of the edge-set E_G.

k-factorization of a graph G: a factorization of G into k-factors.

independence number of a graph G: the maximum cardinality of an independent set of vertices, denoted by $\alpha(G)$.

independent set of edges: a set of mutually non-adjacent edges.

independent set of vertices: a set of mutually non-adjacent vertices.

join of two graphs G and H: the graph $G + H$ obtained from the graph union $G \cup H$ by adding an edge between each vertex of G and each vertex of H.

Kempe i-j chain for a vertex-coloring of a graph: a component of the subgraph induced on the set of all vertices colored either i or j.

Kempe i-j edge-chain in an edge-colored graph: a component of the subgraph induced on all the i-colored and j-colored edges.

leaf block of a graph G: a block that contains exactly one cut-vertex of G.

line graph of a graph G: the graph $L(\mathrm{G})$ whose vertices are the edges of G, such that edges with a common endpoint in G are adjacent in $L(G)$.

map on a surface: an imbedding of a graph on that surface.

map-coloring for an imbedding of a graph: a function from the set of faces to a set whose elements are regarded as *colors.*

matching in a graph G: a subset of edges of G that are mutually non-adjacent.

—, maximum: a matching with the maximum number of edges.

multipartite graph: a loopless graph whose vertices can be partitioned into k independent sets, which are sometimes called the **partite sets**, is said to be k-partite.

multiplicity $\mu(G)$ of a graph: the maximum number of edges joining two vertices.

neighbor of an edge e: another edge that shares one or both of its endpoints with edge e.

obstruction to k-chromaticity: a graph whose presence as a subgraph forces the chromatic number to exceed k.

odd component of a graph: a component with an odd number of vertices.

odd wheel: a wheel graph $W_n = K_1 + C_n$ such that n is odd.

optimum k-edge coloring: a k-edge-coloring with the highest total chromatic incidence among all k-edge-colorings.

partite sets: see *multipartite graph.*

perfect graph: a graph G such that every induced subgraph H has its chromatic number $\chi(H)$ equal to its clique number $\omega(H)$.

proper edge-coloring of a graph: an edge-coloring such that if two edges have a common endpoint, then they are assigned two different colors.

proper map-coloring: a coloring such that if two regions meet at an edge, then they are colored differently.

proper vertex-coloring: a coloring such that the endpoints of each edge are assigned two different colors.

total chromatic incidence of an edge-coloring on a graph: the sum of the chromatic incidences at the vertices.

Tutte's condition on a graph G: the condition that for every subset $S \subset V_G$, the number of odd components of $G - S$ does not exceed $|S|$.

vertex k-colorable graph: a graph that has a proper vertex k-coloring.

vertex-coloring: a function from the vertex-set of a graph to a set whose members are called *colors*.

vertex-k-coloring: a vertex-coloring that uses exactly k different colors.

(vertex) k-chromatic graph: a graph G with $\chi(G) = k$.

(vertex) chromatic number of a graph G, denoted $\chi(G)$: the minimum number of different colors required for a proper vertex-coloring of G.

Chapter 9

SPECIAL DIGRAPH MODELS

INTRODUCTION

Many of the methods and algorithms for digraphs are just like their counterparts for undirected graphs in Chapters 1 through 6, with little or no modification. For instance, the algorithm given in §6.1 for constructing an Eulerian tour in an undirected graph works equally well on a digraph. On the other hand, the non-reversibility of directed paths creates complications in a digraph-connectivity algorithm.

The primary focus of this chapter is on some uniquely digraphic models that arise in selected problems and applications. Some of these applications require acyclic digraphs (e.g., *task scheduling* and *tournaments*), whereas others use a general digraph model (e.g., *Markov chains* and *transitive closure*).

Although depth-first search of a digraph and of a graph work the same way, the use of depth-first search in finding the strong components of a digraph is considerably more intricate than in the analogous problem of finding the components of an undirected graph.

9.1   DIRECTED PATHS AND MUTUAL REACHABILITY

Clearly, assigning directions to edges can affect the notions of reachability and distance in a graph, but we shall see throughout the chapter how assigning directions adapts a graph to modeling processes, relationships, task organization, and operations analysis.

The term *arc* is used throughout this chapter instead of its synonym *directed edge*. Some definitions from earlier chapters are now repeated for immediate reference.

REVIEW FROM §1.4:

- A **connected** digraph is a digraph whose underlying graph is connected. Some authors use the term *weakly connected* to describe such digraphs.

- Let u and v be vertices in a digraph D. Then u and v are said to be **mutually reachable** in D if D contains both a directed u-v walk and a directed v-u walk Every vertex is regarded as reachable from itself (by the trivial walk).

- A digraph D is **strongly connected** if every two vertices are mutually reachable in D.

REVIEW FROM §3.2:

- A **directed tree** is a digraph whose underlying graph is a tree. Sometimes, when the context is clear, the word "directed" is dropped.

- A **rooted tree** is a directed tree having a distinguished vertex r, called the **root**, such that for every other vertex v, there is a directed r-v path. (Since the underlying graph of a rooted tree is acyclic, the directed r-v path is unique.)

Remark: Designating a root in a directed tree does *not* necessarily make it a rooted tree.

Strong Components

DEFINITION: A **strong component** of a digraph D is a maximal strongly connected subdigraph of D. Equivalently, a strong component is a subdigraph induced on a maximal set of mutually reachable vertices.

Example 9.1.1: Figure 9.1.1 below shows a digraph D and its four strong components.

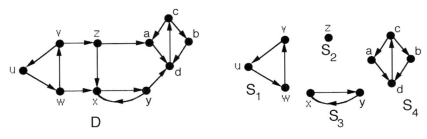

Figure 9.1.1   A digraph and its four strong components.

Notice that the vertex-sets of the strong components of D partition the vertex-set of D and that the edge-sets of the strong components do *not* include all the edges of D. This is in sharp contrast to the situation for an undirected graph G, in which the edge-sets of the components of G partition E_G. The effect that one-way edges have on the study of digraph connectivity is especially apparent when the relatively complicated strong component-finding in §9.5 is compared to the very simple tree-growing algorithm in §4.1, which finds the components of an undirected graph.

Remark: The analogy between a component of an undirected graph and a strong component of a digraph carries over to viewing each as an induced subgraph on an equivalence class of vertices. In particular, the components of an undirected graph G are the subgraphs induced on the equivalence classes of the reachability relation on V_G (§2.3), and the strong components of a digraph D are the subdigraphs induced on the equivalence classes of the *mutual-reachability* relation on V_D.

Proposition 9.1.1: *Let D be a digraph. Then the mutual-reachability relation is an equivalence relation on V_D, and the strong components of digraph D are the subdigraphs induced on the equivalence classes of this relation.*   ◇ *(Exercises)*

Corresponding to any digraph D, there is a new digraph whose definition is based on the strong components of D.

DEFINITION: Let $S_1, S_2, \ldots, S_r$ be the strong components of a digraph D. The **condensation** of D is the simple digraph D' with vertex set $V_{D'} = \{s_1, s_2, \ldots, s_r\}$, such that there is an arc in digraph D' from vertex s_i to vertex s_j if and only if there is an arc in digraph D from a vertex in component S_i to a vertex in component S_j.

Example 9.1.2: Figure 9.1.2 shows the digraph D from the previous example and its condensation D'. Notice that the condensation D' is an acyclic digraph (i.e., it has no *directed* cycles).

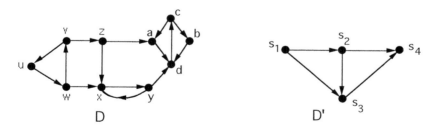

Figure 9.1.2   A digraph D and its condensation D'.

Proposition 9.1.2: *The condensation of any digraph is an acyclic digraph.*

Proof: A directed cycle in the condensation would contradict the maximality of any strong component that corresponds to a vertex on that cycle. The details are left as an exercise.   ◇ *(Exercises)*

TERMINOLOGY: An acyclic digraph is often called a **dag** (as an acronym for *directed acyclic graph*).

Tree-Growing Revisited

Algorithm 9.1.1, shown below, is simply the basic tree-growing algorithm (Algorithm 4.1.1), recast for digraphs. Its output is a rooted tree whose vertices are reachable from the starting vertex. But because the paths to these vertices are directed (i.e., one-way), the vertices in this *output tree* need not be mutually reachable from one another.

If the digraph D is strongly connected, then the output tree is a spanning tree of D, regardless of the starting vertex. But for a general digraph D, the vertex-set of the output tree depends on the choice of a starting vertex, as the example following Algorithm 9.1.1 illustrates.

REVIEW FROM §4.1:

- A **frontier arc** for a rooted tree T in a digraph is an arc whose tail is in T and whose head is not in T.

Algorithm 9.1.1:   Basic Tree-Growing in a Digraph
Input: a digraph D and a starting vertex $v \in V_D$.
Output: a rooted tree T with root v and a standard vertex-labeling of T.
    Initialize tree T as vertex v.
    Write label 0 on vertex v.
    Initialize label counter $i := 1$
    While there is at least one frontier arc for tree T
        Choose a frontier arc e for tree T.
        Let w be $head(e)$ (which lies outside of T).
        Add arc e and vertex w to tree T.
        Write label i on vertex w.
        $i := i + 1$
    Return tree T and vertex-labeling of T.

COMPUTATIONAL NOTE: We assume that there is some implicit *default priority* for choosing vertices or edges, which is invoked whenever there is more than one frontier arc from which to choose.

Example 9.1.3: Figure 9.1.3 shows a digraph and all possible output trees that could result for each of the different starting vertices and each possible default priority. Two opposite extremes for possible output trees are represented here. When the algorithm starts at vertex u, the output tree spans the digraph. The other extreme occurs when the algorithm starts at vertex x, because x has outdegree 0.

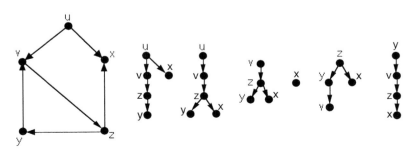

Figure 9.1.3   A digraph and all possible output trees.

Notice that any two output trees in Figure 9.1.3 with the same vertex-set have roots that are mutually reachable. This property holds in general, as the following proposition asserts.

Proposition 9.1.3: *Let u and v be two vertices of a digraph D. Then u and v are in the same strong component of D if and only if the output trees that result from starting Algorithm 9.1.1 at vertex u and at vertex v have the same vertex-set.*   ◇ *(Exercises)*

It is easy enough to modify Dijkstra's algorithm (Algorithm 4.3.2) to find least-cost (or shortest) paths in directed graphs. The following application illustrates how finding least-cost directed paths might be useful in economic planning.

Application 9.1.1:   *Equipment-Replacement Policy*   Suppose that today's price for a new car is \$16,000, and that the price will increase by \$500 for each of the next four years. The projected annual operating cost and resale value of this kind of car are shown in the table below. To simplify the setting, assume that these data do not change for the next five years. Starting with a new car today, determine a replacement policy that minimizes the net cost of owning and operating a car for the next five years.

Annual Operating Cost	Resale Value
\$600 (for 1st year of car)	\$13,000 (for a 1-year-old car)
\$900 (for 2nd year of car)	\$11,000 (for a 2-year-old car)
\$1200 (for 3rd year of car)	\$9,000 (for a 3-year-old car)
\$1600 (for 4th year of car)	\$8,000 (for a 4-year-old car)
\$2100 (for 5th year of car)	\$6,000 (for a 5-year-old car)

The digraph model has six vertices, labeled 1 through 6, representing the *beginning* of years 1 through 6. The beginning of year 6 signifies the end of the planning period. For each i and j with $i < j$, an arc is drawn from vertex i to vertex j and is assigned a weight c_{ij}, where c_{ij} is the total net cost of purchasing a new car at the beginning of year i and keeping it until the beginning of year j. Thus,

$$C_{ij} = \text{price of new car at beginning of year } i$$
$$+ \text{ sum of operating costs for years } i, i+1, \ldots, j-1$$
$$- \text{ resale value at beginning of year } j$$

For example, the net cost of buying a car at the beginning of year 2 and selling it at the beginning of year 4 is $c_{24} = 16,500 + 600 + 900 - 11,000 = \$7,000$. Figure 9.1.4 shows the resulting digraph with seven of its 15 arcs drawn. The arc-weights are in units of \$100.

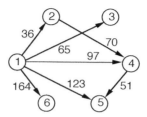

Figure 9.1.4   Part of the digraph model for a car-replacement problem.

The problem of determining the optimal replacement policy is reduced to finding the shortest (least-cost) path from vertex 1 to vertex 6. This is a simple task for Dijkstra's algorithm, even for much larger instances of this kind of problem.

Characterization of Strongly Orientable Graphs

A classical theorem of Herbert Robbins arose as an application of digraph models to a whimsical traffic problem. Seeing how the one-way streets of New York sometimes necessitated roundabout driving patterns, Robbins determined the class of road-networks in which it was possible to make every street one-way, without eliminating mutual reachability between any two locations.

REVIEW FROM §1.4: A graph G is **strongly orientable** if there exists an assignment of directions to the edge-set of G such that the resulting digraph is strongly connected.

REVIEW FROM §5.1 AND §5.2:

- A graph G is **2-edge-connected** if G is connected and has no cut-edges.

- A **path addition** to a graph G is the addition to G of a path between two existing vertices of G, such that the edges and internal vertices of the path are not in G.

- A **cycle addition** is the addition to G of a cycle that has exactly one vertex in common with G.

- A **Whitney-Robbins synthesis** of a graph G from a graph H is a sequence of graphs, $G_0, G_1, \ldots, G_l$, where $G_0 = H, G_l = G$, and G_i is the result of a path or cycle addition to G_{i-1}, for $i = 1, ldots, l$.

- **Theorem 5.2.4:** A graph G is 2-edge-connected if and only if G is a cycle or a Whitney Robbins synthesis from a cycle.

Example 9.1.4: Of the three graphs shown in Figure 9.1.5, only graph G_2 is strongly orientable.

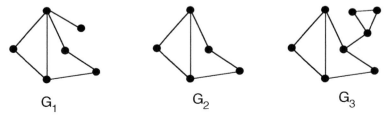

G_1 G_2 G_3

Figure 9.1.5   Only one of these graphs is strongly orientable.

Notice that G_2 is the only graph in the example that does not have a cut-edge. In fact, as Robbins proved, the absence of cut-edges is a necessary and sufficient condition for a graph to be strongly orientable.

Theorem 9.1.4: [Robbins, 1939] *A connected graph G is strongly orientable if and only if G has no cut-edges.*

Proof: ($\Rightarrow$) Arguing by contrapositive, suppose that graph G has a cut-edge e joining vertices u and v. Then the only u-v path in G and the only v-u path is edge e itself (see Figure 9.1.6). Thus, for any assignment of directions to the edges of G, $tail(e)$ will not be reachable from $head(e)$.

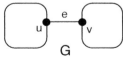

Figure 9.1.6

($\Leftarrow$) Suppose that G is a connected graph with no cut-edges. By Theorem 5.2.4, G is a cycle or a Whitney-Robbins synthesis from a cycle. Since a cycle is obviously strongly orientable, and strong orientability is clearly preserved under path or cycle addition, it follows that G is strongly orientable. $\diamondsuit$

Markov Chains

The concept of a *Markov process* was mentioned briefly in §1.3 and is reintroduced here. Several details are omitted in order to get to the presentation of the digraph model as quickly as possible. In such a digraph, there is at most one arc from any vertex to any other vertex, but some vertices may have self-arcs.

The topic of Markov processes is part of a more general area known as *stochastic processes*, a branch of mathematics and operations research with far-ranging applications and theoretical challenges. The reader may consult any of the standard texts in this subject for a formal presentation of these concepts (e.g., [Ci75]).

Suppose that some characteristic of a phenomenon is being observed at discrete instants, labeled $t = 0, 1, 2, dots$. Let X_t be the value (or *state*) of the characteristic at each such t, and suppose that $S = \{s_1, s_2, \ldots, s_n\}$ is the set of possible values (called the *state space*).

DEFINITION: A *(discrete-time) Markov chain* is a phenomenon whose behavior can be modeled by a sequence $\{X_t\}, t = 0, 1, 2, \ldots$, such that the *(one-step) transition probability* p_{ij} that $X_{t+1} = s_j$, given that $X_t = s_i$, does *not* depend on any earlier terms in the sequence $\{X_t\}$ and does not depend on t. Thus, p_{ij} is the *conditional probability* given by

$$p_{ij} = prob(X_{t+1} = s_j \mid X_t = s_i), \quad \text{for } t = 1, 2, \ldots$$

TERMINOLOGY NOTE: We are calling a *Markov chain* what experts in this area are likely to consider a special kind of Markov chain, called a *finite discrete-time stationary Markov chain*.

DEFINITION: The *transition matrix* of a Markov chain is the matrix whose ij^{th} entry is the transition probability p_{ij}.

DEFINITION: The **Markov diagram** (or **Markov digraph**) of a given Markov chain is a digraph such that each vertex i corresponds to state s_i, and an arc directed from vertex i to vertex j corresponds to the transition from state s_i to state s_j and is assigned the weight P_{ij}.

Thus, the sum of the weights on the out-directed arcs from any vertex of a Markov digraph must be equal to 1.

Application 9.1.2:   *A Gambler's Problem*   Suppose that a gambler starts with \$3 and agrees to play the following game. Two coins are tossed. If both come up heads, then he wins \$3; otherwise, he loses \$1. He agrees to play until either he loses all his money or he reaches a total of at least \$5. Let X_t be the amount of money he has after t plays, with $X_0 = 3$. The state space is $S = \{0, 1, 2, 3, 4, \geq 5\}$, and the sequence $\{X_t\}$ is a discrete-time stochastic process. Moreover, the value of X_{t+1} depends only on the amount of money the gambler has after the tth play and not on any other part of the past history. Thus, the sequence $\{X_t\}$ is a Markov chain. The transition matrix and Markov diagram for this Markov chain are shown in Figure 9.1.7.

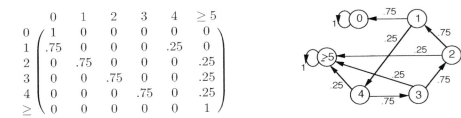

$$
\begin{array}{c}
\quad\;\; 0 \quad 1 \quad 2 \quad 3 \quad 4 \;\; \geq 5 \\
\begin{array}{c}
0 \\ 1 \\ 2 \\ 3 \\ 4 \\ \geq
\end{array}
\left(
\begin{array}{cccccc}
1 & 0 & 0 & 0 & 0 & 0 \\
.75 & 0 & 0 & 0 & .25 & 0 \\
0 & .75 & 0 & 0 & 0 & .25 \\
0 & 0 & .75 & 0 & 0 & .25 \\
0 & 0 & 0 & .75 & 0 & .25 \\
0 & 0 & 0 & 0 & 0 & 1
\end{array}
\right)
\end{array}
$$

Figure 9.1.7   **Gambler's transition matrix and Markov diagram.**

k-Step Transition Probability

DEFINITION: Let $\{X_t\}$ be a Markov chain with state space $S = \{s_1, s_2, \ldots, s_n\}$. The conditional probability that $X_{t+k} = s_j$, given that $X_t = s_i$, is called the k-**step transition probability** and is denoted $p_{ij}^{[k]}$. The matrix whose ij^{th} entry is $p_{ij}^{[k]}$ is called the k-**step transition matrix** and is denoted $P^{[k]}$.

The k-step transition probability has a simple interpretation in terms of the Markov digraph, and it is easily calculated from the transition matrix of the Markov chain.

Example 9.1.5: A company has three copier machines. During any given day, each machine that is working at the beginning of that day has a 0.1 chance of breaking down. If a machine breaks down during the day, it is sent to a repair center and will be working at the beginning of the second day after the day it broke down. Let X_t be the number of machines that are in working order at the start of day t. If at the start of a particular day, there is exactly one copier working, what is the probability that exactly two copiers are working three days later?

It is easy to verify that the transition matrix and Markov digraph are as shown in Figure 9.1.8 (see Exercises). For instance, if two are working, then the probability that exactly one of them breaks down is $.1 \times .9 + .9 \times .1 = .18$.

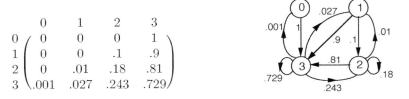

$$
\begin{array}{c}
\quad\;\; 0 \quad\;\; 1 \quad\;\; 2 \quad\;\; 3 \\
\begin{array}{c}
0 \\ 1 \\ 2 \\ 3
\end{array}
\left(
\begin{array}{cccc}
0 & 0 & 0 & 1 \\
0 & 0 & .1 & .9 \\
0 & .01 & .18 & .81 \\
.001 & .027 & .243 & .729
\end{array}
\right)
\end{array}
$$

Figure 9.1.8   **Copier-machine transition matrix and Markov diagram.**

The problem is to calculate $p_{1,2}^{[3]}$, the 3-step transition probability of going from state 1 to state 2. Each directed 1-2 walk of length 3 in the digraph represents one such 3-step probability, and the product of the arc-steps on that walk is the probability of that particular 3-step sequence of transitions. Thus, the 3-step transition probability is the sum of the products of the arc-steps of all possible walks of length 3 from vertex 1 to vertex 2. There are six different 1-2 paths of length 3, and their arc probabilities lead to the following calculation.

$$p_{1,2}^{[3]} = (.9 \times .729 \times .243) + (.9 \times .027 \times .1) + (.9 \times .243 \times .18)$$
$$+ (.1 \times .18 \times .18) + (.1 \times .81 \times .243) + (.1 \times .01 \times .1)$$
$$\approx .224$$

In general, an entry of the transition matrix of a Markov chain is nonzero if and only if the corresponding entry in the adjacency matrix of the Markov diagram is nonzero. Because of this relationship, the k-step transition probabilities can be obtained by simple matrix multiplication, in much the same way that matrix multiplication can be used to determine the number of v_i-v_j walks of length k in an undirected graph (§2.6).

Proposition 9.1.5: *Let P be the transition matrix of a Markov chain. Then the k-step transition matrix is the kth power of matrix P, i.e., $P^{[k]} = P^k$.*

Proof: By definition of the transition matrix P, the assertion is clearly true for $k = 1$, which establishes the base of an induction on the number of steps k. Next, each i-j path of length $k + 1$ in the Markov diagram is the concatenation of an i-l path of length k with an arc from vertex l to vertex j, for some vertex l. Thus, $p_{ij}^{[k+1]} = \sum_l p_{il}^{[k]} \times p_{lj}$, from which the induction step follows, by the definition of matrix multiplication.   ◇

Classification of States in a Markov Chain

The digraph model is helpful in classifying the different types of states that can occur in a Markov chain. To illustrate, we introduce the following standard terminology for Markov chains.

DEFINITION: A state s_j is **reachable** from state s_j in a Markov chain if there is a sequence of transitions, each having nonzero probability, from state s_i to state s_j.

Thus, the term *reachable* for Markov chains coincides with the digraph term applied to the corresponding Markov diagram.

DEFINITION: Two states are said to **communicate** if they are mutually reachable in the corresponding Markov diagram.

DEFINITION: A **communication class** in a Markov chain is a maximal subset of states that communicate with one another.

Thus, a communication class is the vertex-set of one of the strong components of the Markov diagram.

DEFINITION: A Markov chain is **irreducible** if every two states communicate, that is, if the Markov diagram is strongly connected.

DEFINITION: A state s_i is an **absorbing state** if $p_{ii} = 1$.

DEFINITION: A state s_i is an *a* **transient state** if there exists a state s_j that is reachable from state s_i, but state s_i is not reachable from state s_j.

Thus, if the current state is s_i, there is a nonzero probability of leaving s_i and never returning.

DEFINITION: A state s_j is a **recurrent state** if it is not a transient state. Observe that an absorbing state is trivially recurrent.

Example 9.1.6: The copiers Markov chain of Example 9.1.5 is irreducible, and every state is recurrent.

Example 9.1.7: In the Gambler's Markov chain of Application 9.1.2, the 0 state and the ≥ 5 state are absorbing states, and all other states are transient.

EXERCISES for Section 9.1

For Exercises 9.1.1 through 9.1.6, identify the strong components and draw the condensation of the given digraph.

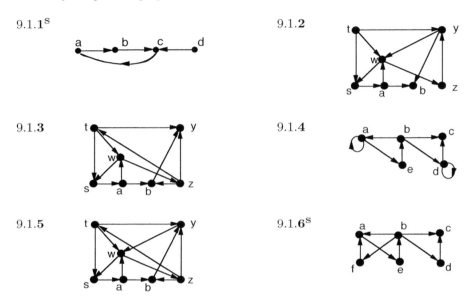

9.1.1^S

9.1.2

9.1.3

9.1.4

9.1.5

9.1.6^S

For Exercises 9.1.7 through 9.1.12, apply Basic Tree-Growing Algorithm 9.1.1 to the specified digraph, and resolve ties (if any) using alphabetical order of vertices. Keep restarting the algorithm until all vertices are labeled.

9.1.**7** The graph of Exercise 9.1.1.

9.1.**8** The graph of Exercise 9.1.2.

9.1.**9** The graph of Exercise 9.1.3.

9.1.**10** The graph of Exercise 9.1.4.

9.1.**11**^S The graph of Exercise 9.1.5.

9.1.**12** The graph of Exercise 9.1.6.

9.1.**13**^S Complete the digraph model for the equipment-replacement problem of Application 9.1.1, and apply Dijkstra's algorithm to determine the optimal replacement policy.

9.1.**14** Referring to Application 9.1.1, suppose that the new-car price increases by $1,000 each year. Also, assume that the resale value of a car in a given year is $500 more than a car of the same age the previous year and that that car's operating cost is $100 more than a comparably aged car the previous year. Modify the arc-weights accordingly and solve this version of the problem.

9.1.15 Suppose that a new fax machine has just been purchased for $200, and assume that the price stays at $200 for the next six years. Also assume that the machine must be replaced by a new machine after five years, but that it can be replaced sooner if desired. The estimated maintenance cost for each year of operation is $80 for the first year, $120 for the second, $160 for the third, $240 for the fourth, and $280 for the fifth year. Determine a replacement policy to minimize the total cost of purchasing and operating a fax machine for the next six years.

For Exercises 9.1.16 through 9.1.21, find a strong orientation of the given graph.

9.1.16   **9.1.17**

9.1.18^S **9.1.19**

9.1.20   **9.1.21**

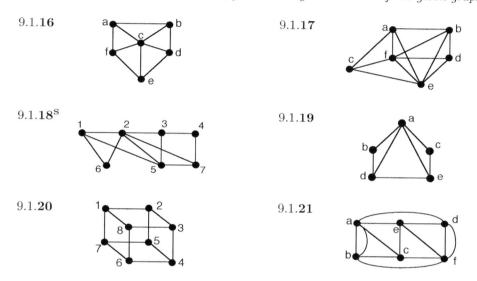

9.1.22^S Find an orientation of the complete graph K_5 that is a dag.

9.1.23 Complete the proof of Proposition 9.1.2.

9.1.24^S Characterize those digraphs that are the same as their condensation.

9.1.25 Prove Proposition 9.1.1.

9.1.26 Prove Proposition 9.1.3.

9.1.27 Prove that two vertices are in the same strong component of a given digraph if and only if every output tree that contains one of them also contains the other.

9.1.28^S Prove or disprove: Let G be a graph with k cut-edges, and let D be a digraph that results from assigning a direction to each edge of G. Then digraph D has at least $k+1$ strong components.

9.1.29 Verify that the transition matrix and Markov diagram for the Gambler's problem of Application 9.1.2 are correct.

9.1.30^S Determine the 2-step transition matrix for the Markov chain of Application 9.1.1, by working directly with the Markov diagram. Check your result by applying Proposition 9.1.5.

9.1.31 Verify that the transition matrix and Markov diagram for the copier-machine Markov chain of Example 9.1.5 are correct.

9.1.32 Referring to Example 9.1.5, calculate $p_{2,1}^{[3]}$ by working directly with the Markov diagram.

9.1.33 Determine the 2-step transition matrix for the Markov chain of Example 9.1.5 by working directly with the Markov diagram. Check your result by applying Proposition 9.1.5.

9.1.34[S] Consider the Markov chain whose Markov diagram is shown.

    a. Classify each of the states.
    b. Calculate $p_{34}^{[4]}$.

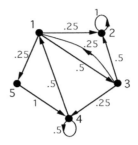

9.1.35 Find the 2-step transition matrix for the Markov diagram in the previous exercise, by working directly with the digraph. Check your results by applying Proposition 9.1.5.

Application 9.1.3:   *Random Walks*   Suppose that a person who has trouble making up his mind is standing on a straight road between two locations, A and B, 400 yards apart. After each minute he chooses from one of three options of what to do during the next minute. With probability $\frac{1}{2}$, he walks 100 yards toward A; with probability $\frac{1}{3}$, he walks 100 yards toward B; and with probability $\frac{1}{6}$, he stays where he is and thinks about it for the next minute. His position after each minute is one of five possible locations, as shown in the figure. Assume that if he arrives at either A or B, he remains there forever.

9.1.36 Consider the situation described in Application 9.1.3.

    a. Draw the Markov diagram.
    b. Determine the probability that he is at position 3 after two minutes, given that he starts there.
    c. Find the 2-step transition matrix by working directly with the digraph, and then check it by applying Proposition 9.1.5.
    d. Classify each of the states.

9.1.37[S] Prove that every finite Markov chain has at least one recurrent state.

9.1.38 [*Computer Project*] Implement Basic Tree-Growing Algorithm 9.1.1, and run the program on each of the digraphs in Exercises 9.1.1 through 9.1.6.

9.1.39 [*Computer Project*] Write a program whose input is a graph G with no cut-edges and whose output is a strong orientation of G. (Hint: See the second part of the proof of Theorem 9.1.4.)

9.2   DIGRAPHS AS MODELS FOR RELATIONS

DEFINITION: The **digraph representation of a relation** R on a finite set S is the digraph whose vertices correspond to the elements of S, and whose arcs correspond to the ordered pairs in the relation. That is, an arc is drawn from vertex x to vertex y if $(x, y) \in R$.[†]

Conversely, a digraph D induces a relation R on V_D in a natural way, namely, $(x, y) \in R$ if and only if there is an arc in digraph D from vertex x to vertex y.

Remark: Because the digraph representation of a relation has no multi-arcs, an arc can be specified by the ordered pair it represents. We adopt this convention for this subsection.

Example 9.2.1: Suppose a relation R on the set $S = \{a, b, c, d\}$ is given by

$$\{(a, a), (a, b), (b, c), (c, b), (c, d)\}$$

Then the digraph D representing the relation R is as shown in Figure 9.2.1.

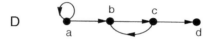

D

Figure 9.2.1   The digraph D representing the relation R.

The Transitive Closure of a Digraph

DEFINITION: A **transitive digraph** is a digraph whose corresponding relation is transitive. That is, if there is an arc from vertex x to vertex y and an arc from y to z, then there is an arc from x to z.

DEFINITION: The **transitive closure** R^* of a binary relation R is the relation R^* defined by $(x, y) \in R^*$ if and only if there exists a sequence $x = v_0, v_1, v_2 \ldots, v_k = y$ such that $k \geq 1$ and $(v_i, v_{i+1}) \in R$, for $i = 0, 1, \cdots, k - 1$.

Equivalently, the transitive closure R^* of the relation R is the smallest transitive relation that contains R.

DEFINITION: Let D be the digraph representing a relation R. Then the digraph $D*$ representing the transitive closure R^* of R is called the **transitive closure of the digraph** D.

Thus, an arc $(x, y), x \neq y$ is in the transitive closure D^* if and only if there is a directed x-y path in D. Similarly, there is a self-loop in digraph D^* at vertex x if and only if there is a directed cycle in digraph D that contains x.

Remark: In many applications, the transitive closure of a digraph is used without making any explicit reference to the underlying relation.

[†] Relations and related terminology appear in Appendix A.2.

Example 9.2.2: The transitive closure of the digraph in Figure 9.2.1 is shown below.

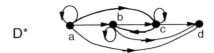

D^*

Figure 9.2.2   The transitive closure D^* of digraph D.

Application 9.2.1:   *Transitive Closure in a Paging Network*   Suppose that the arcs of an n-vertex digraph D represent the one-way direct links between specified pairs of nodes in an n-node paging network. Thus, an arc from vertex i to vertex j indicates that a page call can be transmitted from person i to person j.

To send an alert from person i to person j, it is not necessary to have a direct link from i to j. There need only be a directed i-j path. The transitive closure D^* of digraph D specifies all pairs i, j of vertices for which there exists a directed i-j path in D.

Constructing the Transitive Closure of a Digraph: Warshall's Algorithm

Let D be an n-vertex digraph with vertices $v_1, v_2, \ldots, v_n$. A computationally efficient algorithm, due to Warshall, constructs a sequence of digraphs, $D_0, D_1, \ldots, D_n$, such that $D_0 = D$, D_{i-1} is a subgraph of D_i, $i = 1, \ldots, n$, and such that D_n is the transitive closure of D. Digraph D_i is obtained from digraph D_{i-1} by adding to D_{i-1} an arc (v_j, v_k) (if it is not already in D_{i-1}) if there is a directed path of length 2 in D_{i-1} from v_j to v_k, having v_i as the internal vertex.

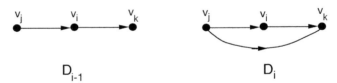

Figure 9.2.3   The arc (v_j, v_k) is added to digraph D_{i-1}.

Algorithm 9.2.1:   **Warshall's Transitive Closure**
Input: an n-vertex digraph D with vertices $v_1, v_2, \ldots, v_n$.
Output: the transitive closure of digraph D.
    Initialize digraph D_0 to be digraph D.
    For $j = 1$ to n
        For $j = 1$ to n
            If (v_j, v_i) is an arc in digraph D_{i-1}
                For $k = 1$ to n
                    If (v_i, v_k) is an arc in digraph D_{j-1}
                        Add arc (v_j, v_k) to D_{i-1} (if it is not already there).
    Return digraph D_n.

COMPUTATIONAL NOTE: An efficient implementation of the algorithm uses the *adjacency matrix* A_D of the digraph D (§2.6), where the ij^{th} element of A_D is given by

$$a_{jk} = \begin{cases} 1, & \text{if } (v_j, v_k) \text{ is an arc in } D \\ 0, & \text{otherwise} \end{cases}$$

If the values 0 and 1 are interpreted as *false* and *true*, respectively, then the innermost loop body can be replaced by $a_{jk} := a_{jk} \vee (a_{ji} \wedge a_{ik})$.

The correctness of Algorithm 9.2.1 is a consequence of the following proposition.

Proposition 9.2.1: *Let D be an n-vertex digraph with vertices $v_1, v_2, \ldots, v_n$, and let D_i be the digraph that results from the ith iteration of Algorithm 9.2.1. Let x and y be any two vertices in D for which there exists a directed x-y path in D whose internal vertices are from the set $\{v_1, v_2, \ldots, v_i\}$ Then digraph D_i contains the arc (x, y).*

Proof: The assertion is true for $i = 1$, because in the first iteration of the algorithm, an arc is added to digraph $D_0(= D)$ If D_0 contains the arcs $(x, v_1$ and (v_1, y). By way of induction, assume that the assertion is true for D_k, for all $k < i$.

Next, let x and y be vertices in D such that there exists a directed x-y path P whose internal vertices are from the set $\{v_1, v_2, \ldots, v_i\}$. If vertex v_i is not an internal vertex of path P, then it follows from the induction hypothesis that digraph D_{i-1} contains arc (x, y), and, hence, so does digraph D_i. If vertex v_i is an internal vertex of path P, then by the induction hypothesis, digraph D_{i-1} contains arcs (x, v_i) and (v_i, y), Hence, arc (x, y) is added during the ith iteration.   ◇

Partial Orders, Hasse Diagrams, and Linear Extensions

NOTATION: Throughout the rest of this subsection, the symbol $\prec$ is used to denote the relation under discussion.

FROM APPENDIX A.2: A relation $\prec$ on a set S is called a **partial order** if $\prec$ is reflexive, antisymmetric, and transitive.

TERMINOLOGY: The pair $(S, \prec)$ is called a **partially ordered set** or **poset**.

Example 9.2.3: Let $S = \{2, 3, 4, 6, 8, 12\}$, and let $\prec$ be the relation on set S given by $x \prec y \Leftrightarrow x$ divides y (i.e., $\frac{y}{x}$ is an integer). Then $\prec$ is a partial order on set S, and its digraph representation is shown in Figure 9.2.4.

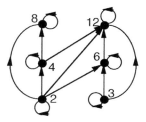

Figure 9.2.4   Representing the *divides* relation.

A partial order on a finite set is often represented by a simplified graph drawing called a *Hasse diagram*.

DEFINITION: Let $\prec$ be a partial order on a finite set S. A **Hasse diagram** representing the poset $(S, \prec)$ is a graph drawing defined as follows:

- Each element of set S is represented by a vertex.
- If $x \prec y$ for distinct elements x and y, then the vertex for y is positioned higher than the vertex for x; and if there is no w different from both x and y such that $x \prec w$ and $w \prec y$, then an edge is drawn from vertex x upward to vertex y.

The effect of the first condition is to allow arcs to be represented by undirected edges, since all arcs are directed upward. The last condition avoids redundant lines that are implied by transitivity. Notice also that the self-loops of the poset are absent from the Hasse diagram.

Example 9.2.4: The digraph representation and its Hasse diagram for the *divides* poset of the previous example are shown in Figure 9.2.5.

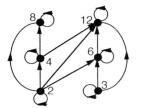 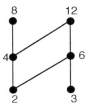

Figure 9.2.5   Digraph and Hasse diagram for the *divides* relation.

Notice that if $\vec{H}$ denotes the digraph obtained by replacing each edge of the Hasse diagram for the example poset by an arc directed upward, then the transitive closure of $\vec{H}$ recaptures the digraph representation of the poset, except for the self-loops. This property holds for all posets, by the definition of Hasse diagram.

DEFINITION: Two elements x and y of a poset $(S, \prec)$ are said to be **comparable** if either $x \prec y$ or $y \prec x$. Otherwise, x and y are **incomparable**.

DEFINITION: A **total order** on a set S is a partial order on S such that every two elements are comparable.

Example 9.2.5: *Lexicographic order* is a total order on the set of words in a dictionary.

Example 9.2.6: The *divides* partial order shown in Figure 9.2.5 is *not* a total order, because there are several incomparable elements (e.g., 3 and 8 are not comparable).

Remark: Because the Hasse diagram of a finite total order can be drawn as a path graph, it should not be surprising that the term *linear order* is a synonym for total order.

DEFINITION: A partial order $\prec'$ on a set S is an **extension of a partial order** $\prec$ on S (or is **compatible with** that partial order) if $x \prec y \Rightarrow x \prec' y$.

DEFINITION: A **linear extension of a poset** $(S, \prec)$ is a total order $\prec^t$ on S that is compatible with $\prec$.

DEFINITION: A **topological sort** is a process that produces a linear extension of a poset. It also refers informally to the resulting linear extension.

Application 9.2.2:   *Job Sequencing on a Single Machine*   Suppose a procedure, consisting of several operations, must be performed on a single machine. There is a natural partial order defined on the set of operations. For any two operations x and y,

$$x \prec y \quad \Longleftrightarrow \quad \text{operation } x \text{ cannot occur after operation } y$$

A linear extension of this poset would solve the problem of sequencing the operations on the machine.

Illustration: Suppose the Hasse diagram shown in Figure 9.2.6 represents the precedence relations among the set $T = \{a, b, c, d, e, j, g\}$ of operations required to perform a particular procedure.

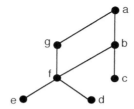

Figure 9.2.6   Hasse diagram for a set of operations.

There is a simple algorithm to construct a linear extension of a poset. The idea is always to choose a *minimal element* at each step.

DEFINITION: A **minimal element** of a poset $(S, \prec)$ is an element $m \in S$ such that for all other $x \in S$, $x \not\prec m$.

In terms of a Hasse diagram, a minimal element is a vertex that has no vertex positioned below it and joined to it by an edge.

Algorithm 9.2.2:   **Linear Extension of a Poset**
Input: an n-element poset $(S, \prec)$.
Output: a permutation $\langle s_1, s_2, \ldots, s_n \rangle$ of the elements of S that represents a linear extension of poset $(S, \prec)$.
    For $i = 1$ to n
        Let s_i be a minimal element of poset $(S, \prec)$.
        $S := S - \{s_i\}$
Return $(s_1, s_2, \ldots, s_n)$.

COMPUTATIONAL NOTE: At any iteration of Algorithm 9.2.2, there may be more than one minimal element from which to choose. Assume that there is some default ordering of the elements of S to resolve ties.

Example 9.2.7: The sequence of Hasse diagrams shown below illustrates Algorithm 9.2.2 for the poset whose Hasse diagram is shown in Figure 9.2.6. Below each Hasse diagram is the total order constructed so far, and the final iteration produces the total order $\langle c, d, e, f, b, g, a \rangle$. Ties are resolved alphabetically.

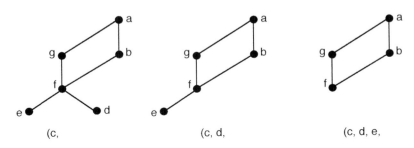

Figure 9.2.7   Iterations 1 through 3.

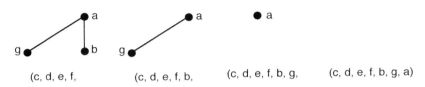

Figure 9.2.8   Iterations 4 through 7.

EXERCISES for Section 9.2

For Exercises 9.2.1 through 9.2.6, draw the transitive closure of the given digraph.

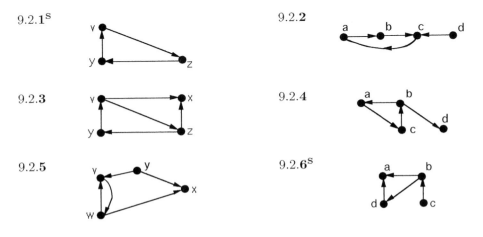

For Exercises 9.2.7 through 9.2.12, apply Warshall's Algorithm 9.2.1 to the given digraph.

9.2.7 The graph of Exercise 9.2.1.   **9.2.8** The graph of Exercise 9.2.2.

9.2.9 The graph of Exercise 9.2.3.   **9.2.10**^{S} The graph of Exercise 9.2.4.

9.2.11 The graph of Exercise 9.2.5.   **9.2.12** The graph of Exercise 9.2.6.

For Exercises 9.2.13 through 9.2.15, draw the Hasse diagram for the given nonreflexive poset and identify all the minimal elements.

9.2.13 $(S, \prec)$, where $S = \{1, 2, 3, 4, 5, 6\}$ and $x \prec y \Leftrightarrow x > y$.

9.2.14^S $(S, \prec)$, where $S = \{2, 3, 4, 5, 6, 8, 9, 72\}$ and $x \prec y \Leftrightarrow x$ properly divides y.

9.2.15 $(S, \prec)$, where S is the set of all subsets of $A = \{1, 2, 3\}$ and $x \prec y \Leftrightarrow x$ is a proper subset of y.

For Exercises 9.2.16 through 9.2.18, construct a linear extension for the specified poset.

9.2.16 The poset of Exercise 9.2.13 .

9.2.17^S The poset of Exercise 9.2.14.

9.2.18 The poset of Exercise 9.2.15.

9.2.19 Construct a linear extension for the poset whose Hasse diagram is shown below. (Resolve ties alphabetically.)

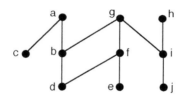

9.2.20 [*Computer Project*] Implement Warshall's Algorithm 9.2.1 and test the program on each of the digraphs in Exercises 9.2.1 through 9.2.6.

9.3   TOURNAMENTS

The applications discussed in this section use models with no multi-arcs or self-arcs, and it will sometimes be convenient for discussion purposes to refer to an arc by its ordered pair of endpoints. However, the more flexible computer representations of digraphs are unlikely to represent arcs in this way.

DEFINITION: A **tournament** is a simple digraph whose underlying graph is complete. Thus, a tournament is obtained by assigning a direction to each edge of a complete graph.

Transitive Tournaments

Much of the interest in tournament models is to determine the winner of a competition. This pursuit sometimes leads to paradoxes.

DEFINITION: A **transitive tournament** is a tournament that is transitive as a digraph.

Example 9.3.1: Each participant in a *round-robin* tennis competition plays exactly one match with every other player. If each player v is represented by a vertex v, and if an arc from u to v means that player u beat player v, then the outcome of the complete round-robin competition is represented by a tournament. Figure 9.3.1 below shows the outcome of a 5-person round-robin for which the representing digraph is a transitive tournament.

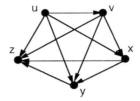

Figure 9.3.1   A 5-vertex transitive tournament.

Proposition 9.3.1: *A tournament is transitive if and only if it is acyclic.*

Proof: Suppose that a tournament T is transitive, and by way of contradiction, assume that T contains a directed cycle $C = \langle v_1, v_2, \cdots, v_k, v_1 \rangle$. By transitivity, there is an arc from vertex v_1 to vertex v_k, which contradicts the existence of the arc from v_k to v_1.

Conversely, let T be an acyclic tournament, and suppose that there is an arc from vertex x to vertex y and an arc from y to z. If the arc between vertex x and vertex z were directed from z to x, then there would be a directed cycle in T. Hence, that arc must be directed from x to z.   ◇

Example 9.3.2: In a transitive tennis tournament, there are no ambiguities regarding the relative strengths of the players. That is, if player x beat y and player y beat z, then player x will have beaten z. The winner beat everyone, and the second place finisher beat everyone except the winner, etc.

TERMINOLOGY: In a tournament, if there is an arc directed from x to y, we say x **dominates** (or **beats**) y.

Notice that the number of matches that a player x wins is equal to $outdegree(x)$. The next proposition shows that the player rankings from a transitive tournament are clear-cut and indisputable.

Score Sequence

DEFINITION: The **score sequence** of a tournament is the outdegree sequence of the digraph.

Proposition 9.3.2: *An n-vertex tournament T is transitive if and only if the score sequence of T is $\langle 0, 1, \ldots, n-1 \rangle$.*

Proof: ($\Rightarrow$) Suppose that tournament T is transitive. Since T is a simple digraph, it suffices to show that all of its n vertices have different outdegrees. Let x and y be any two vertices in T, and assume, without loss of generality, that (x, y) is an arc in T.

Next, let U be the set of all vertices in T that are adjacent *from* vertex y. That is $U = \{u \in V_T | (y, u)$ is an arc in $T\}$, from which it follows that $outdegree\ (y) = |U|$. For each $u \in U$, the transitivity of tournament T implies that (x, u) is an arc in T. Thus, $outdegree(x) \geq |U| + 1 > outdegree(y)$.

($\Leftarrow$) Suppose that the score sequence of tournament T is $\langle 0, 1, \ldots, n-1 \rangle$. Relabel the vertices of T so that $outdegree(v_i) = i$ for $i = 0, 1, \ldots, n-1$. Then there is an arc from vertex v_{n-1} to each of the vertices $v_0, v_1, \ldots, v_{n-2}$, there is an arc from vertex v_{n-2} to each of the vertices $v_0, v_1, \ldots, v_{n-3}$, and so on. In other words, there is an arc from vertex v_i to vertex v_j if and only if $i > j$, from which transitivity follows. $\Diamond$

Corollary 9.3.3: *Every transitive tournament contains exactly one Hamiltonian path.*

Proof: The relabeling of the vertices given in the proof of Proposition 9.3.2 produces the vertex sequence of a Hamiltonian path. Its uniqueness is left as an exercise (see Exercises).

$\Diamond$

Remark: This last result reaffirms the notion that a transitive tournament has an indisputable ranking. Although there is no clear-cut ranking for non-transitive tournaments, one can use additional criteria to choose the "best" of the Hamiltonian paths. The next result guarantees that there is at least one from which to choose.

Proposition 9.3.4: [*Rédei, 1934*] *Every tournament contains a Hamiltonian path.*

Proof: Let T be an n-vertex tournament, and let $P = \langle v_1, v_2, \ldots, v_l \rangle$ be a path in T of maximum length. By way of contradiction, suppose that $l < n$. Then there is some vertex w not on path P. By the maximality of P, the arc between vertices w and v_1 must be directed from v_1 to w, and, similarly, (w, v_l) must be an arc in T. If (w, v_2) were an arc in T, then the path $\langle v_1, w, v_2, \ldots, v_l \rangle$ would contradict the maximality of P. It follows that (v_2, w) is an arc, and similarly, that (v_j, w) is an arc in T, for all $j = 1, \ldots, l-1$. However, this implies that the vertex sequence $\langle v_1, v_2, \ldots, v_{l-1}, w, v_l \rangle$ forms a path of length $l+1$, which contradicts the maximality of P. $\Diamond$

Remark: *Tournament Rankings*. Given the results of a round-robin competition, one would like to rank the teams or at least pick a clear-cut winner. Ranking by the order of a Hamiltonian path will not work unless the path is unique. Ranking by score sequence usually results in ties, and a team that beats only a few teams, with those few teams having lots of wins, might deserve a better ranking. This suggests considering the *second-order score sequence*, where each team's score is the sum of the outdegrees of the teams it beats. One can continue by defining the kth-order score sequences recursively. See [M068] for more details and references.

The following result shows that the vertex-sets of the strong components can be ranked.

Proposition 9.3.5: *The condensation of a tournament is transitive.*

Proof: This follows from Propositions 9.1.2 and 9.3.1, since the condensation of a tournament is a tournament. $\Diamond$

Application 9.3.1:   *Partitioning a Sports League into Divisions*   Suppose that an adult soccer league is to consist of twelve teams. To avoid having a few teams dominate the rest of the league, the organizers would like to group the teams into divisions of comparable strength. One way to achieve this is to have a round-robin competition involving the twelve teams and then use the strong components of the outcome to form the divisions. For an overview of round-robin sports competition scheduling and its relationship to edge-coloring, see [BuWeKi04].

Regular Tournaments and Kelly's Conjecture

DEFINITION: A **regular tournament** is a tournament T in which all scores are the same, i.e., there is an integer s such that $outdegree(v) = s$ for all vertices $v \in V_T$.

Example 9.3.3: A regular 7-vertex tournament is shown in Figure 9.3.2 below. Notice that the arc-set of this regular tournament can be partitioned into three Hamiltonian cycles, as shown on the right of the figure.

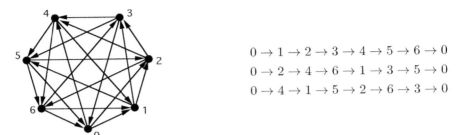

$$0 \to 1 \to 2 \to 3 \to 4 \to 5 \to 6 \to 0$$
$$0 \to 2 \to 4 \to 6 \to 1 \to 3 \to 5 \to 0$$
$$0 \to 4 \to 1 \to 5 \to 2 \to 6 \to 3 \to 0$$

Figure 9.3.2   A regular 7-vertex tournament.

Kelly's Conjecture (see [Mo68]). The arc-set of a regular n-vertex tournament can be partitioned into $(n-1)/2$ subsets, each of which induces a Hamiltonian cycle.

Remark: B. Alspach has shown that the conjecture is true for $n \geq 9$. For this result and other evidence for the conjecture, see [BeTh81], [Zh80], [ThS2], and [Hä93]. Partitioning the arc-set is a special case of *graph factorization*, which is discussed in §8.4.

Tournament Kings

H. G. Landau [La53] developed the idea of a *king* in attempting to determine the "strongest" individuals in certain animal societies in which there exists a pairwise *pecking* relationship.

DEFINITION: A **king** in a tournament T is a vertex x such that every vertex in T is reachable from x by a directed path of length 2 or less.

Proposition 9.3.6: *Every tournament has a king.*

Proof: We use induction on the number of vertices. The lone vertex in a 1-vertex tournament is trivially a king. Assume for some $n \geq 1$ that every n-vertex tournament has a king. Let T be an $(n + 1)$-vertex tournament, and let v be any vertex in T. By the induction hypothesis, the tournament $T - v$ has a king, say x. Let D be the set containing vertex x and all vertices that x dominates. If any vertex in D dominates vertex v, then x is a king in tournament T. Otherwise, v dominates every vertex in D, in which case, v is a king in T.   ◇

Remark: The article by S. Maurer [Ma80] generated early interest in the topic of kings. For a discussion of kings and generalizations in other digraphs, see [Re96].

Voting and the Condorcet Paradox

An n-vertex tournament is a natural way to model a voter's pairwise preferences among n candidates (or alternatives). Each of the n candidates is represented by a vertex, and an arc is directed from x to y if the voter prefers candidate x to candidate y.

DEFINITION: Let T be a set of tournaments, each having the same vertex-set V. The **majority digraph** D of set T has vertex-set V, and there is an arc directed from vertex x to vertex y in D if and only if x dominates y in a majority of the tournaments.

Remark: If each of the tournaments represents one voter's preferences, then the resulting majority digraph represents the voters' preferences under majority voting. If there is an odd number of voters or there are no ties, then the majority digraph is a tournament.

Example 9.3.4: Figure 9.3.3 shows the majority digraph of a set of three transitive 5-vertex tournaments, representing three voters' pairwise preferences for five candidates. Each vertical line of vertices represents a tournament. Each vertex dominates exactly the vertices below it. For example, in the first tournament on the left, vertex C dominates exactly vertices D and E, and vertex D is dominated by exactly vertices A, B, and C.

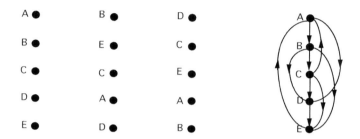

Figure 9.3.3   **The majority digraph of three tournaments.**

DEFINITION: The **Condorcet winner** is a candidate x such that for every other candidate y, x is preferred over y by a majority of the voters.

Thus, a Condorcet winner corresponds to a vertex in the majority digraph that has an arc directed *to* every other vertex.

DEFINITION: The **Condorcet paradox** occurs when the majority digraph of a set of tournaments contains a directed cycle.

Example 9.3.4 continued: The existence of the directed cycle $\langle A, B, C \rangle$ in the majority digraph of Figure 9.3.3 illustrates the Condorcet paradox. We see that a majority of voters prefer A to B, a majority prefer B to C, and yet, a majority prefer C to A. Thus, if candidate A wins, then a majority of the voters would have been happier if C had won, if candidate C wins, then a majority of the voters would have been happier if B had won, and if candidate B wins, then a majority of the voters would have been happier if A had won.

Remark: There is a rich source of literature in voting theory appearing in periodicals such as *Public Choice*, *Social Choice and Welfare*, and *Mathematical Social Sciences*.

EXERCISES for Section 9.3

9.3.1[S] Group the 3-vertex tournaments into isomorphism classes.

9.3.2 Construct a 5-vertex tournament with at least two vertices of maximum outdegree.

9.3.3[S] Determine the number of non-isomorphic 4-vertex tournaments.

9.3.4 Show that every tournament has at most one vertex of outdegree 0.

9.3.5 Prove that every vertex of maximum outdegree in a tournament is a king.

9.3.6 Let T be a tournament having no vertex with indegree 0, and let x be a king. Prove that there is some vertex that dominates x and is also a king.

9.3.7[S] Prove that a transitive tournament with at least two vertices cannot be strongly connected.

9.3.8 Complete the proof of Corollary 9.3.3.

9.3.9 Prove that every tournament is either strongly connected or can be transformed into a strongly connected tournament by reversing the direction on one arc.

9.3.10 Prove that a non-transitive tournament contains at least three Hamiltonian paths.

9.4   PROJECT SCHEDULING

A digraph model can be used as an aid in the scheduling of large complex projects that consist of several interrelated activities. Typically, some of the activities can occur simultaneously, but some activities cannot begin until certain others are completed. For instance, in building a house, the electrical work cannot start until the framing is complete. Two desirable objectives are to schedule the activities so that the project, completion time is minimized and to identify those activities whose delay will delay the overall completion time.

The Critical-Path Method for Project Scheduling

If the duration of each activity in the project is known in advance, then the **criticalpath method (CPM)** can be used to solve this problem. CPM can also be used to determine how long each activity in the project can be delayed without delaying the completion of the project.

One way to represent the precedences among the various activities is to use a digraph called an **AOA (activity-on-arc) network**. Each arc in the digraph represents an activity (or task), with its *head* indicating the direction of progress in the project. The weight of the arc is the duration time of that activity. Each vertex in the AOA network represents an event that signifies the completion of one or more activities and the beginning of new ones.

TERMINOLOGY: Activity A is said to be a **predecessor** of activity B if B cannot begin until A is completed.

Example 9.4.1: The next few figures illustrate how various precedences and events may be depicted in an AOA network. In each figure, the activities are represented by arcs $A, B,$ and C, and the events are represented by vertices 1, 2, and 3.

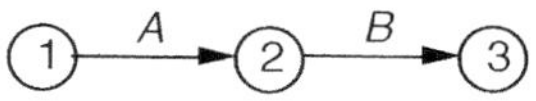

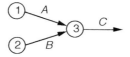

Figure 9.4.1   Event 2 marks the completion of activity A and the start of B.

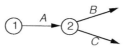

Figure 9.4.2   Activities A and B are predecessors of activity C.

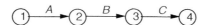

Figure 9.4.3   Activity A is the predecessor of both B and C.

When listing the predecessors of an activity, only the *immediate* predecessors are specified. For instance, in the figure below, activity A is an immediate predecessor of activity B, and activity B is an immediate predecessor of activity C. The predecessor list for C does not include activity A, since its precedence is implied.

Figure 9.4.4   Activity A is not listed as the predecessor of C.

Remark: Notice that if there were a directed cycle in an AOA network, then none of the activities corresponding to the arcs on that cycle could ever begin. Thus, if an AOA network represents a feasible project, it must be an *acyclic* digraph.

Given a list of activities and predecessors, an AOA network representation of a project can be constructed according to the following rules:

1. Vertex 1 represents the event marking the start of the project, and vertex n marks the end of the project.

2. Each activity that has no predecessors is represented by an arc incident from the start vertex.

3. The vertices in the network are numbered so that the vertex representing the completion of an activity (i.e., the *head* of that arc) has a larger number than the vertex that represents the beginning of that activity (the *tail* of that arc).

4. Each activity corresponds to exactly one arc in the network.

5. Two vertices can be joined by at most one arc (i.e., no multi-arcs).

Adhering to these rules sometimes requires using an artificial arc representing a *dummy activity* with zero duration time. For instance, if activities A and B can be performed concurrently, but they are both predecessors of activity C, then their relationship can be represented by introducing a dummy activity as shown in Figure 9.4.5. Notice that according to the figure, activity C cannot begin until both A and B have been completed.

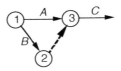

Figure 9.4.5   Using a dummy arc so that A and B precede C.

Application 9.4.1:   *Scheduling the Construction of a House*   Suppose that the tasks (activities) remaining to complete the construction of a house are as shown in the table.

Activity	Predecessors	Duration (days)
$A =$ Wiring	—	6
$B =$ Plumbing	—	9
$C =$ Walls & Ceilings	A, B	8
$D =$ Floors	A, B	7
$E =$ Carpeting	D	10
$F =$ Interior decorating	C, E	12

In the AOA network shown in Figure 12.4.6, the activity and its duration are listed alongside each arc.

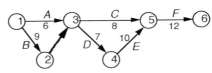

Figure 9.4.6   AOA network for completing the house construction.

The critical-path method requires the calculation of the earliest and latest times at which each event can occur.

NOTATION: Let $wt(i, j)$ denote the weight of arc (i, j), that is, the duration of the corresponding activity.

Calculating the Earliest Event Times

NOTATION: Let $ET(i)$ denote the earliest time at which the event corresponding to vertex i can occur.

TERMINOLOGY: A vertex j is an ***immediate predecessor*** of a vertex i if there is an arc from j to i.

NOTATION: Let $pred(i)$ be the set of immediate predecessors of vertex i.

The calculation of the earliest times for the vertices begins by setting event time $ET(1) = 0$. The value of $ET(i)$ for each of the remaining vertices depends on $ET(j)$ for all immediate predecessors j of vertex i.

In particular, suppose that vertex j is any immediate predecessor of vertex i. Then the event corresponding to vertex i cannot occur before $ET(j) + wt(j, i)$. Since this is true for every immediate predecessor of vertex i, $ET(i)$ is the maximum of these sums taken over all the immediate predecessors of i, as illustrated in Figure 9.4.7.

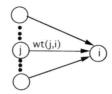

Figure 9.4.7 $ET(i) = \max_{j \in pred(i)} \{ET(j) + wt(j, i)\}$.

Example 9.4.2: In the house-construction example,

$$
\begin{aligned}
ET(2) &= ET(1) + 9 = 9 \\
ET(3) &= \max\{ET(1) + 6 = 6, ET(2) + 0 = 9\} = 9 \\
ET(4) &= ET(3) + 7 = 16 \\
ET(5) &= \max\{ET(3) + 8 = 17, ET(4) + 10 = 26\} = 26 \\
ET(6) &= ET(5) + 12 = 38
\end{aligned}
$$

Thus, the earliest completion time for the construction is 38 days.

Proposition 9.4.1: *Let D be an AOA network with vertices $1, 2, \ldots, n$. Then the earliest time $ET(i)$ for the event i to occur is the length of the longest directed path in the network from vertex 1 to vertex i.*

Proof: Since the digraph is acyclic, the longest path from vertex 1 to itself is 0, which establishes the base of an induction proof on i. The induction step follows from the way $ET(i)$ is computed from the ET values of its immediate predecessors. The details are left as an exercise (see Exercises). ◇

Proposition 9.4.2: *Let D be an acyclic digraph. Then there is a vertex whose indegree is 0, and there is a vertex whose outdegree is 0.*   ◇ (*Exercises*)

The following algorithm, which computes the ET values, is based on the iterative application of Proposition 9.4.2. In each iteration, a vertex v with indegree 0 is selected, and the ET values of its successor vertices are updated accordingly. Vertex v is then deleted from the digraph, and the next iteration begins. An induction argument can be used to show that the algorithm does indeed compute the longest paths from the start vertex to each other vertex (see Exercises).

Algorithm 9.4.1:   **Computation of Earliest Event Time**

Input: an n-vertex AOA network N with vertices $1, 2, \ldots, n$.
Output: the earliest event times $ET(1), ET(2), \ldots, ET(n)$.
    [*Initialization*]
    Initialize a working copy $D := N$.
    For $i = 1$ to n
        $ET(i) := 0$
    [*Begin computation*]
    For $j = 1$ to n
        Let v be a vertex with indegree 0.
        For each arc (v, w)
            Set $ET(w) := \max\{ET(w), ET(v) + wt(v, w)\}$.
        $D := D - v$
    Return $ET(1), ET(2), \ldots, ET(n)$.

Calculating the Latest Event Times

NOTATION: Let $LT(i)$ denote the latest time at which the event corresponding to vertex i can occur without delaying the completion of the project.

TERMINOLOGY: A vertex j is an **immediate successor** of a vertex i if there is an arc from i to j.

NOTATION: Let $succ(i)$ be the set of immediate successors of vertex i.

The calculation of the latest time $LT(i)$ is similar to the ET calculation but works backward from the end-of-project vertex n. It begins by setting $LT(n) = ET(n)$.

Next, suppose that vertex j is an immediate successor of vertex i. If the event corresponding to vertex i occurs after $LT(j) - wt(i, j)$, then event j will occur after $LT(j)$, thereby delaying the completion of the project. Since this is true for any immediate successor of i, $LT(i)$ is the minimum of these differences taken over all immediate successors of i, as illustrated in Figure 9.4.8.

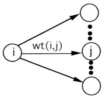

Figure 9.4.8 $LT(i) = \min_{j \in succ(i)}\{\boldsymbol{LT}(j) - \boldsymbol{wt}(i, j)\}.$

Example 9.4.3: Continuing with the house-construction example,

$$
\begin{aligned}
LT(6) &= 38 \\
LT(5) &= LT(6) - 12 = 26 \\
LT(4) &= LT(5) - 10 = 16 \\
LT(3) &= \min\{LT(4) - 7 = 9, LT(5) - 8 = 18\} = 9 \\
LT(2) &= LT(3) - 0 = 9 \\
LT(1) &= \min\{LT(2) - 9 = 0, LT(3) - 6 = 3\} = 0
\end{aligned}
$$

Algorithm 9.4.2, which computes the LT values, is analogous to Algorithm 9.4.1.

> **Algorithm 9.4.2:**   **Computation of Latest Event Time**
>
> *Input*: an n-vertex AOA network N with vertices $1, 2, \ldots, n$ and with earliest project completion time K.
>
> *Output*: the latest event times $LT(1), LT(2), \ldots, LT(n)$.
>
> $\quad$ [*Initialization*]
>
> $\quad$ Initialize a working copy $D := N$.
>
> $\quad$ For $i = 1$ to n
>
> $\qquad$ $LT(i) := K$
>
> $\quad$ [*Begin computation*]
>
> $\quad$ For $j = 1$ to n
>
> $\qquad$ Let v be a vertex with outdegree 0.
>
> $\qquad$ For each arc (w, v)
>
> $\qquad\qquad$ Set $LT(w) := \min\{LT(w), LT(w) - wt(w, v)\}$.
>
> $\qquad$ $D := D - v$
>
> $\quad$ Return $LT(1), LT(2), \ldots, LT(n)$.

Determining the Critical Activities

DEFINITION: The **total float** of an activity corresponding to arc (i, j), denoted $TF(i, j)$, is the amount by which the duration $wt(i, j)$ of activity (i, j) can be increased without delaying the completion of the project.

Equivalently, the total float is the amount by which the starting time of activity (i, j) can be delayed without delaying the completion of the overall project.

The total float is easily calculated once the earliest and latest event times have been determined.

Proposition 9.4.3: *The total float for activity (i, j) is given by $TF(i, j) = LT(j) - ET(i) - wt(i, j)$.*

Proof: Suppose that the occurrence of event i or the duration of activity (i, j) is delayed by k time units. Then activity (i, j) will be completed at time $ET(i) + k + wt(i, j)$. Thus, the largest value for k without incurring a delay in the project is the value that satisfies $ET(i) + k + wt(i, j) = LT(j)$. $\qquad\qquad\qquad\qquad\qquad\qquad\qquad\qquad\qquad\diamond$

Example 9.4.4: The earliest and latest event times for the house-construction example are shown in the table below on the left. The total float for each activity is calculated using Proposition 9.4.3 and is shown on the right.

Vertex	$ET(i)$	$LT(i)$
1	0	0
2	9	9
3	9	9
4	16	16
5	26	26
6	38	38

$TF(A) = TF(1, 3) = LT(3) - ET(1) - 6 = 3$
$TF(B) = TF(1, 2) = LT(2) - ET(1) - 9 = 0$
$TF(C) = TF(3, 5) = LT(5) - ET(3) - 8 = 9$
$TF(D) = TF(3, 4) = LT(4) - ET(3) - 7 = 0$
$TF(E) = TF(4, 5) = LT(5) - ET(4) - 10 = 0$
$TF(F) = TF(5, 6) = LT(6) - ET(5) - 12 = 0$

Notice that there is some leeway for activities A and C but none for the other activities. Identifying the activities that have this kind of leeway would enable a contractor to shift workers from such an activity to one that is threatening the delay of the overall project completion.

DEFINITION: A **critical activity** is an activity whose total float equals zero.

DEFINITION: A **critical path** is a directed path from the start vertex to the finish vertex such that each arc on the path corresponds to a critical activity.

Example 9.4.5: For the house-construction project, the path $1 \to 2 \to 3 \to 4 \to 5 \to 6$ is a critical path. This is the only critical path in this example, but in general, the critical path is *not unique*.

Proposition 9.4.4: *Let D be an AOA network. A directed path from the start vertex to the finish vertex is a critical path if and only if it is a longest path from start to finish.*
                                                                    ◇ *(Exercises)*

EXERCISES for Section 9.4

In Exercises 9.4.1 through 9.4.3, a table listing the tasks of a project, along with their duration times and immediate predecessors is given. Apply CPM to determine the earliest starting time of each task, the earliest possible completion time of the entire project, and the critical tasks.

9.4.1^S

Task	a	b	c	d	e	f	g
Duration	10	7	5	3	2	1	14
Predecessors	—	—	a	c	d	b,e	e,f

9.4.**2**

Task	a	b	c	d	e	f	g	h
Duration	10	5	3	4	5	6	5	5
Predecessors	—	—	b	a,c	a,c	d	e	f,g

9.4.**3**

Task	a	b	c	d	e	f	g	h	i	j
Duration	7	3	2	8	4	6	1	10	5	9
Predecessors	—	a	a	c	b,d	e	d	g	f,g	h,i

9.4.4 A student must complete 10 courses before he or she can graduate in applied mathematics. The courses and their prerequisites are listed in the following table. Apply CPM to determine the minimum number of semesters needed to graduate.

Course	Prerequisites
C1 = Calculus 1	None
C2 = Calculus 2	C1
DM = Discrete Math	C1
C3 = Calculus 3	C2
A = Algorithms	C1, DM
GT = Graph Theory	DM
DE = Differential Equations	C2
S = Statistics	C1
P = Probability	C2, DM
LA = Linear Algebra	DM, C3

9.4.**5** Complete the proof of Proposition 9.4.1.

9.4.**6**^S Prove Proposition 9.4.2.

9.4.**7** Prove that the Earliest-Time Algorithm 9.4.1 is correct.

9.4.**8** Prove that the Latest-Time Algorithm 9.4.2 is correct.

9.4.9$^\mathrm{S}$ Prove Proposition 9.4.4.

9.4.10 [*Computer Project*] Write a computer program that takes as input an n-vertex AOA network and that produces as output the earliest and latest event times and the total float of each activity.

9.5   FINDING THE STRONG COMPONENTS OF A DIGRAPH

The backbone of the algorithm for finding the strong components of a digraph is *depth-first search*, just as it was for the block-finding algorithm of §5.4. The strategy for the one-pass algorithm given here can also be used to convert the block-finding algorithm into a one-pass algorithm (see Exercises).

Depth-First Search Revisited

TERMINOLOGY: An output tree produced by executing a depth-first search on a digraph is sometimes referred to as a **dfs-tree**.

TERMINOLOGY: A **non-tree arc** of a dfs-tree T is an arc that is not part of tree T but whose endpoints are.

Remark: Although the depth-first search algorithm in §4.2 works equally well for digraphs, the version presented in this section is better suited for finding strong components. The vertex stack used in Algorithm 9.5.1 makes it easier to examine and process the non-tree arcs, which is essential for Algorithm 9.5.2.

TERMINOLOGY: A tree vertex v is said to be **completely processed** if all arcs incident from v and all arcs incident from descendants of v have been examined.

TERMINOLOGY: If an arc incident from a vertex v is currently being examined, then the vertex v is said to be **active**.

NOTATION: Recall from §4.2 that the integer-label assigned to a vertex w during depth-first search is denoted *dfnumber*(w).

During the execution of Algorithm 9.5.1, there are two possibilities that can occur when an arc is examined. If the arc is a frontier arc (i.e., its head is not yet in the dfs-tree), then (i) its head vertex is assigned the next *dfnumber*, (ii) the arc and vertex are added to the growing dfs-tree, and (iii) the vertex becomes the active vertex. The second possibility is that the head vertex of the arc is already in the tree, in which case the arc does not get added to the tree (i.e., it is a non-tree arc), and the current vertex stays active.

To distinguish tree vertices from non-tree vertices, Algorithm 9.5.1 initializes *dfnumber*(w) at -1. The function *examined* keeps track of which edges have been examined by initializing *examined*(e) at FALSE, for each edge e, and setting it to TRUE when e is examined.

Algorithm 9.5.1:   **Depth-First Search of a Digraph**

Input: a digraph D and a starting vertex v.

Output: a rooted tree T with root v and a *dfnumber* for each $v \in v_T$.

[*Initialization*]

For each vertex $w \in V_D$

  $dfnumber(w) := -1$

For each arc $e \in E_D$

  $examined(e) := \text{FALSE}$

Initialize *vertexstack* to contain vertex v

Initialize tree T as vertex v.

$dfnumber(v) := 0$

Initialize *dfnumber* counter $dfnum := 1$

[*Begin processing*]

While *vertexstack* is not empty

  Let vertex t be $top(vertexstack)$

  While there are unexamined arcs incident from vertex t

    Let e be an unexamined arc incident from vertex t.

    $examined(e) := \text{TRUE}$

    Let vertex w be $head(e)$.

    If $dfnumber(w) = -1$ [*vertex w is not yet in tree T*]

      $dfnumber(w) := dfnum$

      $dfnum := dfnum + 1$

      Add edge e (and vertex w) to tree T.

      Push vertex w onto *vertexstack*.

      [*begin processing of vertex w*]

      $t := w$

  [*processing of vertex t is complete; remove it from vertexstack*]

Return tree T with its *dfnumbers*.

Classifying the Non-Tree Arcs of a DFS-Tree

REVIEW FROM §4.1: For tree-growing in digraphs, the non-tree arcs fall into three categories according to how their endpoints are related in the dfs-tree.

- A **back-arc** is directed from a vertex to one of its ancestors.

- A **forward-arc** is directed from a vertex to one of its descendants.

- A **cross-arc** is directed from a vertex to another vertex that is neither an ancestor nor a descendant.

Example 9.5.1: Figure 9.5.1 shows one possible result of executing Algorithm 9.5.1 on the digraph D shown on the left, starting at vertex u. The non-tree arcs of the dfs-tree T on the right are drawn as dashed lines. Arc a is a back-arc, b is a forward-arc, and c and d are both cross-arcs.

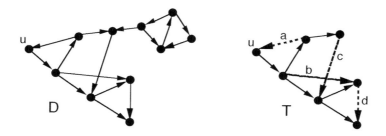

Figure 9.5.1   A digraph D and the non-tree arcs of an output tree T.

Each type of non-tree arc of a dfs-tree can be described in terms of the *dfnumbers* of the vertices. It follows directly from the definitions that a forward-arc is directed from a smaller *dfnumber* to a larger one, and a back-arc is directed from a larger *dfnumber* to a smaller one. The following proposition shows that a cross-arc is always directed from a larger *dfnumber* to a smaller one.

Proposition 9.5.1: *Let e be a cross-arc of a depth-first search tree from vertex x to vertex y. Then dfnumber(x) > dfnumber(y).*

Proof: By way of contradiction, suppose that *dfnumber(x) $\leq$ dfnumber(y)*. This means that vertex x was added to the search tree before vertex y. Immediately after x was added to the tree, arc e became a frontier arc. Since e is a non-tree arc, vertex y must have been labeled via some other frontier arc. But in each iteration after vertex x was labeled, until and including the iteration in which vertex y was labeled, the chosen frontier arc was directed either from vertex x or from a descendant of x. In either case, vertex y is a descendant of vertex x, which contradicts the definition of cross-arc.   ◇

Additional Preliminaries for the Strong-Component-Finding Algorithm

The strategy for finding mutually reachable vertices is to keep track of how the non-tree arcs connect the tree vertices. In §4.4 and in §5.4, a function *low* was used in determining the cut-vertices and the blocks, respectively, of an undirected graph, and it plays a similar role for Algorithm 9.5.2. The presentation given here is similar to those given in [ThSw92] and [Ba83], and additional details, including proofs of correctness, may be found in either reference. A somewhat different approach may be found in [St93].

TERMINOLOGY: Let T be a dfs-tree for a digraph, and let $v \in V_T$. Then *low(v)* is the smallest *dfnumber* among all vertices in V_T that are known to be in the same strong component as vertex v.

The examination of an arc directed from the active vertex t results either in assigning a *dfnumber* to a new vertex w and making w active or recognizing that the arc is a non-tree arc and updating *low(t)* appropriately.

Additional Variables and Initialization for Algorithm 9.5.2

placed(w): is TRUE if vertex w has been placed in a strong component, and is FALSE otherwise.

parent(v): the parent of vertex v in the dfs-tree.

holdstack: a stack that holds the vertices that have been processed but not yet placed in a strong component.

Initialization for Algorithm 9.5.2
   For each vertex $w \in V_D$
    $dfnumber(w) := -1$
    $low(w) := -1$
    $placed(w) :=$ FALSE
   For each arc $e \in E_D$
    $examined(e) :=$ FALSE
   Initialize *vertexstack* to contain vertex v
   Initialize tree T as vertex v.
   $dfnumber(v) := 0$
   Initialize *dfnumber* counter $dfnum := 1$
   Initialize strong component counter $k := 0$
   Initialize *holdstack* as empty.

Computation of *low*

When the algorithm assigns a *dfnumber* to a vertex w, it also initializes $low(w)$ to be that *dfnumber*. The *low* values change as the algorithm proceeds. After vertex v is completely processed, $low(v) = dfnumber(v)$ if and only if vertex v is the root of a subtree whose vertices form a strong component.

The value of $low(v)$ is updated according to the following rules. Their justification follows from the properties of the non-tree arcs, including the one established in Proposition 9.5.1.

1. If the processing of vertex v is complete and if $low(v) < dfnumber(v)$, then the value of $low(v)$ is carried back up to the parent of v. That is,
   $low(parent(v)) := \min\{low(parent(v)), low(v)\}$.

2. If vertex v is active and the arc being examined is a back-arc from v to w, then
   $low(v) := \min\{low(v), dfnumber(w)\}$.

3. If vertex v is active and the arc being examined is a cross-arc from v to w, then set
   $low(v) := \min\{low(v), dfnumber(w)\}$ if and only if w has not already been assigned to a strong component.

Algorithm 9.5.2:   **Strong-Component Finding**

Input: a digraph D and a starting vertex v.

Output: a dfs-tree T with root v;

        vertex-sets $S_1, S_2, \ldots, S_k$ of the strong components of tree T

Initialize variables as prescribed above.

[*Begin processing*]

While *vertexstack* is not empty

    Let vertex t be *top(vertexstack)*.

    While there are unexamined arcs incident from vertex t

        Let e be an unexamined arc incident from vertex t.

        examined(e) := *TRUE*

        Let vertex w be *head(e)*.

        If *dfnumber(w)* $= -1$ [vertex w is not yet in tree T]

            dfnumber(w) := *dfnum*

            dfnum := *dfnum* $+ 1$

            low(w) := *dfnumber(w)*

            Push vertex w onto *vertexstack*.

            $t := w$

        Else If *placed(w)* $=$ FALSE

            [vertex w has not yet been placed in a strong component]

            low(t) := min$\{$*low(t), dfnumber(w)*$\}$

    [processing of vertex t has been completed]

    If *low(t)* $=$ *dfnumber(t)*

        [vertex t is the root of a subtree of T whose vertices form a

           strong component]

        $k := k + 1$

        placed(t) := *TRUE*

        Initialize set S_k := $\{t\}$. [place vertex t in strong component S_k]

        While *holdstack* $\neq \emptyset$ AND *low(top(holdstack))* $\geq$ *dfnumber(t)*

            $z :=$ *top(holdstack)*

            placed(z) := TRUE [z and t are mutually reachable]

                $S_k ::= S_k \cup \{z\}$

            Pop *holdstack*.

        Pop vertex t from *vertexstack*.

    Else

        Push vertex t onto *holdstack*

        Pop *vertexstack*.

        If *parent(t)* exists (i.e., t is not the root of T)

            low(parent(t)) := min$\{$*low(parent(t)), low(t)*$\}$

    [depth-first search backs up to *parent(t)*]

Return tree T vertex-sets $S_1, S_2, \ldots, S_k$.

Remark: For a digraph D, the vertex-set of a dfs-tree T is not necessarily the vertex-set of one of the strong components of D, but it is always the union of the vertex-sets of one or more of the strong components of D (see Exercises). Algorithm 9.5.2 produces each of these vertex-sets. If the dfs-tree does not include all the vertices of digraph D, then Algorithm 9.5.2 will have found the vertex-sets of only some of the strong components. However, the algorithm can be restarted with the deletion subdigraph $D' = D - V_T$ to find additional strong components of D. The process can continue until all the components of digraph D are found.

EXERCISES for Section 9.5

For Exercises 9.5.1 through 9.5.4, draw the dfs-trees that result when Depth-First Search Algorithm 9.5.1 is applied to the given digraph using each of the specified vertices as the starting vertex. Whenever there is more than one frontier arc from which to choose, choose the one that is directed to the lexicographically smallest vertex. Also, classify all the non-tree arcs for each dfs-tree.

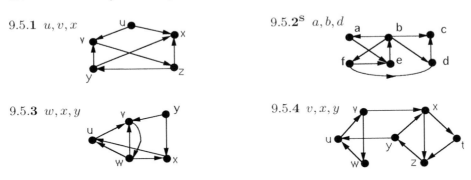

9.5.1 u, v, x

9.5.2^{S} a, b, d

9.5.3 w, x, y

9.5.4 v, x, y

For Exercises 9.5.5 through 9.5.8, verify the assertion of Proposition 9.5.1 for each of the cross-arcs in each of the dfs-trees from the specified exercise.

9.5.5 The graph of Exercise 9.5.1.

9.5.6^{S} The graph of Exercise 9.5.2.

9.5.7 The graph of Exercise 9.5.3.

9.5.8^{S} The graph of Exercise 9.5.4.

For Exercises 9.5.9 through 9.5.12, apply Strong-Component-Finding Algorithm 9.5.2 to the specified digraph. If there remain unlabeled vertices, reapply the algorithm on the subdigraph induced on these unlabeled vertices. Continue until all strong components have been found.

9.5.9 The graph of Exercise 9.5.1.

9.5.10^{S} The graph of Exercise 9.5.2.

9.5.11 The graph of Exercise 9.5.3.

9.5.12 The graph of Exercise 9.5.4.

9.5.13 Apply Strong-Component-Finding Algorithm 9.5.2 to the digraph in Figure 9.1.1 If there remain unlabeled vertices, reapply the algorithm on the subdigraph induced on these unlabeled vertices. Continue until all strong components have been found.

9.5.14^{S} Characterize those digraphs for which one application of Strong-Component Finding Algorithm 9.5.2 is guaranteed to find *all* of the strong components, regardless of where the algorithm starts.

9.5.15 Prove that the vertex-set of a dfs-tree for a digraph D is partitioned by the vertex-sets of one or more of the strong components of D.

9.5.16 Adapt the strategy in Strong-Component-Finding Algorithm 9.5.2 to convert the block-finding algorithm of §5.4 to a one-pass algorithm.

9.5.17 [*Computer Project*] Implement a modified version of Strong-Component-Finding Algorithm 9.5.2 that finds *all* of the strong components. Test the program on each of the digraphs in Exercises 9.5.1 through 9.5.4.

9.6   SUPPLEMENTARY EXERCISES

9.6.1 Draw all isomorphism types of simple digraphs with 4 vertices and 3 arcs.

9.6.2 Draw all isomorphism types of acyclic simple digraphs with 4 vertices and 4 arcs.

9.6.3 Draw all isomorphism types of acyclic simple digraphs with 5 vertices and 3 arcs.

9.6.4 Draw all isomorphism types of Eulerian simple digraphs with 4 vertices.

DEFINITION: A **self-converse digraph** is a digraph D such that the result of reversing every arc direction is isomorphic to D.

9.6.5 Draw all isomorphism types of self-converse simple digraphs on 3 vertices.

9.6.6 Draw all isomorphism types of self-converse simple digraphs on 4 vertices and 4 arcs.

DEFINITION: The **arc-complement of a simple digraph** D is the digraph $\overline{D}$ with the same vertex set V_D, such that $uv \in E_{\overline{D}}$ if and only if $uv \notin E_D$.

DEFINITION: A **self-complementary digraph** is a simple digraph D such that $\overline{D} \cong D$.

9.6.7 Draw all the isomorphism types of self-complementary simple digraphs on 3 vertices.

9.6.8 Draw all the isomorphism types of self-complementary simple digraphs on 4 vertices.

9.6.9 Let G be an undirected graph G without self-loops. Prove that there is an orientation of G that is acyclic.

9.6.10 Prove that a digraph is acyclic if and only if it is isomorphic to its condensation.

9.6.11 Prove that the condensation of any digraph has at least one vertex with indegree 0 and at least one with outdegree 0.

9.6.12 Draw a Hasse diagram for a 5-element poset that has exactly four linear extensions.

9.6.13 Draw all the isomorphism types of tournaments with 4 vertices.

9.6.14 Draw all the isomorphism types of tournaments with 5 vertices.

9.6.15 Do there exist n-vertex tournaments, for $n \geq 4$, that have a constant score sequence? If they do not exist, then explain why not, and if they do exist, then say as much as you can about such a tournament.

9.6.16 Draw a 5-vertex tournament so that every vertex is a king.

9.6.17 Prove or disprove: There exists an n-vertex tournament with $indegree(v) = outdegree(v)$ at each vertex v if and only if n is odd.

9.6.18 Let T be an n-vertex tournament. Prove that the following statements are equivalent.

(1) T is transitive.
(2) T is acyclic.
(3) T contains a unique Hamiltonian path.

Glossary

absorbing state in a Markov chain: a state s_i with $p_{ii} = 1$.

active vertex during a depth-first search: the tail of the arc that is currently being examined.

AOA (activity-on-arc) network: a digraph model of a project such that each arc in the digraph represents an activity (or task). The weight of the arc is the duration time of that activity. Each vertex represents an *event* that signifies the completion of one or more activities and the beginning of new ones.

arc-complement of a digraph D**:** the digraph $\overline{D}$ with the same vertex set V_D, such that $uv \in E_{\overline{D}}$ if and only if $uv \notin E_D$.

back-arc in a dfs-tree T: an arc whose tail is a descendant of its head.

communicating states in Markov chain: two states that are mutually reachable in the corresponding Markov diagram.

communication class in a Markov chain: a maximal subset of states that communicate with one another.

comparable elements in a poset $(S, \prec)$: two elements x and y such that either $x \prec y$ or $y \prec x$.

compatible total order with respect to a given partial order $\prec$ on S: a total order $\prec^t$ on set S such that $x \prec y \Rightarrow x \prec^t y$.

completely processed tree vertex in a depth-first search: a vertex v such that all arcs incident from v and all arcs incident from descendants of v have been examined.

condensation of a digraph D: the simple digraph D' with vertex-set $V_{D'} = \{s_1, s_2, \ldots, s_r\}$, such that there is an arc in digraph D' from vertex s_i to vertex S_j if and only if there is an arc in digraph D from a vertex in strong component S_i to a vertex in strong component S_j.

Condorcet winner: a candidate x such that for every other candidate y, x is preferred over y by a majority of the voters. Thus, a Condorcet winner corresponds to a vertex x in the majority digraph that has an arc directed *to* every other vertex.

Condorcet paradox: an occurrence of a directed cycle in the majority digraph of a set of tournaments (see §9.3).

connected digraph: a digraph whose underlying graph is connected. Some authors use the term *weakly connected* to describe such digraphs.

critical activity in an AOA network: an activity whose total float equals zero; any delay will delay the overall project completion.

critical path in an AOA network: a directed path from the start vertex to the finish vertex such that each arc on the path corresponds to a critical activity.

cross-arc in a dfs-tree T: an arc whose tail is neither an ancestor nor a descendant of its head.

cycle addition to a graph G: the addition of a cycle that has exactly one vertex in common with G.

dag: a mnemonic for *directed acyclic digraph.*

dfs-tree: an output tree resulting from a depth-first search.

digraph representation of a relation R on a finite set S: the digraph whose vertices correspond to the elements of S and whose arcs correspond to the ordered pairs in the relation. That is, an arc is drawn from vertex x to vertex y if $(x, y) \in R$.

directed tree: a digraph whose underlying graph is a tree.

dominates (or **beats**) a vertex y in a tournament: said of any vertex x that has an outgoing arc directed to y.

2-edge-connected graph: a connected graph that has no cut-edges.

forward-arc in a dfs-tree T: an arc whose tail is an ancestor of its head.

frontier arc for a rooted tree T: an arc whose tail is in T and whose head is not in T.

Hasse diagram representing the poset $(S, \prec)$: a graph drawing defined as follows:

(1) Each element of set S is represented by a vertex.

(2) If $x \prec y$, then the vertex for y is positioned higher than the vertex for x; and if there is no w such that $x \prec w$ and $w \prec y$, then an edge is drawn from vertex x upward to vertex y.

immediate predecessor of a vertex i in an AOA network: a vertex j such that there is an arc from j to i.

immediate successor of a vertex i in an AOA network: a vertex j such that there is an arc from i to j.

king in a tournament T: a vertex x such that every vertex in T is reachable from x by a directed path of length 2 or less.

linear extension of a poset $(S, \prec)$: a total order $\prec^t$ on S that is compatible with $\prec$.

majority digraph of a set $\mathcal{T}$ of tournaments, each having the same vertex-set V: a digraph D with vertex-set V, and such that there is an arc directed from vertex x to vertex y in D if and only if x dominates y in a majority of the tournaments.

Markov chain, discrete-time: a phenomenon whose behavior can be modeled by a sequence $\{X_t\}, t = 0, 1, 2, \ldots$, such that the **(one-step) transition probability** P_{ij} is the *conditional probability* given by

$$p_{ij} = prob(X_{t+1} = s_j \,|\, X_t = s_i), \quad \text{for } t = 1, 2, \ldots$$

—, irreducible: a Markov chain whose Markov diagram is strongly connected.

—, stationary: a Markov chain in which the transition probabilities are independent of time t.

Markov diagram of a given Markov chain: the digraph whose vertices represent the states and whose arcs represent the transitions from one state to the next. The arc from vertex i to vertex j is assigned the weight p_{ij}.

minimal element of a poset $(S, \prec)$: an element $m \in S$ such that for all other $x \in S, x \not\prec m$.

mutually reachable vertices u and v in a digraph D: there are both a directed u-v path and a directed v-u path in digraph D. A single vertex is regarded as reachable from itself.

non-tree arc of a dfs-tree T: an arc that is not part of tree T but whose endpoints are.

partial order: a relation that is reflexive, antisymmetric, and transitive.

partially ordered set: a pair (S, R), where R is a partial order on set S.

path addition to a graph G: the addition to G of a path between two existing vertices of G, such that the edges and internal vertices of the path are new.

poset: synonym for partially ordered set.

predecessor of an activity B in a project: an activity A that must be completed before B can begin.

—, immediate of a vertex i in an AOA network: a vertex j such that there is an arc from j to i.

reachability among states in a Markov chain: coincides with the digraph term applied to the corresponding Markov diagram (digraph).

recurrent state in a Markov chain: a state that is not a transient state.

root of a rooted tree: see *rooted tree*.

rooted tree: a directed tree having a distinguished vertex r, called the root, such that for every other vertex v, there is a (unique) directed r-v path.

score sequence of a tournament: the outdegree sequence of the tournament.

self-complementary digraph: a digraph D such that $\overline{D} \cong D$.

self-converse digraph: a digraph D such that the result of reversing every arc direction is isomorphic to D.

strong component of a digraph D: a maximal strongly connected subdigraph of D; a subdigraph induced on a maximal set of mutually reachable vertices.

strongly connected digraph D: every two vertices are mutually reachable in D.

strongly orientable graph: a graph for whose edge-set there exists an assignment of directions such that the resulting digraph is strongly connected.

topological sort: a process that produces a linear extension of a poset, or (informally) the resulting linear extension.

total float $TF(i, j)$ of an activity corresponding to arc (i, j) in an AOA network: the amount by which the duration $wt\,(i, j)$ of activity (i, j) can be increased without delaying the completion of the project.

total order on a set S: a partial order on S such that every two elements are comparable.

tournament: a digraph whose underlying graph is a complete graph.

—, **regular:** a tournament T in which there is an integer s such that $outdegree(v) = s$ for all vertices $v \in V_T$.

—, **transitive:** a tournament that is transitive as a digraph.

transient state in a Markov chain: a state s_i such that there exists a state s_j that is reachable from state s_i, but s_i is not reachable from s_j.

transition matrix for a Markov chain: a matrix whose ijth entry is the transition probability p_{ij}.

—, k-**step** $P^{[k]}$: the matrix whose ijth entry is the k-step transition probability $p_{ij}^{[k]}$.

transition probability in a Markov chain: the conditional probability p_{ij} that $X_{t+1} = s_j$, given that $X_t = s_i$.

—, k-**step** $p_{ij}^{[k]}$: the conditional probability that $X_{t+k} = s_j$, given that $X_t = s_i$.

transitive closure of a digraph D: the smallest transitive superdigraph of D.

transitive digraph: a digraph whose corresponding relation is transitive. That is, if there is an arc from vertex x to vertex y and an arc from y to z, then there is an arc from x to z.

Whitney-Robbins synthesis of a graph G from a graph H: a sequence of graphs, $G_0, G_1, \ldots, G_l$, where $G_0 = H, G_l = G$, and G_i is the result of a path or cycle addition to G_{i-1}, for $i = 1, \ldots, l$.

Chapter 10

NETWORK FLOWS AND APPLICATIONS

INTRODUCTION

Flow in a network can mean literally the flow of oil or water through a system of pipelines. More often in scientific writing, it refers to the flow of electricity, phone calls, email messages, commodities being transported across truck routes, or other such kinds of flow. Indeed, the richness of the network-flow model extends well beyond these applications.

The classical theory of network flows bridges several diverse and seemingly unrelated areas of combinatorial optimization. The equivalences among the celebrated *Max-Flow Min-Cut Theorem* of Ford and Fulkerson, the connectivity theorems of Menger, and the *Marriage Theorem* of Phillip Hall have led to the development of efficient algorithms for a number of practical problems. These include calculating the edge- and vertex-connectivity of a graph and finding *matchings*, which are used to solve various scheduling and assignment problems and which have applications in other areas of operations research, computer science, and engineering.

10.1   FLOWS AND CUTS IN NETWORKS

A pipeline network for transporting oil from a single source to a single sink is one prototype of a network model. Each arc represents a section of pipeline, and the endpoints of an arc correspond to the junctures at the ends of that section. The arc capacity is the maximum amount of oil that can flow through the corresponding section per unit of time. A network could just as naturally represent a system of truck routes for transporting commodities from one point to another, or it could represent a network of phone lines from one distribution center to another.

Single Source–Single Sink Capacitated Networks

DEFINITION: A *single source–single sink network* is a connected digraph that has a distinguished vertex called the *source*, with nonzero outdegree, and a distinguished vertex called the *sink*, with nonzero indegree.

TERMINOLOGY: A single source–single sink network with source s and sink (or *target*) t is often referred to as an *s-t network*.

DEFINITION: A *capacitated network* is a connected digraph such that each arc e is assigned a nonnegative weight $cap(e)$, called the *capacity* of arc e.

Remark: Later in this chapter, several applications with no apparent connection to networks are approached by transforming them into network problems, thereby demonstrating the robustness of the network model.

TERMINOLOGY: All of the networks discussed in this chapter are assumed to be capacitated *s-t* networks, even when one or both of the modifiers are dropped.

DEFINITION: Let v be a vertex in a digraph N.

- The *out-set* of v, denoted $Out(v)$, is the set of all arcs that are directed *from* vertex v. That is,
$$Out(v) = \{e \in E_N | tail(e) = v\}$$

- The *in-set* of v, denoted $In(v)$, is the set of all arcs that are directed *to* vertex v. That is,
$$In(v) = \{e \in E_N | head(e) = v\}$$

NOTATION: For any two vertex subsets X and Y of a digraph N, let $\langle X, Y \rangle$ denote the set of all arcs in N that are directed *from* a vertex in X to a vertex in Y. That is,

$$\langle X, Y \rangle = \{e \in E_N | tail(e) \in X \text{ and } head(e) \in Y\}$$

Remark: The examples and applications throughout this chapter involve networks with integer capacities, which simplifies the exposition. There is no additional complication if capacities are non-integer rational numbers. Such a network can be transformed into an "equivalent" one whose capacities are integers by multiplying each capacity by the least common multiple of the denominators of the capacities (see Exercises).

Example 10.1.1: A 5-vertex capacitated s-t network is shown in Figure 10.1.1. If $X = \{x, v\}$ and $Y = \{w, t\}$, then the elements of arc set $\langle X, Y \rangle$ are the arc directed from vertex x to vertex w and the arc directed from vertex v to sink t. The only element in arc set $\langle Y, X \rangle$ is the arc directed from vertex w to vertex v.

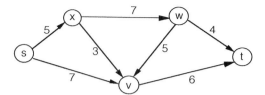

Figure 10.1.1   A 5-vertex capacitated network with source s and sink t.

Feasible Flows

DEFINITION: Let N be a capacitated s-t network. A (**feasible**) **flow** f in N is a function $f : E_N \rightarrow R^+$ that assigns a nonnegative real number $f(e)$ to each arc e such that:

1. (*capacity constraints*) $f(e) \leq cap(e)$, for every arc e in network N.

2. (*conservation constraints*) $\sum_{e \in In(v)} f(e) = \sum_{e \in Out(v)} f(e)$, for every vertex v in network N, other than source s and sink t.

TERMINOLOGY: Property 2 in the flow definition is called the **conservation-of-flow** condition. For an oil pipeline, it says that the total flow of oil going into any juncture (vertex) in the pipeline must equal the total flow leaving that juncture.

NOTATION: To distinguish visually between the flow and the capacity of an arc, we adopt the convention in drawings that when both numbers appear, the capacity will always be in bold and to the left of the flow.

Example 10.1.2: Figure 10.1.2 shows a feasible flow for the capacitated network of Example 10.1.1. Notice that the total amount of flow leaving source s equals 6, which is also the net flow entering sink t. The conservation of flow at every internal vertex in the network is intuitively consistent with this phenomenon. Later in this section, we establish the general result that outflow from the source equals inflow to the sink.

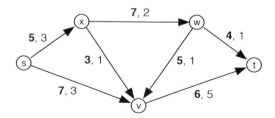

Figure 10.1.2   A flow for the example network.

DEFINITION: The **value of flow** f in a capacitated network, denoted $val(f)$, is the net flow leaving the source s, that is,

$$val(f) = \sum_{e \in Out(s)} f(e) - \sum_{e \in In(s)} f(e)$$

DEFINITION: A **maximum flow** f^* in a capacitated network N is a flow in N having the maximum value, i.e., $val(f) \leq val(f^*)$, for every flow f in N.

Cuts in s-t Networks

By definition, any nonzero flow must use at least one of the arcs in $Out(s)$. In other words, if all of the arcs in $Out(s)$ were deleted from network N, then no flow could get from source s to sink t. This is a special case of the following definition, which combines the concepts of *partition-cut* (from §4.5) and *s-t separating set* (from §5.3).

DEFINITION: Let N be an s-t network, and let V_s and V_t form a partition of V_N such that source $s \in V_s$ and sink $t \in V_t$. Then the set of all arcs that are directed from a vertex in set V_s to a vertex in set V_t is called an *s-t* **cut** of network N and is denoted (V_s, V_t).

Remark: Notice that the arc sets $Out(s)$ and $In(t)$ for an s-t network N are the s-t cuts $\langle\{s\}, V_N - \{s\}\rangle$ and $\langle V_N - \{t\}, \{t\}\rangle$, respectively.

Example 10.1.3: Figure 10.1.3 portrays the arc sets $Out(s)$ and $In(t)$ as s-t cuts, where $Out(s) = \langle\{s\}, \{x, v, w, t\}\rangle$ and $In(t) = \langle\{s, x, v, w\}, \{t\}\rangle$.

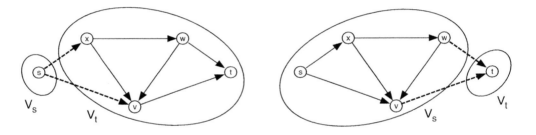

Figure 10.1.3   **Arc sets $Out(s)$ and $In(t)$ shown as s-t cuts.**

Example 10.1.4: A more general s-t cut $\langle V_s, V_t \rangle$ is shown in Figure 10.1.4, where $V_s = \{s, x, v\}$ and $V_t = \{w, t\}$.

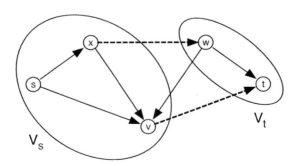

Figure 10.1.4   **The s-t cut $\langle\{s, x, v\}, \{w, t\}\rangle$.**

Proposition 10.1.1: *Let $\langle V_s, V_t \rangle$ be an s-t cut of a network N. Then every directed s-t path in N contains at least one arc in $\langle V_s, V_t \rangle$.*

Proof: Let $P = \langle s = v_0, v_1, v_2, \ldots, v_l = t \rangle$ be the vertex sequence of a directed s-t path in network N. Since $s \in V_s$ and $t \in V_t$, there must be a first vertex v_j on this path that is in set V_t (see Figure 10.1.5). Then the arc from vertex V_{j-1} to v_j is in $\langle V_s, V_t \rangle$.   ◇

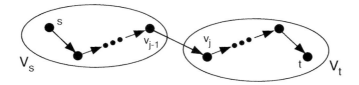

Figure 10.1.5   A directed s-t path with an arc in $\langle V_s, V_t \rangle$.

Relationship between Flows and Cuts

Similar to viewing the set $Out(s)$ of arcs directed from source s as the s-t cut $\langle \{s\}, V_N - \{s\} \rangle$, the set $In(s)$ may be regarded as the set of "backward" arcs relative to this cut, namely, the arc set $\langle V_N - \{s\}, \{s\} \rangle$. From this perspective, the definition of $val(f)$ may be rewritten as

$$val(f) = \sum_{e \in \langle \{s\}, V_N - \{s\} \rangle} f(e) - \sum_{e \in \langle V_N - \{s\}, \{s\} \rangle} f(e)$$

In other words, the value of any flow equals the total flow across the arcs of the cut $\langle \{s\}, V_N - \{s\} \rangle$ minus the flow across the arcs of $\langle V_N - \{s\}, \{s\} \rangle$. The next proposition generalizes this property to all s-t cuts. Its proof uses the following simple consequence of the definitions.

Lemma 10.1.2: *Let $\langle V_s, V_t \rangle$ be any s-t cut of an s-t network N. Then*

$$\bigcup_{v \in V_s} Out(v) = \langle V_s, V_s \rangle \cup \langle V_s, V_t \rangle \quad \text{and} \quad \bigcup_{v \in V_s} In(v) = \langle V_s, V_s \rangle \cup \langle V_t, V_s \rangle$$

Proof: For any vertex $v \in V_s$, each arc directed from v is either in $\langle V_s, V_s \rangle$ or in $\langle V_s, V_t \rangle$. (Figure 10.1.6 illustrates for a vertex v, the partition of $Out(v)$ into a 4-element subset of $\langle V_s, V_s \rangle$ and a 3-element subset of $\langle V_s, V_t \rangle$.) Similarly, each arc directed to vertex v is either in $\langle V_s, V_s \rangle$ or in $\langle V_t, V_s \rangle$.   ◇

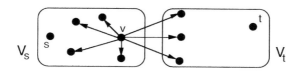

Figure 10.1.6 $\bigcup_{v \in V_s} Out(v) = \langle V_s, V_s \rangle \cup \langle V_s, V_t \rangle$.

Proposition 10.1.3: *Let f be a flow in an s-t network N, and let $\langle V_s, V_t \rangle$ be any s-t cut of N. Then*

$$val(f) = \sum_{e \in \langle V_s, V_t \rangle} f(e) - \sum_{e \in \langle V_t, V_s \rangle} f(e)$$

Proof: By definition, $val(f) = \sum_{e \in Out(s)} f(e) - \sum_{e \in In(s)} f(e)$, and by the conservation of

flow, $\sum_{e \in Out(v)} f(e) - \sum_{e \in In(v)} f(e) = 0$ for every $v \in V_s$ other than s. Thus,

$$val(f) = \sum_{v \in V_s} \left(\sum_{e \in Out(v)} f(e) - \sum_{e \in In(v)} f(e) \right) = \sum_{v \in V_s} \sum_{e \in Out(v)} f(e) - \sum_{v \in V_s} \sum_{e \in In(v)} f(e)$$

By Lemma 10.1.2,

$$\sum_{v \in V_s} \sum_{e \in Out(v)} f(e) = \sum_{e \in \langle V_s, V_s \rangle} f(e) + \sum_{e \in \langle V_s, V_t \rangle} f(e)$$

and

$$\sum_{v \in V_s} \sum_{e \in In(v)} f(e) = \sum_{e \in \langle V_s, V_s \rangle} f(e) + \sum_{e \in \langle V_t, V_s \rangle} f(e)$$

The result follows by substitution.   ◇

Example 10.1.5: The flow f and cut $\langle \{s, x, v\}, \{w, t\} \rangle$, shown in Figure 10.1.7, illustrate Proposition 10.1.3.

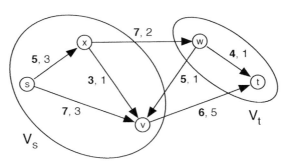

Figure 10.1.7 $6 = val(f) = \displaystyle\sum_{e \in \langle \{s,x,v\}, \{w,t\} \rangle} f(e) - \sum_{e \in \langle \{w,t\}, \{s,x,v\} \rangle} f(e) = 7 - 1.$

The next result confirms what was apparent earlier from intuition, namely, that the net flow out of source s equals the net flow into sink t.

Corollary 10.1.4: *Let f be a flow in an s-t network. Then*

$$val(f) = \sum_{e \in In(t)} f(e) - \sum_{e \in Out(t)} f(e)$$

Proof: Apply Proposition 10.1.3 to the s-t cut $In(t) = \langle V_N - \{t\}, \{t\} \rangle$.   ◇

DEFINITION: The **capacity of a cut** $\langle V_s, V_t \rangle$, denoted $cap\langle V_s, V_t \rangle$, is the sum of the capacities of the arcs in cut $\langle V_s, V_t \rangle$. That is,

$$cap\langle V_s, V_t \rangle = \sum_{e \in \langle V_s, V_t \rangle} cap(e)$$

DEFINITION: A ***minimum cut*** of a network N is a cut with the minimum capacity.

Example 10.1.6: The capacity of the cut shown in Figure 10.1.7 is 13, and the cut $\langle\{s,x,v,w\},\{t\}\rangle$, with capacity 10, is the only minimum cut.

The Maximum-Flow Problem and the Minimum-Cut Problem

The next few results demonstrate that the problems of finding the maximum flow in a capacitated network N and finding a minimum cut in N are closely related. In fact, these two optimization problems form a *max-min pair*, much like the max-min pair that appears in §5.3.

The first result provides an upper bound for the maximum-flow problem.

Proposition 10.1.5: *Let f be any flow in an s-t network N, and let $\langle V_s, V_t\rangle$ be any s-t cut. Then*

$$val(f) \leq cap\langle V_s, V_t\rangle$$

Proof: The following chain of inequalities, which begins with the assertion of Proposition 10.1.3, establishes the result.

$$
\begin{aligned}
val(f) &= \sum_{e\in\langle V_s,V_t\rangle} f(e) - \sum_{e\in\langle V_t,V_s\rangle} f(e) &&\text{(by Proposition 10.1.3)}\\[4pt]
&\leq \sum_{e\in\langle V_s,V_t\rangle} cap(e) - \sum_{e\in\langle V_t,V_s\rangle} f(e) &&\text{(by capacity constraints)}\\[4pt]
&= cap\langle V_s,V_t\rangle - \sum_{e\in\langle V_t,V_s\rangle} f(e) &&\text{(by definition of } cap\langle V_s,V_t\rangle)\\[4pt]
&\leq cap\langle V_s,V_t\rangle &&\text{(since each } f(e) \text{ is nonnegative)}
\end{aligned}
$$

$\diamond$

The following corollary resembles the weak-duality result in §5.3

Corollary 10.1.6: [***Weak Duality for Flows***] *Let f^* be a maximum flow in an s-t network N, and let K^* be a minimum s-t cut in N. Then*

$$val(f^*) \leq cap(K^*)$$

Proof: This is an immediate consequence of Proposition 10.1.5. $\diamond$

Corollary 10.1.7: [***Certificate of Optimality for Flows***] *Let f be a flow in an s-t network N and K an s-t cut, and suppose that $val(f) = cap(K)$. Then flow f is a maximum flow in network N, and cut K is a minimum cut.*

Proof: Let $\hat{f}$ be any feasible flow in network N. The maximality of flow f is implied by Proposition 10.1.5 and the premise, as follows.

$$val(\hat{f}) \leq cap(K) = val(f)$$

The minimality of cut K can be established with a similar argument (see Exercises). $\diamond$

Example 10.1.7: The flow for the example network shown in Figure 10.1.8 has value 10, which also is the capacity of the *s-t* cut $\langle\{s, x, v, w\}, \{t\}\rangle$. By Corollary 10.1.7, both the flow and the cut are optimal for their respective problems.

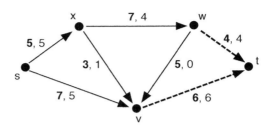

Figure 10.1.8   A maximum flow and minimum cut.

The final result of this section is used in the next section to establish the correctness of a classical algorithm for finding the maximum flow in a network.

Corollary 10.1.8: Let $\langle V_s, V_t \rangle$ be an *s-t* cut in a network N, and suppose that f is a flow such that

$$f(e) = \begin{cases} cap(e) & \text{if } e \in \langle V_s, V_t \rangle \\ 0 & \text{if } e \in \langle V_t, V_s \rangle \end{cases}$$

Then f is a maximum flow in N, and $\langle V_s, V_t \rangle$ is a minimum cut. $\Diamond$ *(Exercises)*

EXERCISES for Section 10.1

For Exercises 10.1.1 through 10.1.4, use trial and error and Certificate-of-Optimality Corollary 10.1.7 to find a maximum flow and a minimum cut in the given network.

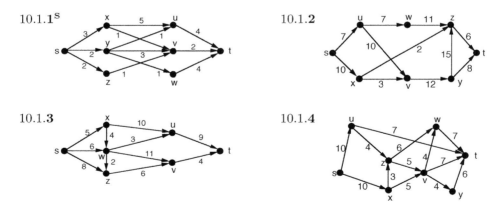

10.1.1$^\text{S}$

10.1.2

10.1.3

10.1.4

For Exercises 10.1.5 and 10.1.6, find all s-t cuts and identify all the minimum ones for the capacitated s-t network shown.

10.1.5$^\text{S}$

10.1.6

10.1.7 Show that if there exists no directed s-t path in an s-t network, then there does not exist a nonzero flow.

10.1.8 Some authors define an s-t network such that the source s must have in degree 0 and the sink t must have outdegree 0. Describe how a network whose source and sink do not satisfy these additional requirements can be transformed by graphical operations into one whose source and sink do meet the requirements.

10.1.9[S] a. Describe how a multi-source, multi-sink network can be transformed into a network with a single source and single sink, so that the methods in this section can be used to analyze it.

b. Use your transformation to solve the problem of finding the maximum total flow from the two sources to the three sinks in the network shown. Use Certificate-of-Optimality Corollary 10.1.7 to confirm your result.

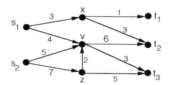

10.1.10 Use the transformation given in Exercise 10.1.9 to find the maximum total flow from the two sources to the two sinks in the network shown. Use Certificate-of-Optimality Corollary 10.1.7 to confirm your result.

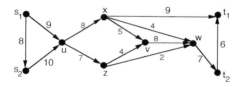

10.1.11 Use the transformation given in 10.1.9 to find the maximum total flow from the three sources X, Y, and Z to the three sinks A, B, and C in the network shown. Use Certificate-of-Optimality Corollary 10.1.7 to confirm your result.

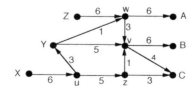

10.1.12 Find a maximum flow and a minimum cut for the network shown, by first transforming the network to one whose capacities are all integers. Use Certificate-of-Optimality Corollary 10.1.7 to confirm your result.

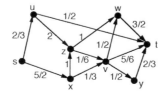

10.1.13^S a. Suppose that some of the vertices in an *s-t* network have capacities on the amount of flow that can pass through them. Describe how to transform such a network to the standard one used in this section.

b. Apply the transformation in part (a) to find the maximum flow in the network shown. The circled numbers represent vertex capacities. Apply Certificate-of-Optimality Corollary 10.1.7 to the transformed network to confirm your result.

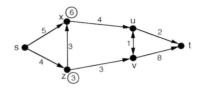

10.1.14 Let A be a subset of arcs in an *s-t* network N, and suppose that every directed *s-t* path contains at least one arc of set A.

a. Show that there exists an *s-t* cut $\langle V_s, V_t \rangle$ in network N such that $\langle V_s, V_t \rangle \subseteq A$.

b. Give an example where the *s-t* cut is a proper subset of A.

10.1.15 Complete the proof of Certificate-of-Optimality Corollary 10.1.7.

10.1.16^S Prove Corollary 10.1.8.

10.2   SOLVING THE MAXIMUM-FLOW PROBLEM

The basic idea of the algorithm presented in this section, originated by Ford and Fulkerson [FoFu62], is to increase the flow in a network iteratively until it cannot be increased any further. Each increase is based on a suitably chosen alternating sequence of vertices and arcs, called an *augmenting flow path*. Before introducing the most general such path, we consider the simplest and most obvious one.

Using Directed Paths to Augment Flow

Suppose that f is a flow in a capacitated *s-t* network N, and suppose that there exists a directed *s-t* path

$$P = \langle s, e, v_1, e_2, \ldots, e_k, t \rangle$$

in N, such that $f(e_i) < cap(e_i)$ for $i = 1, \ldots, k$. Then considering arc capacities only, the flow on each arc e_i can be increased by as much as $cap(e_i) - f(e_i)$. But to maintain the conservation-of-flow property at each of the vertices v_i, the increases on all of the arcs of path P must be equal. Thus, if Δ_P denotes this increase, then the largest possible value for Δ_P is $\min\{cap(e_i) - f(e_i)\}$.

Example 10.2.1: In the example network shown on the left in Figure 10.2.1, the value of the current flow is 6. Consider the directed s-t path $P = \langle s, x, w, t \rangle$. The flow on each arc of path P can increase by $\Delta p = 2$, and the resulting flow, which has value 8, is shown on the right.

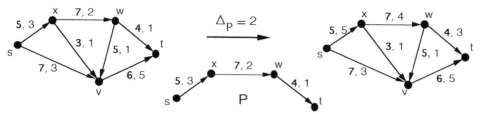

Figure 10.2.1   Using directed path P to increase the *flow* from 6 to 8.

Using the directed path $\langle s, v, t \rangle$, the flow can then be increased to 9. The resulting flow is shown in Figure 10.2.2. At this point, the flow cannot be increased any further along *directed s-t* paths, because each such path must use either the arc directed from source s to vertex x or the one from vertex v to sink t, and both these arcs have flow at capacity.

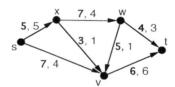

Figure 10.2.2   Current flow cannot be increased along a directed path.

However, the flow can be increased further. One way to accomplish this is to increase the flow on the arc from source s to vertex v by one unit, *decrease* the flow on the arc from w to v by one unit, and increase the flow on the arc from w to t by one unit. The decrease in the flow on the arc from w to v has the effect of redirecting one unit of the flow from vertex w, so that instead of going to vertex v, it goes directly to sink t. This makes room for the extra unit of flow on the arc from source s to vertex v.

Remark: Notice that in increasing the flow from 9 to 10 in the example network, the arcs whose flow have changed form an s-t path *if directions are ignored*. A generalization of a directed path helps in replacing the ad hoc strategy used above with a systematic one.

f-Augmenting Paths

DEFINITION: An s-t **quasi-path** in a network N is an alternating sequence

$$\langle s = v_0, e_1, v_1, \ldots, v_{k-1}, e_k, v_k = t \rangle$$

of vertices and arcs that forms an s-t path in the underlying graph of N.

TERMINOLOGY: For a given s-t quasi-path

$$Q = \langle s = v_o, e_1, v_1, \ldots, v_{k-1}, e_k, v_k = t \rangle$$

arc e_i is called a **forward-arc** if it is directed from vertex v_{i-1} to vertex v_i, and arc e_i is called a **backward-arc** if it is directed from v_i to v_{i-1}.

Remark: Clearly, a directed s-t path is a quasi-path whose arcs are all forward.

TERMINOLOGY NOTE: Some authors use the term *semipath* instead of *quasi-path*.

Example 10.2.2: On the *s-t* quasi-path shown in Figure 10.2.3, arcs a and b are backward, and the three other arcs are forward.

Figure 10.2.3   A quasi-path with two backward arcs.

DEFINITION: Let f be a flow in an *s-t* network N. An *f-augmenting path* Q is an *s-t* quasi-path in N such that the flow on each forward arc can be increased, and the flow on each backward arc can be decreased.

Thus, for each arc e on an *f*-augmenting path Q,

$$f(e) < cap(e), \quad \text{if } e \text{ is a forward arc}$$
$$f(e) > 0, \qquad \text{if } e \text{ is a backward arc}$$

NOTATION: For each arc e on a given *f*-augmenting path, let Δ_e be the quantity given by

$$\Delta_e = \begin{cases} cap(e) - f(e), & \text{if } e \text{ is a forward arc} \\ f(e), & \text{if } e \text{ is a backward arc} \end{cases}$$

TERMINOLOGY: The quantity Δ_e is called the **slack on arc** e. Its value on a forward arc is the largest possible increase in the flow, and on a backward arc, the largest possible decrease in the flow, *disregarding conservation of flow*.

Remark: Conservation of flow requires that the change in the flow on the arcs of an augmenting flow path be of equal magnitude. Thus, the maximum allowable change in the flow on an arc of quasipath Q is Δ_Q, where

$$\Delta_Q = \min_{e \in Q}\{\Delta_e\}$$

Notice that this definition of Δ_Q coincides with that of Δ_P, defined earlier, whenever the quasi-path Q is a directed path.

Example 10.2.3: For the example network in Figure 10.2.4, the current flow f has value 9, and the quasi-path $Q = \langle s, v, w, t \rangle$ is an *f*-augmenting path, with $\Delta_Q = 1$.

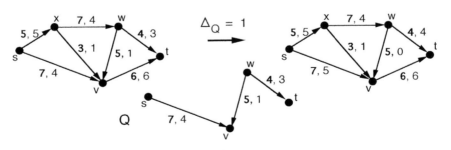

Figure 10.2.4   Using an *f*-augmenting path Q to increase the flow by $\Delta_Q = 1$.

TERMINOLOGY NOTE: To simplify the terminology, the prefix *quasi-* was not used in the definition of *f-augmenting path*. But to underscore that an *f*-augmenting path is not necessarily a directed path, the letter "Q" is often used for the name of these quasi-paths.

The following proposition summarizes how an f-augmenting path is used to increase the flow f in a network.

Proposition 10.2.1: [**Flow Augmentation**] *Let f be a flow in a network N, and let Q be an f-augmenting path with minimum slack Δ_Q on its arcs. Then the augmented flow $\hat{f}$ given by*

$$\hat{f}(e) = \begin{cases} f(e) + \Delta_Q, & \text{if } e \text{ is a forward arc of } Q \\ f(e) - \Delta_Q, & \text{if } e \text{ is a backward arc of } Q \\ f(e), & \text{otherwise} \end{cases}$$

is a feasible flow in network N, and $val(\hat{f}) = val(f) + \Delta_Q$.

Proof: Clearly $0 \le \hat{f}(e) \le cap(e)$, by the definition of Δ_Q. The only vertices through which the net flow may have changed are those vertices on the augmenting path Q. Thus, to verify that $\hat{f}$ satisfies conservation of flow, only the internal vertices of Q need to be checked.

For a given vertex v on augmenting path Q, the two arcs of Q that are incident on v are configured in one of four ways, as illustrated in Figure 10.2.5. In each case, the net flow into or out of vertex v does not change, thereby preserving the conservation-of-flow property.

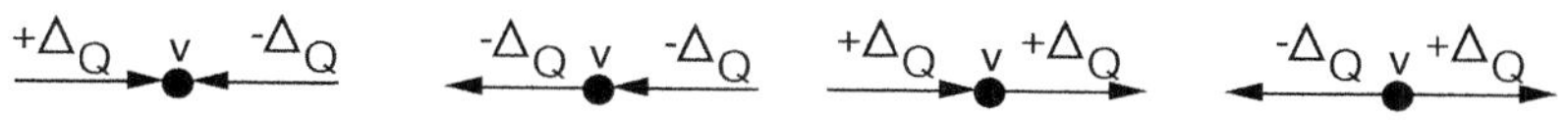

Figure 10.2.5   The four possibilities for an internal vertex v of a quasi-path.

It remains to be shown that the flow has increased by Δ_Q. The only arc incident on source s whose flow has changed is the first arc e_1 of augmenting path Q. If e_1 is a forward arc, then $\hat{f}(e_1) = f(e_1) - \Delta_Q$ and if e_1 is a backward arc, then $\hat{f}(e_1) = f(e_1) - \Delta_Q$. In either case,

$$val(\hat{f}) = \sum_{e \in Out(s)} \hat{f}(e) - \sum_{e \in In(s)} \hat{f}(e) = \Delta_Q + val(f)$$

$\diamondsuit$

Corollary 10.2.2: *Let f be an integer-valued flow in a network whose arc capacities are integers. Then the flow that results from each successive flow augmentation will be integer-valued.*

Max-Flow Min-Cut Theorem

Theorem 10.2.3: [**Characterization of Maximum Flow**] *Let f be a flow in a network N. Then f is a maximum flow in network N if and only if there does not exist an f-augmenting path in N.*

Proof: *Necessity* ($\Rightarrow$) Suppose that f is a maximum flow in network N. Then by Proposition 10.2.1, there is no f-augmenting path.

Sufficiency ($\Leftarrow$) Suppose that there does not exist an f-augmenting path in network N. Consider the collection of all quasi-paths that start at vertex s and have the following property: each forward arc on the quasi-path has positive slack (i.e., it can be increased), and each backward arc on the quasi-path has positive flow (i.e., it can be decreased). Let V_s be the union of the vertex-sets of these quasi-paths.

Since there is no f-augmenting path, it follows that sink $t \notin V_s$. Let $V_t = V_N - V_s$. Then $\langle V_s, V_t \rangle$ is an s-t cut of network N. Moreover, by definition of the sets V_s and V_t,

$$f(e) = \begin{cases} cap(e) & \text{if } e \in \langle V_s, V_t \rangle \\ 0 & \text{if } e \in \langle V_t, V_s \rangle \end{cases}$$

Hence, f is a maximum flow, by Corollary 10.1.8. $\diamondsuit$

Theorem 10.2.4: [*Max-Flow Min-Cut*] *For a given network, the value of a maximum flow is equal to the capacity of a minimum cut.*

Proof: The s-t cut constructed in the proof of Theorem 10.2.3 has capacity equal to value of the maximum flow. $\diamondsuit$

The outline of an algorithm (shown below) for maximizing the flow in a network emerges from Proposition 10.2.1 and Theorem 10.2.3.

Algorithm 10.2.1:   **Maximum Flow Outline**

Input: an s-t network N.
Output: a maximum flow f^* in network N.
    [*Initialization*]
    For each arc e in network N
      $f^*(e) := 0$
    [*Flow Augmentation*]
    While there exists an f^*-augmenting path in network N
       Find an f^*-augmenting path Q.
       Let $\Delta_Q = \min_{e \in Q} \{\Delta_e\}$.
       For each arc e of augmenting path Q
         If e is a forward arc
           $f^*(e) := f^*(e) + \Delta_Q$
         Else [e is a backward arc]
           $f^*(e) := f^*(e) - \Delta_Q$
    Return flow f^*.

Finding an f-Augmenting Path

The discussion of f-augmenting paths culminating in the flow-augmentation Proposition 10.2.1 provides the basis of a vertex-labeling strategy, due to Ford and Fulkerson, that finds an f-augmenting path, when one exists. Their labeling scheme is essentially basic *tree-growing* (§4.1). Here, the idea is to grow a tree of quasi-paths, each starting at source s. If the flow on each arc of these quasi-paths can be increased or decreased, according to whether that arc is *forward* or *backward*, then an f-augmenting path is obtained as soon as the sink t is labeled.

REVIEW FROM §4.1: A **_frontier arc_** is an arc e directed from a labeled endpoint v to an unlabeled endpoint w.

For constructing an f-augmenting path, frontier arc e is allowed to be backward (directed from vertex w to vertex v), and it can be added to the tree as long as it has slack $\Delta_e > 0$.

TERMINOLOGY: At any stage during tree-growing for constructing an f-augmenting path, let e be a frontier arc of tree T, with endpoints v and w. Then arc e is said to be **_usable_** if, for the current flow f, either

   e is directed from vertex v to vertex w and $f(e) < cap(e)$, or

   e is directed from vertex w to vertex v and $f(e) > 0$.

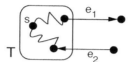

Figure 10.2.6   Frontier arcs e_1 and e_2 are _usable_ if $f(e_1) < cap(e_1)$ and $f(e_2) > 0$.

Remark: From this vertex-labeling scheme, any of the existing f-augmenting paths could result. But the efficiency of Algorithm 10.2.1 hinges on being able to find "good" f-augmenting paths. In fact, Ford and Fulkerson showed that if the arc capacities are irrational numbers, then an algorithm using their labeling scheme may not terminate (strictly speaking, with irrational labels, it is not an algorithm).

   But even when flows and capacities are restricted to be integers, problems concerning efficiency still exist. For instance, if each flow augmentation were to increase the flow by only one unit, then the number of augmentations required for maximization would equal the capacity of a minimum cut. Such an algorithm would depend on the size of the arc capacities instead of on the size of the network.

Example 10.2.4: For the network shown in Figure 10.2.7, the arc from vertex v to vertex w has flow capacity 1, and the other arcs have capacity M, which could be made arbitrarily large. If the choice of augmenting flow path at each iteration were to alternate between the directed path $\langle s, v, w, t\rangle$ and the quasi-path $\langle s, w, v, t\rangle$, then the flow would increase by only one unit at each iteration. Thus, it could take as many as $2M$ iterations to obtain the maximum flow.

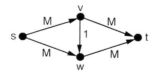

Figure 10.2.7   This network could require as many as $2M$ augmentations.

   Edmonds and Karp avoid these problems with the following algorithm, which is a slight refinement of the Ford-Fulkerson labeling scheme [EdKa72]. It uses _breadth-first search_ to find an f-augmenting path with the least number of arcs. Algorithm 10.2.2 either returns an f-augmenting path or returns a minimum cut that indicates the current flow f is a maximum flow.

Algorithm 10.2.2:   **Finding an Augmenting Path**

Input: a flow f in an s-t network N.

Output: an f-augmenting path Q or a minimum s-t cut with capacity $val(f)$.

    Initialize vertex set $V_s := \{s\}$.

    Write label 0 on vertex s.

    Initialize label counter $i := 1$

    While vertex set V_s does not contain sink t

        If there are usable arcs

            Let e be a usable arc whose labeled endpoint v has the

                smallest possible label.

            Let w be the unlabeled endpoint of arc e.

            Set $backpoint(w) := v$.

            Write label i on vertex w.

            $V_s := V_s \cup \{w\}$

            $i := i + 1$

        Else

            Return s-t cut $\langle V_s, V_N - V_s \rangle$.

    Reconstruct the f-augmenting path Q by following backpointers,

        starting from sink t.

    Return f-augmenting path Q.

Algorithm 10.2.3 combines Algorithms 10.2.1 and 10.2.2. The "FFEK" in the algorithm name refers to Ford, Fulkerson, Edmonds, and Karp.

Algorithm 10.2.3:   **FFEK-Maximum Flow**

Input: an s-t network N.

Output: a maximum flow f^* in network N.

    [*Initialization*]

    For each arc e in network N

        $f^*(e) := 0$

    [*Flow Augmentation*]

    Repeat

        Apply Algorithm 10.2.2 to find an f^*-augmenting path Q.

        Let $\Delta_Q = \min_{e \in Q}\{\Delta_e\}$.

        For each arc e of augmenting path Q

            If e is a forward arc

                $f^*(e) := f^*(e) + \Delta_Q$

            Else [e is a backward arc]

                $f^*(e) := f^*(e) - \Delta_Q$

    Until an f^*-augmenting path cannot be found in network N.

    Return flow f^*.

Example 10.2.5: The following sequence of figures illustrates Algorithm 10.2.3 for the example network shown, starting with zero flow. Each of the first three figures shows the current flow in the network and a shortest f-augmenting path (in number of arcs) for that flow, which is then used to get the flow shown in the subsequent figure. The final figure in this example shows an s-t cut with capacity equal to the value of the current flow, thereby establishing optimality.

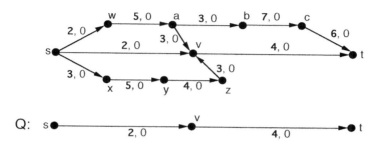

Figure 10.2.8   Iteration 0: $val(f) = 0$; augmenting path Q has $\Delta_Q = 2$.

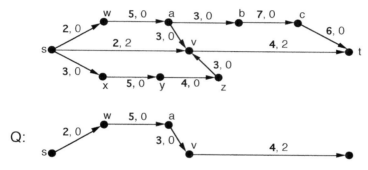

Figure 10.2.9   Iteration 1: $val(f) = 2$; augmenting path Q has $\Delta_Q = 2$.

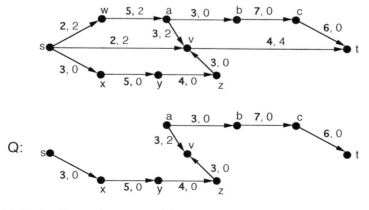

Figure 10.2.10   Iteration 2: $val(f) = 4$; augmenting path Q has $\Delta_Q = 2$.

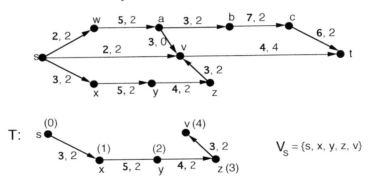

Figure 10.2.11   Final iteration: $val(f) = 6 = cap\langle\{s, x, y, z, v\}, \{w, a, b, c, t\}\rangle.$

Notice that at the end of the final iteration, the arc directed from source s to vertex w and the arc directed from vertex v to sink t are the only frontier arcs of tree T, but neither is usable. These two arcs form the minimum cut $\langle\{s, x, y, z, v\}, \{w, a, b, c, t\}\rangle$. This illustrates the s-t cut that was constructed in the proof of Theorem 10.2.3.

COMPUTATIONAL NOTE: Edmonds and Karp [EdKa72] show that Algorithm 10.2.3 finds a maximum flow in no more than $\frac{|E_N|(|V_N|+2)}{2}$ augmentations. Since the computation for breadth-first search is $O(|E_N|)$, it follows that the overall computation of Algorithm 10.2.3 is $O(|E_N|^2|V_N|)$. There are more efficient algorithms that are considerably more complicated and outside the scope of this book (*see* [Mi78] or [ThSw92]).

EXERCISES for Section 10.2

10.2.1^S Starting with the flow of value 9 given in Example 10.2.3, obtain a different maximum flow by using a different f-augmenting path.

For Exercises 10.2.2 through 10.2.6, apply Maximum-Flow Algorithm 10.2.3 to find a maximum flow and minimum cut for the given network.

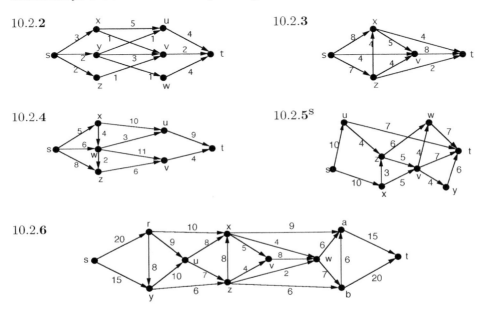

10.2.2

10.2.3

10.2.4

10.2.5^S

10.2.6

10.2.7 A company maintains three warehouses, X, Y, and Z, and three stores, A, B, and C. The warehouses have, respectively, 500, 500, and 900 lawn mowers in stock. There is an immediate demand from the stores for 700, 600, and 600 mowers, respectively. In the graph model shown, the arc capacities represent upper bounds on the number of mowers that can be shipped in a single day on that truck route segment. The intermediate vertices may be regarded as *transshipment points*, where trucks are checked, refueled, maintained, etc. Can all the demand be met? If not, how close can the company come to satisfying the demand?

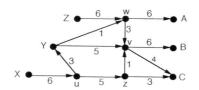

10.2.8^S The Johnson, Pate, Shargaa-Sears, and Ward families are going to the Winter Park Sidewalk Art Festival. Four cars are available to transport the families to the show. The cars can carry the following numbers of people: car 1, four; car 2, three; car 3, three; and car 4, four. There are four people in each family, and no car can carry more than two people from any one family. Formulate the problem of transporting the maximum possible number of people to the festival as a maximum-flow problem.

10.2.9 Seven types of chemicals are to be shipped in five trucks. There are three containers storing each type of chemical, and the capacities of the five trucks are 6, 4, 5, 4, and 3 containers, respectively. For security reasons, no truck can carry more than one container of the same chemical. Determine whether it is possible to ship all 21 containers in the five trucks.

10.2.10 Let N be a slightly more general s-t network, in which a nonnegative lower bound is specified for the flow on each arc. Prove that there exists a feasible flow in network N if and only if every arc e satisfies one of the following conditions:
   a. arc e is on a directed circuit of N;
   b. arc e is on a directed path from source s to sink t;
   c. arc e is on a directed path from sink t to source s.

10.2.11 Modify Maximum-Flow Algorithm 10.2.3 so that it finds a maximum flow in a network that has multiple sources and multiple sinks (see Exercise 10.2.9).

10.2.12 Modify Maximum-Flow Algorithm 10.2.3 so that it finds a maximum flow in a network in which a nonnegative lower bound is specified for the flow on each arc.

10.2.13 Design an algorithm whose input is an s-t network N and whose output is an arc with the property that increasing its capacity will increase the maximum flow in network N.

10.2.14 [*Computer Project*] Implement Maximum-Flow Algorithm 10.2.3 and test the program on the networks in Exercises 10.2.2 through 10.2.6.

10.3   FLOWS AND CONNECTIVITY

The theory of network flows is used in this section to give constructive proofs of Menger's Theorems, which were introduced in §5.3. These proofs lead directly to algorithms for determining the edge-connectivity and vertex-connectivity of a graph.

The strategy behind using network flows to prove Menger's Theorems is based on properties of networks whose arcs all have unit capacity. These *0-1 networks* are constructed from the original graph. The next few results are used to establish certain properties of 0-1 networks needed later.

Lemma 10.3.1: Let N be an s-t network such that $outdegree(s) > indegree(s)$, $indegree(t) > outdegree(t)$, and $outdegree(v) = indegree(v)$ for all other vertices v. Then there exists a directed s-t path in network N.

Proof: Let W be a longest directed trail in network N that starts at source s, and let z be its terminal vertex. If vertex z were not the sink, then there would be an arc not in trail W that is directed from z (since $indegree(z) = outdegree(z)$). But this would contradict the maximality of trail W. Thus, W is a directed trail from source s to sink t. If W has a repeated vertex, then part of W determines a directed cycle, which can be deleted from W to obtain a shorter directed s-t trail. This deletion step can be repeated until no repeated vertices remain, at which point, the resulting directed trail is an s-t path. ◇

Proposition 10.3.2: Let N be an s-t network such that
$$outdegree(s) - indegree(s) = m = indegree(t) - outdegree(t), \text{ and}$$
$$outdegree(v) = indegree(v), \text{ for all vertices } v \neq s, t.$$
Then there exist m arc-disjoint directed s-t paths in network N.

Proof: If $m = 1$, then there exists an open Eulerian directed trail T from source s to sink t, by Theorem 6.1.3. By Theorem 1.5.2, trail T is either an s-t directed path or can be reduced to an s-t path.

By way of induction, assume that the assertion is true for $m = k$, for some $k \geq 1$, and consider a network N for which the condition holds for $m = k+1$. There exists a directed s-t path P by Lemma 10.3.1. If the arcs of path P are deleted from network N, then the resulting network $\hat{N}$ satisfies the condition of the proposition for $m = k$. By the induction hypothesis, there exist k arc-disjoint directed s-t paths in network $\hat{N}$. These k paths together with path P form a collection of $k + 1$ arc-disjoint directed s-t paths in network N. ◇

Two Basic Properties of 0-1 Networks

DEFINITION: A **0-1 network** is a capacitated network whose arc capacities are either 0 or 1.

Proposition 10.3.3: *Let N be an s-t network such that $cap(e) = 1$ for every arc e. Then the value of a maximum flow in network N equals the maximum number of arc-disjoint directed s-t paths in N.*

Proof: Let f^* be a maximum flow in network N, and let r be the maximum number of arc-disjoint directed s-t paths in N. Consider the network N^*, obtained by deleting from N all arcs e such that $f^*(e) = 0$. Then $f^*(e) = 1$ for all arcs e in network N^*. It follows from the definitions that for every vertex v in network N^*,

$$\sum_{e \in Out(v)} f^*(e) = |Out(v)| = outdegree(v) \text{ and } \sum_{e \in In(v)} f^*(e) = |In(v)| = indegree(v)$$

Thus, by the definition of $val(f^*)$ and by the conservation-of-flow property,

$$outdegree(s) - indegree(s) = val(f^*) = indegree(t) - outdegree(t), \text{ and}$$
$$outdegree(v) = indegree(v), \text{ for all vertices } v \neq s, t.$$

By Proposition 10.3.2, there are $val(f^*)$ arc-disjoint directed s-t paths in network N^* and, hence, also in N, which implies that $val(f^*) \leq r$.

To obtain the reverse inequality, let $\{P_1, P_2, \ldots, P_r\}$ be a largest collection of arc-disjoint directed s-t paths in N, and consider the function $f : E_N \to R^+$ defined by

$$f(e) = \begin{cases} 1, & \text{if some path } P_i \text{ uses arc } e \\ 0, & \text{otherwise.} \end{cases}$$

Then f is a feasible flow in network N, with $val(f) = r$ (see Exercises). It follows that $val(f^*) \geq r$.   ◊

Separating Sets and Cuts

The next proposition is stated in terms of a digraph analogue of an *s-t separating edge set* (introduced in Chapter 5).

REVIEW FROM §5.3: Let s and t be distinct vertices in a graph G. An s-t **separating edge set** in G is a set of edges whose removal destroys all s-t paths in G, i.e., an edge subset that contains at least one edge of every s-t path in G.

DEFINITION: Let s and t be distinct vertices in a digraph D. An s-t **separating arc set** in D is a set of arcs whose removal destroys all directed s-t paths in D.

Thus, an s-t separating arc set in D is an arc subset of E_D that contains at least one arc of every directed s-t path in digraph D.

Remark: For the degenerate case in which the original graph or digraph has no s-t paths, the empty set is regarded as an s-t separating set.

Proposition 10.3.4: *Let N be an s-t network such that $cap(e) = 1$ for every arc e. Then the capacity of a minimum s-t cut in network N equals the minimum number of arcs in an s-t separating arc set in N.*

Proof: Let $K^* = \langle V_s, V_t \rangle$ be a minimum s-t cut in network N, and let q be the minimum number of arcs in an s-t separating arc set in N. Since K^* is an s-t cut, it is also an s-t separating arc set. Thus, $cap(K^*) \geq q$.

To obtain the reverse inequality, let S be an s-t separating arc set in network N containing q arcs, and let R be the set of all vertices in N that are reachable from source s by a directed path that contains no arc from set S. Then, by the definitions of arc set S and vertex set R, $t \notin R$, which means that $\langle R, V_N - R \rangle$ is an s-t cut. Moreover, $\langle R, V_N - R \rangle \subseteq S$. Therefore,

$$\begin{aligned}
cap(K^*) &\leq cap\langle R, V_N - R \rangle &&\text{(since } K^* \text{is a minimum s-t cut)} \\
&= |\langle R, V_N - R \rangle| &&\text{(since all capacities are 1)} \\
&\leq |S| &&\text{(since} \langle R, V_N - R \rangle \subseteq S) \\
&= q
\end{aligned}$$

which completes the proof.   ◇

Arc and Edge Versions of Menger's Theorem Revisited

Theorem 10.3.5: *[**Arc Form of Menger's Theorem**] Let s and t be distinct vertices in a digraph D. Then the maximum number of arc-disjoint directed s-t paths in D is equal to the minimum number of arcs in an s-t separating arc set of D.*

Proof: Let N be the s-t network obtained by assigning a unit capacity to each arc of digraph D. Then the result follows from Propositions 10.3.3 and 10.3.4, together with the Max-Flow Min-Cut Theorem (Theorem 10.2.4).   ◇

Remark: Conversely, the arc form of Menger's Theorem together with Propositions 10.3.3 and 10.3.4 can be used to prove the Max-Flow Min-Cut Theorem (see Exercises).

The edge form of Menger's Theorem for undirected graphs follows directly from the next two assertions concerning the relationship between a graph G and the digraph $\overset{\leftrightarrow}{G}$ obtained by replacing each edge e of graph G with a pair of oppositely directed arcs having the same endpoints as edge e. Each of these assertions follows directly from the definitions.

Assertion 10.3.6: *Let s and t be distinct vertices of a graph G, and let $\overset{\leftrightarrow}{G}$ be the digraph obtained by replacing each edge e of G with a pair of oppositely directed arcs having the same endpoints as edge e. Then there are at least as many edge-disjoint s-t paths in G as there are arc-disjoint s-t paths in $\overset{\leftrightarrow}{G}$.*   ◇ *(Exercises)*

Lemma 10.3.7: *Let S be a minimal s-t separating arc set of a digraph D. Then for every pair of vertices x, y in D, at most one of the arcs $a = xy$ and $b = yx$ is in S.*

Proof: Suppose not. Let $P_a = \langle s, v_1, v_2, \ldots, v_j, x, y, v_{j+1}, v_{j+2}, \ldots, t \rangle$ be (the vertex sequence of) an s-t directed path in D whose only arc in S is $a = xy$, and let $P_b = \langle s, w_1, w_2, \ldots, w_k, y, x, w_{k+1}, w_{k+2}, \ldots, t \rangle$ be an s-t directed path in D whose only arc in S is $b = yx$. (The minimality of S guarantees the existence of two such paths.) Then $P = \langle s, v_1, v_2, \ldots, v_j, x, w_{k+1}, w_{k+2}, \ldots, t \rangle$ is an s-t directed path in D that contains neither arc a nor arc b. Thus, P (and hence P_a and P_b) must contain some other arc in S, which contradicts the definitions of paths P_a and P_b.   ◇

Assertion 10.3.8: *Let s and t be distinct vertices of a graph G, and let $\overset{\leftrightarrow}{G}$ be the digraph obtained by replacing each edge e of G with a pair of oppositely directed arcs having the same endpoints as edge e. Then the minimum number of edges in an s-t separating edge set of graph G is equal to the minimum number of arcs in an s-t separating arc set of digraph $\overset{\leftrightarrow}{G}$.*

Proof: Let m be the size of a minimum s-t separating edge set S of G, and let $\overset{\leftrightarrow}{m}$ be the size of a minimum s-t separating arc set of $\overset{\leftrightarrow}{G}$. Let $\overset{\leftrightarrow}{S}$ be the arc set of $\overset{\leftrightarrow}{G}$ that results when each edge in S is replaced by its two oppositely directed arcs. Then $\overset{\leftrightarrow}{S}$ is an s-t separating arc set of $\overset{\leftrightarrow}{G}$ of size $2m$. Let $\overset{\leftrightarrow}{S_*}$ be a minimal s-t separating arc set of $\overset{\leftrightarrow}{G}$ contained in $\overset{\leftrightarrow}{S}$. Lemma 10.3.7 implies that $\left|\overset{\leftrightarrow}{S_*}\right| \leq |S|$, and hence, $\overset{\leftrightarrow}{m} \leq m$. Now let $\overset{\leftrightarrow}{T_*}$ be a minimum s-t separating arc set of $\overset{\leftrightarrow}{G}$. Then the corresponding edge set T that results from ignoring directions is an s-t separating edge set of G, from which it follows that $m \leq \overset{\leftrightarrow}{m}$.   ◇

Theorem 10.3.9: **[Edge Form of Menger's Theorem]** *Let s and t be distinct vertices in a graph G. Then the maximum number of edge-disjoint s-t paths in G equals the minimum number of edges in an s-t separating edge set of graph G.*

Proof: Let m and M be the sizes of a minimum s-t separating edge set and a maximum set of edge-disjoint s-t paths, respectively, in graph G, and let $\overset{\leftrightarrow}{m}$ and $\overset{\leftrightarrow}{M}$ be the sizes of a minimum s-t separating arc set and a maximum set of arc-disjoint s-t directed paths, respectively, in digraph $\overset{\leftrightarrow}{G}$.

If $\overset{\leftrightarrow}{F_*}$ is a maximum set of arc-disjoint s-t directed paths in $\overset{\leftrightarrow}{G}$, then by Assertion 10.3.6, there exists a set F of s-t paths in G with $\left|\overset{\leftrightarrow}{F_*}\right| \leq |F|$, which implies that

$$\overset{\leftrightarrow}{M} = \left|\overset{\leftrightarrow}{F_*}\right| \leq |F| \leq M$$

By Assertion 10.3.8 and Theorem 110.3.5, we have

$$\overset{\leftrightarrow}{m} = \overset{\leftrightarrow}{M} \leq M \leq m = \overset{\leftrightarrow}{m}$$

where the second inequality follows because it takes at least m edges to destroy m edge-disjoint paths.   ◇

Determining Edge-Connectivity Using Network Flows

REVIEW FROM §5.1: The **edge-connectivity** $\kappa_e(G)$ is the size of a smallest edge-cut in graph G.

DEFINITION: The **local edge-connectivity** $\kappa_e(s,t)$ between distinct vertices s and t in a graph G is the minimum number of edges in an s-t separating edge set in G.

The following proposition shows that the edge-connectivity of a graph can be expressed in terms of the local edge-connectivity between all pairs of vertices. The assertion and its proof are analogous to Lemma 5.3.5 and its proof in §5.3.

Proposition 10.3.10: *The edge-connectivity of a graph G is equal to the minimum of the local edge-connectivities, taken over all pairs of vertices s and t. That is,*

$$\kappa_e(G) = \min_{s,t \in V_G} \{\kappa_e(s,t)\}$$

◇ (Exercises)

Proposition 10.3.10 and the results used in the proof of Theorem 10.3.9 suggest an algorithm for determining the edge-connectivity $\kappa_e(G)$ of an arbitrary graph G. The algorithm calculates the local edge-connectivity between each pair of vertices in G, by solving an appropriate maximum-flow problem in the network $\overset{\leftrightarrow}{G}$ (referred to in Assertions 10.3.6 and 10.3.8). In fact, it is not necessary to calculate the local edge-connectivity between every pair of vertices, as the next two results show. Both results make use of the relationship between local edge-connectivity and *partition-cuts*.

REVIEW FROM §4.5: Let G be a graph, and let V_1 and V_2 form a partition of V_G. The set of all edges of G having one endpoint in vertex subset V_1 and the other endpoint in vertex subset V_2 is called a **partition-cut** of G and is denoted $\langle V_1, V_2 \rangle$.

Proposition 10.3.11: *Let $\langle V_1, V_2 \rangle$ be a partition-cut of minimum cardinality in a graph G, and let v_1 and v_2 be any vertices in V_1 and V_2, respectively. Then the edge-connectivity $\kappa_e(G)$ equals the local edge-connectivity $\kappa_e(v_1, v_2)$.*

Proof: Suppose that the minimum local edge-connectivity is achieved between vertices x and y. Then $\kappa_e(G) = \kappa_e(x, y)$ (by Proposition 10.3.10). It suffices to show that $\kappa_e(v_1, v_2) \leq \kappa_e(x, y)$.

Let $\overset{\leftrightarrow}{G}$ be the digraph obtained by replacing each edge of graph G with two oppositely directed arcs. Then digraph $\overset{\leftrightarrow}{G}$ can be regarded as a v_1-v_2 capacitated network $\overset{\leftrightarrow}{G}_{v_1 v_2}$ and as an x-y capacitated network $\overset{\leftrightarrow}{G}_{xy}$, where each arc is assigned unit capacity. Let K^* be a minimum v_1-v_2 cut in network $\overset{\leftrightarrow}{G}_{v_1 v_2}$ It follows that $cap(K^*) \leq cap\langle V_1, V_2 \rangle$, since the partition-cut $\langle V_1, V_2 \rangle$ corresponds to a v_1-v_2 cut in network $\overset{\leftrightarrow}{G}_{v_1 v_2}$.

Next, let f^* be a maximum flow and $\langle V_x, V_y \rangle$ a minimum x-y cut in x-y network $\overset{\leftrightarrow}{G}_{xy}$, so that $cap\langle V_x, V_y \rangle = val(f^*)$. Then the following chain of inequalities establishes the desired inequality.

$$
\begin{aligned}
\kappa_e(v_1, v_2) &= cap(K^*) && \text{(Proposition 10.3.4 and Assertion 10.3.8)} \\
&\leq cap\langle V_1, V_2 \rangle && (|\langle V_1, V_2 \rangle| \text{ corresponds to a } v_1 - v_2 \text{ cut in } \overset{\leftrightarrow}{G}_{v_1 v_2}) \\
&= |\langle V_1, V_2 \rangle| && \text{(all arcs have unit capacity)} \\
&\leq |\langle V_x, V_y \rangle| && (\langle V_x, V_y \rangle \text{ corresponds to a partition-cut in } G) \\
&= cap\langle V_x, V_y \rangle && \text{(all arcs have unit capacity)} \\
&= val(f^*) && \text{(Max-Flow Min-Cut Theorem 10.2.4)} \\
&= \kappa_e(x, y) && \text{(Proposition 10.3.3 and Theorem 10.3.5)}
\end{aligned}
$$

$\diamondsuit$

Corollary 10.3.12: *Let s be any vertex in a graph G. Then*

$$
\kappa_e(G) = \min_{t \in V_G - \{s\}} \{\kappa_e(s, t)\}
$$

Proof: Let $\langle V_1, V_2 \rangle$ be a partition-cut of minimum cardinality, and suppose, without loss of generality, that vertex $s \in V_1$. There must be some vertex $t \in V_2$ (otherwise, $E_G = \emptyset$, and the assertion would be trivially true). By Proposition 10.3.11, it follows that $\kappa_e(G) = \kappa_e(s, t)$. $\diamondsuit$

The variable κ_e, used in the next algorithm, represents the edge-connectivity of graph G and is initialized with the sufficiently large positive integer $|E_G|$.

Algorithm 10.3.1:   **Edge-Connectivity Calculation**

Input: a graph G.

Output: the edge-connectivity κ_e of graph G.

> Construct digraph $\overset{\leftrightarrow}{G}$.
> Let s be an arbitrary vertex of graph G.
> Initialize edge-connectivity $\kappa_e := |E_G|$.
> For each vertex $t \in V_G - \{s\}$
>> Apply Algorithm 10.2.3 to s-t network $\overset{\leftrightarrow}{G}$ to obtain maximum flow f^*
>> If $val(f^*) < \kappa_e$
>>> $\kappa_e := val(f^*)$
>
> Return κ_e

COMPUTATIONAL NOTE: Algorithm 10.3.1 requires $O(n)$ iterations, and since Algorithm 10.2.3 requires $O(n|E|^2)$ computations, the overall computation of Algorithm 10.3.1 is $O(n^2|E|^2)$. This can be reduced if one of the more efficient maximum-flow algorithms cited earlier is used instead of Algorithm 10.2.3.

Using Network Flows to Prove the Vertex Forms of Menger's Theorem

The vertex form of Menger's Theorem for digraphs and graphs can be proved using the arc and edge forms (Theorems 10.3.5 and 10.3.9). The proofs are based on the following construction.

Construction of Digraph N^D from a Digraph D:

Let s and t be any pair of non-adjacent vertices in a digraph D. The digraph N^D is obtained from digraph D as follows:

- Each vertex $x \in V_D - \{s, t\}$ corresponds to two vertices x' and x'' in digraph N^D and an arc directed from x' to x''.

- Each arc in digraph D that is directed from vertex s to a vertex $x \in V_D - \{s, t\}$ corresponds to an arc in digraph N^D directed from s to x'.

- Each arc in D that is directed from a vertex $x \in V_D - \{s, t\}$ to vertex t corresponds to an arc in N^D directed from x'' to t.

- Each arc in D that is directed from a vertex $x \in V_D - \{s, t\}$ to a vertex $y \in V_D - \{s, t\}$ corresponds to an arc in N^D directed from x'' to y'.

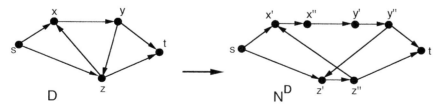

Figure 10.3.1   The digraph N^D corresponding to a digraph D.

REVIEW FROM §5.3: Let s and t be a pair of non-adjacent vertices in a graph G (or digraph D). An *s-t separating vertex set* in G (or D) is a set of vertices whose removal destroys all *s-t* paths in G (or all directed *s-t* paths in D).

Thus, an *s-t* separating vertex set is a set of vertices that contains at least one internal vertex of every (directed) *s-t* path.

DEFINITION: Two (directed) *s-t* paths in a digraph are **internally disjoint** if they have no internal vertices in common.

Relationships between Digraphs D and N^D

The vertex forms of Menger's Theorem are easily established from the following four relationships between a digraph D and its corresponding digraph N^D. Each of these follows from the definitions.

Assertion 10.3.13: *There is a one-to-one correspondence between directed s-t paths in digraph D and directed s-t paths in digraph N^D.* ◇ *(Exercises)*

Assertion 10.3.14: *Two directed s-t paths in D are internally disjoint if and only if their corresponding s-t directed paths in N^D are arc-disjoint.* ◇ *(Exercises)*

Assertion 10.3.15: *The maximum number of internally disjoint directed s-t paths in D is equal to the maximum number of arc-disjoint directed s-t paths in N^D.* ◇ *(Exercises)*

Assertion 10.3.16: *The minimum number of vertices in an s-t separating vertex set in digraph D is equal to the minimum number of arcs in an s-t separating arc set in N^D.* ◇ *(Exercises)*

Theorem 10.3.17: [**Vertex Form of Menger for Digraphs**] *Let s and t be a pair of non-adjacent vertices in a digraph D. Then the maximum number of internally disjoint directed s-t paths in D is equal to the minimum number of vertices in an s-t separating vertex set in D.*

Proof: This follows from Assertions 10.3.13 through 10.3.16, together with the arc form of Menger's Theorem (Theorem 10.3.5).   ◇

Theorem 10.3.18: [**Vertex Form of Menger for Undirected Graphs**] *Let s and t be a pair of non-adjacent vertices in a graph G. Then the maximum number of internally disjoint s-t paths in G is equal to the minimum number of vertices in an s-t separating vertex set in G.*

Proof: Let $\overleftrightarrow{G}$ be the digraph obtained by replacing each edge e of G with a pair of oppositely directed arcs having the same endpoints as edge e. The result follows by Theorem 10.3.17 and Assertions 10.3.13 and 10.3.16.   ◇

Determining Vertex-Connectivity Using Network Flows

The proof of the vertex form of Menger's Theorem based on the theory of network flows leads to an algorithm for determining the vertex-connectivity of a graph, in much the same way as the edge form of Menger's Theorem leads to an algorithm for edge-connectivity.

REVIEW FROM §5.3:

- Let s and t be non-adjacent vertices of a connected graph. Then the **local vertex-connectivity** between s and t, denoted $\kappa_v(s,t)$, is the minimum number of vertices in an s-t separating vertex set.
- **Lemma 5.3.5:** Let G be a connected graph containing at least one pair of non-adjacent vertices. Then the vertex-connectivity $\kappa_v(G)$ is the minimum of the local vertex-connectivity $\kappa_v(s,t)$, taken over all pairs of non-adjacent vertices s and t.

Remark: The following algorithm, which has $O(n|E_G|^3)$ time-complexity, computes the vertex-connectivity of an n-vertex graph by calculating the local vertex-connectivity between various pairs of non-adjacent vertices.

Algorithm 10.3.2:   **Vertex-Connectivity Calculation**
Input: a graph G with $V_G = \{V_1, V_2, \dots v_n\}$.
Output: the vertex-connectivity κ_v of graph G.

    Construct digraph $\overset{\leftrightarrow}{G}$.
    Initialize vertex-connectivity $\kappa_v := |V_G|$.
    Initialize index $k := 0$.
    While $k \leq \kappa_v$
        $k := k + 1$
        For $j = k + 1$ to n
            If vertices v_k and v_j are not adjacent
                Construct digraph $N^{\overset{\leftrightarrow}{G}}$.
                Assign unit capacity to each arc in digraph $N^{\overset{\leftrightarrow}{G}}$.
                Apply Algorithm 10.2.3 to network $N^{\overset{\leftrightarrow}{G}}$ with source v_k
                    and sink v_j to obtain maximum flow f^*.
                If $val(f^*) < \kappa_v$
                    $\kappa_v := val(f^*)$
    Return κ_v.

As in Algorithm 10.3.1, it is not necessary to calculate the local vertex-connectivity between every pair. Verification of its correctness and its time-complexity are omitted but may be found in [Ev79].

The variable κ_v, appearing in Algorithm 10.3.2, represents the vertex-connectivity of graph G and is initialized with the sufficiently large positive integer $|V_G|$.

EXERCISES for Section 10.3

For Exercises 10.3.1 through 10.3.4, find the local edge-connectivity between the pair of solid vertices for the graph shown, by finding the maximum flow in an appropriate network.

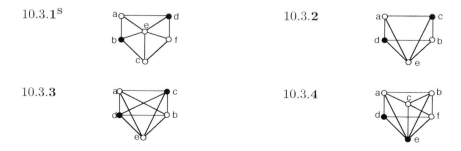

10.3.1$^\text{S}$

10.3.2

10.3.3

10.3.4

For Exercises 10.3.5 through 10.3.8, apply Algorithm 10.3.1 to find the edge-connectivity of the given graph.

10.3.5$^\text{S}$ The graph of Exercise 10.3.1.

10.3.6 The graph of Exercise 10.3.2.

10.3.7 The graph of Exercise 10.3.3.

10.3.8 The graph of Exercise 10.3.4.

For Exercises 10.3.9 through 10.3.12, find the local vertex-connectivity between the pair of solid vertices for the graph shown, by finding the maximum flow in an appropriate network.

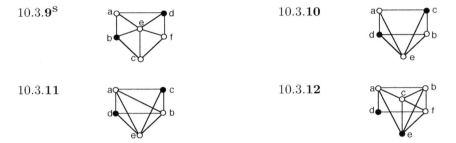

10.3.9$^\text{S}$

10.3.10

10.3.11

10.3.12

For Exercises 10.3.13 through 10.3.16, apply Algorithm 10.3.2 to find the vertex-connectivity of the given graph.

10.3.13$^\text{S}$ The graph of Exercise 10.3.9.

10.3.14 The graph of Exercise 10.3.10.

10.3.15 The graph of Exercise 10.3.11.

10.3.16 The graph of Exercise 10.3.12.

10.3.17 Supply the details that were omitted in the proof of Lemma 10.3.1.

10.3.18$^\text{S}$ Suppose that $\{P_1, P_2, \ldots, P_r\}$ is a collection of r arc-disjoint directed s-t paths in a network N whose arc capacities are all ≥ 1. Show that the function $f : E_N \to R^+$ defined by

$$f(e) = \begin{cases} 1, & \text{if some path } P_i \text{ uses arc } e \\ 0, & \text{otherwise} \end{cases}$$

is a flow in network N, with $val(f) = r$.

10.3.19 Prove Assertion 10.3.6.

10.3.20 Prove Proposition 10.3.10.

10.3.21$^\text{S}$ Prove Assertion 10.3.13.

10.3.22 Prove Assertion 10.3.14.

10.3.23 Prove Assertion 10.3.15.

10.3.24 Prove Assertion 10.3.16.

10.3.25$^\text{S}$ Prove Max-Flow-Min-Cut Theorem 10.2.4 using Arc-Form-of-Menger's Theorem 10.3.5 together with Propositions 10.3.3 and 10.3.4.

10.4   MATCHINGS, TRANSVERSALS, AND VERTEX COVERS

The link between Menger's theorems and the theory of network flows, which was established in the last section, extend to several other combinatorial problems discussed here.

Matchings in a Graph

REVIEW FROM §9.3: A set M of edges in a graph G is called a **matching** in G if no two edges in set M have an endpoint in common.

TERMINOLOGY: If e, with endpoints x and y, is an edge in a matching M, then M **matches vertex x with vertex y**, or x is **matched with y by** M.

DEFINITION: A **maximum (-cardinality) matching** is a matching with the greatest number of edges.

Remark: A *maximal* matching in a graph G is a matching that is not a proper subset of any other matching in G. Clearly, every maximum matching is a maximal matching, but a maximal matching need not be a maximum one.

Example 10.4.1: Consider the graph G shown in Figure 10.4.1. The singleton edge sets $\{b\}$, $\{c\}$, $\{d\}$ and $\{e\}$ are the nonempty matchings in graph G that are *not* maximal. The edge set $\{a\}$ is a maximal matching in G that is not maximum, and the edge sets $\{b,d\}$, $\{c,e\}$ are the maximum matchings in G.

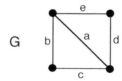

Figure 10.4.1   Two of the three maximal matchings in G are maximum.

Maximum Matching in a Bipartite Graph

REVIEW FROM §1.2: A graph G is a **bipartite graph** if there exists a vertex bipartition $\{X,Y\}$ of V_G such that every edge of G has one endpoint in vertex set X and one endpoint in vertex set Y.

Application 10.4.1:   *The Personnel-Assignment Problem*   Suppose that a company requires a number of different types of jobs, and suppose each worker is suited for some of these jobs, but not others. Assuming that each person can perform at most one job at a time, how should the jobs be assigned so that the maximum number of jobs can be performed simultaneously?

The bipartite graph of Figure 10.4.2 has vertex bipartition $\{V_J, V_W\}$, where the solid vertices represent the jobs, and the hollow vertices represent the workers. Each edge corresponds to a suitable job assignment. The maximum matching whose edges appear in bold shows that all five jobs can be assigned to workers.

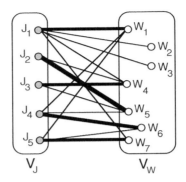

Figure 10.4.2   A maximum matching that solves the assignment problem.

Converting a Maximum-Matching Problem into a Maximum-Flow Problem

Let G be a bipartite graph with vertex bipartition $\{X, Y\}$. Then the problem of finding a maximum matching in graph G can be transformed into a maximum-flow problem in an s-t network $\vec{G}_{st}$, constructed from graph G as follows:

(1) The vertex-set of network $\vec{G}_{st}$ is $V_G \cup \{s, t\}$, where s and t are two new vertices that are the source and sink, respectively, of $\vec{G}_{st}$.

(2) Each edge in graph G between a vertex $x \in X$ and a vertex $y \in Y$ corresponds to an arc in network $\vec{G}_{st}$ directed from vertex x to vertex y.

(3) For each vertex $x \in X$, an arc in network $\vec{G}_{st}$ is drawn from source s to vertex x.

(4) For each vertex $y \in Y$, an arc in network $\vec{G}_{st}$ is drawn from vertex y to sink t.

(5) Each arc e in network $\vec{G}_{st}$ is assigned a capacity $cap(e) = 1$.

Example 10.4.2: Figure 10.4.3 shows the bipartite graph from Application 10.4.1 and the corresponding s-t network $\vec{G}_{st}$.

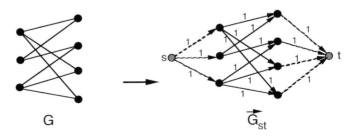

Figure 10.4.3   A bipartite graph G and its corresponding network $\vec{G}_{st}$.

Relationship between Matchings in G and Flows in $\vec{G}_{st}$

The following proposition and its corollary show that a maximum-matching problem can be solved as a maximum-flow problem.

Proposition 10.4.1: *Let G be a bipartite graph and $\vec{G}_{st}$ the s-t network constructed from G. Then there is a one-to-one correspondence between the integral flows of network $\vec{G}_{st}$ and the matchings in graph G.*

Proof: Let f be an integer-valued flow in the network $\vec{G}_{st}$. Then the unit capacities imply that $f(e) = 0$ or 1, for each arc e in network $\vec{G}_{st}$. Let M be the set of edges in graph G whose corresponding arcs in network $\vec{G}_{st}$ have unit flow. It follows from the structure of $\vec{G}_{st}$ and the conservation-of-flow property that edge set M is a matching in graph G (see Exercises).

Conversely, let M be a matching in graph G, and for any edge e in graph G, let $\vec{e}$ denote the corresponding arc in network $\vec{G}_{st}$. Then for every arc $a \in \vec{G}_{st}$, the function defined by

$$f(a) = \begin{cases} 1, & \text{if } a = \vec{e}, \text{ for some edge } e \in M \\ 1, & \text{if arc } a \text{ is adjacent to } \vec{e}, \text{ for some edge } e \in M \\ 0, & \text{otherwise} \end{cases}$$

is a flow in network $\vec{G}_{st}$ (see Exercises).   ◇

Corollary 10.4.2: *Let M be a matching in a bipartite graph G and f the corresponding flow in network $\vec{G}_{st}$. Then $val(f) = |M|$, and f is a maximum flow if and only if M is a maximum matching.*

Proof: This follows from the correspondence established in Proposition 10.4.1.   ◇

Example 10.4.3: On the left in Figure 10.4.4 below is a maximum matching of size 3 in a bipartite graph G. Its corresponding maximum flow in network $\vec{G}_{st}$ has unit flow on each arc represented as a dashed line.

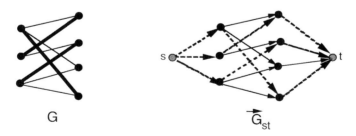

$$G \qquad\qquad\qquad \vec{G}_{st}$$

Figure 10.4.4   A maximum matching and its corresponding maximum flow.

Algorithm 10.4.1 finds a maximum matching in a bipartite graph G by finding a maximum flow in the corresponding network $\vec{G}_{st}$, as prescribed by Proposition 10.4.1. As in the proposition, $\vec{e}$ denotes the arc in network $\vec{G}_{st}$ that corresponds to edge e in graph G.

Algorithm 10.4.1:   **Maximum Bipartite Matching**
Input: a bipartite graph G with vertex bipartition $\{X, Y\}$.
Output: a maximum matching M of graph G.
       Initialize edge set $M := 0$.
       Construct the s-t network $\vec{G}_{st}$ that corresponds to bipartite graph G.
       Apply Algorithm 10.2.3 to network $\vec{G}_{st}$ to obtain maximum flow f^*.
       For each arc $\vec{e}$ in network $\vec{G}_{st}$
          If $f^*(\vec{e}) = 1$
              $M := M \cup \{e\}$
       Return edge set M.

Remark: One can view the algorithm above as iteratively augmenting a matching M by finding an *M-augmenting path*, analogous to the flow-augmenting paths of the maximum-flow algorithm. Edmonds [Ed65b] extended this notion by designing an algorithm for finding a maximum matching in a general graph. The Edmonds algorithm is more complicated than a bipartite-matching algorithm, but it and its more recent improvements have running time not much beyond that needed for bipartite matching. The following application illustrates the need for a more general algorithm.

Application 10.4.2:   *Pairing Volunteers for a Rescue Mission*   Suppose that several first-aid workers from around the world have volunteered for a rescue mission in some country that has been struck by a disaster. The volunteers are to be divided into two-person teams, but the members of each team must speak the same language. What is the maximum number of pairs that can be sent out on a rescue mission?

A graph model for this problem has a vertex corresponding to each volunteer, and an edge exists between a pair of vertices if the corresponding volunteers speak the same language. Then a maximum matching in the graph corresponds to a maximum set of two-person teams.

Transversals for Families of Subsets

DEFINITION: Let A be a nonempty finite set, and let $\mathcal{F} = \{S_1, S_2, \ldots, S_r\}$ be a family of (not necessarily distinct) nonempty subsets of set A. Then a ***transversal*** (or ***system of distinct representatives***) of family $\mathcal{F}$ is a sequence $T = \langle a_1, a_2, \ldots, a_r \rangle$ of r distinct elements of set A, such that $a_i \in S_i$, for each $i = 1, \ldots, r$.

Example 10.4.4: Let $A = \{a, b, c, d, e\}$, and suppose that $\mathcal{F} = \{S_1 S_2, \ldots, S_5\}$, where

$$S_1 = \{a, b\}; \quad S_2 = \{b, c, d\}; \quad S_3 = \{c, d, e\}; \quad S_4 = \{d, e\}; \quad \text{and } S_5 = \{e, a, b\}$$

Then $T = \langle b, c, e, d, a \rangle$ is a transversal for family $\mathcal{F}$.

Application 10.4.3:   *Pairing Interns with Hospitals*   Suppose that r medical school graduates have applied for internships at various hospitals. For the ith medical school graduate, let S_i be the set of all hospitals that find applicant i acceptable. Then a transversal for the family $\mathcal{F} = \{S_l, S_2, \ldots, S_r\}$ would assign each potential intern to a hospital that is willing to take him or her, such that no hospital gets assigned more than one intern.

Relationship between Bipartite Matchings and Transversals

The problem of finding a transversal for a given family of subsets can be formulated as a matching problem in an associated bipartite graph. The statement of the matching problem uses the following definition.

DEFINITION: Let G be a bipartite graph with vertex bipartition $\{X, Y\}$, and let edge set M be a matching in G. Then the matching M is said to be X-**saturating** if each vertex in set X is an endpoint of an edge in M, and to be Y-**saturating** if each vertex in Y is an endpoint of an edge in M.

Notice that every X-saturating matching and every Y-saturating matching must be a maximum matching.

Example 10.4.5: The maximum matching shown on the left in Figure 10.4.5 is X-saturating, but the one on the right is not.

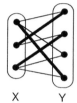 

$$X \quad Y \qquad\qquad X \quad Y$$

Figure 10.4.5   The matching on the left is X-saturating.

TERMINOLOGY NOTE: Some authors refer to an X-saturating matching as a *complete matching from X to Y*.

Finding a Transversal by Finding an X-Saturating Matching

Let $A = \{a_1, a_2, \ldots, a_n\}$ be a set of n elements and $\mathcal{F} = \{S_1, S_2, \ldots, S_r\}$ be a family of (not necessarily distinct) subsets of set A. Consider the following bipartite graph G constructed from $\mathcal{F}$. The vertex bipartition $\{X, Y\}$ of graph G consists of two vertex sets

$$X = \{x_1, x_2, \ldots, x_r\} \text{ and } Y = \{y_1, y_2, \ldots, y_n\}$$

where each vertex $x_i \in X$ corresponds to the subset S_i in family $\mathcal{F}$, and each vertex $y_j \in Y$ corresponds to the element $a_j \in A$. An edge exists between vertex x_i and vertex y_j if $a_j \in S_i$. Then a transversal for family $\mathcal{F}$ corresponds to an X-saturating matching of graph G.

Example 10.4.6: Figure 10.4.6 shows the X-saturating matching that corresponds to the transversal $T = \langle b, c, e, d, a \rangle$ from Example 10.4.4.

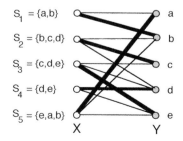

Figure 10.4.6   Transversal $T = \langle b, c, e, d, a \rangle$ and corresponding matching.

Hall's Theorem

The next theorem, known as *Hall's Theorem*, gives a necessary and sufficient condition for a bipartite graph to have an X-saturating matching. Its corollary, which is simply a restatement in terms of a family of subsets, characterizes those families of subsets that have a transversal. The graph version uses the following notation.

NOTATION: Let G be a bipartite graph with vertex bipartition $\{X, Y\}$, and let W be a subset of vertex set X. Then $N(W)$ denotes the subset of Y consisting of all vertices in Y that are adjacent to at least one vertex in set W. In other words, $N(W)$ is the set of *neighbors* of set W.

DEFINITION: A bipartite graph with vertex bipartition $\{X, Y\}$ is said to satisfy **Hall's Condition** if $|W| \leq |N(W)|$ for every subset W of X.

Theorem 10.4.3: [***Hall's Theorem for Bipartite Graphs, 1935***] *Let G be a bipartite graph with vertex bipartition $\{X, Y\}$. Then graph G has an X-saturating matching if and only if for each subset W of X, $|W| \leq |N(W)|$.*

Proof: *Necessity* ($\Rightarrow$) Suppose that there exists an X-saturating matching M in graph G, and let W be any subset of X. Then every vertex $w \in W$ is matched to a vertex $y \in Y$ by an edge in matching M. Each such Y is in the neighbor set $N(W)$ of W, and, thus, $|W| \leq |N(W)|$.

Sufficiency ($\Leftarrow$) Suppose that Hall's Condition holds, that is, $|W\| \leq |N(W)|$, for every subset W of X. Consider the s-t network $\vec{G}_{st}$ corresponding to the bipartite graph G (as in Proposition 10.4.1). An X-saturating matching in graph G would be a maximum matching in G and would correspond to a maximum flow in network $\vec{G}_{st}$ with value $|X|$. Thus, by the Max-Flow Min-Cut Theorem (Theorem 10.2.4), it suffices to show that a minimum s-t cut in network $\vec{G}_{st}$ has capacity $|X|$. The s-t cut $\langle \{s\}, X \cup Y \cup \{t\}\rangle$ has capacity $|X|$, so it remains to show that every other s-t cut K has capacity $cap(K) \geq |X|$.

Let $\langle V_s, V_t\rangle$ be any s-t cut in network $\vec{G}_{st}$, and let $W = V_s \cap X$. Then the cut $\langle V_s, V_t\rangle$ can be expressed as the disjoint union of three arc sets (depicted as dashed lines in Figure 10.4.7), that is,

$$\langle V_s, V_t\rangle = \langle \{s\}, V_t \cap X\rangle \cup \langle W, V_t \cap Y\rangle \cup \langle V_s \cap Y, \{t\}\rangle$$

where $\langle A, B\rangle$ denotes the set of arcs directed from a vertex in A to a vertex in B.

The following chain of inequalities completes the proof.

$$
\begin{aligned}
cap\langle V_s, V_t\rangle &= |\langle \{s\}, X - W\rangle| + |\langle W, V_t \cap Y\rangle| + |\langle V_s \cap Y, \{t\}\rangle| &&\text{(all capacities = 1)}\\
&= |X - W| + |\langle W, V_t \cap Y\rangle| + |V_s \cap Y| &&\text{(by the construction of } \vec{G}_{st})\\
&\geq |X - W| + |V_t \cap N(W)| + |V_s \cap Y| &&\text{(by definition of } N(W)\\
&= |X - W| + |N(W)| - |V_s \cap N(W)| + |V_s \cap Y| &&\text{(see Figure 10.4.7)}\\
&\geq |X - W| + |N(W)| - |V_s \cap Y| + |V_s \cap Y| &&\text{(since } N(w) \subseteq Y)\\
&= |X - W| + |N(W)|\\
&\geq |X - W| + |W| &&\text{(by Hall's Condition)}\\
&= |X|
\end{aligned}
$$

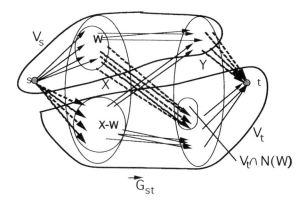

Figure 10.4.7 $\langle V_s, V_t \rangle = \langle \{s\}, X - W \rangle \cup \langle W, V_t \cap Y \rangle \cup \langle V_s \cap Y, \{t\} \rangle$

Corollary 10.4.4: [**Hall's Theorem for Transversals**] *Let A be a nonempty finite set, and let $\mathcal{F} = \{S_1, S_2, \ldots, S_r\}$ be a family of nonempty subsets of set A. Then family $\mathcal{F}$ has a transversal if and only if the union of any k of the subsets S_i contains at least k elements of set A ($1 \le k \le r$).*

Proof: This is Hall's Theorem for Bipartite Graphs, restated in terms of transversals.

$\diamondsuit$

Remark: One of the earliest incarnations of Hall's Theorem appeared as a solution to the following problem, known as the **Marriage Problem**: *Given a set of women, each of whom knows a subset of men, under what conditions can each of the women marry a man whom she knows?* (See Exercises.)

Two Graph Factorization Theorems

REVIEW FROM §8.4:

- A *factor* of a graph is a spanning subgraph.
- A *factorization* of a graph G is a set of factors whose edge-sets form a partition of the edge-set E_G.
- A *k-factor* of a graph G is a k-regular factor of G.
- A *k-factorization* of a graph is a factorization into k-factors.

Theorem 10.4.5: [**König's 1-Factorization Theorem**] *[Kö16] Every r-regular bipartite graph G with $r > 0$ is 1-factorable.*

Proof: Suppose that X and Y are the two partite sets, and let $W \subseteq X$. The number of edges from W to Y is $r |W|$. Since each vertex of Y has degree r, at most r of these edges are incident on any one vertex of Y, from which it follows (by the generalized pigeonhole principle) that

$$N(W) \ge \left\lfloor \frac{r |W|}{r} \right\rfloor = |W|$$

By Hall's Theorem for Bipartite Graphs, the graph G has an X-saturating matching. Since $|X| = |Y|$, such a matching is a 1-factor in G. The graph obtained from G by deleting the edges of this 1-factor is an $(r - 1)$-regular bipartite graph, and the result follows by induction.

$\diamondsuit$

REVIEW FROM §1.5 AND §4.5:

- An **Eulerian trail** in a graph is a trail that contains every edge of that graph.

- An **Eulerian tour** is a closed Eulerian trail.

- An **Eulerian graph** is a graph that has an Eulerian tour.

- Theorem 4.5.11 [**Eulerian-Graph Characterization**]: A connected graph G is Eulerian if and only if the degree of every vertex in G is even.

Theorem 10.4.6: [**Petersen's 2-Factorization Theorem**] *[Pe1891] Every regular graph G of even degree is 2-factorable.*

Proof: We may assume that G is connected since a graph is factorable if and only if each of its components is factorable. Let G be a $2r$-regular graph with vertices $v_1, \ldots, v_n$, and let C be a closed Eulerian trail in G, whose existence is guaranteed by the Eulerian-Graph Characterization, cited above. We define a new bipartite graph H with partite sets $U = \{u_1, \ldots, u_n\}$ and $W = \{w_1, \ldots, w_n\}$ such that u_i and w_j are adjacent if vertex v_j immediately follows vertex v_i on the Eulerian trail C. Graph H is r-regular, because the trail C enters and leaves each vertex of G exactly r times.

Since H is a regular bipartite graph, it follows from Theorem 10.4.5 that H has a 1-factor F. By the construction of H, the edge $e \in F$ with endpoints u_i and w_j corresponds to edge e' of G with endpoints v_i and v_j. Moreover, for each subscript $k = 1, \ldots, n$, the vertices u_k and w_k both occur exactly once in the 1-factor F, and hence, the edge set

$$F' = \{e' | e \in F\}$$

contains exactly two edges of G that are incident on the vertex v_k. It follows that F' is a 2-factor of G. The graph obtained from G by deleting the edges of this 2-factor is a $(2r - 2)$-regular graph, and the result follows by induction. $\diamond$

Maximum Matchings and Minimum Vertex Covers

The theme of *max-min* pairs of optimization problems, seen earlier in this chapter and in Chapter 5, appears once again in the context of *vertex covers*.

DEFINITION: Let G be a graph, and let C be a subset of the vertices of G. Then set C is a **vertex cover** of graph G if every edge of G is incident on at least one vertex in C.

DEFINITION: A **minimum vertex cover** is a vertex cover with the least number of vertices.

Example 10.4.7: For the bipartite graph shown in Figure 10.4.8, a maximum matching (the bold edges) and minimum vertex cover (the solid vertices) both have cardinality 5.

Figure 10.4.8 A maximum matching and minimum vertex cover.

Proposition 10.4.7: [**Weak Duality for Matchings**] *Let M be a matching in a graph G, and let C be a vertex cover of G. Then $|M| \leq |C|$.* $\diamond$ *(Exercises)*

Corollary 10.4.8: [**Certificate of Optimality for Matchings**] *Let M be a matching in a graph G, and let C be a vertex cover of G such that $|M| = |C|$. Then M is a maximum matching and C is a minimum vertex cover.* $\diamond$ *(Exercises)*

Remark: The converse of Corollary 10.4.8 does not hold in general (see Exercises); however, it does hold for bipartite graphs.

NOTATION: The neighbor set of a given subset W of vertices in a graph G is denoted $N_G(W)$ (instead of the usual $N(W)$) when there is more than one graph involved.

Theorem 10.4.9: [**König, 1931**] *Let G be a bipartite graph. Then the number of edges in a maximum matching in G is equal to the number of vertices in a minimum vertex cover of G.*

Proof: Let $\{X, Y\}$ be the vertex bipartition of bipartite graph G, and let C^* be a minimum vertex cover of G. Then C^* is the disjoint union of its set of X-vertices, $C_X^* = C^* \cap X$, and its set of Y-vertices, $C_Y^* = C^* \cap Y$, as illustrated in Figure 10.4.9.

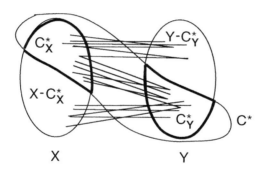

Figure 10.4.9   Minimum vertex cover $C^* = C_X^* \cup C_Y^*$.

Consider the bipartite subgraph G_1 of graph G induced on the vertex bipartition $\{C_X^*, Y - C_Y^*\}$. Let W be any subset of C_X^*. If $|W| > |N_{G_1}(W)|$, then there would exist $w \in W$ such that $N_{G_1}(W - \{w\}) = N_{G_1}(W)$. But this would imply that $(C_X^* - \{W\}) \cup C_Y^* = C^*$ is a vertex cover of graph G, contradicting the minimality of C^*. Thus, the bipartite graph G_1 satisfies Hall's Condition, and by Hall's Theorem (Theorem 10.4.3), G_1 has a C_X^*-saturating matching M_1^*, with $|M_1^*| = |C_X^*|$.

Next, let G_2 be the bipartite subgraph induced on the vertex bipartition $\{X - C_X^*, C_Y^*\}$. Then a similar argument applied to graph G_2 shows that it has a C_Y^*-saturating matching M_2^*, with $|M_2^*| = |C_Y^*|$. The edge set $M = M_1^* \cup M_2^*$ is clearly a matching in graph G, and $|M| = |C_X^*| + |C_Y^*| = |C^*|$. Thus, by Corollary 10.4.8, M is a maximum matching.

0-1 Matrices and the König-Egerváry Theorem

An interesting interpretation of this last theorem involves *0-1 matrices*, which are matrices each of whose entries is 0 or 1.

Theorem 10.4.10: [*König-Egerváry, 1931*] *Let A be a 0-1 matrix. Then the maximum number of 1's in matrix A, no two of which lie in the same row or column, is equal to the minimum number of rows and columns that together contain all the 1's in A.*

Proof: Let G be a bipartite graph with vertex bipartition $\{X, Y\}$, where X and Y correspond to the rows and columns of matrix A, respectively, and whose adjacencies correspond to the 1's in A. The result follows by applying Theorem 10.4.9 (see Exercises).

$\diamondsuit$

Application 10.4.4: *The Bottleneck Problem* Suppose that a manufacturing process consists of five operations that are performed simultaneously on five machines. The time in minutes that each operation takes when executed on each machine is given in the table below. Determine whether it is possible to assign the operations so that the process is completed within 4 minutes.

	M1	M2	M3	M4	M5
Op1	4	5	3	6	4
Op2	5	6	2	3	5
Op3	3	4	5	2	4
Op4	4	8	3	2	7
Op5	2	6	6	4	5

Let M be the 5×5 matrix whose ij^{th} entry a_{ij} equals 1 if operation i takes no more than 4 minutes when performed on machine j, and equals 0 otherwise. Thus,

$$M = \begin{pmatrix} 1 & 0 & 1 & 0 & 1 \\ 0 & 0 & 1 & 1 & 0 \\ 1 & 1 & 0 & 1 & 1 \\ 1 & 0 & 1 & 1 & 0 \\ 1 & 0 & 0 & 1 & 0 \end{pmatrix}$$

It is easy to check that this matrix contains five 1's no two of which are in the same row or column, which implies that it is possible to complete the process within 4 minutes.

Summary of Equivalences among Theorems in Chapter 10

The occurrence of various max-min pairs of problems and the similarities among many of the proofs in this chapter suggest strong connections among the theorems involved, some of which have already been established. In fact, any one of the following theorems can be used to prove the others.

- Menger's Theorem (Theorem 10.3.18)
- Max-Flow Min-Cut Theorem (Theorem 10.2.4)
- König's Theorem (Theorem 10.4.9)
- Hall's Theorem (Theorem 10.4.3)
- König-Egerváry Theorem (Theorem 10.4.10)

The following is an outline of some of these connections.

Max-Flow Min-Cut $\Rightarrow$ Menger (Theorems 10.3.5, 10.3.9, 10.3.17, 10.3.18)

Max-Flow Min-Cut $\Rightarrow$ Hall (Theorem 10.4.3)

Hall $\Rightarrow$ König (Theorem 10.4.9)

König $\Rightarrow$ König-Egerváry (Theorem 10.4.10)

König-Egerváry $\Rightarrow$ Hall (Exercise 10.4.32)

Menger $\Rightarrow$ Hall (Exercise 10.4.31)

Menger $\Rightarrow$ Max-Flow Min-Cut (Exercise 10.3.25)

EXERCISES for Section 10.4

For Exercises 10.4.1 through 10.4.4, use Maximum-Bipartite-Matching Algorithm 10.4.1 to find a maximum matching for the given bipartite graph.

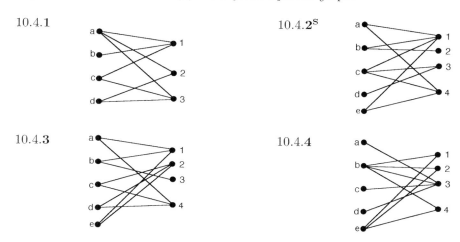

10.4.1

10.4.2^S

10.4.3

10.4.4

10.4.5 Five men and five women are attending a dance. Katie will dance only with Gary or Harry, Barbara will dance only with Fred or Ignatz, Carol will dance only with Harry or Jack, Donna will dance only with Fred or Gary, and Emma will dance only with Gary or Ignatz. Is it possible for all ten people to dance the last dance so that each woman dances with someone she finds acceptable?

10.4.6^S The Art History Department would like to offer six courses during the fall semester. There are seven professors in the department, each of whom is willing to teach certain courses, as shown in the table. Is there an assignment of professors to courses so that no professor teaches more than one course?

Course	Professor
Greek & Roman	Shargaa, Ward, Johnson, Pate
Renaissance	Maupin, Shargaa, Margeson
Baroque	Maupin, Shargaa
Impressionism	Maupin, Margeson
Early Modern	Vigorito, Johnson, Margeson
Contemporary	Shargaa, Margeson

For Exercises 10.4.7 through 10.4.10, determine whether Hall's Condition for the existence of a transversal is met by the given family of subsets. If the condition is met, then find a transversal; otherwise, show how the condition is violated.

10.4.7 $\{1,2,4\},\{2,4\},\{2,3\},\{1,2,3\}$.

10.4.8 $\{1,2,5\},\{1,5\},\{1,2\},\{2,5\}$.

10.4.9 $\{1,2,3\},\{1,2,4\},\{1,3,4\},\{1,2,3,4\},\{2,3,4\}$.

10.4.10^S $\{1,2\},\{2,3\},\{5\},\{1,3\},\{4,5\},\{4,5\}$.

For Exercises 10.4.11 through 10.4.14, find a maximum matching and a minimum vertex cover for the given graph.

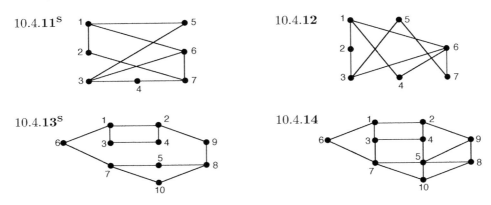

10.4.15 Let G be a bipartite graph and $\vec{G}_{st}$ be its corresponding s-t network (referred to in Proposition 10.4.1). Prove that if f is an integral flow in network $\vec{G}_{st}$, then the set M of edges of graph G whose corresponding arcs in $\vec{G}_{st}$ have unit flow is a matching in G.

10.4.16 Let M be a matching in a bipartite graph G. For any edge e in G, let $\vec{e}$ denote the corresponding arc in network $\vec{G}_{st}$ (described in the subsection preceding Proposition 10.4.1). Consider the function f defined by

$$f(a) = \begin{cases} 1, & \text{if } a = \vec{e}, \text{for some edge } e \in M \\ 1, & \text{if arc } a \text{ is adjacent to } \vec{e}, \text{some edge} e \in M \\ 0, & \text{otherwise} \end{cases}$$

for every arc a in network $\vec{G}_{st}$. Show that f is a feasible flow in $\vec{G}_{st}$.

10.4.17^S Let G be a bipartite graph with vertex bipartition $\{X,Y\}$, and let δ_X be the minimum degree of the vertices in X, and Δ_Y be the maximum degree of the vertices in Y. Prove that if $\delta_X \geq \Delta_Y$, then there exists an X-saturating matching in graph G.

10.4.18 Suppose that there are n workers and n jobs to be performed. Each worker is qualified to perform exactly k jobs, $k \geq 1$, and each job can be performed by exactly k workers. Prove that each job can be assigned to a different worker who is qualified for that job.

10.4.19^S Let G be a bipartite graph with vertex bipartition $\{X,Y\}$ such that $|X| = |Y| = m$ and every vertex has degree $k, k \geq 1$. Prove that there exists an X-saturating matching in graph G.

10.4.20 Let $\mathcal{F} = \{S_1, S_2, \ldots, S_r,\}$ be a family of subsets of the set $\{1, 2, \ldots, r\}$, each of size $m, m \geq 1$. Prove that there exists a transversal of $\mathcal{F}$.

10.4.21 Let G be a graph, and let W be a subset of V_G. Prove that W is an *independent set* of graph G (§2.3) if and only if the set $V_G - W$ is a vertex cover of G.

For Exercises 10.4.22 through 10.4.24, verify König-Egerváry Theorem 10.4.10 for the given 0-1 matrix.

10.4.22
$$\begin{pmatrix} 1 & 1 & 0 & 0 & 1 \\ 0 & 1 & 1 & 1 & 1 \\ 1 & 0 & 1 & 0 & 0 \\ 1 & 1 & 0 & 0 & 1 \end{pmatrix}$$

10.4.23
$$\begin{pmatrix} 0 & 0 & 1 & 0 & 1 \\ 1 & 0 & 1 & 1 & 1 \\ 0 & 1 & 1 & 0 & 0 \\ 0 & 0 & 0 & 0 & 1 \end{pmatrix}$$

10.4.24
$$\begin{pmatrix} 1 & 0 & 1 & 0 & 1 \\ 1 & 0 & 1 & 1 & 1 \\ 0 & 1 & 1 & 0 & 0 \\ 0 & 0 & 1 & 1 & 0 \\ 1 & 1 & 0 & 0 & 1 \end{pmatrix}$$

10.4.25^S Suppose that there are four jobs to be performed and five workers who are qualified for every job. The time in hours each worker needs to do each job is given in the table below. Is it possible to assign each job to a different worker so that each worker starts at the same time, and all four jobs are completed within four hours?

	Job1	Job2	Job3	Job4
W1	3	7	5	8
W2	6	3	2	3
W3	3	5	8	6
W4	5	8	6	4
W5	6	5	7	3

10.4.26 Prove the weak duality result (Proposition 10.4.7) for matchings and vertex covers.

10.4.27^S Prove the certificate of optimality (Corollary 10.4.8) for matchings and vertex covers.

10.4.28 Show that Theorem 10.4.9 of König does not necessarily hold for a nonbipartite graph.

10.4.29^S Express Hall's Theorem as a solution to the *Marriage Problem* (see the remark following Corollary 10.4.8).

10.4.**30** Fill in the details of the proof of König-Egerváry Theorem 10.4.10 for 0-1 matrices.

10.4.**31** Derive Hall's Theorem 10.4.3 from Menger's Theorem 10.3.18.

10.4.**32** Derive Hall's Theorem 10.4.3 from König-Egerváry Theorem 10.4.10.

10.5   SUPPLEMENTARY EXERCISES

10.5.1 Let G be a bipartite graph with vertex bipartition $\{X, Y\}$ such that $|X| = |Y| = m$ and every vertex has degree k, $k \geq 1$. Prove that there exists a partition $\{M_1, M_2, \ldots, M_k\}$ of E_G such that each edge set M_i is a matching in graph G. (Hint: See Exercise 10.4.19.)

10.5.2 At a certain job fair, there are 10 prospective employers and 10 job seekers. Each employer would like to interview exactly three job seekers, and each job seeker would like to be interviewed by exactly three employers. By prior arrangement, each of the interviews has been mutually agreed upon. In other words, an employer wants to interview a job seeker if and only if that job seeker also wants to be interviewed by that employer. Prove that it is possible to arrange all 30 interviews over the course of three days, such that all 20 people interview exactly once each day.

10.5.3 Let M be an $n \times n$ matrix of 0's and 1's such that every row and every column contain exactly s 1's, $s \geq 1$. Prove that there are n 1's such that no two lie in the same row or column.

10.5.4 Prove or disprove: A graph G is non-bipartite if and only if the number of edges in a maximum matching is unequal to the number of vertices in a minimum vertex cover.

10.5.5 Let G be a simple n-vertex graph. Let i^* be the number of vertices in a largest independent set of G, and let c^* be the number of vertices in a minimum vertex cover of G. Prove that $i^* + c^* = n$. (Hint: See Exercise 10.4.21.)

DEFINITION: An $m \times n$ **Latin rectangle** is an $m \times n$ matrix L whose entries are integers between 1 and n such that no two entries in any row or in any column are equal.

DEFINITION: A **Latin square** is a Latin rectangle with an equal number of rows and columns.

10.5.6 Application 10.5.1:   *Extending Latin Rectangles to Latin Squares*   Use Hall's Theorem to prove that any $m \times n$ Latin rectangle L with $m < n$ can be extended to a Latin square by the addition of $n - m$ new rows. (Hint: Let $A = \{1, 2, \ldots, n\}$ and $\mathcal{F} = \{S_1, S_2, \ldots, S_n\}$, where S_i is the subset of A of elements that do not occur in the ith column of rectangle L. Then show that L can be extended to an $(m+1) \times n$ Latin rectangle by showing that family $\mathcal{F}$ satisfies Hall's Condition.)

10.5.7 Use Hall's Theorem to show there is no perfect matching in the following graph. (Hint: Think strategically; don't fight the 2^8 subsets.) Also, find a maximum matching.

10.5.8 Determine whether the following collection of subsets of $\{1, 2, 3, 4, 5, 6, 7, 8\}$ has a system of distinct representatives? $B = \{5, 6, 7\}$, $C = \{4, 5\}$, $J = \{2, 3, 6, 7, 8\}$, $K = \{1, 4\}$, $L = \{1, 3, 4\}$, $Q = \{2, 5, 6, 8\}$, $T = \{1, 5\}$, $U = \{3, 5\}$. Hint: If your approach would require exponential time, think of a polynomial-time method.

Glossary

f-augmenting path Q for a given flow f in a network N: an s-t quasi-path in N such that the flow on each forward arc can be increased, and the flow on each backward arc can be decreased.

backward arc on an s-t quasi-path $Q = \langle s = v_0, e_1, v_1, \ldots, v_{k-1}, e_k, v_k = t \rangle$: an arc e_i that is directed from vertex v_i to vertex v_{i-1}.

bipartite graph: a graph for which there exists a vertex bipartition $\{X, Y\}$ of V_G such that every edge of G has one endpoint in vertex set X and one endpoint in vertex set Y.

capacitated network: a connected digraph such that each arc e is assigned a non-negative weight $cap(e)$, called the *capacity* of arc e.

capacity of an arc: see *capacitated network*.

capacity of an s-t cut $\langle V_s, V_t \rangle$: the sum of the capacities of the arcs in cut $\langle V_s, V_t \rangle$. That is, $cap\langle V_s, V_t \rangle = \sum_{e \in \langle V_s, V_i \rangle} cap(e)$.

conservation-of-flow property for a flow f in a network N: one of two conditions that a feasible flow must satisfy; $\sum_{e \in In(v)} f(e) = \sum_{e \in Out(v)} f(e)$ for every vertex v in network N, excluding source s and sink t.

s-t cut $\langle V_s, V_t \rangle$ in an s-t network N: the set of all arcs from a vertex subset V_s to a vertex subset V_t, where $\{V_s, V_t\}$ is a partition of V_N with source $s \in V_s$ and sink $t \in V_t$.

edge-connectivity $\kappa_e(G)$ of a graph G: the size of a minimum edge-cut in G.

—, local $\kappa_e(w, t)$ between distinct vertices s and t: the minimum number of edges in an s-t separating edge set in G.

Eulerian graph: a graph that has an Eulerian tour.

Eulerian tour in a graph G: a closed Eulerian trail in G.

Eulerian trail in a graph G: a trail that contains every edge of G.

factor of a graph G: a spanning subgraph of G.

—, k-: a k-regular factor of G.

factorization of a graph G: a set of factors whose edge-sets form a partition of the edge-set E_G.

—, k-: a factorization of G into k-factors.

(feasible) flow f in a capacitated network N: a function $f : E_N \to R^+$ that assigns a nonnegative real number $f(e)$ to each arc e such that:
1. (*capacity constraints*) $f(e) \le cap(e)$, for every arc e in network N.
2. (*conservation constraints*) $\sum_{e \in In(v)} f(e) = \sum_{e \in Out(v)} f(e)$, for every vertex v in network N, other than source s and sink t.

—, **maximum:** a flow in network N having the maximum value.

flow augmenting path: see (*f-*)*augmenting path.*

forward arc on an *s-t* quasi-path $Q = \langle s = v_0, e_l, v_1, \ldots, v_{k-1}, e_k, v_k = t \rangle$: an arc e_i that is directed from vertex v_{i-1} to vertex v_i.

frontier arc: an arc e directed from a tree vertex v to a non-tree vertex w.

Hall's Condition for a bipartite graph G with vertex bipartition $\{X, Y\}$: the condition $|W| \le |N(W)|$ for every subset W of X.

in-set for a vertex v in a digraph N, denoted $In(v)$: the set of all arcs e that are directed *to* v, i.e., $In(v) = \{e \in E_N | head(e) = v\}$.

independent set of vertices of G: a subset of V_G such that no two vertices in W are adjacent in graph G.

internally disjoint (directed) paths in a (di)graph: (directed) paths that have no internal vertices in common.

$m \times n$ **Latin rectangle:** an $m \times n$ matrix L whose entries are integers between 1 and n such that no two entries in any row or in any column are equal.

Latin square: a Latin rectangle with an equal number of rows and columns.

local edge-connectivity $\kappa_e(s, t)$ between vertices s and t in a graph G: the minimum number of edges in an *s-t* separating edge set in G.

local vertex-connectivity $\kappa_v(s, t)$ between non-adjacent vertices s and t in a graph G: the minimum number of vertices in an *s-t* separating vertex set in G.

matched with a vertex y by a matching M: said of a vertex x that is joined to y by an edge in M.

matching in a graph G: a set of edges of G such that no two of them have an endpoint in common.

—, **maximum:** a matching with the greatest number of edges.

maximum flow f^* in a network N: a flow in network N having the maximum value.

maximum matching in a graph: see *matching, maximum.*

minimum cut of a network N: a cut with the minimum capacity.

minimum vertex cover: a vertex cover with the least number of vertices.

network: shortened name used in this chapter for *single source–single sink, capacitated* network.

s-t **network:** a single source–single sink network with source s and sink t.

0-1 network: a network such that each arc has capacity 0 or 1.

out-set for a vertex v in a digraph N, denoted $Out(v)$: the set of all arcs e that are directed *from* v, i.e., $Out(v) = \{e \in E_N | tail(e) = v\}$.

partition-cut $\langle V_1, V_2 \rangle$ in a graph G: the set of all edges of G having one endpoint in vertex subset V_1 and the other endpoint in vertex subset V_2, where $\{V_1, V_2\}$ form a partition of V_G.

s-t **quasi-path** in a network N: an alternating sequence $\langle s = v_0, e_1, v_1, \ldots, v_{k-1}, e_k, v_k = t \rangle$ of vertices and arcs that forms an *s-t* path in the underlying graph of N.

X-**saturating** matching M in a bipartite graph G with vertex bipartition $\{X, Y\}$: a matching M such that each vertex in set X is an endpoint of an edge in M.

s-t **separating arc set** in a digraph D: a set of arcs whose removal destroys all directed *s-t* paths in D; an arc subset of E_D that contains at least one arc of every directed *s-t* path in digraph D.

s-t **separating edge set** in a graph G: a set of edges whose removal destroys all *s-t* paths in G; an edge subset of E_G that contains at least one edge of every *s-t* path in G.

s-t **separating vertex set** in a graph G (or digraph D): a set of vertices whose removal destroys all *s-t* paths in G (or all directed *s-t* paths in D); a set of vertices that contains at least one internal vertex of every (directed) *s-t* path.

single source–single sink network: a connected digraph that has a distinguished vertex called the **source** with nonzero outdegree, and a distinguished vertex called the **sink** with nonzero indegree.

sink see *single source–single sink network*.

slack Δ_e **for arc** e on a given f-augmenting path Q: the quantity $cap(e) - f(e)$ if e is a forward arc, and the quantity $f(e)$ if e is a backward arc.

source: see *single source–single sink network*.

system of distinct representatives: synonym for *transversal*.

transversal of a family $\mathcal{F} = \{S_1, S_2, \ldots, S_r\}$ of (not necessarily distinct) nonempty subsets of a nonempty finite set A: a sequence $T = \langle a_1, a_2, \ldots, a_r \rangle$ of r distinct elements of set A, such that $a_i \in S_i$ for each $i = 1, \ldots, r$.

value of flow f in a network N, denoted $val(f)$: the net flow leaving the source s; $val(f) = \sum_{e \in Out(s)} f(e) - \sum_{e \in In(s)} f(e)$.

vertex cover of a graph G: a subset C of the vertices of G such that every edge of G is incident on at least one vertex in C.

—, **minimum:** a vertex cover with the least number of vertices.

Chapter 11

GRAPH COLORINGS AND SYMMETRY

INTRODUCTION

This chapter explores the interplay between a graph's symmetry and the number of different colorings of that graph. For example, if one coloring of a graph can be obtained from another coloring by a simple rotation of the graph, then the two colorings are equivalent. The symmetries of a graph are precisely defined using the graph's automorphism group, along with its vertex- and edge-permutations. The equivalence classes under these group actions determine when colorings are equivalent. We develop the basic mathematical concepts and tools to enumerate these equivalence classes.

Although we restrict our attention mainly to simple graphs, the methods studied in this chapter are applicable to general graphs.

11.1   AUTOMORPHISMS OF SIMPLE GRAPHS

The concept of graph automorphism was introduced in §2.2. For convenience, its definition is presented below.

REVIEW FROM §2.1 AND §2.2:

- A vertex bijection $f_V : V_G \to V_H$ between the vertex-sets of simple graphs G and H is **structure-preserving** if it preserves adjacency and non-adjacency, that is, if for every pair of vertices in G,

$$u \text{ and } v \text{ are adjacent in } G \Longleftrightarrow f_V(u) \text{ and } f_V(v) \text{ are adjacent in } H$$

- Every structure-preserving vertex bijection $f_V : V_G \to V_H$ of simple graphs induces an edge bijection $f_E : E_G \to E_H$, given by the rule $f_E(uv) = f_V(u)f_V(v)$.

- A **graph isomorphism** $f : G \to H$ of simple graphs is completely specified by a structure-preserving vertex bijection $f_V : V_G \to V_H$. Formally, it consists of the vertex bijection f_V paired with the induced edge bijection f_E.

- An **automorphism** π of a graph G is an isomorphism from G to itself. Thus, the vertex bijection π_V is a permutation on V_G, and the edge bijection π_E is a permutation on E_G.

NOTATION: The set of automorphisms of a graph G is denoted $\mathcal{A}ut(G)$, and the corresponding sets of vertex-permutations and of edge-permutations are denoted $\mathcal{A}ut_V(G)$ and $\mathcal{A}ut_E(G)$, respectively.

The Sets $\mathcal{A}ut_V(G)$ and $\mathcal{A}ut_E(G)$ Are Permutation Groups

DEFINITION: A collection P of permutations on the same set of objects is **closed under composition** if for every pair $\pi_1, \pi_2 \in P$, the composition $\pi_1\pi_2$ is in P.

DEFINITION: Let P be a nonempty collection of permutations of the finite set of objects Y such that P is closed under composition. Then the set P together with set Y is called a **permutation group** and is denoted $[P : Y]$.

TERMINOLOGY: The permutation group is said to **act on the set** Y.

DEFINITION: The **full symmetric group** Σ_Y on a set Y is the collection of *all* permutations on Y.

Theorem 11.1.1: *The set $Aut(G)$ of all automorphisms of a simple graph G acts as a permutation group on V_G and on E_G.*

Proof: It is straightforward to verify that the composition of two structure-preserving vertex-permutations is a structure-preserving vertex-permutation (see Exercises). Thus, $\mathcal{A}ut_V(G)$ and $\mathcal{A}ut_E(G)$ are permutation groups. $\diamond$

DEFINITION: Let $[P : Y]$ be a permutation group. The **orbit** of an object $y \in Y$ is the set of all objects $z \in Y$ such that $\pi(y) = z$ for some permutation $\pi \in P$.

- The action of the automorphism group $\mathcal{A}ut(G)$ on a graph G partitions V_G into **vertex orbits**. That is, the vertices u and v are in the same orbit if there exists an automorphism π such that $\pi(u) = v$. Similarly, $\mathcal{A}ut(G)$ partitions E_G into **edge orbits**.

- A graph G is **vertex-transitive** if all the vertices are in a single orbit. Similarly, G is **edge-transitive** if all the edges are in a single orbit.

Remark: In this chapter, permutations are represented in *disjoint-cycle form* (introduced in §2.2). Permutation groups and abstract groups are discussed in Appendix A.4.

Example 11.1.1: The graph $K_{1,3}$, shown below, has six automorphisms, and the vertex- and edge-permutation groups corresponding to $\mathcal{A}ut(K_{1,3})$ are given by

$$\mathcal{A}ut_V(K_{1,3}) = \{(u)(v)(w)(x),\ (x)(u\ v\ w),\ (x)(u\ w\ v),\ (x)(u)(v\ w),$$
$$(x)(v)(u\ w),\ (x)(w)(u\ v)\}$$

$$\mathcal{A}ut_E(K_{1,3}) = \{(a)(b)(c),\ (a\ b\ c),\ (a\ c\ b),\ (a)(b\ c),\ (b)(a\ c),\ (c)(a\ b)\}$$

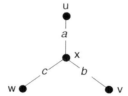

Figure 11.1.1   The graph $K_{1,3}$.

The vertex-orbits are $\{u, v, w\}$ and $\{x\}$. The only edge-orbit is $\{a, b, c\}$. Therefore $K_{1,3}$ is edge-transitive, but not vertex-transitive.

Remark: Figure 11.1.1 gives a drawing of $K_{1,3}$ whose geometric symmetry captures all of the automorphisms of that graph. However, this is not always possible. The Petersen graph has no such drawing, as shown in Example 2.2.4.

Automorphism Groups of Some Other Simple Graphs

The methods used above to determine $\mathcal{A}ut(K_{1,3})$ are now applied to calculating the automorphism groups of some other graphs.

Example 11.1.2: $\mathcal{A}ut(K_n)$ Each of the $n!$ permutations on the vertex-set of K_n is structure-preserving, because every pair of vertices is joined by an edge. Thus, every vertex-permutation specifies a different automorphism of K_n.

Figure 11.1.2   Automorphism action on an edge and its endpoints.

The complete graph K_4 can be represented as the 1-skeleton of a regular tetrahedron in Euclidean 3-space. All 24 automorphisms of K_4 can be realized by rotations and reflections of a regular tetrahedron. To generalize this geometric viewpoint to larger values of n, represent the complete graph K_n as the 1-skeleton of a regular n-simplex in Euclidean n-space. Then all $n!$ automorphisms can be realized by combinations of rotations and reflections of a regular n-simplex.

Example 11.1.3: $Aut(C_n)$ The cycle graph C_n can be represented as the 1-skeleton of a regular n-gon in the plane, as illustrated in Figure 11.1.3 for $n = 5$. All of its automorphisms can be realized as rotations and reflections of a regular n-gon. For instance, rotation 72° clockwise corresponds to the automorphism of C_5 whose vertex-permutation is $(u\ v\ w\ x\ y)$ and whose edge-permutation is $(a\ b\ c\ d\ e)$. Also, reflection through the vertical axis of symmetry corresponds to the automorphism whose vertex-permutation is $(u)(v\ y)(w\ x)$ and whose edge-permutation is $(a\ e)(b\ d)(c)$. In general, a regular n-gon has n rotations and n reflections, which gives the graph C_n a total of $2n$ automorphisms.

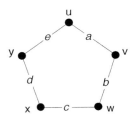

Figure 11.1.3   The cycle graph C_5 has 5 rotations and 5 reflections.

Example 11.1.4: $Aut(W_n)$ A geometric representation of the n-spoked wheel graph W_n can be constructed by placing a vertex at the center of a regular n-gon in the plane, and joining it to every vertex of the n-gon, as illustrated for $n = 5$ in Figure 11.1.4. All of its automorphisms can be realized as rotations and reflections of the n-gon. For instance, rotation 72° clockwise corresponds to the automorphism of W_5 whose vertex-permutation is $(t)(u\ v\ w\ x\ y)$ and whose edge-permutation is $(a\ b\ c\ d\ e)(f\ g\ h\ i\ j)$. Also, reflection through the vertical axis of symmetry corresponds to the automorphism whose vertex-permutation is $(t)(u)(v\ y)(w\ x)$ and whose edge-permutation is $(a\ e)(b\ d)(c)(f\ i)(g\ h)(j)$. Thus, the wheel graph W_n has n rotations and n reflections, for a total of $2n$ automorphisms, except when $n = 3$. In the special case $W_3 \cong K_4$, in which the hub vertex has the same degree as the rim vertices, there are 24 automorphisms.

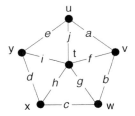

Figure 11.1.4   The wheel graph W_5 has 5 rotations and 5 reflections.

Example 11.1.5: $\mathcal{Aut}(K_{m,n})$ Let $K_{m,n}$ have vertex bipartition $R \cup S$, with $m = |R|$ and $n = |S|$, and let $\tau \in \mathcal{Aut}(K_{m,n})$. Then, since τ preserves non-adjacency, two vertices on one side of the bipartition must map to two vertices on that same side or to two vertices on the other side. It follows that either $\tau(R) = R$ and $\tau(S) = S$, or else the bijection τ swaps R and S (i.e., $\tau(R) = S$ and $\tau(S) = R$).

If $m \neq n$, then $\tau(R) = R$ and $\tau(S) = S$ is the only possibility. Thus, τ permutes the elements of set R and permutes the elements of set S. That is, $\tau = \rho \oplus \sigma$, where $\rho \in \Sigma_R$ and $\sigma \in \Sigma_S$, and

$$(\rho \oplus \sigma)(u) = \left\{ \begin{array}{ll} \rho(u), & \text{if } u \in R \\ \sigma(u), & \text{if } u \in S \end{array} \right.$$

Conversely, for each pair of permutations $\rho \in \Sigma_R$ and $\sigma \in \Sigma_S$, $\rho \oplus \sigma$ is a structure-preserving vertex-permutation of $K_{m,n}$. Therefore, if $m \neq n$, then $|\mathcal{Aut}_V(K_{m,n})| = m!n!$.

If $m = n$, then $\tau \in \mathcal{Aut}(K_{n,n})$ is either a permutation of the form $\rho \oplus \sigma$ or is a vertex bijection that swaps R and S. Moreover, each of the $(n!)^2$ vertex bijections that swap R and S is structure-preserving, since every vertex in R is adjacent to every vertex in S. Thus, $|\mathcal{Aut}(K_{n,n})| = 2(n!)^2$.

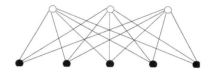

Figure 11.1.5   The bipartite graph $K_{3,5}$ has $3!5! = 720$ automorphisms.

EXERCISES for Section 11.1

In Exercises 11.1.1 through 11.1.8, determine the vertex-permutation and the edge-permutation for every automorphism of the graph shown. Also list the vertex-orbits and the edge-orbits.

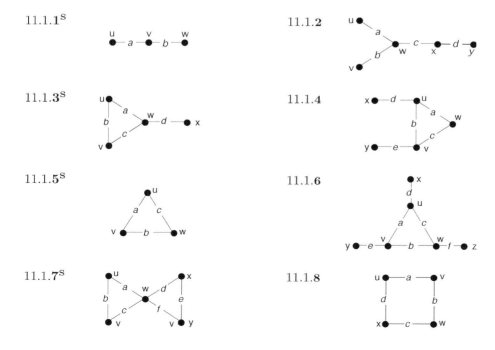

11.1.1^S

11.1.2

11.1.3^S

11.1.4

11.1.5^S

11.1.6

11.1.7^S

11.1.8

In Exercises 11.1.9 through 11.1.12, redraw the graph so that as many automorphisms as possible are represented by symmetries of the drawing.

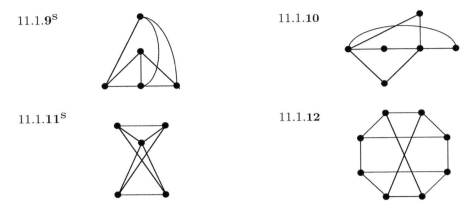

11.1.9$^\text{S}$ 11.1.10

11.1.11$^\text{S}$ 11.1.12

11.1.13 How many automorphisms are in the group $\mathcal{A}ut(K_{1,4})$?

$K_{1,4}$

11.1.14 Suppose that K_3 and K_4 are amalgamated at a vertex. How many automorphisms does the resulting graph have? How many vertex-orbits and edge-orbits?

11.1.15 Suppose that an edge is deleted from K_5. How many automorphisms does the resulting graph have? How many vertex-orbits and edge-orbits?

11.1.16 Suppose that the complete graph K_3 is joined to the edgeless graph on 5 vertices. How many automorphisms does the resulting graph have? How many vertex-orbits and edge-orbits?

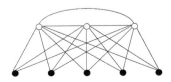

11.1.17 How many automorphisms are in the group $\mathcal{A}ut(K_{2,3,4})$, where $K_{2,3,4}$ is the *complete tripartite graph* with tripartition consisting of a 2-vertex, 3-vertex, and 4-vertex subset? How many vertex-orbits and edge-orbits?

11.1.18 How many automorphisms are in the group $\mathcal{A}ut(CL_n)$, where CL_n is the n-vertex circular ladder graph (§1.2)? How many vertex-orbits and edge-orbits?

11.2   EQUIVALENCE CLASSES OF COLORINGS

The symmetry of a graph G has a substantial effect on how we determine the number of *different* vertex-colorings and edge-colorings. What we actually count is the number of equivalence classes of its vertex- and edge-colorings, induced by the automorphism group $\mathcal{A}ut(G)$, and this section gives a precise description of these equivalence classes.

Coloring a Set Subject to the Action of a Permutation Group

Counting equivalence classes of graph colorings lies within the general context of counting equivalent colorings of a set acted upon by a permutation group.

DEFINITION: A k-**coloring** of a set Y is a mapping f from Y *onto* the set $\{1, 2, \ldots, k\}$, in which the value $f(y)$ is called the **color** of y. Any k-coloring of Y is also called a **coloring**.

DEFINITION: A $(\leq k)$-**coloring** of a set Y is a coloring that uses k or *fewer* colors, formally a mapping f from Y onto any set $\{1, 2, \ldots, t\}$ with $t \leq k$.

NOTATION: The set of all $(\leq k)$-colorings of the elements of a set Y is denoted $Col_k(Y)$.

Proposition 11.2.1: *Let Y be a set. Then $|Col_k(Y)| = k^{|Y|}$.*

Proof: This is a direct application of the Rule of Product, a familiar counting principle in elementary discrete mathematics (see Appendix A.3).   ◇

DEFINITION: Let $\mathcal{P} = [P : Y]$ be a permutation group acting on a set Y, and let f and g be $(\leq k)$-colorings of the objects in Y. Then the coloring f is $\mathcal{P}$-**equivalent** to the coloring g if there is a permutation $\pi \in P$ such that $g = f\pi$, that is, if for every object $y \in Y$, the color $g(y)$ is the same as the color $f(\pi(y))$.

TERMINOLOGY: The $\mathcal{P}$-equivalence classes are also called **coloring classes**, or $\mathcal{P}$-**orbits**. That is, any two $\mathcal{P}$-equivalent colorings are in the same $\mathcal{P}$-orbit.

NOTATION: The set of $\mathcal{P}$-orbits of $Col_k(Y)$ is denoted $\{Col_k(Y)\}_{\mathcal{P}}$. Thus, $|\{Col_k(Y)\}_{\mathcal{P}}|$ equals the number of non-equivalent $(\leq k)$-colorings.

NOTATION: The identity permutation on a set Y is often denoted ε.

- For any $y \in Y$, $\varepsilon(y) = y$.

- For any permutation in $\pi \in \Sigma_Y$, $\varepsilon\pi = \pi = \pi\varepsilon$.

Example 11.2.1: Let $Y = \{x, y, z\}$ be the vertex-set V_{C_3} of the 3-cycle graph C_3, shown on the left in Figure 11.2.1, and let $P = \{\varepsilon, (x\ y\ z), (x\ z\ y)\}$ be the group of vertex-permutations corresponding, respectively, to the $0°$, $120°$, and $240°$ clockwise rotations of C_3. The set $Col_2(V_{C_3})$, which consists of eight (≤ 2)-colorings, is shown on the right in Figure 11.2.1. Each (≤ 2)-coloring $f : V_{C_3} \to \{1, 2\}$ is represented graphically using white (color 1) and black (color 2) vertices, and also as the string $f(x)f(y)f(z)$.

These eight vertex-colorings are partitioned into four $\mathcal{P}$-orbits of $Col_2(V_{C_3})$ given by $[Col_2(V_{C_3})]_{\mathcal{P}} = \{\{111\}, \{211, 121, 112\}, \{122, 212, 221\}, \{222\}\}$.

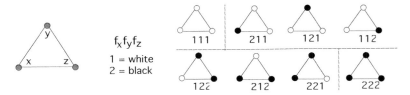

Figure 11.2.1   The four coloring classes of V_{C_3} under rotational equivalence.

Remark: Intuitively, the reason why the colorings in one class, e.g., $\{211, 121, 112\}$, are equivalent is because rotating a drawing of the graph superimposes one of these colorings automorphically onto another with correct color matching. This intuitive notion is what is formally represented by the condition $g = f\pi$.

Induced Permutation Actions

In Example 11.2.1, we saw how the permutations acting on the three vertices of C_3 *induced* another action on the set of eight vertex-(≤ 2)-colorings. More generally, a permutation group acting on a set of objects also acts on the set of colorings of those objects, as we now show.

Proposition 11.2.2: *Let $\mathcal{P} = [P : Y]$ be a permutation group acting on a set Y. Let $f \in Col_k(Y)$ be a ($\leq k$)-coloring of Y, and let $\pi \in P$ be a permutation in $\mathcal{P}$. Then the composition $f\pi$ of permutation π followed by coloring f is a coloring in $Col_k(Y)$.*

Proof: The composition $f\pi$ is a coloring of Y because it assigns to each object $y \in Y$ whatever color f assigns to the object $\pi(y)$. $\Diamond$

Corollary 11.2.3: *Let $\mathcal{P} = [P : Y]$ be a permutation group acting on a set Y, and let $\pi \in P$. Then the mapping $\pi_{YC} : Col_k(Y) \to Col_k(Y)$ defined by*

$$\pi_{YC} : (f) = f\pi, \text{ for every coloring } f \in Col_k(Y)$$

is a permutation on the set $Col_k(Y)$.

Proof: The mapping π_{YC} is a bijection, because it has an inverse, namely, the rule $g \mapsto g\pi^{-1}$. $\Diamond$

DEFINITION: The mapping $\pi_{YC} : Col(Y) \to Col_k(Y)$ defined (in Corollary 11.2.3) by the rule $f \mapsto f\pi$ is called the **induced permutation action** of π on $Col_k(Y)$.

NOTATION: To distinguish between the action of a permutation π on a set Y and its induced action on the set $Col_k(Y)$ of colorings of Y, we let π_Y denote its action on Y and π_{YC} its action on $Col_k(Y)$. When there is no risk of confusion, the subscripts Y and YC may both be omitted.

DEFINITION: Let $\mathcal{P} = [P : Y]$ be a permutation group acting on a set Y. The collection $\mathcal{P}_{YC} = [P : Col_k(Y)]$ of induced permutations on $Col_k(Y)$ is called the **induced permutation group**.

NOTATION: When it is necessary to distinguish between the group that acts on the set Y and the group that acts on the set $Col_k(Y)$, the respective notations $\mathcal{P}_Y$ and $\mathcal{P}_{YC}$ are used.

Example 14.2.1, continued: Figure 11.2.2 below depicts the disjoint-cycle form of each permutation π_Y and also of the corresponding induced permutation π_{YC} acting on the set $Col_2(Y) = \{111, 112, 121, 122, 211, 212, 221, 222\}$.

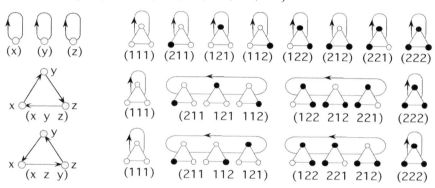

Figure 11.2.2   Correspondence of permutations and induced permutations.

Notice that two colorings are $\mathcal{P}$-equivalent (i.e., in the same $\mathcal{P}$-orbit) if the two colorings appear in a cycle of at least one induced permutation.

Equivalent Colorings of a Graph G under $\mathcal{A}ut(G)$

If the symmetries of a graph are ignored, then counting vertex- or edge-colorings is trivial; using k or fewer colors, there are $k^{|V_G|}$ vertex-colorings and $k^{|E_G|}$ edge-colorings (by Proposition 11.2.1). However, taking symmetry into account, by counting orbits of colorings, is more complicated.

TERMINOLOGY: $Col_k(V_G)$ is sometimes called the **full set of vertex-($\leq k$)-colorings** of a graph G, and $Col_k(E_G)$ the **full set of edge-($\leq k$)-colorings**.

DEFINITION: Let G be a graph with automorphism group $\mathcal{A}ut(G)$. **Equivalent vertex-colorings** are vertex-colorings that are $\mathcal{A}ut_V(G)$-equivalent, and **equivalent edge-colorings** are edge-colorings that are $\mathcal{A}ut_E(G)$-equivalent.

Thus, equivalent vertex-colorings are in the same $\mathcal{A}ut_V(G)$-orbit, and equivalent edge-colorings are in the same $\mathcal{A}ut_E(G)$-orbit.

NOTATION: The set of all $\mathcal{A}ut_V(G)$-orbits (coloring classes of $Col_k(V_G)$) is denoted $\{Col_k(V_G)\}_{\mathcal{A}ut_V(G)}$. Similarly, $\{Col_k(E_G)\}_{\mathcal{A}ut_E(G)}$ denotes the set of all $\mathcal{A}ut_E(G)$-orbits of $Col_k(E_G)$.

Example 11.2.2: The two vertex-colorings in Figure 11.2.3 are equivalent, because a 180°-rotation or a reflection of one graph drawing through its horizontal axis corresponds to a graph automorphism.

Figure 11.2.3   Two equivalent vertex-colorings of the graph $K_4 - K_2$.

Example 11.2.3: The two edge-colorings in Figure 11.2.4 are equivalent. Reflection of the graph through its vertical or horizontal axis corresponds to an automorphism that maps one coloring onto the other.

Figure 11.2.4   Two equivalent edge-colorings of the graph $K_4 - K_2$.

Counting Vertex- and Edge-Coloring Orbits One by One

For a small graph and a small number of colors, it is possible to count the orbits of vertex- and edge-colorings by drawing a list of representatives of those classes. Using graph automorphism invariants like vertex degree simplifies this kind of counting.

Example 11.2.4: An automorphism on the graph $K_4 - K_2$ must either fix the two 2-valent vertices or swap them, and it must either fix the two 3-valent vertices or swap them. Thus, $Aut(K_4 - K_2)$ has the following representations as a group of vertex-permutations and as a group of edge-permutations.[†]

Symmetry	$\pi \in \mathcal{Aut}_V(G)$	$\pi \in \mathcal{Aut}_E(G)$
identity	$(u)(v)(w)(x)$	$(a)(b)(c)(d)(e)$
refl. thru vert. axis	$(u)(w)(v\ x)$	$(e)(a\ b)(c\ d)$
refl. thru horiz. axis	$(v)(x)(u\ w)$	$(e)(a\ b)(b\ d)$
180° rotation	$(v\ w)(v\ x)$	$(e)(a\ d)(b\ c)$

Figure 11.2.5 shows how the (full) set $Col_2(V_{K_4-K_2})$ of 16 vertex-(≤ 2)-colorings is partitioned into nine $\mathcal{Aut}_V(K_4 - K_2)$-orbits.

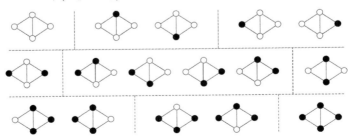

Figure 11.2.5   The $\mathcal{Aut}_V(K_4 - K_2)$-orbits of $Col_2(V_{K_4-K_2})$.

[†]The reader familiar with group theory will notice that the group $\mathcal{Aut}(K_4 - K_2)$ is abstractly isomorphic to $Z_4 \times Z_2$.

Example 11.2.4, continued: Since the graph $K_4 - K_2$ has five edges, there are, ignoring equivalences, $2^5 = 32$ edge-colorings that use two or fewer colors. Figure 11.2.6 shows a representative of each of the 14 $Aut_E(K_4 - K_2)$-orbits of $Col_2(E_{K_4 - K_2})$. Next to each representative edge-(≤ 2)-coloring of $K_4 - K_2$ in the figure is the number of colorings in its orbit.

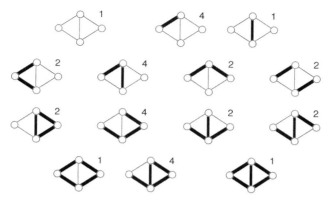

Figure 11.2.6   Representatives and sizes of the 14 orbits of $Col_2(E_{K_4 - K_2})$.

Elementary Application of Symmetries in Itemizing Colorings

As the size of the graph and the number of colors increase, it becomes progressively less practical to count orbits by ordinary itemization. However, systematic exploitation of graph symmetries often reduces the work.

Example 11.2.5: A systematic approach to counting the $Aut_V(P_5)$-orbits (coloring classes) of vertex-(≤ 2)-colorings of the path graph P_5 may begin with the observation that there is only one vertex-(≤ 2)-coloring with all vertices white. There are three coloring classes with four white vertices and one black, depending on whether the black vertex is at an end, next to an end, or in the middle. There are six coloring classes with three white and two black, as shown in Figure 11.2.7.

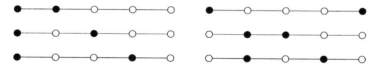

Figure 11.2.7   The six vertex-colorings of P_5 with 3 white and 2 black.

By symmetry between the colors white and black, there are also six coloring classes with three black vertices and two white, three coloring classes with four black and one white, and one class with all five vertices black, for a total of 20 $Aut_V(P_5)$-orbits of colorings. In other words, there are 20 non-equivalent vertex-(≤ 2)-colorings.

Example 11.2.6: The number of coloring classes of edge-(≤ 2)-colorings of P_5 is 10, according to the following inventory:

$$10 \text{ total} = \begin{cases} 1 & \text{with all four edges light} \\ 1 & \text{with all four edges dark} \\ 2 & \text{with three edges light and one edge dark} \\ 2 & \text{with one edge light and three edges dark} \\ 4 & \text{with two edges light and two edges dark} \end{cases}$$

Example 11.2.7: The number of (non-equivalent) edge-(≤ 3)-colorings of P_5 can be derived from the number of edge-(≤ 2)-colorings. We begin by calculating the number of edge-3-colorings. The three different edge colors are depicted by *bold*, *plain*, and *dashed* edges. Figure 11.2.8 shows the six kinds of edge-3-colorings of P_5 with the color *bold* used twice.

Figure 11.2.8   Six of the non-equivalent edge-3-colorings of P_5.

Since there are three choices for the color that is used twice, there are 18 edge-3-colorings. Since P_5 has eight edge-2-colorings with exactly two colors, according to the inventory in Example 11.2.6, and there are three choices for which two of the three colors are used, it follows that there are 24 edge-(≤ 3)-colorings that use exactly two of the three available colors. Finally, there are three ways to color all the edges with exactly one of the three available colors. The following inventory summarizes the number of non-equivalent edge-(≤ 3)-colorings of P_5.

$$45 \text{ total} = \begin{cases} 18 & \text{using all three colors} \\ 24 & \text{using exactly two of the three colors} \\ 3 & \text{using only one color} \end{cases}$$

If equivalences were ignored, the total would be $3^4 = 81$.

EXERCISES for Section 11.2

In Exercises 11.2.1 through 11.2.7, group the full set of vertex-(≤ 2)-colorings of the given graph G into $\mathcal{A}ut_V(G)$-orbits, as in Figure 11.2.5.

11.2.1 K_3.   **11.2.2^S** P_4.

11.2.3 $K_{1,3}$.   **11.2.4**

11.2.5 $K_{2,3}$.   **11.2.6** $K_5 - K_2$.

11.2.7 CL_3.

In Exercises 11.2.8 through 11.2.14, determine the number of $\mathcal{A}ut_V(G)$-orbits of the full set of vertex-(≤ 3)-colorings of the given graph G. Itemize by inventory as in Example 11.2.7.

11.2.8 K_3.   **11.2.9^S** P_4.

11.2.10 The graph of Exercise 11.2.3.   **11.2.11** $K_{1,3}$.

11.2.12 $K_{2,3}$.   **11.2.13** $K_5 - K_2$.

11.2.14 CL_3.

In Exercises 11.2.15 through 11.2.21, determine the number of $Aut_E(G)$-orbits of the full set of edge-(≤ 2)-colorings of the given graph. Itemize by inventory as in Example 11.2.6.

11.2.15 K_3.   **11.2.16^S** P_4.

11.2.17 The graph of Exercise 11.2.3.   **11.2.18** $K_{1,3}$.

11.2.19 $K_{2,3}$.   **11.2.20** $K_5 - K_2$.

11.2.21 CL_3.

In Exercises 11.2.22 through 11.2.28, determine the number of $Aut_E(G)$-orbits of the full set of edge-(≤ 3)-colorings of the given graph. Itemize by inventory as in Example 11.2.7.

11.2.22 K_3.   **11.2.23^S** P_4.

11.2.24 The graph of Exercise 11.2.3.   **11.2.25** $K_{1,3}$.

11.2.26 $K_{2,3}$.   **11.2.27** $K_5 - K_2$.

11.2.28 CL_3.

11.3   SUPPLEMENTARY EXERCISES

In Exercises 11.3.1 through 11.3.4, count the number of different vertex- and edge-colorings with 2 or fewer colors for each of the given graphs. That is, for each graph G, determine the number of $Aut_V(G)$-orbits in $Col_2(V_G)$ and the number of $Aut_E(G)$-orbits in $Col_2(E_G)$. Draw a representative (≤ 2)-coloring for each orbit.

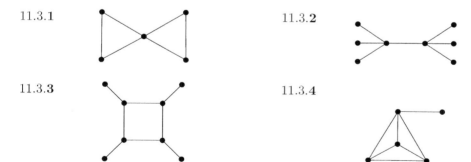

11.3.1

11.3.2

11.3.3

11.3.4

Glossary

automorphism of a graph G: an isomorphism from the graph G to itself.

automorphism group $Aut(G)$ **of a graph** G: the permutation group of all automorphisms of the graph G.

closed collection of permutations: a collection P of permutations on the same set of objects, such that for every pair $\pi_1, \pi_2 \in P$, the composition $\pi_1 \pi_2$ is in P.

coloring of a set Y: a mapping f from Y *onto* a set $\{1, 2, \ldots, k\}$ of integers, in which the number $f(y)$ is called the *color* of y.

$(\leq k)$-coloring of a set Y: a coloring that uses k or *fewer* colors. The set of all $(\leq k)$-colorings of Y is denoted $Col_k(Y)$.

coloring class for a set with a coloring: a subset of all objects of like color.

edge-automorphism group $Aut_E(G)$ **of a graph** G: the permutation group whose object set is E_G and whose permutations are the edge functions of the automorphisms of the graph G.

$\mathcal{P}$-equivalent colorings of a set Y **under a permutation group** $\mathcal{P} = [P : Y]$: colorings f and g for which there is a permutation $\pi \in P$ such that $g = f\pi$. That is, for every object $y \in Y$, the color $g(y)$ is the same as the color $f(\pi(y))$. The set of $\mathcal{P}$-equivalence classes of $Col_k(Y)$ is denoted $\{Col_k(Y)\}_{\mathcal{P}}$.

equivalent edge-colorings on a graph G: edge-colorings f_1 and f_2 on G that are $Aut_E(G)$-equivalent.

equivalent vertex-colorings on a graph G: vertex-colorings f_1 and f_2 on G that are $Aut_V(G)$-equivalent.

full symmetric group Σ_Y: the collection of *all* permutations on a set Y.

induced permutation π_{YC}: given a permutation $\pi \in \mathcal{P} = [P : Y]$, the rule $f \mapsto f\pi$ that permutes the set $Col_k(Y)$ of colorings of Y.

induced permutation group $\mathcal{P}_k$: given a permutation group $\mathcal{P} = [P : Y]$, the group of induced permutations on $Col_k(Y)$.

orbit of an object $y \in Y$ **under a permutation group** P: the set $\{\pi(y) | \pi \in P\}$.

permutation group $[P : Y]$: a mathematical structure such that P is a nonempty closed collection of permutations on the same finite set of objects Y.

symmetric group Σ_Y **on a set** Y: the collection of *all* permutations on Y.

vertex-automorphism group $Aut_V(G)$ **of a graph** G: the permutation group whose object set is V_G and whose permutations are the vertex functions of the automorphisms of the graph G.

APPENDIX

A.1 LOGIC FUNDAMENTALS

Propositional Logic

DEFINITION: Let p and q be propositions. The **conditional proposition** $p \to q$ is a proposition that is false when p is true and q is false and is true otherwise. It is read "if p then q."

TERMINOLOGY: The following are equivalent statements.

> $p \to q$
>
> if p then q
>
> p is a **sufficient condition** for q
>
> q is a **necessary condition** for p
>
> p only if q
>
> q if p

DEFINITION: The **converse** of $p \to q$ is the conditional proposition $q \to p$.

DEFINITION: The **contrapositive** of $p \to q$ is the conditional proposition $\neg q \to \neg p$.

Proposition A.1.1: *The proposition $p \to q$ and its contrapositive $\neg q \to \neg p$ are logically equivalent.*

DEFINITION: The statement p **if and only if** q means that both conditionals $p \to q$ and $q \to p$ are true.

Types of Proof

TERMINOLOGY: When an assertion is in the form $p \to q$, the proposition p is often referred to as the **antecedent** and q the **consequent**.

DEFINITION: A **direct proof** of the conditional $p \to q$ consists of showing that statement q is true under the assumption that statement p is true.

DEFINITION: A **proof by contrapositive** of the conditional $p \to q$ is a direct proof of the contrapositive $\neg q \to \neg p$.

Example A.1.1: A proof by contrapositive of the assertion "If n^2 is odd, then n is odd" takes the form: Assume n is even. Then $n = 2k$, for some integer k. Hence, $n^2 = (2k)^2 = 2(2k^2)$, which is even.

DEFINITION: A **proof by contradiction** of an assertion S consists of assuming that S is false and deriving some contradiction.

Example A.1.2: Prove: There is no largest prime number.

Proof by contradiction: Assume that P is the largest prime number. Then the product $r = 2 \times 3 \times \cdots \times P + 1$ is not divisible by any prime and, hence, is itself prime, which contradicts the maximality of P.

Mathematical Induction

DEFINITION: **Axiom of Mathematical Induction – First Form**: Statement $P(n)$ is true for all integers $n \geq b$ if both of the following are true.

Base: $P(b)$

Inductive step: For all $k \geq b, P(k) \to P(k+1)$

Remark: Typically (but not always), the inductive step is demonstrated by a direct proof of the statement $P(k) \to P(k+1)$.

Example A.1.3: Prove: For all $n \geq 1, \sum_{i=1}^{n} i = \frac{n}{2}(n+1)$.

Proof: The equation holds for $n = 1$ since its left and right sides are both equal to 1 when $n = 1$. Assume that the equation holds for some $k \geq 1$. Then

$$\sum_{i=1}^{k+1} i = \left(\sum_{i=1}^{k} i\right) + (k+1) = \left(\frac{k}{2}(k+1)\right) + (k+1) = \frac{k+1}{2}(k+2)$$

$\diamond$

DEFINITION: **Axiom of Mathematical Induction — Second Form**: Statement $P(n)$ is true for all integers $n \geq b$ if both of the following are true.

Base: $P(b)$

Inductive step: For all $k \geq b, (P(b) \wedge P(b+1) \wedge \ldots \wedge P(k)) \to P(k+1)$

Thus, a direct proof of this inductive step is to show that $P(k+1)$ is true under the assumption that the statements $P(b), P(b+1), \ldots, P(k)$ are true.

A.2   RELATIONS AND FUNCTIONS

Relations

DEFINITION: The **cartesian product** of two sets A and B is the set

$$A \times B = \{(a,b) \mid a \in A \text{ and } b \in B\}$$

DEFINITION: A **binary relation** R from a set S to a set B is a subset of $A \times B$. Alternatively, a binary relation may be regarded as a function from $A \times B$ to the Boolean set $\{true, false\}$.

TERMINOLOGY: Let R be a relation from set A to set B. If $(x, y) \in R$, then x is said to be **related to** y (by R). This is often denoted xRy.

DEFINITION: A **(binary) relation** on a set A is a relation from A to A.

DEFINITION: A relation R on a set A is **reflexive** if for all $x \in A$,

$$xRx \in R$$

DEFINITION: A relation R on a set A is **symmetric** if for all $x, y \in A$,

$$xRy \Rightarrow yRx$$

DEFINITION: A relation R on a set A is **transitive** if for all $x, y, z \in A$,

$$(xRy \wedge yRz) \Rightarrow xRz$$

DEFINITION: A relation R on a set A is an **equivalence relation** if R is reflexive, symmetric, and transitive.

Example A.2.1: Let R be the relation on the set $\mathbb{Z}$ of all integers, given by

$$xRy \Leftrightarrow x - y = 3k, \text{ for some } k \in \mathbb{Z}$$

Then R is an equivalence relation.

DEFINITION: Let R be an equivalence relation on a set A, and let $x \in A$. The **equivalence class** of a, denoted $[a]$, is given by

$$[x] = \{a \in A \mid aRx\}$$

Example A.2.2: For the equivalence relation defined in Example A.2.1,

$$[0] = \{3k \mid k \in \mathbb{Z}\}$$
$$[1] = \{3k + 1 \mid k \in \mathbb{Z}\}$$
$$[2] = \{3k + 2 \mid k \in \mathbb{Z}\}$$

Observe that the equivalence class of any other element is one of these three classes.

DEFINITION: A collection of distinct non-empty subsets $\{S_1, S_2, \ldots, S_l\}$ of a set A is a **partition** of A if both of the following conditions are satisfied.

- $S_i \cap S_j = \emptyset$, for all $1 \le i < j \le l$
- $\bigcup_{i=1}^{l} S_i = A$

Proposition A.2.1: *Let R be an equivalence relation on a set A, and let $x, y \in A$. Then the following three statements are logically equivalent.*

1. xRy
2. $[x] = [y]$
3. $[x] \cap [y] \neq \emptyset$

Corollary A.2.2: *Let R be an equivalence relation on a set A. Then the distinct equivalence classes of R partition A.*

Example A.2.3: The three equivalence classes $[0]$, $[1]$, and $[2]$ of Example A.2.2 partition the set $\mathbb{Z}$ of integers.

Partial Orderings and Posets

DEFINITION: A relation R on a set A is **antisymmetric** if for all $x, y \in A$,

$$x R y \wedge y R x \Rightarrow x = y$$

DEFINITION: A relation R on a set A is a **partial ordering** (or **partial order**) if R is reflexive, antisymmetric, and transitive.

Example A.2.4: Let R be the *divisibility relation* on the set $\mathbb{Z}^+$ of positive integers, given by $x R y \Leftrightarrow x$ divides y. Then R is a partial ordering.

NOTATION: The symbol $\prec$ is often used to denote a partial ordering.

DEFINITION: Let $\prec$ be a partial ordering on a set A. The pair $\{A, \prec\}$ is called a **partially ordered set** (or **poset**).

Functions

DEFINITION: Let A and B be two sets. A **function** f from A to B, denoted $f : A \to B$, is an assignment of exactly one element of B to each element of A. We also say that f **maps** A to B.

ALTERNATIVE DEFINITION: A **function** f from A to B, denoted $f : A \to B$, is a relation from A to B, such that each element of set A appears as a first component in exactly one of the ordered pairs of f.

TERMINOLOGY: Let $f : A \to B$ be a function from set A to set B. Then A is called the **domain** of f, and B is called the **codomain** of f.

TERMINOLOGY: Let $f : A \to B$ be a function from set A to B, and let $a \in A$. The element b assigned to a by f, denoted $b = f(a)$, is called the **image of** a **under** f, and element a is called a **preimage of** b **under** f.

DEFINITION: Let $f : A \to B$ be a function from set A to set B, and let $S \subseteq A$. Then the set $f(S) = \{f(s) \mid s \in S\}$ is the **image of** S **under** f.

DEFINITION: Let $f : A \to B$ be a function from set A to set B, and let $S \subseteq B$. Then the set $f^{-1}(S) = \{x \in A \mid f(x) \in S\}$ is the **inverse image of** S **under** f.

DEFINITION: A function $f : A \to B$ is **one-to-one** (or **injective**) if for all $x, y \in A$, $x \neq y \Rightarrow f(x) \neq f(y)$.

Thus, a function $f : A \to B$ is one-to-one if and only if each element of set B has at most one preimage under f.

DEFINITION: A function $f : A \to B$ is **onto** (or **surjective**) if for all $b \in B$, there exists at least one $a \in A$ such that $f(a) = b$.

Thus, a function $f : A \to B$ is onto if and only if each element of set B has at least one preimage under f.

DEFINITION: A function $f : A \to B$ is a **bijection** if f is both injective and surjective.

Example A.2.5: In a standard (0-based) vertex-labeling of an n-vertex graph G, each vertex of G is assigned an integer label from the set $\{0, 1, \ldots, n - 1\}$, such that no two vertices get the same label. Thus, the labeling is a bijection

$$f : \{0, 1, \ldots, n - 1\} \to V_G$$

It is common to say that the integers from 0 to $n-1$ **have been bijectively assigned** to the vertices of G.

Theorem A.2.3: [**Pigeonhole Principle**] *Let* $f : A \to B$ *be a function from a finite set A to a finite set B. Let any two of the following three statements be true.*

1. *f is one-to-one.*

2. *f is onto.*

3. *$|A| = |B|$.*

Then the third statement is also true.

The following corollary uses familiar pigeonhole jargon.

Corollary A.2.4: *If each member of a flock of pigeons flies into a pigeonhole, and if there are more pigeons than pigeonholes, then there must be at least one pigeonhole that contains more than one pigeon.*

DEFINITION: Let $f : A \to B$ be a function from a set A to a set B, and let $g : B' \to C$ be a function from a set B' to a set C, where $B \subseteq B'$. Then the **composition of f and g** is the function $g \circ f : A \to C$, given by $(g \circ f)(a) = g(f(a))$.

DEFINITION: Let $f : A \to B$ be a function from a set A onto a set B, and let $g : B \to A$ be a function from a set B to a set A. Then g is the **inverse function** of f if for all $a \in A$, $(g \circ f)(a) = a$.

Proposition A.2.5: *Let* $f : A \to B$ *be a function from a set A onto a set B, and let $g : B \to A$ be a function from a set B onto a set A. Then g is the inverse function f if and only if f is the inverse function of g.*

Proposition A.2.6: *A function $f : A \to B$ has an inverse if and only if f is a bijection.*

A.3   SOME BASIC COMBINATORICS

Proposition A.3.1: [**Rule of Product**] *Let A and B be two finite sets. Then*

$$|A \times B| = |A| \cdot |B|$$

DEFINITION: A **permutation** of an n-element set S is a bijection $\pi : S \to S$.

DEFINITION: A **k-permutation** of an n-element set S is a permutation of a k-element subset of S.

A consequence of the Rule of Product is the following simple formula for the number of k-permutations.

Proposition A.3.2: *The number of k-permutations of an n-element set is*

$$\frac{n!}{(n-k)!}$$

DEFINITION: A **k-combination** of a set S is a k-element subset of S.

NOTATION: The number of k-combinations of an n-element set is denoted $\binom{n}{k}$ or, alternatively, $C(n, k)$.

Proposition A.3.3:

$$\binom{n}{k} = \frac{n!}{k!(n-k)!}$$

Theorem A.3.4: [***Binomial Theorem***] *For every nonnegative integer* n *and any numbers* x *and* y,

$$(x+y)^n = \sum_{k=0}^{n} \binom{n}{k} x^{n-k} y^k$$

DEFINITION: Let $\langle a_1, a_2, \dots, a_k, \dots \rangle$ be a sequence of real numbers. The **generating function** for the sequence $\{a_n\}$ is the power series

$$g(x) = \sum_{k=0}^{\infty} a_k x^k = a_0 + a_1 x + a_2 x^2 + \cdots + a_k x^k + \cdots$$

Example A.3.1: The generating function for the constant sequence $\langle 1, 1, \dots \rangle$ is

$$\sum_{k=0}^{\infty} x^k = \frac{1}{1-x}$$

Example A.3.2: The generating function for the sequence $\langle 1, 2, 3, \dots \rangle$ is

$$\sum_{k=0}^{\infty} k x^{k-1} = \sum_{k=0}^{\infty} \frac{d}{dx} x^k = \frac{d}{dx} \left[\sum_{k=0}^{\infty} x^k \right] = \frac{d}{dx} \left(\frac{1}{1-x} \right) = \frac{1}{(1-x)^2}$$

A.4   ALGEBRAIC STRUCTURES

Groups

DEFINITION: A **binary operation** $*$ on a nonempty set A is a function $f : A \times A \to A$, given by $f((a,b)) = a * b$.

NOTATION: The set A together with a binary operation $*$ is denoted $(A, *)$.

Example A.4.1: Let $\mathbb{Z}_{10} = \{0, 1, 2, \dots, 9\}$, and let $\oplus$ and $\odot$ be the operations of addition and multiplication mod 10. For instance, $6 \oplus 6 = 12 \,(\text{mod } 10) = 2$ and $6 \odot 6 = 36 \,(\text{mod } 10) = 6$. Then $\oplus$ and $\odot$ are binary operations on the set $\mathbb{Z}_{10}$.

DEFINITION: The binary operation $*$ on set A is **associative** if for all $a, b, c \in A$, $(a * b) * c = a * (b * c)$.

DEFINITION: The binary operation $*$ on set A is **commutative** if for all $a, b \in A$, $a * b = b * a$.

DEFINITION: Let $*$ be a binary operation on set A. An element $e \in A$, is an **identity element** of set A under $*$ if for all $a \in A$,

$$a * e = e * a = a$$

DEFINITION: Let $*$ be a binary operation on set A, and let e be an identity element of $(A, *)$. For $a \in A$, an element $a' \in A$ is an **inverse** of a in $(A, *)$ if

$$a * a' = a' * a = e$$

DEFINITION: A **group** $\mathcal{G} = (G, *)$ is a nonempty set G and a binary operation $*$ that satisfy the following conditions.

1. The operation $*$ is associative.
2. $\mathcal{G}$ has an identity element.
3. Each $g \in G$ has an inverse in $(G, *)$.

NOTATION: Each element g in a group $\mathcal{G}$ has a *unique* inverse, which is denoted g^{-1}.

DEFINITION: If $\mathcal{G} = (G, *)$ is a group, the set G is the domain (set) of the group $\mathcal{G}$.

NOTATION: When convenient, and when there is no ambiguity, we sometimes write "$g \in \mathcal{G}$" to mean that g is an element of the domain set G.

DEFINITION: An **abelian** group is a group whose operation is commutative.

NOTATION: In an abelian group, the operation is commonly denoted "+" and called "sum."

TERMINOLOGY: For an abstract group $(G, *)$, the term "product" is used generically to refer to the element $x * y$, and its use does not imply anything about the type of group operation.

Example A.4.2: The group $\mathbb{Z}_n = (\mathbb{Z}_n, +)$, which generalizes $(\mathbb{Z}_{10}, \oplus)$ in Example A.4.1, is abelian and is denoted $\mathbb{Z}_n$.

Example A.4.3: Observe that $(\mathbb{Z}_n, \odot)$ *is not* a group, since the element 0 has no inverse under $\odot$.

Example A.4.4: Let S be an n-element set, and let P be the set of all permutations of set S. Then $(P, \circ)$, where $\circ$ is the *composition* operation, is a group.

DEFINITION: Let $\mathcal{G} = (G, *)$ be a finite group. A subset X of G is a generating set for group $\mathcal{G}$ if every element of $\mathcal{G}$ can be expressed as a product of elements of set X.

Example A.4.5: The set $\{2, 5\}$ is a generating set for $\mathbb{Z}_{10}$.

DEFINITION: A **cyclic group** is a group that has a 1-element generating set.

NOTATION: If $\{a\}$ is a generating set for a cyclic group $\mathcal{G}$, then we may write $\mathcal{G} = \langle a \rangle$.

Example A.4.6: $\mathbb{Z}_n = (\mathbb{Z}_n, +) = \langle 1 \rangle$.

Example A.4.7: $(\mathbb{Z}_7 - \{0\}, \odot) = \langle 3 \rangle = \langle 5 \rangle$.

DEFINITION: Let $\mathcal{G}_1 = (G_1, *_1)$ and $\mathcal{G}_2 = (G_2, *_2)$ be two groups. The **direct sum** (or **direct product**) $\mathcal{G} \times \mathcal{G}_2$ is the group $(G_1 \times G_2, *)$ whose domain set is the Cartesian product of the domain sets G_1 and G_2 and whose operation is defined by

$$(x_1, x_2) * (y_1, y_2) = (x_1 *_1 y_1, x_2 *_2 y_2)$$

DEFINITION: The n-fold direct sum of groups $\mathcal{G}_1 = (G_1, *_1), \mathcal{G}_2 = (G_2, *_2), \ldots, \mathcal{G}_n = (G_2, *_n)$ is defined recursively as

$$\mathcal{G}_1 \times \mathcal{G}_2 \times \ldots \mathcal{G}_n = (\mathcal{G}_1 \times \mathcal{G}_2 \times \ldots \mathcal{G}_{n-1}) \times \mathcal{G}_n$$

Theorem A.4.1: [**Fundamental Theorem of Finite Abelian Groups**] *Let $\mathcal{G}$ be a finite abelian group. Then $\mathcal{G}$ can be expressed as the direct sum of cyclic groups.*

Remark: In fact, the cyclic groups in the direct-sum decomposition of an abelian group have prime-power order, and the direct sum is *essentially* unique, but the above assertion is all that we need in this book.

Order of an Element in a Group

NOTATION: Let x be an element of a group $\mathcal{G} = (G, *)$. The element $x*x$ may be denoted x^2, and the product of r factors of element x may be denoted x^r. In an abelian group, the element $x + x$ may be denoted $2x$, and the sum of r occurrences of element x may be denoted $r \cdot x$ or sometimes rx.

DEFINITION: Let x be an element of a group with identity e. The **order** of x is the smallest positive integer m such that $x^m = e$.

DEFINITION: The **order** of a group $\mathcal{G} = (G, *)$ is the cardinality $|G|$ of its domain set.

Proposition A.4.2: *In a finite group, every element has finite order.*

Example A.4.8: For $2 \in \mathbb{Z}_{10}$, we have $5 \cdot 2 = 2 + 2 + 2 + 2 + 2 = 0 \pmod{10}$. Thus, the order of the element $2 \in \mathbb{Z}_{10}$ is 5.

Permutation Groups

DEFINITION: The group of all permutations of a set S with n elements is called the **full symmetric group** on S and is denoted Σ_S or Σ_n.

TERMINOLOGY: The elements of the set on which a permutation group acts are called **objects**.

NOTATION: One way of representing a permutation π of the set $\{1, 2, \ldots, n\}$ is to use a $2 \times n$ matrix, where the first row contains the n integers in order, and the second row contains the images, so that $\pi(j)$ appears below j.

Example A.4.9: The group Σ_3 consists of the six permutations of $S = \{1, 2, 3\}$. They are as follows:

$$\pi_1 = \begin{pmatrix} 1 & 2 & 3 \\ 1 & 2 & 3 \end{pmatrix}; \quad \pi_2 = \begin{pmatrix} 1 & 2 & 3 \\ 1 & 3 & 2 \end{pmatrix}; \quad \pi_3 = \begin{pmatrix} 1 & 2 & 3 \\ 2 & 1 & 3 \end{pmatrix};$$

$$\pi_4 = \begin{pmatrix} 1 & 2 & 3 \\ 3 & 2 & 1 \end{pmatrix}; \quad \pi_5 = \begin{pmatrix} 1 & 2 & 3 \\ 2 & 3 & 1 \end{pmatrix}; \quad \pi_3 = \begin{pmatrix} 1 & 2 & 3 \\ 3 & 1 & 2 \end{pmatrix};$$

The permutation π_1 is the identity element of Σ_3. Remembering that the composition $f \circ g$ is applied *right-to-left*, we have the following sample calculations:

$$\pi_2 \circ \pi_2 = \pi_3 \circ \pi_3 = \pi_4 \circ \pi_4 = \pi_1; \quad \pi_5 \circ \pi_5 = \pi_6; \quad \pi_4 \circ \pi_5 = \pi_3; \quad \pi_5 \circ \pi_4 = \pi_2$$

The last two calculations show that the full symmetric group is *non-abelian*. Additional calculation shows that the set $\{\pi_4, \pi_5\}$ is a generating set for Σ_3. Moreover, $\pi_5^3 = \pi_6 \circ \pi_5 = \pi_1$. Thus, the order of the permutation π_5 is 3.

DEFINITION: A permutation π is an **involution** if $\pi = \pi^{-1}$.

Cycle Notation for a Permutation

Each object of $\{1, 2, 3\}$ in Example A.4.9 *cycles* back to itself after three applications of the permutation $\pi_5 = \begin{pmatrix} 1 & 2 & 3 \\ 2 & 3 & 1 \end{pmatrix}$. That is,

$$1 \xrightarrow{\pi_5} 2 \xrightarrow{\pi_5} 3 \xrightarrow{\pi_5} 1 \qquad 2 \xrightarrow{\pi_5} 3 \xrightarrow{\pi_5} 1 \xrightarrow{\pi_5} 2 \qquad 3 \xrightarrow{\pi_5} 1 \xrightarrow{\pi_5} 2 \xrightarrow{\pi_5} 3$$

The *cycling* of the images under successive applications of a given permutation is compactly represented by enclosing the chain of images in parentheses so that the image of each element appears to the right of that element, with *wraparound* for the rightmost element.

NOTATION: If S is a set and $a_1, a_2, \ldots, a_k \in S$, then $(a_1 a_2 \ldots a_k)$ denotes the permutation π of S for which

$$\pi(a_i) = \begin{cases} a_{i+1} & \text{for } i = 1, \ldots, k-1 \\ a_1 & \text{for } i = k \end{cases}$$

and

$$\pi(x) = x \quad \text{for all other } x \in S$$

Example A.4.10: The cycle representation of permutation π_5 from Example A.4.9 can be written in any one of three equivalent ways: (1 2 3), (2 3 1), or (3 1 2).

NOTATION: This cycle notation can be extended to any permutation by writing that permutation as a **product of disjoint cycles**. The elements that appear in each cycle form the chain of images under the action of that permutation.

Example A.4.11: Consider the permutation $\alpha \in \Sigma_9$ given by

$$\alpha = \begin{pmatrix} 1 & 2 & 3 & 4 & 5 & 6 & 7 & 8 & 9 \\ 7 & 2 & 6 & 1 & 8 & 3 & 4 & 9 & 5 \end{pmatrix}$$

Then the disjoint cycle representation of α is

$$\alpha = (1\ 7\ 4)(2)(3\ 6)(5\ 8\ 9) \quad \text{or} \quad (1\ 7\ 4)(3\ 6)(5\ 8\ 9)$$

Fields

DEFINITION: A **field** $\mathcal{F} = (F, +, \cdot)$ is a set F together with two operations, $+$ and $\cdot$, (called generically *addition* and *multiplication*), such that each of the following conditions is satisfied:

1. $(F, +)$ is an abelian group.
2. $(F - \{0\}, \cdot)$ is an abelian group, where 0 is the additive identity.
3. $a \cdot (b + c) = (a \cdot b) + (a \cdot c)$.

Finite Field $GF(2)$

DEFINITION: The **finite field** $GF(2)$ consists of the set $\mathbb{Z}_2 = \{0, 1\}$ together with the mod 2 operations $\oplus$ and $\odot$. Thus,

$$0 \oplus 0 = 1 \oplus 1 = 0; \quad 0 \oplus 1 = 1 \oplus 0 = 1; \quad 0 \odot 0 = 1 \odot 0 = 0 \odot 1 = 0; \quad 1 \odot 1 = 1$$

Vector Spaces

DEFINITION: A **_vector space_** over a field (of **_scalars_**) $\mathcal{F}$ is a set V (of **_vectors_**) together with a binary operation $+$ on V and a mapping, called **_scalar multiplication_**, from the cartesian product $F \times V$ to V : $(a, v) \mapsto av$, such that the following conditions are satisfied for all scalars $a, b \in F$ and all vectors $v, w \in V$.

1. $(V, +)$ is an abelian group (the symbol "$+$" is being used to denote two different operations, addition in the field $\mathcal{F}$ and addition in the set V)
2. $(a \cdot b)v = a(bv)$
3. $(a + b)v = av + bv$
4. $a(v + w) = av + aw$
5. $ev = v$, where e is the multiplicative identity of field $\mathcal{F}$

Example A.4.12: Let $W = \{(\alpha_1, \alpha_2, \alpha_3) \,|\, \alpha_i \in GF(2)\}$, and consider componentwise addition $\oplus$ (mod 2) on the elements of W. For each scalar $\alpha \in GF(2)$ and element $w = (\alpha_1, \alpha_2, \alpha_3)$, let scalar multiplication be defined by

$$\alpha w = (\alpha \odot \alpha_1, \alpha \odot \alpha_2 \alpha \odot \alpha_3)$$

Then it is easy to verify that W with these operations forms a vector space over $GF(2)$.

Remark: The vector space defined in Example A.4.12 is a special case of the vector space $\mathcal{F}^n$ of all n-tuples whose components are elements in a field $\mathcal{F}$.

Independence, Subspaces, and Dimension

The remaining definitions in this section all assume that V is a vector space over a field $\mathcal{F}$. The examples refer to the vector space W defined in Example A.4.12.

DEFINITION: A vector v is a **_linear combination_** of vectors $v_1, v_2, \ldots, v_m$ if $v = \alpha_1 v_1 + \alpha_2 v_2 + \cdots + \alpha_m v_m$ for scalars $\alpha_i \in \mathcal{F}$, $i = 1, \ldots, m$.

Example A.4.13: In vector space W, the vector $(1, 0, 1)$ is a linear combination of vectors $(1, 1, 0)$ and $(0, 1, 1)$, given by $(1, 0, 1) = 1(1, 1, 0) + 1(0, 1, 1)$. This last equation also implies that $(1, 0, 1)$ is a linear combination of any superset of $\{(1, 1, 0), (0, 1, 1)\}$, where the scalar multipliers for the other vectors in the superset are 0.

DEFINITION: A set $\{v_1, v_2, \ldots, v_m\}$ of vectors is **_linearly independent_** if

$$\alpha_1 v_1 + \alpha_2 v_2 + \cdots + \alpha_m v_m = \mathbf{0} \text{ only if } \alpha_1 = \alpha_2 = \cdots = \alpha_m = 0$$

Thus, nonzero vectors $v_1, v_2, \ldots, v_m$ are **_linearly independent_** if none of them is expressible as a linear combination of the remaining ones.

Example A.4.14: It is easy to check that $\{(1, 1, 0), (0, 1, 0), (0, 1, 1)\}$ is a set of linearly independent vectors in vector space W.

DEFINITION: A set S of vectors **_spans_** V if every vector in V is expressible as a linear combination of vectors in S.

Example A.4.15: The three vectors in Example A.4.14 span W. A more obvious spanning set is $\{(1, 0, 0), (0, 1, 0), (0, 0, 1)\}$.

DEFINITION: A set B is a **_basis_** for V if the vectors in B span V and are linearly independent.

Proposition A.4.3: *All the bases of a given vector space have the same number of vectors.*

DEFINITION: The **dimension** of a vector space V, denoted $\dim(V)$, is the number of vectors in a basis of V.

DEFINITION: A subset U of vector space V is a **subspace** of V if U is itself a vector space with respect to the addition and scalar multiplication of V.

Proposition A.4.4: *Let V be a vector space over a field $\mathcal{F}$, and let U be a subset of V. Then U is a subspace of V if and only if*

1. *$v + w \in U$ for all $v, w \in U$, and*
2. *$\alpha v \in U$ for all $\alpha \in \mathcal{F}$, $v \in U$.*

Example A.4.16: Either of the 3-element sets given in Example A.4.15 shows that W is a 3-dimensional vector space. The set $U = \{(0,0,0),(1,1,0),(0,1,1),(1,0,1)\}$ is a 2-dimensional subspace of W.

Proposition A.4.5: *Let U_1 and U_2 be two subspaces of a vector space V. Then the intersection $U_1 \cap U_2$ is a subspace of V.*

DEFINITION: The **direct sum** $U_1 \oplus U_2$ of two subspaces U_1 and U_2 of V is the set of all vectors of the form $v_1 + v_2$, where $v_1 \in V_1$ and $v_2 \in V_2$.

Proposition A.4.6: *Let U_1 and U_2 be two subspaces of a vector space V. Then the direct sum $U_1 \oplus U_2$ is a subspace of V, and*

$$\dim(U_1 \oplus U_2) = \dim(U_1) + \dim(U_2) - \dim(U_1 \cap U_2)$$

A.5   ALGORITHMIC COMPLEXITY

One approach to the analysis and comparison of algorithms is to consider the *worst-case* performance, ignoring constant factors that are often outside a given programmer's control. These factors include the computer on which the algorithm is run, the operating system, and the machine code translation of the programming language in which the algorithm is written. The objective of this approach is to determine the functional dependence of the running time (or of some other measure) on the number n of inputs (or on some other such variable). A first step toward this goal is to make the notion of "proportional to" mathematically precise.

Big-O Notation

DEFINITION: Let $f : Z^+ \to \mathbb{R}$ and $g : Z^+ \to \mathbb{R}$ be functions from the set of positive integers to the set of real numbers. The function $f(n)$ is in the family $O(g(n))$ (read "big-O" of $g(n))$ *if there are constants c and N such that*

$$|f(n)| \leq c|g(n)| \text{ for all } n > N$$

TERMINOLOGY: If $f(n)$ is in $O(g(n))$, then $f(n)$ is said to be **asymptotically dominated** by $g(n)$. In other words, for sufficiently large n and a sufficiently large constant c, $cg(n)$ is larger than $f(n)$.

TERMINOLOGY NOTE: The usage "f is $O(g(n))$" (omitting the preposition "in") is quite common.

Remark: It follows from the definition that $f(n)$ is in $O(g(n))$ if and only if there is a constant c such that $\lim_{n\to\infty} \frac{f(n)}{g(n)} = c$.

Example A.5.1: The function $f(n) = \sum_{i=1}^{n} i$ is in $O(n^2)$. To see this, observe that for all $n > 0$

$$f(n) = \frac{n^2 + n}{2} < n^2 + n \leq n^2 + n^2 = 2n^2$$

Thus, the condition is satisfied for $c = 2$ and $N = 0$. It is also true that $f(n)$ is in $O(n^r)$ for any $r \geq 2$, but $O(n^2)$ is the *sharpest* among these. In fact, letting $c = 2$ and $N = 0$, the following chain of inequalities shows that n^2 is in $O(f(n))$.

$$n^2 < n^2 + n = 2f(n)$$

The next example illustrates the effect of information-structure representation of a graph on algorithmic computation.

Example A.5.2: Given a simple graph G, find the degree of each vertex $v \in V_G$. For each vertex, we must count the number of edges incident on that vertex. The number of steps required depends on how the input graph is represented in the computer. We consider two of the possibilities. Let n be the number of vertices and m the number of edges.

Representation by adjacency matrix Because the adjacency matrix is symmetric, only the upper (or lower) triangular submatrix (excluding the main diagonal) needs to be examined. There are $\sum_{i+1}^{n-1} i$ cells in this triangular submatrix, and each cell must be checked to see if it is nonzero. Hence, the number of checks is $O(n^2)$.

Representation by adjacency lists of edges For each vertex, the list of edges incident with that vertex is traversed. Thus, the number of add operations is $O(m)$.

DEFINITION: Functions $f(n)$ and $g(n)$ are **big-O equivalent** if $f(n)$ is in $O(g(n))$ and $g(n)$ is in $O(f(n))$.

Proposition A.5.1: *Functions $f(n)$ and $g(n)$ are big-O equivalent if and only if* $\lim_{n\to\infty} \frac{f(n)}{g(n)} = c$ *for some nonzero constant c.*

Proposition A.5.2: *Two polynomial functions are big-O equivalent if and only if they have the same degree.*

DEFINITION: A **polynomial-time** algorithm is an algorithm whose running time (as a function on the input size n) is in $O(p(n))$ for some polynomial $p(n)$.

TERMINOLOGY: A *linear-time* algorithm is an algorithm whose running time is in $O(n)$. A *quadratic-time* algorithm is an algorithm whose running time is in $O(n^2)$.

Decision Problems

DEFINITION: A **decision problem** is a problem that requires only a *yes* or *no* answer regarding whether some element of its domain has a particular property.

DEFINITION: A decision problem belongs to the **class P** if there is a polynomial-time algorithm to solve the problem.

DEFINITION: A decision problem belongs to the **class NP** if there is a way to provide evidence of the correctness of a *yes* answer so that it can be confirmed by a polynomial-time algorithm.

Example A.5.3: The decision problem "Is this graph bipartite?" is in class P.

Example A.5.4: It is not known whether the decision problem "Can the vertices of this graph be colored with three colors such that adjacent vertices get different colors?" is in class P. However, a *yes* answer could be supported by supplying a proposed 3-coloring of the vertices. Checking the correctness of this 3-coloring would require counting the number of different colors used on the vertices, which can be accomplished in linear time, and checking for each pair of vertices, whether the two vertices have received the same color and whether they are adjacent, which can be performed in quadratic time. Hence, this decision problem is in class NP.

Remark: The letters NP stand for *nondeterministic polynomial*. This terminology derives from the fact that problems in class NP would be answered in polynomial time if there was a nondeterministic (limitlessly parallel) machine that could check all possible evidence sets simultaneously. This philosophically creative notion provides a meaningful theoretical construct for categorizing problems.

It might appear that the class of problems that could be solved by a nondeterministic computer should "obviously" be larger than those that can be solved by strictly sequential polynomial-time algorithms. However, no one has been able to find a problem in class NP that is provably not in class P. Determining whether such a decision problem exists is among the foremost unsolved problems in theoretical computer science.

The Problem Classes NP-Complete and NP-Hard

There is a large collection of problems (called *NP-complete*), including the vertex-coloring problem and the traveling salesman problem, for which no polynomial-time algorithms have been found, despite considerable effort over the last several decades. It remains open whether there exist such algorithms for these problems. In turns out that if one can find a polynomial-time algorithm for an NP-complete problem, then polynomial-time algorithms can be found for all NP-complete problems. Our goal here is to provide a brief and informal introduction to this topic. A precise, detailed description of this class of problems may be found, for instance, in [Ev79], [GaJo79], or [Wi86].

DEFINITION: A decision problem R is **polynomially reducible** to Q if there is a polynomial-time transformation of each instance I_R of problem R to an instance I_Q of problem Q, such that instances I_R and I_Q have the same answer (*yes* or *no*).

DEFINITION: A decision problem is NP-**hard** if every problem in class NP is polynomially reducible to it.

DEFINITION: An NP-hard problem R is **NP-complete** if R is in class NP.

DEFINITION: Let $X = \{x_1, x_2, \ldots, x_n\}$ be a set of n Boolean variables. The **satisfiability problem** is to decide, for a propositional form $S(x, \ldots, x_n)$ in the variables $x_i \in X$, whether there is a set of truth values $\{a_1, \ldots, a_n\}$ such that $S(a_1, \ldots, a_n)$ is true.

Stephen Cook ([Co71]) showed that the class NP-complete is nonempty by showing that the satisfiability problem is NP-complete. Since that time, the list has grown rapidly.

A.6   SUPPLEMENTARY READING

For discrete mathematics: [PoSt90], [Ro84], [Ro03], [St93], [Tu95].

For algebra: [Ga90], [MaBi67].

For algorithmics: [AhHoUl83], [Ba88], [Se88], [Wi86].

BIBLIOGRAPHY

B.1   GENERAL READING

The first part of this bibliography is a list of books of general interest on graph theory and on background topics for graph theory, at a level consistent with our present text, grouped according to topic.

ALGEBRA and ENUMERATION

[Ga90] J. A. Gallian, *Contemporary Abstract Algebra*, Second Edition, D. C. Heath, 1990.

[GrKnPa94] R. L. Graham, D. E. Knuth, and O. Patashnik, *Concrete Mathematics: A Foundation for Computer Science*, Second Edition, Addison-Wesley, 1994.

[MaBi67] S. Maclane and G. Birkhoff, *Algebra*, Macmillan, 1967.

ALGORITHMS and COMPUTATION

[AhHoUl83] A. V. Aho, J. E. Hopcroft, and J. D. Ullman, *Data Structures and Algorithms*, Addison-Wesley, 1983.

[Ba83] S. Baase, *Computer Algorithms: Introduction to Design and Analysis*, Addison Wesley, 1983.

[Ev79] S. Even, *Graph Algorithms*, Computer Science Press, 1979.

[GaJo79] M. R. Garey and D. S. Johnson, *Computers and Intractability: A Guide to the Theory of NP-Completeness*, W. H. Freeman, 1979.

[Gr97] J. Gruska, *Foundations of Computing*, Thomson, 1997.

[La76] E. L. Lawler, *Combinatorial Optimization: Networks and Matroids*, Holt, Rinehart and Winston, 1976.

[Se88] R. Sedgewick, *Algorithms*, Addison-Wesley, 1988.

[Wi86] H. S. Wilf, *Algorithms and Complexity*, Prentice-Hall, 1986.

COMBINATORIAL MATHEMATICS

[Bo00] K. P. Bogart, *Introductory Combinatorics*, Third Edition, Academic Press, 2000.

[Br04b] R. A. Brualdi, *Introductory Combinatorics*, Fourth Edition, Prentice Hall, 2004.

[Gr08] J. L. Gross, *Combinatorial Methods with Computer Applications*, CRC Press, 2008.

[MiRo91] J. G. Michaels and K. H. Rosen (Eds.), *Applications of Discrete Mathematics*, McGraw-Hill, 1991.

[PoSt90] A. Polimeni and H. J. Straight, *Foundations of Discrete Mathematics*, Brooks/Cole, 1990.

[Ro84] F. S. Roberts, *Applied Combinatorics*, Prentice-Hall, 1984.

[Ro03] K. H. Rosen, *Discrete Mathematics and Its Applications*, Fifth Edition, McGraw-Hill, 2003.

[St93] H. J. Straight, *Combinatorics: An Invitation*, Brooks/Cole, 1993.

[Tu01] A. Tucker, *Applied Combinatorics*, Fourth Edition, John Wiley & Sons, 2001.

GRAPH THEORY

[Be85] C. Berge, *Graphs*, North-Holland, 1985.

[BiLlWi86] N. L. Biggs, E. K. Lloyd, and R. J. Wilson, *Graph Theory 1736-1936*, Oxford, 1986.

[Bo98] B. Bollobás, *Modern Graph Theory*, Springer, 1998.

[BoMu76] J. A. Bondy and U. S. R. Murty, *Graph Theory with Applications*, American Elsevier, 1976.

[ChLe04] G. Chartrand and L. Lesniak, *Graphs & Digraphs*, Fourth Edition, CRC Press, 2004.

[ChZh05] G. Chartrand and P. Zhang, *Introduction to Graph Theory*, McGraw Hill, 2005.

[Di00] R. Diestel, *Graph Theory*, Springer-Verlag, Second Edition, 2000.

[Gi85] A. Gibbons, *Algorithmic Graph Theory*, Cambridge University Press, 1985.

[Go80] M. C. Golumbic, *Algorithmic Graph Theory and Perfect Graphs*, Academic Press, 1980.

[Go88] R. Gould, *Graph Theory*, Benjamin/Cummings, 1988.

[GrYe14] J. L. Gross and J. Yellen, eds, *Handbook of Graph Theory*, CRC Press, Second Edition, 2014.

[Ha69] F. Harary, *Graph Theory*, Addison-Wesley, 1969.

[ThSw92] K. Thulasiraman and M. N. S. Swamy, *Graphs: Theory and Algorithms*, John Wiley & Sons, 1992.

[We01] D. B. West, *Introduction to Graph Theory*, Second Edition, Prentice-Hall, 2001.

[Wi96] R. J. Wilson, *Introduction to Graph Theory*, Addison Wesley Longman, 1996.

[WiWa90] R. J. Wilson and J. J. Watkins, *Graphs: An Introductory Approach*, John Wiley & Sons, 1990.

SURFACES and TOPOLOGICAL GRAPH THEORY

[GrTu87] J. L. Gross and T. W. Tucker, *Topological Graph Theory*, Dover, 2001. (First Edition, Wiley-Interscience, 1987.)

[Ma67] W. S. Massey, *Algebraic Topology: An Introduction*, Harbrace, 1967.

[Ri74] G. Ringel, *Map Color Theorem*, Springer, 1974.

[Wh84] A. T. White, *Graphs, Groups and Surfaces*, Revised Edition, North-Holland, 1984.

B.2   REFERENCES

All the references cited in the text are listed below. Some additional references are also included for supplementary reading.

Chapter 1: Introduction to Graph Models

[Ro84] F. S. Roberts, *Applied Combinatorics*, Prentice-Hall, 1984.

[Wi04] R. J. Wilson, History of Graph Theory, §1.3 *in Handbook of Graph Theory*, eds., J. L. Gross and J. Yellen, CRC Press, 2014.

[WiWa90] R. J. Wilson and J. J. Watkins, *Graphs: An Introductory Approach*, John Wiley & Sons, 1990.

Chapter 2: Structure and Representation

[Mc77] B. D. McKay, Computer reconstruction of small graphs, *J. Graph Theory* 1 (1977), 281–283.

[Ni77] A. Nijenhuis, Note on the unique determination of graphs by proper subgraphs, *Notices Amer. Math. Soc.* 24 (1977), A-290.

Chapter 3: Trees

[AhHoUl83] A. V. Aho, J. E. Hopcroft, and J. D. Ullman, *Data Structures and Algorithms*, Addison-Wesley, 1983.

[Hu52] D. A. Huffman, A method for the construction of minimum-redundancy codes, *Proc. IRE* 40 (1952), 1098–1101.

Chapter 4: Spanning Trees

[AhHoUl83] A. V. Aho, J. E. Hopcroft, and J. D. Ullman, *Data Structures and Algorithms*, Addison-Wesley, 1983.

[Ba83] S. Baase, *Computer Algorithms: Introduction to Design and Analysis*, Addison-Wesley, 1983.

[Di59] E. W. Dijkstra, A note on two problems in connexion with graphs, *Numerische Mathematik* 1 (1959), 269–271.

[Fl62] R. W. Floyd, Algorithm 97: Shortest path, *Comm. ACM* 5:6 (1962), 345.

[Kr56] J. B. Kruskal Jr., On the shortest spanning subtree of a graph and the traveling salesman problem, *Proceedings AMS* 7, 1 (1956).

[La76] E. L. Lawler, *Combinatorial Optimization: Networks and Matroids*, Holt, Rinehart and Winston, 1976.

[Ox92] J. G. Oxley, *Matroid Theory*, Oxford, 1992.

[Pr57] R. C. Prim, Shortest connection networks and some generalizations, *Bell System Technical Journal* 36 (1957).

[Se88] R. Sedgewick, *Algorithms*, Addison-Wesley, 1988.

[ThSw92] K. Thulasiraman and M. N. S. Swamy, *Graphs: Theory and Algorithms*, John Wiley & Sons, 1992.

[We76] D. J. A. Welsh, *Matroid Theory*, Academic Press, 1976.

[Wi96] R. J. Wilson, *Introduction to Graph Theory*, Addison Wesley Longman, 1996.

Chapter 5: Connectivity

[AhHoUl83] A. V. Aho, J. E. Hopcroft, and J. D. Ullman, *Data Structures and Algorithms*, Addison-Wesley, 1983.

[Ba83] S. Baase, *Computer Algorithms: Introduction to Design and Analysis*, Addison-Wesley, 1983.

[ChHa68] G. Chartrand and F. Harary, Graphs with described connectivity, *Theory of Graphs* [P. Erdös and G. Katona, eds.], Academic Press, New York, 1968, pp. 61–63.

[FoFu56] L. R. Ford and D. R. Fulkerson, Maximal flow through a network, *Canad. J. Math.* 8 (1956), 399–404.

[Ha62] F. Harary, The maximum connectivity of a graph, *Proc. Natl. Acad. Sci., U.S.* 48 (1962), 1142–1146.

[ThSw92] K. Thulasiraman and M. N. S. Swamy, *Graphs: Theory and Algorithms*, John Wiley & Sons, 1992.

[Tu61] W. T. Tutte, A theory of 3-connected graphs, *Indag. Math.* 23 (1961), 441–455.

Chapter 6: Optimal Graph Traversals

[CrPa80] H. Crowder and M. W. Padberg, Solving large-scale symmetric traveling salesman problem to optimality, *Management Science* 26 (1980), 495–509.

[DaFuJo54] G. B. Dantzig, D. R. Fulkerson, and S. M. Johnson, Solution of a large-scale traveling salesman problem, *Oper. Res.* 2 (1954), 393–410.

[DeEh51] N. G. deBruijn and T. Ehrenfest, Circuits and trees in oriented graphs, *Simon Stevin* 28 (1951), 203–217.

[Di52] G. A. Dirac, Some theorems on abstract graphs, *Proc. London Math. Soc.* 2 (1952), 69–81.

[Ed65a] J. Edmonds, The Chinese postman problem, *Oper. Res.* 13-1 (1965), 373.

[EdJo73] J. Edmonds and E. L. Johnson, Matching Euler tours and the Chinese postman, *Math. Programming* 5 (1973), 88–124.

[Fl90] H. Fleischner, *Eulerian Graphs and Related Topics, Part 1, Vol. 1*, Ann. Discrete Math **45** North-Holland, Amsterdam, 1990.

[Fl91] H. Fleischner, *Eulerian Graphs and Related Topics, Part 1, Vol. 2*, Ann. Discrete Math **50** North-Holland, Amsterdam, 1991.

[Fl04] H. Fleischner, Eulerian Graphs, §4.2 *in Handbook of Graph Theory*, editors, J. L. Gross and J. Yellen, CRC Press, Boca Raton, Fl, 2004.

[Fr79] G. N. Frederickson, Approximation algorithms for some postman problems, *JACM* 26 (1979), 538–554.

[Gu62] M. Guan, Graphic programming using odd and even points, *Chinese Math.* 1 (1962), 273–277.

[Hu69] G. Hutchinson, Evaluation of polymer sequence fragment data using graph theory, *Bull. Math. Biophys.* 31 (1969), 541–562.

[HuWi75] J. P. Hutchinson and H. S. Wilf, On Eulerian circuits and words with prescribed adjacency patterns, *J. Comb. Theory* 18 (1975), 80–87.

[LaLeKaSh85] E. L. Lawler, J. K. Lenstra, A. H. G. Rinnooy Kan, and D. Shmoys, *The Traveling Salesman Problem: A Guided Tour of Combinatorial Optimization*, John Wiley, 1985.

[Or60] O. Ore, Note on Hamilton circuits, *Amer. Math. Monthly* 67 (1960), 55.

[Pa76] C. H. Papadimitriou, On the complexity of edge traversing, *JACM* 23 (1976), 544–554.

[PaSt82] C. H. Papadimitriou and K Stieglitz, *Combinatorial Optimization: Algorithms and Complexity*, Prentice-Hall, 1982.

[ReTa81] E. M. Reingold and R. E. Tarjan, On a greedy heuristic for complete matching, *SIAM J. Computing* 10 (1981), 676–681.

[Ro99] H. Robinson, Graph Theory Techniques in Model-Based Testing, 1999 International Conference on Testing Computer Software.

[RoStLe77] D. J. Rosenkrantz, R. E. Stearns, and P. M. Lewis, An analysis of several heuristics for the travelling salesman problem, *SIAM J. Comput.* 6 (1977), 563–581.

[SaGo76] S. Sahni and T. Gonzalez, P-complete approximation problems, *JACM* 23 (1976), 555–565.

[SmTu41] C. A. B. Smith and W. T. Tutte, On unicursal paths in a network of degree 4, *Amer. Math. Monthly* 48 (1941), 233–237.

[TuBo83] A. C. Tucker and L. Bodin, A model for municipal street-sweeping operations, in W. F. Lucas, F. S. Roberts, and R. M. Thrall (Eds.), *Discrete and System Models* 3, *of Modules in Applied Mathematics*, Springer-Verlag (1983), 76–111.

[Wo72] D. R. Woodall, Sufficient conditions for circuits in graphs, *Proc. London Math. Soc.* 24 (1972), 739–755.

Chapter 7: Planarity and Kuratowski's Theorem

[BoMu76] J. A. Bondy and U. S. R. Murty, *Graph Theory with Applications*, American Elsevier, 1976.

[DeMaPe64] G. Demoucron, Y. Malgrange, and R. Pertuiset, Graphes planaires: reconnaissance et construction de representations planaires topologiques, *Rev. Francaise Recherche Operationelle* 8 (1964), 33–47.

[Fa48] I. Fary, On the straight-line representations of planar graphs, *Acta Sci. Math.* 11 (1948), 229–233.

[GrRo79] J. L. Gross and R. H. Rosen, A linear-time planarity algorithm for 2-complexes, *J. Assoc. Comput. Mach.* 20 (1979), 611–617.

[GrRo81] J. L. Gross and R. H. Rosen, A combinatorial characterization of planar 2-complexes, *Colloq. Math.* 44 (1981), 241–247.

[Ha69] F. Harary, *Graph Theory*, Addison-Wesley, 1969.

[HoTa74] J. Hopcroft and R. E. Tarjan, Efficient Planarity Testing, *JACM* 21 (1974), 549–568.

[Ne54] M. H. A. Newman, *Elements of the Topology of Plane Sets of Points*, Cambridge University Press, 1954.

[Th80] C. Thomassen, Planarity and duality of finite and infinite graphs, J. *Combin. Theory Ser. B* 29 (1980), 244–271.

[Th81] C. Thomassen, Kuratowski's theorem, *J. Graph Theory* 5 (1981), 225–241.

[Wa36] K. Wagner, Bemerkungen zum Vierfarbenproblem, *Jber. Deutch. Math. Verien.* 46 (1936), 21–22.

[We01] D. B. West, *Introduction to Graph Theory*, Second Edition, Prentice-Hall, 2001.

[Wi04] R. J. Wilson, History of Graph Theory, §1.3 *in Handbook of Graph Theory*, eds., J. L. Gross and J. Yellen, CRC Press, 2004.

Chapter 8: Graph Colorings

[ApHa76] K. I. Appel and W. Haken, Every planar map is four-colorable, *Bull. Amer. Math. Soc.* 82 (1976), 711–712.

[Bä38] F. Bäbler, Über die Zerlegung regulärer Streckencomplexe ungerader Ordnung, *Comment. Math. Helv.* 10 (1938), 275–287.

[Be68] L. W. Beineke, Derived graphs and digraphs, in *Beitrage zur Graphentheorie* Tuebner, 1968.

[Br79] D. Brelaz, New methods to color the vertices of a graph, *Commun. ACM* 22 (1979), 251–256.

[Ca86] M. W. Carter, A survey of practical applications of examination timetabling algorithms, *Oper. Res.* 34 (1986), 193–202.

[GaJo79] M. R. Garey and D. S. Johnson, *Computers and Intractability: A Guide to the Theory of NP-Completeness*, W. H. Freeman, 1979.

[Ho81] I. Holyer, The NP-completeness of edge-coloring, *SIAM J. Computing* 10 (1981), 718–720.

[KiYe92] L. Kiaer and J. Yellen, Weighted graphs and university course timetabling, *Computers and Oper. Res.* 19 (1992), 59–67.

[Kö16] D. König, Über Graphen und ihre Andwendung auf Determinantentheorie und Mengenlehre, *Math. Ann.* 77 (1916), 453–465.

[Lo75] L. Lovász, Three short proofs in graph theory, *J. Combin. Theory Ser. B* 19 (1975), 269–271.

[Ma81] B. Manvel, Coloring large graphs, *Proceedings of the 1981 Southeastern Conference on Graph Theory, Combinatorics and Computer Science*, 1981.

[Pe1891] J. Petersen, Die Theorie der regulären Graphen, *Acta Math.* 15 (1891), 193–220.

[Pl04] M. Plummer, Factors and Factorization, §5.4 *in Handbook of Graph Theory*, eds., J. L. Gross and J. Yellen, CRC press, 2004.

[RoSaSeTh97] N. Robertson, D. P. Sanders, P. Seymour, and R. Thomas, The four-colour theorem, *J. Combin. Theory Ser. B* 70 (1997), 166–183.

[WeYe14] A. Wehrer and J. Yellen, The design and implementation of an interactive course-timetabling system, *Annals of Oper. Res.* 218, (2014), 327–345.

Chapter 9: Special Digraph Models

[Ba83] S. Baase, *Computer Algorithms: Introduction to Design and Analysis*, Addison-Wesley, 1983.

[BeTh81] J.-C. Bermond and C. Thomassen, Cycles in digraphs: A survey, *J. Graph Theory* 5 (1981) 1–43.

[BuWeKi04] E. Burke, D. de Werra, and J. Kingston, Applications to Timetabling, §5.6 *in Handbook of Graph Theory*, eds., J. L. Gross and J. Yellen, CRC Press, 2004.

[Ci75] E. Cinlar, *An Introduction to Stochastic Processes*, Prentice-Hall, 1975.

[Hä93] R. Häggkvist, Hamiltonian cycles in oriented graphs, *Combin. Probab. Comput.* 2 (1993), 25–32.

[La53] H. G. Landau, On dominance relations and the structure of animal societies. III. The condition for a score structure, *Bull. Math. Biophys.* 15 (1953), 143–148.

[Ma04] S.B. Maurer, Directed Acyclic Graphs, §3.2 *in Handbook of Graph Theory*, eds., J. L. Gross and J. Yellen, CRC Press, 2004.

[Ma80] S. Maurer, The king chicken theorems, *Math. Mag.* 53 (1980), 67–80.

[Mo68] J. W. Moon, *Topics on Tournaments*, Holt, Rinehart, and Winston, 1968.

[Re04] K. B. Reid, Tournaments, §3.3 *in Handbook of Graph Theory*, eds., J. L. Gross and J. Yellen, CRC Press, 2004.

[Re96] K. B. Reid, Tournaments: Scores, kings, generalizations and special topics, *Congressus Numer.* 115 (1996), 171–211.

[St93] H. J. Straight, *Combinatorics: An Invitation*, Brooks/Cole, 1993.

[Th82] C. Thomassen, Edge-disjoint Hamiltonian paths and cycles in tournaments, *Proc. London Math. Soc.* 45 (1982), 151–168.

[ThSw92] K. Thulasiraman and M. N. S. Swamy, *Graphs: Theory and Algorithms*, John Wiley & Sons, 1992.

[Zh80] C. Q. Zhang, Every regular tournament has two arc-disjoint Hamiltonian cycles, *J. Qufu Normal College, Special Issue Oper. Res.* (1980), 70–81.

Chapter 10: Network Flows and Related Problems

[Ed65b] J. Edmonds, Paths, trees, and flowers, *Canad. J. Math.* 17 (1965), 449–467.

[EdKa72] J. Edmonds and R. M. Karp, Theoretical improvements in algorithmic efficiency for network flow problems, *JACM* 19 (1972), 248–264.

[Ev79] S. Even, *Graph Algorithms*, Computer Science Press, 1979.

[FoFu62] L. R. Ford and D. R. Fulkerson, *Flows in Networks*, Princeton University Press, 1962.

[Kö16] D. König, Über Graphen und ihre Andwendung auf Determinantentheorie und Mengenlehre, *Math. Ann.* 77 (1916), 453–465.

[Mi78] E. Minieka, *Optimisation Algorithms for Networks and Graphs*, Marcel Dekker, 1978.

[Pe1891] J. Petersen, Die Theorie der regularen Graphen, *Acta Math.* 15 (1891), 193–220.

[ThSw92] K. Thulasiraman and M. N. S. Swamy, *Graphs: Theory and Algorithms*, John Wiley & Sons, 1992.

Chapter 11: Graphical Enumeration

[Bo00] K. P. Bogart, *Introductory Combinatorics*, Third Edition, Academic Press, 2000.

[Br04b] R. A. Brualdi, *Introductory Combinatorics*, Fourth Edition, Prentice Hall, 2004.

[Gr08] J. L. Gross, *Combinatorial Methods with Computer Applications*, CRC Press, 2008.

[HaPa73] F. Harary and E. M. Palmer, *Graphical Enumeration*, Academic Press, 1973.

Appendix

[Co71] S. A. Cook, The complexity of theorem proving procedures, *Proc. 3rd Annual ACM Symposium on Theory of Computing (STOC 71)* (1971), 151–158.

[Ev79] S. Even, *Graph Algorithms*, Computer Science Press, 1979.

[GaJo79] M. R. Garey and D. S. Johnson, *Computers and Intractability: A Guide to the Theory of NP-Completeness*, W. H. Freeman, 1979.

[Wi86] H. S. Wilf, *Algorithms and Complexity*, Prentice-Hall, 1986.

SOLUTIONS AND HINTS

Chapter 1 Introduction to Graph Models

1.1 Graphs and Digraphs

1.1.1:

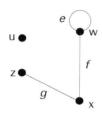

edge	e	f	g
endpts	w	w	x
	w	x	z

$\langle 3, 2, 1, 0 \rangle$

1.1.4:

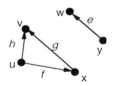

1.1.7:

edge	a	b	c	d
endpts	y^h	y	z	x
	x	x^h	z^h	z^h

1.1.10:

edge	a	b	c	d
endpts	y	y	z	x
	x	x	z	z

1.1.15: a. Does not exist.   b. Does not exist.

1.1.22: No.

1.1.24: The *endpts* function must be one-to-one, and none of its images is a one-element set.

1.1.32: No, the conditions imply an even number of vertices.

1.1.36: The minimum number is 5.

1.1.39: The maximum number is 6.

1.1.42: The maximum number is 5.

1.1.45: A minimum dominating set has four vertices.

1.2 Common Families of Graphs

1.2.1: a. $\binom{n}{2} = \frac{n^2 - n}{2}$ b. mn

1.2.4: a. $n \leq 2$ b. All even n.   c. All n.

1.2.6: $U = \{x,v\}$ and $W = \{u,z\}$

1.2.12: Use Euler's Degree-Sum Theorem.

1.2.15: $2^{\frac{n^2 - n}{2}}$

1.2.25: Let $S_k = (0, k)$ for $k = 1, \dots, n$.

1.2.27: With a minus sign denoting BC, the following collection of intervals assigned to $a_1, \dots, a_5$, respectively, shows that the given graph is an interval graph.

$$(-100, 50) \quad (250, 400) \quad (100, 300) \quad (0, 200) \quad (150, 300)$$

1.3   Graph Modeling Applications

1.3.1: A 4-coloring is easy to construct, and a 3-coloring is impossible because vertices G, C, D, E are mutually adjacent. Hence four is the minimum number of frequencies.

1.3.3: .048

1.3.5: All binary strings that have an odd number of ones.

1.3.6: Vertices correspond to the volunteers, and two vertices are adjacent if the corresponding volunteers have a language in common. Then a maximum-size collection of vertex-disjoint edges is desired.

1.3.13: d. No two vertices have a pair of oppositely directed arcs between them.

1.4   Walks and Distance

1.4.1: a. yes   b. yes   c. no   d. yes

1.4.4: Only the first two represent connected graphs.

1.4.6: Not strongly connected. For example, vertex u is not reachable from vertex x.

1.4.10: Any digraph that has a directed closed walk.

1.4.12: The distance equals 4.

1.4.15: Any connected graph that contains no closed walks.

1.4.18: $diam(G) = 3$, $rad(G) = 2$, and the central vertices are v and z.

1.4.32: For the second inequality, let x and y be two vertices such that $d(x, y) = diam(G)$, and let w be a central vertex. Then apply the triangle inequality.

1.4.34: Rigid

1.4.38: Choose any vertex $x_1 \in V_G - \{u, v\}$, and let W_1 be a u-x_1 walk. Choose $x_2 \in V_G - \{u, v, x_1\}$, and let W_2 be an x_1-x_2 walk. Continue until the last vertex, x_m, is chosen, and let W_{m+1} be an x_m-v walk. Then form the concatenation of walks W_1 through W_{m+1}.

1.4.41: There are two walks of length 2 and seven walks of length 3.

1.4.46: Let x and y be any two non-adjacent vertices of graph G. Since G is a simple graph, the condition implies that the set of vertices adjacent to vertex x and the set of vertices adjacent to y have at least one vertex in common.

1.5   Paths, Cycles, and Trees

1.5.1: a. It is a walk of length 4, but it is not a path.

1.5.3: b. $\langle r, s, n \rangle$

1.5.5: The three w-v paths can be concatenated with the three v-r paths. Hence there are nine different w-r paths.

1.5.8: Let e be any edge with endpoints x and y, and consider the walk $\langle x,\, e,\, y,\, e,\, x\rangle$.

1.5.11: None of length 2 or 4, and four paths of length 3.

1.5.15: The girth equals 4.

1.5.30: Consider a nontrivial closed subtrail of minimal length that contains vertex v.

1.5.32: Repeated vertices would imply the existence of a directed cycle, which could then be deleted, leading to a shorter directed walk.

1.5.40: The root represents the empty board. Each child represents the configuration that results from placing an x in one of the positions, etc.

1.5.43: Start at any vertex. Since that vertex has nonzero outdegree, there is an arc directed from it to some vertex. Now use the finiteness of the vertex set to argue that eventually a vertex must be repeated.

1.6   Vertex and Edge Attributes: More Applications

1.6.2: $(n-1)!$

1.6.3: The n vertices of a graph represent the exams. When it is undesirable to schedule two exams in overlapping timeslots, the two corresponding vertices are joined by an edge, and that edge is assigned a weight indicating how undesirable a conflict would be. The objective would then be to assign one of k possible *colors* to each of the n vertices so that the sum of the weights of all edges whose endpoints have the same color is minimized. Vertex coloring is discussed in §8.1.

1.6.5: Modify the model used in Application 1.3.1 by assigning a weight to each edge. The objective is to find a maximum-weight matching (see §10.4).

1.6.7: Some variation of a network-flow model (see §10.1).

1.6.8: This is a type of vehicle-routing problem that is notoriously difficult computationally. One approach is to first partition the bus stops into m subsets and then solve m separate traveling salesman problems to assess the cost (mileage) of that partition.

Chapter 2 Structure and Representation

2.1   Graph Isomorphism

2.1.1:

2.1.7:

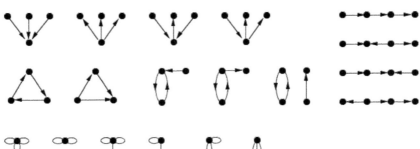

2.1.11:

2.1.15: $v \to r; w \to s; u \to t; x \to z$ or $v \to r; w \to s; u \to z; x \to t$.

2.2   Automorphisms and Symmetry

2.2.1: identity automorphism and $(u)(x)(vw)$

2.2.5: identity automorphism and $\{f_V = (u)(x)(v)(w), f_E = (a)(b)(cd)\}$

2.2.9: For any two vertices of K_n, the vertex-bijection that swaps them and fixes all the others is adjacency-preserving. It is its own inverse, so the inverse bijection is also adjacency-preserving.

2.2.14: $circ(9 : 1, m)$ is edge-transitive unless $m = 3$ or 6 (in which case a 3-edge or a 6-edge would lie on a 3-cycle, but a 1-edge does not).

2.2.18: vertex orbits $\{u\}, \{x\}, \{v, w\}$; edge orbits $\{ux\}, \{vw\}, \{vx,\ wx\}$

2.3   Subgraphs

2.3.1: a. Yes: $endpoints(q) = \{u, g\} \subset W$; $endpoints(g) = \{u, y\} \subset W$.

2.3.3:

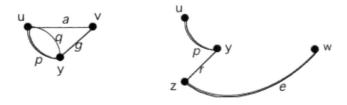

2.3.7:

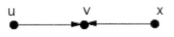

2.3.13:

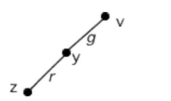

2.3.19: (a) There are 6 cliques; each has 3 vertices. (b) $w = 3$. (c) uyz, uw, xv, xw, xz, rw, rz are maximal independent sets. (d) $\alpha = 3$. (e) x and z are non-central. The center is the subgraph induced by the other five vertices.

2.3.23: Using the solution above to Exercise 19, we see that the largest cardinality among such sets is 2. For instance, $\{w, z\}$ is such a set.

2.3.30: There are four spanning trees that contain edge p and four that do not.

2.3.40: If one tree is K_1, the other tree is one of the three trees with 5 vertices. If one tree is K_2, the other tree is $K_{1,3}$ or P_4. Or both trees are P_3.

2.4   Some Graph Operations

2.4.1:

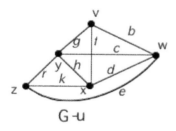

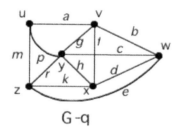

2.4.4:

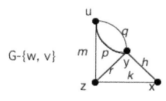

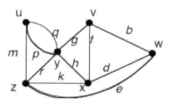

2.4.8: u, v, z are the cutpoints.

2.4.12: 2; $\{b, d\}$.

2.4.16: 4; $\{da, db, du, dc\}$.

2.4.20:

2.4.22: There are 4 vertices and 4 edges. The degree sequence is 3211. There is only one such graph.

2.4.35: False. If G is not connected, then vertex v is a cut-vertex of graph $G + v$. If G is connected, then $G + v$ has no cut-vertices.

2.4.40: Assume that G is not connected, and let x and y be two vertices in $\overline{G}$. If x and y are in different components of G, then they are adjacent in $\overline{G}$. If x and y are in the same component of G, then there is at least one vertex w in some other component of G, and w is adjacent to both x and y in $\overline{G}$.

2.5   Tests for Non-Isomorphism

2.5.1: a. The given graph is isomorphic to graph B.

2.5.2: One graph has two 3-cycles, but the other graph has none.

2.5.8: Non-isomorphic because they have a different degree sequence.

2.5.13: Start with a digraph version of Theorem 2.1.3.

2.5.20: Isomorphic.

2.6   Matrix Representations

2.6.1:
$$
A_G = \begin{array}{c} u \\ v \\ w \\ x \\ y \end{array}
\begin{pmatrix}
\begin{array}{ccccc} u & v & w & x & y \end{array} \\
\begin{array}{ccccc}
0 & 2 & 1 & 0 & 0 \\
2 & 2 & 1 & 1 & 0 \\
1 & 1 & 0 & 0 & 0 \\
0 & 1 & 0 & 0 & 0 \\
0 & 0 & 0 & 0 & 0
\end{array}
\end{pmatrix}
\quad ; \quad
I_G = \begin{array}{c} u \\ v \\ w \\ x \\ y \end{array}
\begin{pmatrix}
\begin{array}{cccccc} a & b & c & d & e & f \end{array} \\
\begin{array}{cccccc}
1 & 0 & 1 & 0 & 0 & 1 \\
1 & 2 & 1 & 1 & 1 & 0 \\
0 & 0 & 0 & 0 & 1 & 1 \\
0 & 0 & 0 & 1 & 0 & 0 \\
0 & 0 & 0 & 0 & 0 & 0
\end{array}
\end{pmatrix}
$$

2.6.4:
$$
A_D = \begin{array}{c} u \\ v \\ w \\ x \\ y \end{array}
\begin{pmatrix}
\begin{array}{ccccc} u & v & w & x & y \end{array} \\
\begin{array}{ccccc}
0 & 1 & 0 & 0 & 0 \\
1 & 2 & 1 & 1 & 0 \\
1 & 0 & 0 & 0 & 0 \\
0 & 0 & 0 & 0 & 0 \\
0 & 0 & 0 & 0 & 0
\end{array}
\end{pmatrix}
\quad ; \quad
I_D = \begin{array}{c} u \\ v \\ w \\ x \\ y \end{array}
\begin{pmatrix}
\begin{array}{cccccc} a & b & c & d & e & f \end{array} \\
\begin{array}{cccccc}
-1 & 0 & 1 & 0 & 0 & 1 \\
1 & 2 & -1 & -1 & -1 & 0 \\
0 & 0 & 0 & 0 & 1 & -1 \\
0 & 0 & 0 & 1 & 0 & 0 \\
0 & 0 & 0 & 0 & 0 & 0
\end{array}
\end{pmatrix}
$$

2.6.9: If the m vertices of one side of the bipartition are listed first, then the adjacency matrix will consist of an $m \times m$ matrix of 0's in the upper left corner, an $m \times n$ matrix of 1's in the upper right corner, an $n \times m$ matrix of 1's in the lower left corner, and an $m \times m$ matrix of 0's in the lower right corner.

2.6.14:

2.6.16: $A_G^2\,[a,a] = (0,\ 1,\ 1,\ 1) \cdot (0,\ 1,\ 1,\ 1) = 3.$

2.6.18: $A_D^2\,[d,a] = (0,\ 0,\ 1,\ 0) \cdot (0,\ 0,\ 0,\ 0) = 0.$

2.6.20:

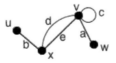

2.6.22:

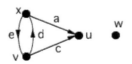

2.7   More Graph Operations

2.7.1:

2.7.7:

2.7.12:

2.7.16:

2.7.20:

Chapter 3 Trees

3.1   Characterizations and Properties of Trees

3.1.3: The only such graph is K_5 plus two isolated vertices.

3.1.9: There is no such graph, because a tree would have eight edges, and each additional edge would create at least one cycle.

3.1.13: The smallest average degree occurs when the connected graph is a tree, so by Euler's Degree-Sum Theorem and Proposition 3.1.3, $\frac{\sum v}{n} = \frac{2|E|}{n} = 2 - \frac{2}{n}$.

3.1.17: Use Part 6 of Theorem 3.1.8 to show it is true.

3.1.20: If H were not connected, then there would be an edge $e \in E_G - E_H$ whose endpoints are in different components of H. But then $H + e$ would contradict the maximality of H. Use Part 6 of Theorem 3.1.8 to prove the reverse implication.

3.1.24: Let x and y be any two non-adjacent vertices of graph G. Since G is a simple graph, the condition implies that the set of vertices adjacent to vertex x and the set of vertices adjacent to y have at least one vertex in common.

3.1.27: All acyclic graphs.

3.2   Rooted Trees, Ordered Trees, and Binary Trees

3.2.1: The two 3-vertex rooted trees shown below are isomorphic as graphs, and the four 4-vertex rooted trees represent two isomorphism types as graphs.

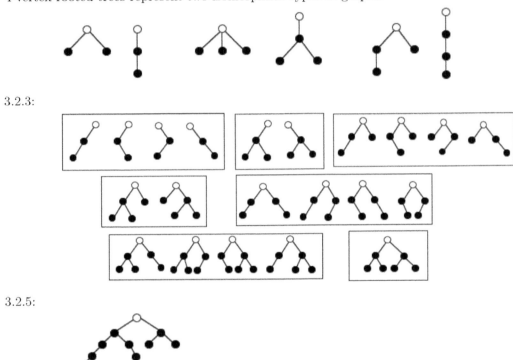

3.2.3:

3.2.5:

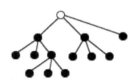

3.2.8: There are several such trees. For instance, start with a path of length 2, and attach 11 children to the last vertex.

3.2.13:

3.2.17: Start by considering the three ternary trees of height 2 whose internal vertices have exactly three children, as shown below.

Then determine the number of different ways three children can be attached to leaves at depth 2. For instance, there is only one way (up to rooted-tree isomorphism) to attach three children to each of two of the leaves of the tree on the left, but there are two different ways of attaching three children to each of two of the leaves of the middle tree.

3.2.21: $-$, 0, 0, 1, 1, 1, 5, 5

3.2.23: The recurrence relation is $f(h) = 2f(h-1) + 1$. By the induction hypothesis, $f(h) = 2(2^h - 1) + 1 = 2^{h+1} - 1$.

3.3   Binary-Tree Traversals

3.3.1: in-order: 1,2,3,4,5,6,7,8,9,10; pre-order: 1,2,3,4,5,6,7,8,9,10; post-order: 7,8,4,5,2,9,10,6,3,1

3.3.5:

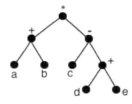

3.4   Binary-Search Trees

3.4.1:

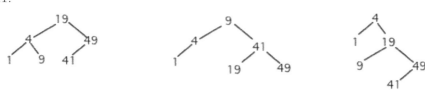

3.4.5: The following tree results after the two insertions and one deletion.

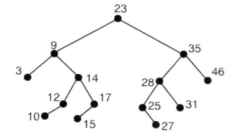

3.4.11: 13, 3, 55, 2, 5, 34, 144, 1, 8, 21, 89

3.4.13: When the keys are in either descending order or in ascending order.

3.5   Huffman Trees and Optimal Prefix Codes

3.5.1: facade

3.5.4:

letter	a	b	c	d	e	f	g	h	av. wt. length
codeword	00	1000	010	011	110	101	111	1001	2.92

defaced = 01111010100010110011;   baggage = 10000011111100111110

3.6   Priority Trees

3.6.2: 2^h

3.6.6:

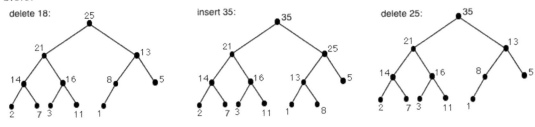

3.6.9:

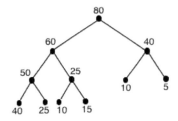

3.6.11: The heap does not represent a priority tree because it puts 22 as a child of 20.

3.7   Counting Labeled Trees: Prüfer Encoding

3.7.1: $\langle 4, \; 2, \; 2, \; 4 \rangle$

3.7.6: $\langle 1, \; 1, \; 3, \; 1, \; 3 \rangle$

3.7.10:

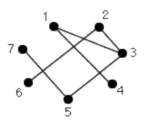

3.7.14: Exactly one of the m vertices on the m-side of the bipartition has degree 2, and each of the other $m - 1$ vertices on that side is joined to exactly one of the two vertices on the other side of the bipartition. Hence, the number of spanning trees is $m \cdot 2^{m-1}$.

3.8   Counting Binary Trees: Catalan Recursion

3.8.1: $b_4 = b_0 b_3 + b_1 b_2 + b_2 b_1 + b_3 b_0 = 14 = \frac{1}{5}\binom{8}{4}$

3.8.4: Each vertex of a binary tree on $n-1$ vertices represents a multiplication, and the lower the depth of the vertex, the earlier that multiplication must be performed. Thus, the number of different ways of multiplying n numbers equals the number of different binary trees on $n-1$ vertices.

3.8.7: Use the fact that

$$(2n)! = (2n)(2n-1)(2n-2)\ldots(3)(2)(1) = 2^n n!(2n-1)(2n-3)\ldots(3)(1)$$

Chapter 4 Spanning Trees

4.1   Tree Growing

4.1.1: $\{f, g, i, j\}$

4.1.4a:

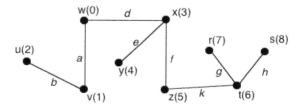

4.1.8b:

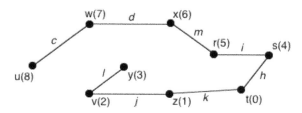

4.1.10:

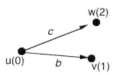

4.1.16: The graph would have to be a path graph, and the starting vertex would have to be at one of the ends of the path.

4.2   Depth-First and Breadth-First Search

4.2.1a:

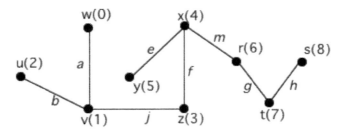

4.2.5b:

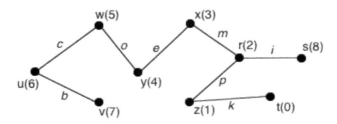

4.2.9a:

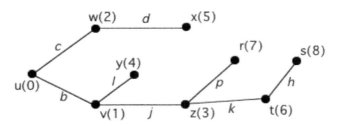

4.2.13b:

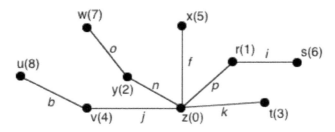

4.2.16: Trees.

4.3   Minimum Spanning Trees and Shortest Paths

4.3.1: total weight = 32

4.3.5:

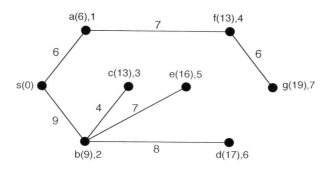

4.3.10:

4.4   Applications of Depth-First Search

4.4.1: According to the depth-first tree shown below, $df number(c) \leq low(b)$ and $df number(g) \leq low(h)$, verifying the assertion of Corollary 4.4.12.

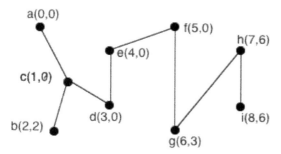

4.4.5: Vertex c is a cut-vertex and is the only non-root in the tree shown below that meets the condition for a cut-vertex given in Corollary 4.4.12.

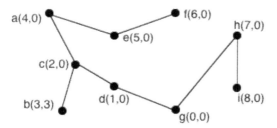

4.5   Cycles, Edge-Cuts, and Spanning Trees

4.5.1:

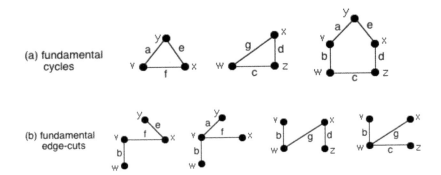

(a) fundamental cycles

(b) fundamental edge-cuts

4.5.6a: For instance, the first fundamental cut in the solution to 4.5.1(a) above has two edges in common with the spanning tree whose edge-set is $\{a, b, d, f\}$, it has one edge in common with $\{a, e, g, d\}$, and it has two edges in common with $\{e, f, c, d\}$ (there are several other spanning trees to be considered).

4.5.6b: For instance, the first fundamental cycle in the solution to 4.5.1(a) above has edge g in common with the relative complement of the spanning tree whose edge-set is $\{a, b, d, f\}$, has edge f in common with the relative complement of the tree whose edge-set is $\{a, e, g, d\}$, and has edges e and f in common with the relative complement of the tree whose edge-set is $\{a, b, g, c\}$.

4.5.6c: The 5-cycle has edges b and e in common with the first fundamental edge-cut listed in the solution to 4.5.1(b) above, it has edges a and b in common with the second edge-cut in the list, it has edges b and d in common with the third one, and it has edges b and c in common with the last one in the list.

4.5.10a: Consider the spanning tree whose edge-set is $\{a, e, c, d\}$, and consider the fundamental cycle with respect to edge f (which is listed first in the solution to 4.5.1a above). The other two edges of this cycle are a and e. The two fundamental edge-cuts with respect to these two edges are the first two listed in the solution to 4.5.1(b), and these are the only two that contain the edge f.

4.5.10b: Consider the spanning tree whose edge-set is $\{a, e, c, d\}$, and consider the fundamental edge-cut with respect to edge e (which is listed first in the solution to 4.5.1(b)). The other two edges of this edge-cut are f and b. The fundamental cycle with respect to I and the fundamental cycle with respect to b are the first and third ones listed in the solution to 4.5.1(a), and these two fundamental cycles are the only ones that contain edge e.

4.5.16: Let v be a vertex adjacent to vertices x and y. If there is an x-y path P that avoids v, then there is a cycle involving P, v, and the two edges joining v to x and y.

4.5.22: Consider the fundamental systems of cycles and edge-cuts in Figure 4.5.2. The edge-cut $\{b, c, e, f\}$ has three edges in common with the 5-cycle of the graph.

4.6   Graphs and Vector Spaces

4.6.1a: $\{b, c, f\}$, $\{a, c, e\}$, $\{b, d, e\}$, $\{a, d, f\}$, $\{a, b, c, d\}$, $\{a, b, e, f\}$, $\{c, d, e, f\}$

4.6.1b: $\{a, c, f\}$, $\{a, d, e\}$, $\{b, c, e\}$, $\{b, d, f\}$, $\{a, b, e, f\}$, $\{a, b, e, f\}$, $\{a, b, c, d\}$, $\{c, d, e, f\}$

4.6.4a: The fundamental system of cycles consists of $\{a, c, e\}$, $\{b, d, e\}$, and $\{a, d, f\}$. Each of the other five non-null elements of $W_C(G)$ (listed in the solution of 4.6.1(a)) is a mod 2 sum of some or all of these three edge-sets. For instance, $\{a, b, c, d\} = \{a, c, e\} + \{b, d, e\}$.

4.6.4b: The fundamental system of edge-cuts consists of $\{b, c, e\}$, $\{b, d, f\}$, and $\{a, c, f\}$. Each of the other four non-null elements of $W_C(G)$ (listed in the solution of 4.6.1(b)) is a mod 2 sum of some or all of these three edge-sets. For instance, $\{a, b, c, d\} = \{a, c, f\} + \{b, d, f\}$.

4.6.10: For the vertex order u, w, x, y, z, the columns of C_D for the edge set $D = \{a, b, c, d\}$ are

$$
\begin{pmatrix} 0 \\ 0 \\ 1 \\ 0 \\ 1 \end{pmatrix}, \quad
\begin{pmatrix} 1 \\ 0 \\ 1 \\ 0 \\ 0 \end{pmatrix}, \quad
\begin{pmatrix} 1 \\ 0 \\ 0 \\ 0 \\ 1 \end{pmatrix}, \quad
\begin{pmatrix} 0 \\ 1 \\ 0 \\ 0 \\ 1 \end{pmatrix}
$$

and their mod 2 sum is not the zero vector. This is consistent with the assertion that edge-set D is neither a cycle nor a disjoint union of cycles. Also, the sum of the first two column vectors equals the third one, which shows that C_D does not form a basis for the column space. This is consistent with the assertion that the edge-set D does not induce a spanning tree of the graph.

4.6.17: The non-null elements of $W_C(G)$ are the three cycles of graph G, and none of these is an element of $W_S(G)$. Thus, by Theorem 4.6.11, the two subspaces are orthogonal complements.

4.7   Matroids and the Greedy Algorithm

4.7.1: Since the edges were not labeled, ties were broken according to the alphabetical order of the edges' endpoints.

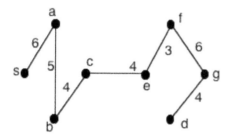

4.7.5: It is an hereditary subset system but is not a matroid. For instance, in the star graph $K_{1,n}$, the vertex of degree n and any two other vertices do not satisfy the augmentation property.

4.7.9: It is not an hereditary subset system because $\emptyset \notin \mathcal{I}$.

4.7.13: Show that $(C_1 \cup C_2) - x$ is a dependent set.

4.7.16: Suppose that the independent set B obtained by the greedy algorithm is not of minimum weight. Then there is some maximal independent set A such that $w(A) < w(B)$. Since A and B are both maximal independent sets, the augmentation property implies that they have the same number of elements. Let $A = \{a_1, a_2, \ldots, a_r\}$, where $w(a_1) \le w(a_2) \le \ldots \le w(a_r)$, and let $B = \{b_1, b_2, \ldots, b_r\}$, where b_i is the element chosen in the ith iteration of the greedy algorithm. Let j be the smallest subscript such that $w(a_j) < w(b_j)$ (j exists since $w(A) < w(B)$). By the augmentation property applied to the independent sets $A_j = \{a_1, \ldots, a_j\}$ and $B_{j-1} = \{b_1, \ldots, b_{j-1}\}$, there is some $a \in A_j - B_{j-1}$ such that $B_{j-1} \cup \{a\}$ is an independent set. But $w(a) \le w(a_j) < w(b_j)$, which contradicts the choice of b_j at the jth iteration.

Chapter 5 Connectivity

5.1   Vertex- and Edge-Connectivity

5.1.1: $\{y,v\}$ and $\{v,w\}$

5.1.5: Corollary 5.1.5 implies no such graph exists.

5.1.9: $\kappa_v(G) = 3$ and $\kappa_e(G) = 4$

5.1.13: Consider three specific pairs of adjacent vertices and three pairs of nonadjacent ones for the cases in which both vertices are on the inner cycle, both are on the outer cycle, and one is on the inner cycle and the other is on the outer cycle, and show none of these pairs can be a vertex-cut. Then conclude, by Corollary 5.1.6, that $\kappa_v(G) = \kappa_e(G) = 3$.

5.1.17: The graph in Figure 5.1.1.

5.1.21: Generalize the graph in Figure 5.1.1.

5.1.23: Since $2 = \kappa_v(G) \le \kappa_e(G)$, every edge must be on that unique cycle. Hence, the graph must be a cycle graph.

5.1.26: See Menger's Theorem 5.3.4.

5.2   Constructing Reliable Networks

5.2.1:

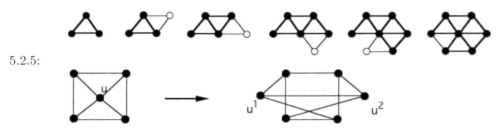

5.2.5:

5.2.7: Use induction. Observe that the Harary graph $H_{3,4}$ is the 4-vertex wheel graph and that $H_{3,n}$ is an n-cycle plus some diameters or quasi-diameters, depending on the parity of n. Then show that if n is odd, $H_{3,n+1}$ can be obtained from $H_{3,n}$ by a single type 2 operation, and if n is even, by a type 1 operation followed by a type 2.

5.2.11:

$H_{6,8}$

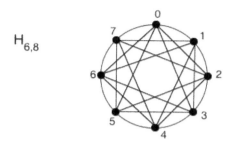

5.2.14: Use Corollary 5.1.6.

5.3   Max-Min Duality and Menger's Theorems

5.3.1: There are several collections of two internally disjoint paths, for instance, $\mathcal{P} = \{\langle u, t, y, v \rangle, \langle u, s, w, z, v \rangle\}$. The u-v separating vertex set $\{s, t\}$ shows that $\mathcal{P}$ is a maximum-size collection.

5.3.5: The u-v separating edge set consisting of the two edges incident on vertex u shows that the two paths given in the solution to 5.3.1 form a maximum-size collection of edge-disjoint u-v paths.

5.3.10: The collection $\mathcal{P} = \{\langle u, t, y, v \rangle, \langle u, s, a, b, v \rangle, \langle u, a, w, x, z, v \rangle\}$ is a maximum-size collection of arc-disjoint directed u-v paths, because the three arcs directed from vertex u form a u-v separating arc set.

5.3.15: Clearly, between each pair of vertices u, v in a wheel graph, there are exactly three internally disjoint u-v paths. For the induction step, a type 1 operation in a Tutte synthesis adds an edge and hence, cannot reduce the number of paths. Let u and v be any two vertices in the graph that results from a type 2 operation, and suppose vertex w is split into vertices w' and w''.

Case 1: $u \neq w'$ and $v \neq w''$. By the induction hypothesis, there are at least k internally disjoint u-v paths in the graph before the operation is performed. If none of these paths uses vertex w, then they are unaffected by the type 2 operation and hence, are still internally disjoint. Suppose that one of these k paths uses vertex w, say $P = \langle u, x_1, x_2, \ldots, x_j, w, x_{j+2}, \ldots, x_l, v \rangle$. After the type 2 operation, vertices x_j and x_{j+2} are either both adjacent to w', both adjacent to w'', or one is adjacent to w' and the other is adjacent to w''. In each of these cases, path P is transformed into a u-v path whose only internal vertices are $x_1, \ldots x_j, x_{j+2}, x_l$ and one or both of w' and w'', and thus, is still internally disjoint from the other $k - 1$ paths.

The two cases $u = w'$ and $v = w''$ and $u = w'$ and $v \neq w''$ are handled in a similar way.

5.4   Block Decompositions

5.4.1: The vertex-sets of the five blocks are $B_1 = \{a, b\}$, $B_2 = \{a, v\}$, $B_3 = \{v, t\}$, $B_4 = \{u, v, w, x, y, z\}$, and $B_5 = \{y, s\}$. The block graph is shown below.

5.4.5:

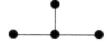

5.4.8: By Theorem 3.1.8, the graph must contain a unique cycle. Therefore, the three blocks must be a 5-cycle, and two K_2's. Either the two K_2's form a path that is joined to one vertex on the cycle, or they are adjacent to the same cycle-vertex, to two adjacent cycle-vertices, or to two non-adjacent cycle-vertices. Thus, there are four non-isomorphic graphs, as shown in the figure below.

5.4.14: The assertion is false even if the graphs are restricted to be simple and connected. For instance, the following two graphs have the same block graph.

5.4.18: For the block-cutpoint graph shown below, the labels on the block vertices correspond to the labels used in the solution to 5.4.1.

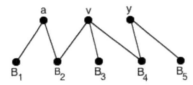

5.4.22: If the block-cutpoint graph had a cycle, then the union of the blocks corresponding to the block vertices on that cycle would be 2-connected, contradicting the maximality of each of those blocks.

Chapter 6 Optimal Graph Traversals

6.1   Eulerian Trails and Tours

6.1.1: K_n is Eulerian for all odd $n \geq 3$.

6.1.5: The octahedral graph is the only platonic graph that is Eulerian.

6.1.7: Starting at the bottom left vertex and using alphabetical order for the default priority, we obtain the following sequence of iterations.

iteration 0: closed trail $T = \langle b, d, e, f \rangle$

iteration 1: closed trail $D = \langle g, a, j, c, h, k, m, i \rangle$;

enlarged trail $T = \langle b, d, g, a, j, c, h, k, m, i, e, f \rangle$

6.1.10: Algorithm 6.1.1a is shown below.

Algorithm 6.1.1a: Constructing an Open Eulerian Trail

Input: a connected graph G that has exactly two vertices of odd degree.

Output: an open Eulerian trail T.

  Let x and y be the two vertices of odd degree.

  Let e^* be a new edge between vertices x and y.

  $G := G + e^*$

  Start at vertex x, and construct a closed trail T in G, starting with edge e^*.

  While there are edges of G not already in trail T

    Choose any vertex w in T that is incident with an unused edge.

    Starting at vertex w, construct a closed trail D of unused edges.

    Enlarge trail T by splicing trail D into T at vertex w.

  Circularly shift the terms of the edge sequence of trail T to the right until
    edge e^* appears first.

  Delete edge e^* from the edge sequence of T.

  Return T.

6.1.13: Edges are chosen using an alphabetical default priority.

  iteration 0: Let x be the odd-valent endpoint of edge e.

        Let y be the odd-valent endpoint of edge n.

        Let e^* be the new edge between x and y, and add e^* to graph G.

        closed trail $T = \langle e^*, c, b, p, n, l, j, k, g, d, m, i, q \rangle$

  iteration 1: closed trail $D = \langle e, a, r, o \rangle$;

        enlarged trail $T = \langle e, a, r, o, e^*, c, b, p, n, l, j, k, g, d, m, i, q \rangle$

  circular shift: $T = \langle e^*, c, b, p, n, l, j, k, g, d, m, i, q, e, a, r, o \rangle$

  delete edge e^*: $T = \langle c, b, p, n, l, j, k, g, d, m, i, q, e, a, r, o \rangle$

6.1.15: Algorithm 6.1.1b is shown below.

Algorithm 6.1.1b: Constructing a Directed Eulerian Tour

Input: an Eulerian digraph G.

Output: an Eulerian tour T.

  Start at any vertex v, and construct a closed directed trail T in G.

  While there are arcs of G not already in trail T

    Choose any vertex w in T that is the tail of an unused arc.

    Starting at vertex w, construct a closed directed trail D of unused arcs.

    Enlarge trail T by splicing trail D into T at vertex w.

  Return T.

6.1.18: This digraph meets the conditions of Theorem 6.1.3. Let x be the vertex incident with arcs e, m, and s, and let y be the vertex incident with arcs a,b, and f. Adding a new arc directed from vertex y to vertex x makes the digraph Eulerian. Assign the label e^* to this new arc. Shown below is the result of starting at vertex y and arc e^* and applying Algorithm 6.1.1b (from the solution to Exercise 6.1.15), using alphabetical order for the default priority.

iteration 0: closed trail $T = \langle e^*, m, f \rangle$

iteration 1: closed trail $D = \langle a, b \rangle$; enlarged trail $T = \langle a, b, e^*, m, f \rangle$

iteration 2: closed trail $D = \langle d, e, s \rangle$; enlarged trail $T = \langle a, d, e, s, b, e^*, m, f \rangle$

iteration 3: closed trail $D = \langle h, g, i, j, k \rangle$;

       enlarged trail $T = \langle a, d, e, s, b, e^*, m, h, g, i, j, k, f \rangle$

Next, analogous to Algorithm 6.1.1a (from the solution to Exercise 6.1.10), shift circularly the arc sequence of trail T so that e^* appears first:

$$T = \langle e^*, m, h, g, i, j, k, f, a, d, e, s, b \rangle$$

Then delete arc e^* from trail T to obtain the following open directed Eulerian trail from x to y.

$$T = \langle m, h, g, i, j, k, f, a, d, e, s, b \rangle$$

6.1.21: If there were no such edge, then the input graph would not be connected.

6.1.25: In graph G, each endpoint of each edge is also an endpoint of an odd number of other edges.

6.1.27: False. For instance, the line graph of the non-Eulerian star graph $K_{1,3}$ is the complete graph K_3, which is Eulerian.

6.2   DeBruijn Sequences and Postman Problems

6.2.1: A different $(2, 4)$-deBruijn sequence is obtained from the Eulerian tour specified by the following sequence of vertices and arcs.

$$000 \xrightarrow{0} 001 \xrightarrow{0} 010 \xrightarrow{0} 101 \xrightarrow{1} 010 \xrightarrow{0} 100 \xrightarrow{1} 001 \xrightarrow{0} 011 \xrightarrow{0} 111$$

$$\xrightarrow{1} 111 \xrightarrow{1} 110 \xrightarrow{1} 101 \xrightarrow{1} 011 \xrightarrow{0} 110 \xrightarrow{1} 100 \xrightarrow{1} 000 \xrightarrow{0} 000$$

The sequence of arc labels is the desired $(2, 4)$-deBruijn sequence: 0001010011110110.

6.2.5: To obtain the line graph $L(G)$ of a digraph G, each arc in G is a vertex in $L(G)$, and there is an arc in $L(G)$ directed from vertex a to vertex b if, in digraph G, $head(a) = tail(b)$. Each vertex of the line graph $L(D_{2,n})$ is labeled with the full label on the corresponding arc in deBruijn digraph $D_{2,n}$. In digraph $D_{2,n}$, the arc directed from vertex $b_1 b_2 \ldots b_{n-1}$ to vertex $b_2 \ldots b_{n-1} b_n$ is labeled $b_1 b_2 \ldots b_n$, and the arc directed from vertex $b_2 \ldots b_{n-1} b_n$ to some vertex $b_3 \ldots b_{n-1} b_n c$ is labeled $b_2 \ldots b_{n-1} b_n c$. Thus, in the line graph $L(D_{2,n})$, there is an arc directed from vertex $b_1 b_2 \ldots b_n$ to vertex $b_2 \ldots b_{n-1} b_n c$. Show that if this arc is labeled $b_1 b_2 \ldots b_{n-1} b_n c$, then the resulting digraph is $D_{2,n+1}$

6.2.8: Each of the nine vertices of $D_{3,3}$ is labeled with a different sequence of length 2 of characters drawn from $\{1, 2, 3\}$. The arc that is directed from vertex $b_1 b_2$ to vertex $b_2 b_3$ is labeled $b_1 b_2 b_3$. The next figure shows some of the vertices and arcs of $D_{3,3}$.

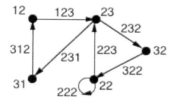

6.2.14: The set S of odd-degree vertices is $S = \{a, d\}$. The shortest a-d path is $P = \langle a, f, e, d \rangle$, which has length 13.

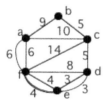

The complete graph K on the odd-degree vertices is simply the vertices a and d and the edge between them has weight 13. This edge is itself the minimum-weight perfect matching, so the edges on the corresponding path P are duplicated, and the following graph is the Eulerian graph shown above. Apply Algorithm 6.1.1 to obtain an Eulerian tour of this graph, which will correspond to an optimal postman tour of length 77 for the original graph.

6.2.16: See Corollary 1.1.3.

6.2.22: 1. The only abnormal fragment is UC, and hence, the chain must end with it. 2. The only irreducible fragment that does not appear elsewhere as an internal subfragment is C, and hence, it must start the chain. 3. The normal fragments with at least two subfragments are: CUG; CAAG; AAGC; GGU. Thus, the associated digraph is as shown below. Only one Eulerian tour in this digraph ends with the arc labeled UC. The corresponding RNA chain is CAAGCUGGUC.

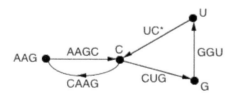

6.2.27: The corresponding digragh is shown below. Notice that the indegree of vertex 2 is two and the outdegree of vertex 1 is two. Therefore, any Eulerian trail would need to start at vertex 2 and end at vertex 1. There are many such trails, including 213231321.

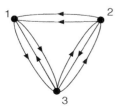

6.3   Hamiltonian Paths and Cycles

6.3.2: For all m and n such that $m = n \geq 2$.

6.3.6:

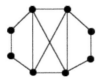

6.3.11: $\langle a,\ j,\ i,\ k,\ f,\ c,\ d,\ h,\ e,\ b,\ g,\ a \rangle$

6.3.13: The graph is not Hamiltonian, because vertices c and w have degree 2, which forces the 4-cycle $\langle c,\ f,\ w,\ b,\ c \rangle$.

6.3.18: $\langle a,\ e,\ f,\ j,\ i,\ m,\ n,\ o,\ p,\ l,\ k,\ g,\ h,\ d,\ c,\ b,\ a \rangle$

6.3.21: a. Hamiltonian graphs. b. Cycle graphs.

6.3.24: The cycle graph on five vertices is a counterexample.

6.4   Gray Codes and Traveling Salesman Problems

6.4.1: First delete the last 000 term of the Gray code given in Example 6.4.1. Then put the reverse of this truncated sequence to the right, and append a 0 to the right of each of the first eight terms and a 1 to the right of each of the second eight terms. The resulting sequence, shown below, is a Gray code of order 4.

$\langle 0000,\ 1000,\ 1100,\ 0100,\ 0110,\ 1110,\ 1010,\ 0010,$

$0011,\ 1011,\ 1111,\ 0111,\ 0101,\ 1101,\ 1001,\ 0001 \rangle$

6.4.3: nearest neighbor output: $\langle a,\ b,\ c,\ b,\ e,\ a \rangle$ with weight 26 double the tree output: $\langle a,\ d,\ c,\ e,\ b,\ a \rangle$ with weight 29

6.4.7: Transformation 1 can be applied to undirected graphs. The graph G^* that results is shown below. The nearest neighbor algorithm with ties resolved alphabetically results in the cycle $\langle a,\ 0,\ b,\ e,\ d,\ c,\ a \rangle$. The open Hamiltonian path that results from deleting vertex 0 is $\langle a,\ b,\ e,\ d,\ c \rangle$, with weight 23. An optimal Hamiltonian path that starts at a is $\langle a,\ d,\ c,\ b,\ e \rangle$ with weight 20.

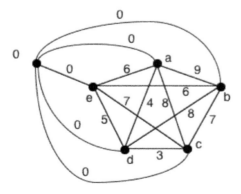

6.4.12: In order to apply Transformation 2, the given graph must be transformed into a digraph by replacing each edge with a pair of oppositely directed arcs that have weight equal to the weight of that edge.

6.4.16: The modified version of the nearest neighbor (Algorithm 6.4.1) would be to take the shortest path to an unvisited vertex, but the internal vertices of this path are allowed to have been visited. For the given graph, the iterations for the modified version are identical to those obtained in Exercise 6.4.3. Thus, the minimum-weight walk is no better than the minimum-weight Hamiltonian cycle.

However, if both edge-weights that are 7 are changed to 9, then the modified version of nearest neighbor will have a different iteration 3. In particular,

iteration 1: $\langle a,\ d \rangle$

iteration 2: $\langle d,\ c \rangle$

iteration 3: $\langle c,\ d,\ e \rangle$

iteration 4: $\langle e,\ b \rangle$

iteration 5: $\langle b,\ a \rangle$

This would result in the closed walk $\langle a,\ d,\ c,\ d,\ e,\ b,\ a \rangle$.

Chapter 7 Planarity and Kuratowski's Theorem

7.1   Planar Drawings and Some Basic Surfaces

7.1.1: Since $\left[\left(\frac{1}{2},\ \frac{1}{2}, 0 \right) - (0, 1, 0) \right] = \left(\frac{1}{2}, \frac{-1}{2}, 0 \right)$, the locus of the line through the points $\left(\frac{1}{2},\ \frac{1}{2}, 0 \right)$ and $(0,1,0)$ is

$$\left\{ (0, 1, 0) + c \left(\frac{1}{2}, \frac{-1}{2}, 0 \right) \middle| -\infty \le c \le \infty \right\}$$

This line meets the plane $y = 0$ when $c = 2$; thus, the point of intersection is $(1, 0, 0)$.

7.1.5: The locus of the line through the points $\left(\frac{\sqrt{3}}{3}, 0, 0 \right)$ and $(0,1,0)$ is

$$\left\{ (0, 1, 0) + c \left(\frac{\sqrt{3}}{3}, -1, 0 \right) \middle| -\infty \le c \le \infty \right\}$$

This line meets the sphere $x^2 + \left(y - \frac{1}{2} \right)^2 + z^2 = \frac{1}{4}$ when $c = \frac{3}{4}$; thus, the point of intersection is $\left(\frac{\sqrt{3}}{4}, \frac{1}{4}, 0 \right)$.

7.2   Subdivision and Homeomorphism

7.2.1: Subdividing two of the edges of D_3 as shown yields a graph that is isomorphic to $K_4 - K_2$.

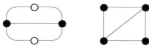

7.2.5: Place 2 subdivision vertices on each of the n self-loops, for a total of $2n$ subdivision vertices.

7.2.7:

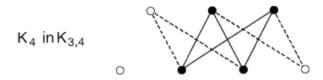

K$_4$ in K$_{3,4}$

7.2.11: No matter how the four vertices of a homeomorphic copy of K_4 could be chosen in $K_{2,3}$, at least two vertex pairs would be non-adjacent. It would be necessary to join each such pair with a path in the homeomorphic copy of K_4, and such paths would have to be mutually internally disjoint. Thus, two or more internally disjoint paths in $K_{2,3}$ would have to go through the one remaining vertex, an impossibility.

7.3   Extending Planar Drawings

7.3.1:

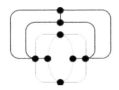

7.3.5: Nonplanar. After pasting, the hub of W_6 is adjacent to all three of the "pasted" vertices. Also, the two nonpasted vertices from W_4 are adjacent to all three pasted vertices. Thus, the amalgamated graph contains $K_{3,3}$.

7.3.9: One appendage has vertex-set $\{u,v,w\}$ and edge-set $\{a,b\}$. The other has vertex-set $\{u,x,w\}$ and edge-set $\{d,e\}$.

7.3.13: The contact points of the appendage with vertex-set $\{u,v,w\}$ and edge-set $\{a,b\}$ are u and w. The contact points of the appendage with vertex-set $\{u,x,w\}$ and edge-set $\{d,e\}$ are also u and w. These two appendages do not overlap.

7.4   Kuratowski's Theorem

7.4.2: In the figure below, the partite sets of a $K_{3,3}$ are represented by three white vertices and three black vertices. The gray vertices are interior vertices on paths joining a pair.

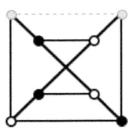

7.4.7: $K_9 - K_{4,5}$ is isomorphic to $K_4 \cup K_5$.

7.5   Algebraic Tests for Planarity

7.5.2:

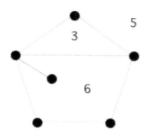

(c) Sum of face-sizes $= 3 + 6 + 5 = 14 = 2 \cdot 7 = 2|E|$.

7.5.6: (a) $girth = 3$. (b) $|F| = 3$. $girth \cdot |F| = 3 \cdot 3 \le 2 \cdot 7 = 2 \cdot |E|$.

7.5.10: (a) $|V| = 6$. $|E| = 7$. $|F| = 3$. (b) $|V| - |E| + |F| = 6 - 7 + 3 = 2$.

7.5.14:

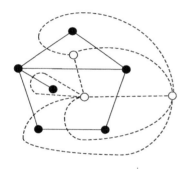

7.5.18:

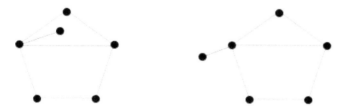

Since the face-size combinations of these two new imbeddings and the original imbedding are mutually distinct, it follows that the dual graphs are mutually distinct.

7.5.22: The graph $K_4 - K_2$ has 4 vertices. By the Euler polyhedral formula, a planar imbedding of it has 3 faces. Thus, the dual graph has 3 vertices. It follows that the dual graph cannot be isomorphic to $K_4 - K_2$.

7.5.29: Each of the four vertices lying on one of the two disjoint (deleted) copies of C_4 is joined to each of the four vertices in the other copy, so the graph contains $K_{4,4}$.

7.5.33: Violates the inequality $|E| \le 3|V| - 6$.

7.5.42: Nonplanar because it contains $K_{3,3}$, However, since $|E| = 15$ and $|V| = 7$, it satisfies the inequality $|E| \le 3|V| - 6$.

7.6   Planarity Algorithm

7.6.2: The blocked appendage in the figure below shows that the input graph is nonplanar. It is blocked since no region contains all four of its contact points.

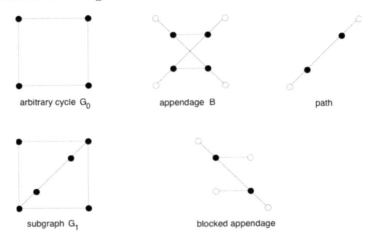

arbitrary cycle G_0 appendage B   path

subgraph G_1 blocked appendage

7.7   Crossing Numbers and Thickness

7.7.1: Apply Theorem 7.7.3.

7.7.5: By Theorem 7.7.5, $v(K_{4,4}) \geq 16 - 2 \cdot 8 + 4 = 4$. This figure establishes that $v(K_{4,4}) \leq 4$.

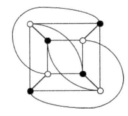

7.7.10: By Theorem 7.7.9, $\theta(K_{6,6}) \geq 2$. This figure establishes that $\theta(K_{6,6}) \leq 2$.

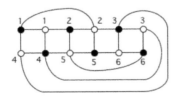 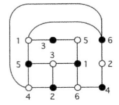

7.7.12:

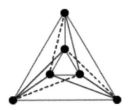

Chapter 8 Graph Colorings

8.1   Vertex-Colorings

8.1.3:

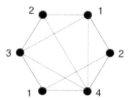

Fewer than 4 colors is impossible because of the K_4.

8.1.7:

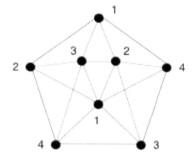

Fewer than 4 colors is impossible because of the wheel W_6 or the K_4.

8.1.10:

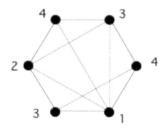

8.1.14: The coloring given above for Exercise 8.1.7 is obtainable by application of the largest-degree-first heuristic.

8.1.19: In each of the two parts of the bipartition, all vertices have the same degree. In the part with larger degree, all vertices are assigned color 1. In the other, all are assigned color 2.

8.1.22: Vertices 1, 2, 4, and 6 get colors 1, 2, 2, and 3, respectively. The algorithm completes the coloring using only these three colors.

8.1.27: No matter what edge is deleted, there remains a 3-cycle.

8.1.31: The independence number is 2. The four vertices of the K_4 are mutually adjacent, and the two other vertices are adjacent.

8.1.35: Finding two non-adjacent vertices is easy, which establishes a lower bound of 2 for the independence number. Since the minimum degree is 4, and since the graph has only eight vertices, it follows that any subset of two non-adjacent vertices has an adjacency to every vertex not in the subset.

8.1.37: The domination number is 1, since the 5-valent vertex is adjacent to all the others.

8.1.41: Any pair of two non-adjacent vertices qualifies as a dominating set.

8.1.45: A subgraph of a bipartite graph is 1-chromatic if it contains no edges, in which case its clique number is 1. Otherwise, it is bipartite and 2-chromatic, and its clique number is 2.

8.2   Map-Colorings

8.2.2: The four regions whose colors are underlined are mutually adjacent.

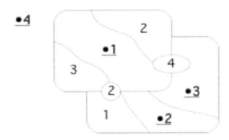

8.2.5: Yes. Here is a minimum example.

8.2.8:

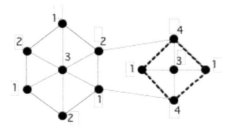

8.2.14: The chromatic number of Africa is 4. Since Africa is planar, 4 is an upper bound. There are several even-order wheels — for instance, with Niger or Mali as the hub — which implies that 4 is also a lower bound.

8.2.16: Six. There are six mutually adjacent regions.

8.3   Edge-Colorings

8.3.2: Since the maximum degree is 3, this 3-edge-coloring must be minimum.

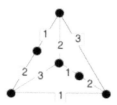

8.3.11: There are 14 edges. Any collection of more than three edges has a common vertex. Thus, no single color class may have more than three edges. Thus, the edge-chromatic number must be 5.

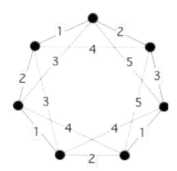

8.3.16: Alternate the colors.

8.3.21: Give the new edge a new color.

8.3.25: Use any edge in an even cycle C_{2n}.

8.3.28: Deleting the cut-edge e from graph G must yield two components each with an odd number of vertices (since smoothing the two resulting 2-valent vertices would yield a 3-regular graph with two components, each necessarily with an even number of vertices). Suppose now, that G were in Class 1. Then deleting the color class of the cut-edge would yield a 2-colorable 2-regular graph, i.e., the disjoint union of some even cycles. However, one sub-union of these even cycles must span one of the components of $G - e$, and the complementary sub-union must span the other component, which contradicts the fact that these components each have an odd number of vertices.

8.4   Factorization

8.4.3: The union of a k-factor of G and a k-factor of H is a k-factor of G + H.

8.4.8: The edge-complement of the graph of Exercise 8.4.1.

8.4.11: $K_2 \times K_2$.

Chapter 9 Special Digraph Models

9.1   Directed Paths and Mutual Reachability

9.1.1:

9.1.6:

9.1.11: This digraph is strongly connected. The output tree is shown below.

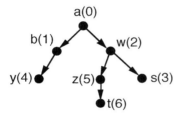

9.1.13: The full model for Application 9.1.1 is shown below. There are two optimal policies for this set of data. Either sell the first car after one year and keep the second for the next four years, or keep the first car for all five years.

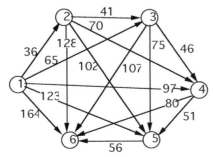

9.1.18:

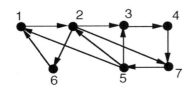

9.1.22:

9.1.24: Dags.

9.1.28: The stronger statement that the resulting digraph has at least $k+1$ strong components can be proved by induction on k. The base for the induction is Theorem 9.1.4.

9.1.30: The 2-step transition matrix is shown below. As an illustration of how these 2-step transition probabilities can be obtained directly from the Markov diagram in Figure 9.1.7, consider $p_{45}^{[2]}$. There are two different paths of length 2 from vertex 4 to vertex 5, namely $4 \rightarrow 3 \rightarrow 5$ and $4 \rightarrow 5 \rightarrow 5$. Multiplying the arc probabilities for each of these two paths and then taking the sum, we obtain $\frac{3}{4} \cdot \frac{1}{4} + \frac{1}{4} \cdot 1 = \frac{7}{16}$, which agrees with the 4,5 entry of the 2-step transition matrix.

$$
\begin{array}{c}
 \\ 0 \\ 1 \\ 2 \\ 3 \\ 4 \\ \geq 5
\end{array}
\begin{array}{cccccc}
0 & 1 & 2 & 3 & 4 & \geq 5 \\
\left(\begin{array}{cccccc}
1 & 0 & 0 & 0 & 0 & 0 \\
.75 & 0 & 0 & .1875 & 0 & .0625 \\
.5625 & 0 & 0 & 0 & .1875 & .0625 \\
0 & .5625 & 0 & 0 & 0 & .4375 \\
0 & 0 & .5625 & 0 & 0 & .4375 \\
0 & 0 & 0 & 0 & 0 & 1
\end{array}\right)
\end{array}
$$

9.1.34: a. All of the states are transient. except state 2, which is absorbing.

b. There are four different paths of length 4 from vertex 3 to vertex 4, namely, $\langle 3, 1, 3, 4, 4 \rangle$; $\langle 3, 1, 5, 4, 4 \rangle$; $\langle 3, 4, 1, 5, 4 \rangle$; and $\langle 3, 4, 4, 4, 4 \rangle$. By multiplying the arc probabilities along each of these four paths and then taking the sum, we obtain the result $p_{34}^{[4]} = \frac{1}{64} + \frac{1}{32} + \frac{1}{32} + \frac{1}{32} = \frac{7}{64}$.

9.1.37: If all of the states were transient, then for each vertex in the Markov digraph, there would exist a vertex that is reachable from it. Show that this would imply the existence of a directed cycle in the digraph, which would then contradict the assumption that all states were transient.

9.2   Digraphs as Models for Relations

9.2.1: Since the given digraph is strongly connected, the transitive closure has both arcs between each pair of vertices and has a self-loop at each vertex.

9.2.6:

9.2.10: The transitive closure is obtained at the end of the second iteration (i.e., $D_4 = D_3 = D_2$).

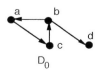

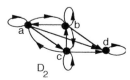

9.2.14: The minimal elements in the Hasse diagram shown below are 2, 3, and 5.

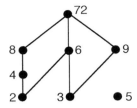

9.2.17: The sequence $\langle 5, 2, 3, 9, 6, 4, 8, 72 \rangle$ represents a compatible total order of the given poset and is the result of the following sequence of iterations of the topological sort.

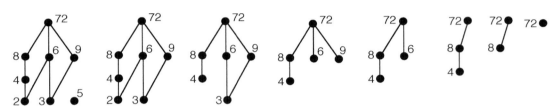

9.3   Tournaments

9.3.1: Since each of the three edges of K_3 can be oriented in one of two ways, there are eight different 3-vertex tournaments. They are shown below, grouped into isomorphism classes.

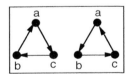   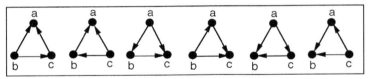

9.3.3: Each 4-vertex tournament is isomorphic to one of the four tournaments shown in the figure below. The first contains a 4-cycle, the second and third each contain exactly one 3-cycle, and the fourth is acyclic.

To see that there is only one tournament, up to isomorphism, that contains a 4-cycle, first arrange the vertices so that the 4-cycle forms a square, as in the figure. Then observe that no matter how the diagonal edges are directed, a rotation of 90°, 180°, or 270° of the tournament will match the one in the figure. For a tournament that contains exactly one cycle, it is not hard to show that the non-cycle vertex must have either 0 indegree or 0 outdegree. Similarly, one can argue that if a 4-vertex tournament has more than one cycle, then it must contain a 4-cycle, putting it in category 1.

Finally, to see that there is only one isomorphism type for the acyclic 4-vertex tournaments, start with the unique Hamiltonian path (whose existence is guaranteed by Corollary 9.3.3) and argue that there is only one way to direct the three arcs that are not on this path without creating a cycle.

9.3.7: If the tournament were strongly connected, then any two vertices would be mutually reachable, which would imply the existence of a directed cycle.

9.4 Project Scheduling and Critical Paths

9.4.1: In the AOA network for the problem shown below at the right, the ET and LT values for each event are as follows:

Event	$ET(i)$	$LT(i)$
1	0	0
2	5	20
3	10	10
4	15	15
5	18	18
6	20	20
7	21	21
8	35	35

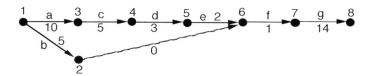

The earliest possible completion time for the project is 35 days, and task b is the only non-critical one. Its total float is $LT(2) - ET(1) - wt(b) = 20 - 0 - 5 = 15$ days.

9.4.6: Suppose that every vertex has nonzero indegree, and consider any vertex v_1. There is an arc directed from some vertex v_2 to vertex v_1. Similarly, there is an arc directed from some vertex v_3 to vertex v_2. Since there are finitely many vertices, this process will eventually produce some $v_k = v_l$ for $k \neq l$, which would contradict the acyclic property. A similar argument can be used to show that there is a vertex whose outdegree is 0.

9.4.9: Suppose that the critical path were not the longest path. Then there would be a subpath P of the critical path from event i to event j such that there is a longer i-j path consisting of non-critical tasks only, as illustrated in the figure below. Work backward from event j and use the definitions of TF, ET, and LT to reach a contradiction.

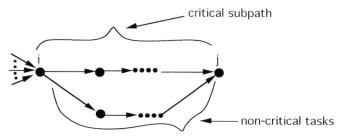

9.5   Finding the Strong Components of a Digraph

9.5.2: The unlabeled vertex in the given digraph has been assigned the label f. The three dfs trees are shown in the figure below. In tree T_a, ba and fe are back arcs, and dc is a cross arc. In tree T_b, f_e and d_c are cross arcs, and eb is a back arc.

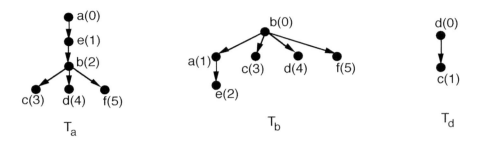

9.5.6: Referring to the dfs trees from the solution to Exercise 9.5.2, the cross arc dc in tree T_a is directed from *dfnumber* 4 to *dfnumber* 3. In tree T_b, cross arc fe is directed from *dfnumber* 5 to *dfnumber* 2, and cross-arc de is directed from *dfnumber* 4 to *dfnumber* 3.

9.5.8: There are no cross arcs.

9.5.10: The unlabeled vertex in the given digraph has been assigned the label f. Starting at vertex a, the dfs tree produced by Algorithm 9.5.2 contains all of the vertices. The vertex-sets of the strong components are: $\{a, b, e, f\}$; $\{c\}$; and $\{d\}$.

9.5.14: Strongly connected digraphs.

Chapter 10 Network Flows and Applications

10.1   Flows and Cuts in Networks

10.1.1: The arc set $\langle \{s, z\}, \{x, y, u, v, w, t\} \rangle$ is an *s-t* cut with capacity 6, and the flow shown in the figure below has value 6. Hence, by Corollary 10.1.7, both are optimal. The number on each arc in the figure indicates the flow across that arc.

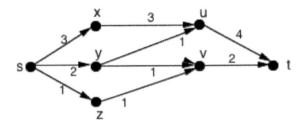

10.1.5: There are eight different *s-t* cuts depending on the subset of $\{x, z, v\}$ into which source s is placed. The two minimum ones, with capacity 14, are $\langle \{s, x, z, v\}, \{t\} \rangle$ and $\langle \{s, z\}, \{x, v, t\} \rangle$.

10.1.9: Add a "super source" S directed to each of the original sources and add a "super sink" T directed from each of the original sinks to obtain the network shown below. For each original sink s_i, assign to the arc from S to s_i a capacity equal to the total capacity of $Out(s_i)$. Similarly, for each original sink t_j, assign to the arc from t_j to T a capacity equal to the total capacity of $In(t_j)$.

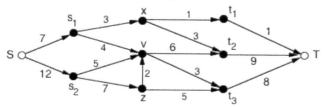

A minimum S-T cut for this network is $\langle \{S, s_1, s_2, v, z\}, \{x, t_1, t_2, t_3\}\rangle$ and has capacity 17. A flow f achieving this value is given by: $f(Ss_1) = 7$; $f(Ss_2) = 10$; $f(s_1x) = 3$; $f(s_1v) = 4$; $f(s_2v) = 3$; $f(s_2z) = 7$; $f(zv) = 2$; $f(zt_3) = 5$; $f(vt_3) = 3$; $f(vt_2) = 6$; $f(xt_2) = 3$; $f(xt_1) = f(t_1T) = 0$; $f(t_2T) = 9$; and $f(t_3T) = 8$. Finally, dropping the arcs from S and the ones to T results in a maximum flow for the original network.

10.1.13: Split each such vertex v into two vertices v_1 and v_2, draw an arc directed from v_1 to v_2, and assign to it a capacity equal to the capacity of vertex v. Then replace each arc in the original network that was directed from a vertex x to vertex v with an arc directed from x to v_1, and replace each arc that was directed from vertex v to a vertex y with an arc from v_2 to y. The network that results when this transformation is applied to the given digraph is as follows.

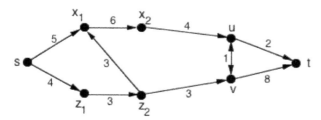

10.1.16: Use Proposition 10.1.3 and Corollary 10.1.7.

10.2   Solving the Maximum-Flow Problem

10.2.1:

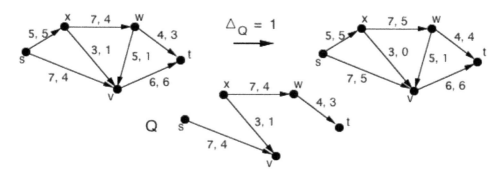

10.2.5: A sequence of quasi-paths and flows is shown below.

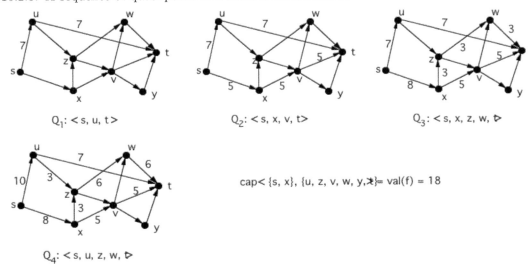

$Q_1: <s, u, t>$ $Q_2: <s, x, v, t>$ $Q_3: <s, x, z, w, \triangleright$

$Q_4: <s, u, z, w, \triangleright$

$cap< \{s, x\}, \{u, z, v, w, y, \ast\}= val(f) = 18$

10.2.8: Solve the following maximum-flow problem.

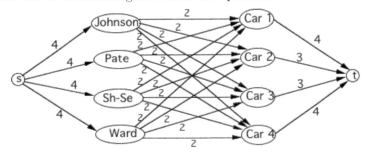

10.3   Flows and Connectivity

10.3.1: Each arc of the digraph $\overleftrightarrow{G}$ below is assumed to have capacity 1. The arcs directed to source b and the arcs directed from sink d are not drawn, since they cannot be part of a directed path from b to d. A maximum flow from b to d of value 3 is shown in the figure and is achieved by assigning a flow of 1 to each arc of the three b-d paths: $\langle b, a, d \rangle$, $\langle b, e, d \rangle$, $\langle b, c, f, d \rangle$, which are arc-disjoint. Thus, the local edge-connectivity between vertices b and d equals 3.

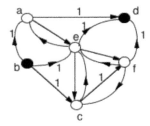

10.3.5: From the solution to Exercise 10.3.1, we know that $\kappa_e(b,d) = 3$. Similarly, $\kappa_e(b,a) = 3$, and by symmetry, $\kappa_e(b,c) = \kappa_e(b,e) = 3$. Thus, by Corollary 10.3.12, $\kappa_e(G) = 3$.

10.3.9: Each arc of the digraph $N^{\overleftrightarrow{G}}$, shown below, is assumed to have capacity 1. The arcs directed to source b and the arcs directed from sink d are not drawn since they cannot be part of a directed path from b to d. A maximum flow of value 3 is obtained by assigning a flow of 1 to each arc of the three arc-disjoint b-d paths $\langle b, a', a'', d \rangle$, $\langle b, e', e'', d \rangle$, $\langle b, c', c'', f', f'', d \rangle$. Thus, $\kappa_v(b, d) = 3$. The three paths correspond to the paths $\langle b, a, d \rangle$, $\langle b, e, d \rangle$, $\langle b, c, f, d \rangle$ in the original graph, which are internally disjoint.

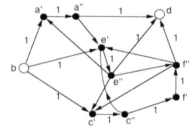

10.3.13: It follows from the solution to Exercise 10.3.9 and symmetry that $\kappa_v(b, f) = 3$, which, by Lemma 5.3.5, implies that $\kappa_v(G) = 3$.

10.3.18: Since the initial arc of each of the r paths has flow 1, we have $val(f) = \sum_{e \in Out(s)} f(e) = r$. Since $f(e) \leq cap(e)$, it remains to show that f satisfies conservation of flow. Let v be any vertex in network N other than s or t. On each of the paths that contain v, the arc directed to v and the arc from v both have flow 1, and hence, the conservation-of-flow property is satisfied.

10.3.21: In digraph N^D, there is only one arc directed from vertex v' to v''. This implies that any s-t directed path in N^D that uses vertex v' must have vertex v'' as the next vertex.

10.3.25: If all the arc capacities are integer, then convert the given network N into a 0-1 network $\hat{N}$ by replacing each arc e in N by k arcs of unit capacity, where $k = cap(e)$. Then use the digraph version of Theorem 5.3.10. If all the arc capacities are rational numbers, then multiply each arc by the least common multiple of the denominators (see Exercise 10.1.12). If some of the capacities are irrational numbers, then the maximum flow can be approximated to any desired accuracy by replacing each irrational by a suitably near rational so that the difference between the value of the flow and the capacity of the cut is sufficiently close to zero.

10.4   Matchings, Transversals, and Vertex Covers

10.4.2: The maximum flow in the network $\vec{G}_{st}$ is shown below. The set of edges in the given graph whose corresponding arcs have unit flow in $\vec{G}_{st}$ is the maximum matching $\{a4, b2, d3, e1\}$.

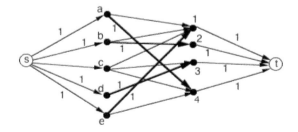

10.4.6: Assume all arcs in the following network have unit capacity and determine whether the maximum flow is 6.

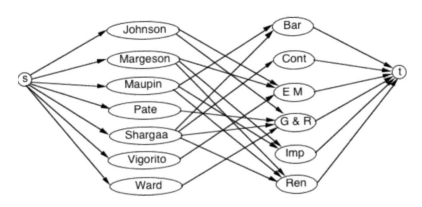

10.4.10: One of several violations of Hall's condition is the set of three subsets $\{\{5\}, \{4, 5\}, \{4, 5\}\}$, since the union contains only two elements.

10.4.13: The edge set given by $\{(6, 1), (3, 4), (2, 9), (8, 10), (5, 7)\}$ is a maximum matching, and the vertex set $\{1, 3, 2, 8, 5\}$ is a minimum vertex cover.

10.4.17: Let W be a subset of X. Since G is bipartite, the degree sum $\sum_{w \in W} deg(w)$ equals the number of edges having one endpoint in W and the other endpoint in $N(W)$. But this number is less than or equal to the total number of edges having an endpoint in $N(W)$. Thus,

$$|W| \cdot \delta_X \leq \sum_{w \in W} deg(w) \leq \sum_{w \in N(W)} deg(w) \leq N(W) \cdot \Delta_Y \leq N(W) \cdot \delta_X$$

which shows that Hall's condition is satisfied.

10.4.19: This is a corollary of the assertion in Exercise 10.4.17, but it can be proved directly with a similar argument. In particular, if W is a subset of X, then since G is bipartite,

$$|W| \cdot k = \sum_{w \in W} deg(w) \leq \sum_{w \in N(W)} deg(w) = N(W) \cdot k$$

10.4.25: Either construct a 0-1 matrix and apply König-Egerváry, or construct a bipartite graph and apply Theorem 10.4.3 or Theorem 10.4.9.

10.4.27: Let $\hat{M}$ be any matching in graph G. Then, by weak duality (Proposition 10.4.7), $\left|\hat{M}\right| \leq |C| = |M|$, which shows that M is a maximum matching. A similar argument shows that C is a minimum vertex cover.

10.4.29: If each subset of women collectively knows at least as many men as there are women in that subset, then each woman can marry a man whom she knows.

Chapter 11 Graph Symmetry and Colorings

11.1   Automorphisms of Simple Graphs

11.1.3:

Symmetry	Vertex permutation	Edge permutation
identity	$(u)(v)(w)(x)$	$(a)(b)(c)(d)$
horiz.refl.	$(x)(w)(u\ v)$	$(b)(d)(a\ c)$

11.1.5:

Vertex permutation	Edge permutation
$(u)(v)(w)$	$(a)(b)(c)$
$(u\ v\ w)$	$(a\ b\ c)$
$(u\ w\ v)$	$(a\ c\ b)$
$(u)(v\ w)$	$(b)(a\ c)$
$(v)(u\ w)$	$(c)(a\ b)$
$(w)(u\ v)$	$(a)(b\ c)$

11.1.9:

11.1.13: 4! automorphisms

11.1.15: 2!3! automorphisms

11.1.17: 2!3!4! automorphisms

11.2   Graph Colorings and Symmetry

11.2.2:

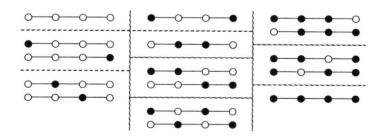

11.2.9:

$$45 \text{ total} = \begin{cases} 18 & \text{using all three colors} \\ 24 & \text{using exactly two of the three colors} \\ 3 & \text{using only one color} \end{cases}$$

11.2.16:

$$6 \text{ total} = \begin{cases} 1 & \text{with all three edges light} \\ 1 & \text{with all three edges dark} \\ 2 & \text{with two edges light and one edge dark} \\ 2 & \text{with one edge light and two edges dark} \end{cases}$$

11.2.23:

$$18 \text{ total} = \begin{cases} 3 & \text{using all three colors} \\ 12 & \text{using exactly two of the three colors} \\ 3 & \text{using only one color} \end{cases}$$

INDEX OF APPLICATIONS

INDEX OF ALGORITHMS

GENERAL INDEX